Particle Physics: Concepts and Applications

Particle Physics: Concepts and Applications

Edited by
Joy Moody

Willford Press
www.willfordpress.com

Published by Willford Press,
118-35 Queens Blvd., Suite 400,
Forest Hills, NY 11375, USA

ISBN: 978-1-64728-361-2

Cataloging-in-Publication Data

Particle physics : concepts and applications / edited by Joy Moody.
p. cm.
Includes bibliographical references and index.
ISBN 978-1-64728-361-2
1. Particles (Nuclear physics). 2. Nuclear physics. 3. Physics. I. Moody, Joy.
QC793.2 .P37 2022
539.72--dc23

For information on all Willford Press publications
visit our website at www.willfordpress.com

Willford Press

Contents

Preface

Over the recent decade, advancements and applications have progressed exponentially. This has led to the increased interest in this field and projects are being conducted to enhance knowledge. The main objective of this book is to present some of the critical challenges and provide insights into possible solutions. This book will answer the varied questions that arise in the field and also provide an increased scope for furthering studies.

The study of the nature of particles that constitute matter and radiation is known as particle physics. This discipline examines the smallest detectable particles and also studies the basic interactions which are necessary to explain particle's behavior. Some of the subatomic particles studied in this field are electrons, protons and neutrons. The dynamics of these particles are studied using quantum mechanics since they exhibit wave-particle duality. Particle physics investigates the standard model and its extensions. Standard model is a theory that describes fundamental particles and fields with their dynamics. Principles and concepts from particle physics are applied in a wide range of fields such as medicine, computing, national security, etc. This book explores all the important aspects of particle physics in the present day scenario. It explains the complex concepts and applications of particle physics in an easy manner. Students, researchers, experts and all associated with this discipline will get benefit alike from this book.

I hope that this book, with its visionary approach, will be a valuable addition and will promote interest among readers. Each of the authors has provided their extraordinary competence in their specific fields by providing different perspectives as they come from diverse nations and regions. I thank them for their contributions.

Editor

Angular Dependence of η Photoproduction in Photon-Induced Reaction

Jun-Zhen Wang and Bao-Chun Li

Department of Physics, Shanxi University, Taiyuan, Shanxi 030006, China

Correspondence should be addressed to Bao-Chun Li; libc2010@163.com

Guest Editor: Sakina Fakhraddin

Photoproduction of η mesons from nucleons can provide valuable information about the excitation spectrum of the nucleons. The angular dependence of η photoproduction in the photon-induced reaction is investigated in the multisource thermal model. The results are compared with experimental data from the $\eta \longrightarrow 3\pi^0 \longrightarrow 6\gamma$ decay mode. They are in good agreement with the experimental data. It is shown that the movement factor increases linearly with the photon beam energies. And the deformation and translation of emission sources are visually given in the formalism.

1. Introduction

The excitation spectrum of nucleons is important to understanding the nonperturbative behavior of the fundamental theory of strong interactions, Quantum Chromodynamics (QCD) [1–4]. The photon-induced meson production off nucleons is mainly used to achieve more information from the excitation spectrum of nucleons. It is very important for missing resonances that the η meson production in photon-induced and hadron-induced reactions on free (and quasi-free) nucleons and on nuclei [5–8]. The advantage of photon-induced reactions is that the electromagnetic couplings can provide valuable information related to the details of the model wave functions. Because the electromagnetic excitations are isospin dependent, we need perform meson-production reactions off the neutron.

Recently, the photoproduction of η mesons from quasi-free protons and neutrons was measured in $\eta \longrightarrow 3\pi^0 \longrightarrow 6\gamma$ decay mode by the CBELSA/TAPS detector at the electron accelerator ELSA in Bonn [9]. At different incident photon energies, the experiments are performed by the incident photon beam on a liquid deuterium target. A great number of η mesons are produced in the photon-induced reaction. The experimental data are regarded as a multiparticle system. And, their angular distributions represent an obvious regularity at different incident photon energies. In order to explain the abundant experimental results, some statistical methods are proposed and developed [10–16]. In this work, we will extend a multi-source thermal model to the statistical investigation of the angular distributions in the photon-induced reaction and try to understand the η photoproduction in the reaction. In our previous wok [17–21], the model was focused on the investigation of the particle production in intermediate-energy and high-energy collisions.

2. η Meson Distribution in the Multi-Source Thermal Model

In the multi-source thermal model [17–21], many emission sources are expected to be formed at the final stage of the photon-induced reaction. Every source emits particles isotropically in the source rest frame. The observed η mesons are from different emission sources. The incident beam direction is defined as an oz axis and the reaction plane is defined as yoz plane. In the source rest frame, the meson momentum p_x, p_y, and p_z obeys a normal distribution. The corresponding transverse momentum $p_T = \sqrt{p_x^2 + p_y^2}$ obeys a Rayleigh distribution:

$$f_{p_T}(p_T) = \frac{p_T}{\sigma^2} e^{-p_T^2/2\sigma^2}, \quad (1)$$

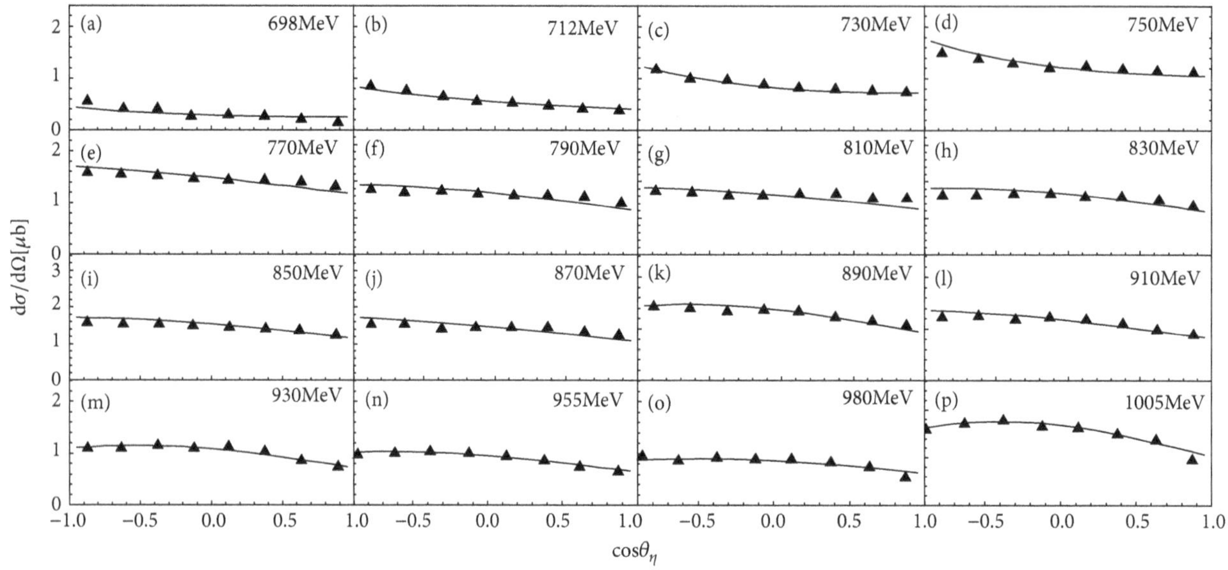

FIGURE 1: Angular distributions for different bins of incident photon energy 698 MeV $\leq E_\gamma \leq$ 1005MeV as a function of $\cos\theta_\eta$ in the beam-target cm system assuming the initial state nucleon at rest. The symbols represent the experimental data from the CBELSA/TAPS detector at the electron accelerator ELSA in Bonn [9]. The results in the multi-source thermal model are shown with the curves.

where σ represents a distribution width. The distribution function of the polar angle θ is

$$f_\theta(\theta) = \frac{1}{2}\sin\theta \tag{2}$$

Because of the interactions with other emission sources, the considered source deforms and translates along the oz axis. Then, the momentum component is revised to

$$p'_z = a_z p_z + b_z \sigma \tag{3}$$

where a_z and b_z represent the coefficients of the source deformation and translation along the oz axis, respectively. The mathematical description of the deformable translational source is formulized simply as a linear relationship between p'_z and p_z, which reflects the mean result of the source interaction. For $a_z \neq 1$ or $b_z \neq 0$, the p'_z distribution of η mesons is anisotropic along the oz axis.

By using Monte Carlo method, p_T and p'_z are given by

$$p_T = \sigma\sqrt{-2\ln r_1}, \tag{4}$$

$$p'_z = a_z\sigma\sqrt{-2\ln r_2}\cos(2\pi r_3) + b_z\sigma, \tag{5}$$

where r_1, r_2, and r_3 are random numbers from 0 to 1. The polar angle θ is revised to

$$\theta' = \arctan\frac{p_T}{p'_z} = \arctan\frac{\sqrt{-2\ln r_1}}{a_z\sqrt{-2\ln r_2}\cos(2\pi r_3) + b_z}. \tag{6}$$

We can calculate a new distribution function of the polar angle by this formula.

3. Angular Dependencies of η Photoproduction in the Photon-Induced Reaction

Figures 1(a)–1(p) show the angular distributions of η mesons for different bins of incident photon energy 698 MeV $\leq E_\gamma \leq$ 1005 MeV as a function of $\cos\theta_\eta$. θ_η is the polar angle of η meson in the beam-target cm system assuming the initial state nucleon at rest. The symbols represent the experimental data from the CBELSA/TAPS detector at the electron accelerator ELSA in Bonn [9]. The results obtained by using the multi-source thermal model are shown with the curves, which behave in the same way as the experimental data in the 16 bins of incident photon energy. By minimizing χ^2 per degree of freedom (χ^2/dof), we determine the corresponding parameters a_z and b_z, which are presented in Table 1. It is found that there is an almost linear relationship between the b_z and E_γ. As representative energies of Figure 1, we give a schematic sketch of these emission sources at the four different energies in Figure 7(a). The deformations and translations can be seen intuitively in the figure.

In Figures 2(a)–2(p) and Figures 3(a)–3(p), we present the angular distributions of η mesons for different bins of incident photon energy 1035 MeV $\leq E_\gamma \leq$ 1835 MeV as a function of $\cos\theta_\eta$. θ_η is the polar angle of η meson in the beam-target cm system assuming the initial state nucleon at rest. Same as Figure 1, the symbols represent the experimental data from the CBELSA/TAPS detector at the electron accelerator ELSA in Bonn [9]. The results obtained by using the multisource thermal model are shown with the curves, which behave in the same way as the experimental data in the 28 bins of incident photon energy. Parameters a_z and b_z are presented in Tables 2 and 3. As the representative energies of Figures 1 and

TABLE 1: Values of a_z and b_z taken in Figure 1 model results.

Figure 1	E_γ (MeV)	a_z	b_z	χ^2/dof
(a)	698	1.050	-0.170	0.118
(b)	712	1.020	-0.210	0.105
(c)	730	1.050	-0.170	0.090
(d)	750	1.040	-0.155	0.134
(e)	770	0.985	-0.105	0.182
(f)	790	0.970	-0.125	0.179
(g)	810	0.979	-0.110	0.194
(h)	830	0.960	-0.120	0.192
(i)	850	0.970	-0.110	0.200
(j)	870	0.980	-0.130	0.211
(k)	890	0.950	-0.120	0.175
(l)	910	0.970	-0.140	0.261
(m)	930	0.940	-0.110	0.179
(n)	955	0.950	-0.120	0.154
(o)	980	0.950	-0.090	0.138
(p)	1005	0.920	-0.110	0.150

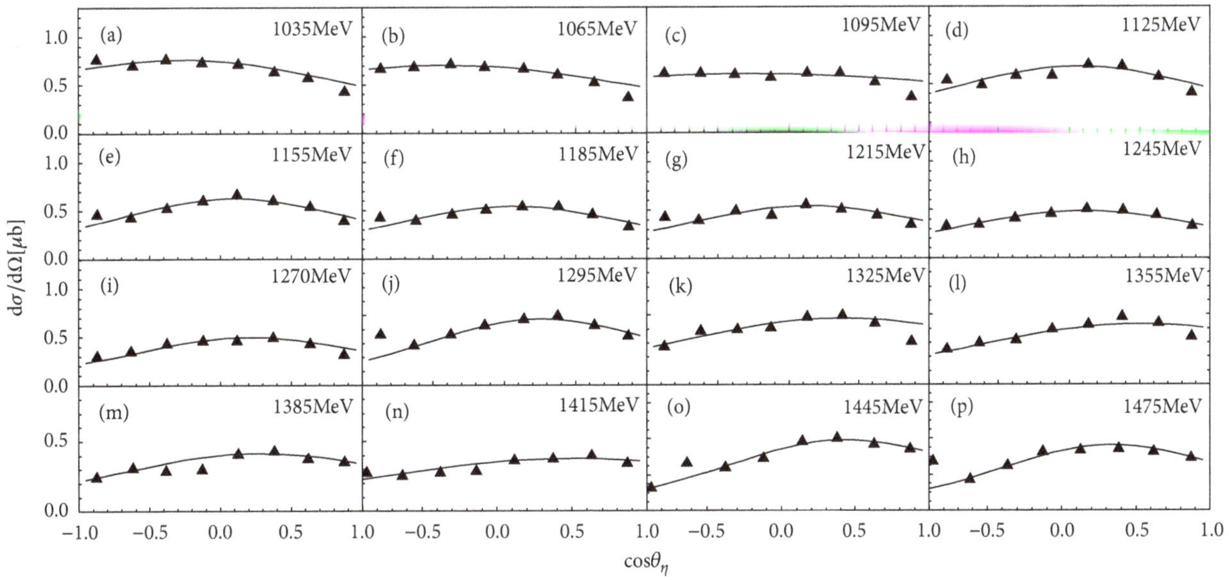

FIGURE 2: Same as Figure 1, but showing angular distributions for different bins of incident photon energy 1035 MeV $\leq E_\gamma \leq$ 1475MeV.

2, the schematic sketches of the emission sources are given at different energies in Figures 7(b) and 7(c).

In Figures 4, 5, and 6, we show angular distributions in the η-nucleon cm system for $\gamma p \longrightarrow p\eta$ reaction for the different bins of final state energy 1488 MeV $\leq W \leq$ 1625 MeV, 1635 MeV $\leq W \leq$ 1830 MeV and 1850 MeV $\leq W \leq$ 2070 MeV, respectively. Same as Figure 1, the model results and experimental data are indicated by the curves and symbols, respectively. The model results can also agree with the experimental data. In the same way, the deformations and translations of these emission sources are given in Tables 4–6 and Figures 7(d)–7(f). All the parameter values taken in the above calculations are also given in Figures 8 and 9. It can be found that a_z keeps almost invariable and fluctuates around 1.0 with the increasing E_γ. The parameter b_z increases linearly with the increasing E_γ and their relationship can be expressed by a linearly function, $b_z = (0.541 \pm 0.005) \times 10^{-3} E_\gamma - (0.622 \pm 0.011)$. There are similar relationships between the parameters and different final state energies W in Figure 9, where the fitting function of b_z is $b_z = (0.808 \pm 0.003) \times 10^{-3} W - (1.322 \pm 0.007)$.

4. Discussion and Conclusions

The excitation spectrum of nucleons can especially help us to understand the strong interaction in the nonperturbative regime. Before, the hadron induced reactions is a main experimental method in the investigation. In the last two

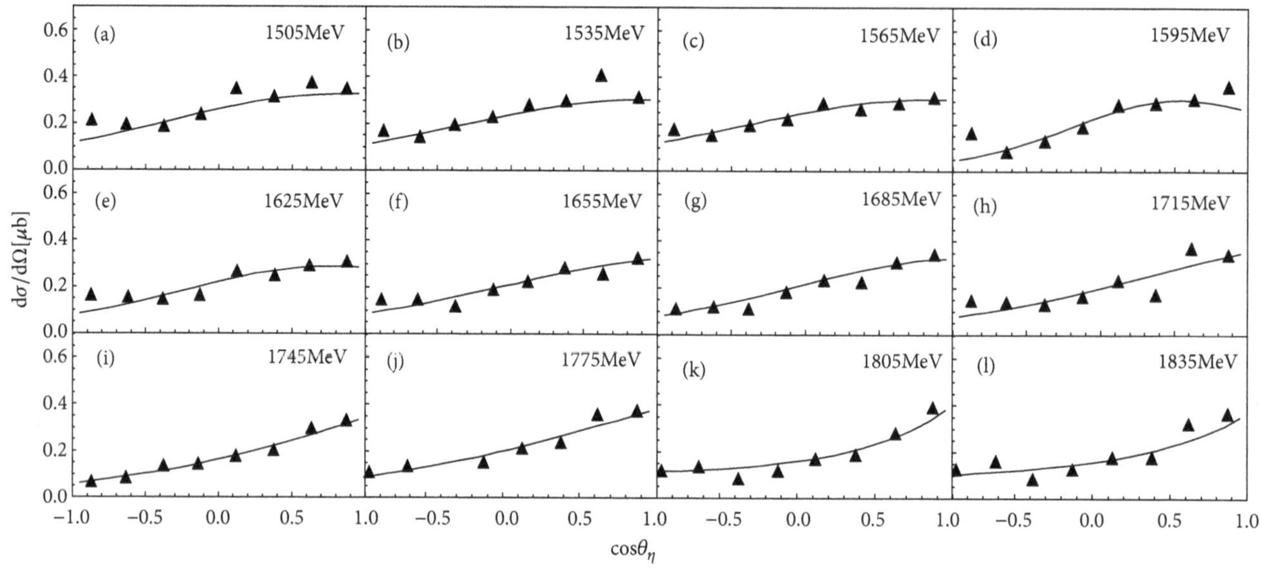

FIGURE 3: Same as Figure 1, but showing angular distributions for different bins of incident photon energy 1505 MeV $\leq E_\gamma \leq$ 1835MeV.

TABLE 2: Values of a_z and b_z taken in Figure 2 model results.

Figure 2	E_γ (MeV)	a_z	b_z	χ^2/dof
(a)	1035	0.920	-0.080	0.124
(b)	1065	0.940	-0.090	0.118
(c)	1095	0.970	-0.030	0.132
(d)	1125	0.870	0.030	0.165
(e)	1155	0.850	0.050	0.170
(f)	1185	0.850	0.030	0.168
(g)	1215	0.860	0.070	0.173
(h)	1245	0.860	0.050	0.122
(i)	1270	0.850	0.110	0.121
(j)	1295	0.830	0.150	0.145
(k)	1325	0.910	0.120	0.147
(l)	1355	0.910	0.160	0.142
(m)	1385	0.890	0.110	0.115
(n)	1415	0.930	0.130	0.106
(o)	1445	0.830	0.250	0.122
(p)	1475	0.810	0.210	0.091

TABLE 3: Values of a_z and b_z taken in Figure 3 model results.

Figure 3	E_γ (MeV)	a_z	b_z	χ^2/dof
(a)	1505	0.915	0.260	0.243
(b)	1535	0.920	0.260	0.190
(c)	1565	0.920	0.240	0.158
(d)	1595	0.800	0.360	0.315
(e)	1625	0.890	0.310	0.179
(f)	1655	0.930	0.340	0.190
(g)	1685	0.910	0.360	0.206
(h)	1715	0.940	0.400	0.235
(i)	1745	0.960	0.460	0.085
(j)	1775	0.960	0.410	0.144
(k)	1805	1.090	0.400	0.132
(l)	1835	1.060	0.390	0.140

TABLE 4: Values of a_z and b_z taken in Figure 4 model results.

Figure 4	W (MeV)	a_z	b_z	χ^2/dof
(a)	1488	0.956	-0.098	0.086
(b)	1492	0.968	-0.069	0.125
(c)	1498	0.966	-0.058	0.101
(d)	1505	0.959	-0.063	0.114
(e)	1515	0.965	-0.052	0.205
(f)	1525	0.952	-0.048	0.192
(g)	1535	0.960	-0.063	0.143
(h)	1545	0.967	-0.066	0.206
(i)	1555	0.948	-0.060	0.174
(j)	1565	0.941	-0.072	0.168
(k)	1575	0.934	-0.059	0.170
(l)	1585	0.955	-0.068	0.235
(m)	1595	0.925	-0.051	0.260
(n)	1605	0.934	-0.039	0.251
(o)	1615	0.942	-0.045	0.307
(p)	1625	0.961	-0.058	0.293

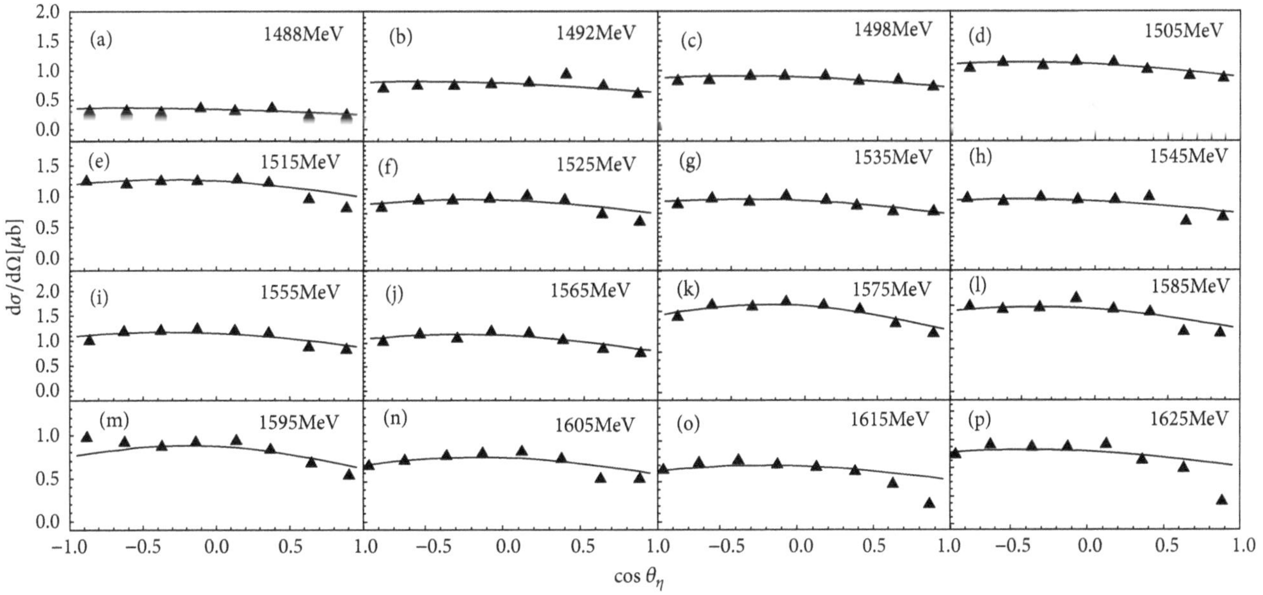

FIGURE 4: Angular distributions in the η-nucleon cm system for the reaction $\gamma p \longrightarrow p\eta$ for the different bins of final state energy 1488 MeV $\leq W \leq$ 1625 MeV. The symbols represent the experimental data from the CBELSA/TAPS detector at the electron accelerator ELSA in Bonn [9]. The results in the multisource thermal model are shown with the curves.

decades, the photon-induced reaction and electron scattering experiment are applied to study the electromagnetic excitation of baryons. Recently, the photoproduction of η mesons from quasi-free protons and neutrons are measured by the CBELSA/TAPS detector. In the paper, we theoretically study the angular distribution of η mesons for different incident photon energies E_γ and for different final state energies W. Then, the results are compared with the experimental data in detail. The deformation coefficient a_z and translation coefficient b_z are extracted by the comparison. a_z is almost independent of incident photon energies and final state energies. b_z is linearly dependent on incident photon energies and final state energies. In particular, we visually give the deformation and translation of the emission sources by schematic sketches. From the patterns, it is intuitive and easy to better understand the motion and configuration of the emission sources.

A great number of η mesons are produced in the photon-induced reaction. These η mesons are regarded as a multiparticle system, which can be analyzed by the statistical method. In recent years, we develop such a model, which is called multisource thermal model. Some emission sources of final-state particles are formed in the reaction. Each emission source emits particles isotropically in the rest frame of the emission

TABLE 5: Values of a_z and b_z taken in Figure 5 model results.

Figure 5	W (MeV)	a_z	b_z	χ^2/dof
(a)	1635	0.958	-0.042	0.305
(b)	1645	0.940	-0.028	0.350
(c)	1655	0.920	-0.021	0.263
(d)	1670	0.925	-0.009	0.187
(e)	1690	0.902	0.022	0.240
(f)	1710	0.899	0.047	0.209
(g)	1730	0.868	0.095	0.181
(h)	1750	0.843	0.102	0.194
(i)	1770	0.853	0.115	0.235
(j)	1790	0.879	0.125	0.170
(k)	1810	0.855	0.142	0.152
(l)	1830	0.867	0.159	0.138

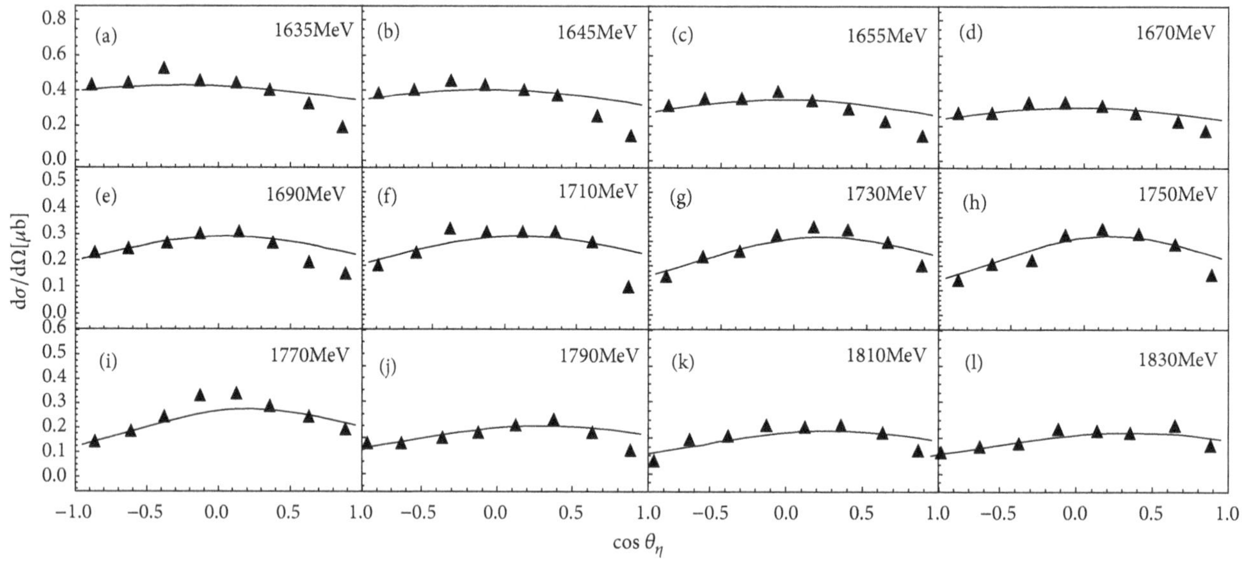

FIGURE 5: Same as Figure 4, but showing angular distributions for the different bins of final state energy 1635 MeV ≤ W ≤ 1830 MeV.

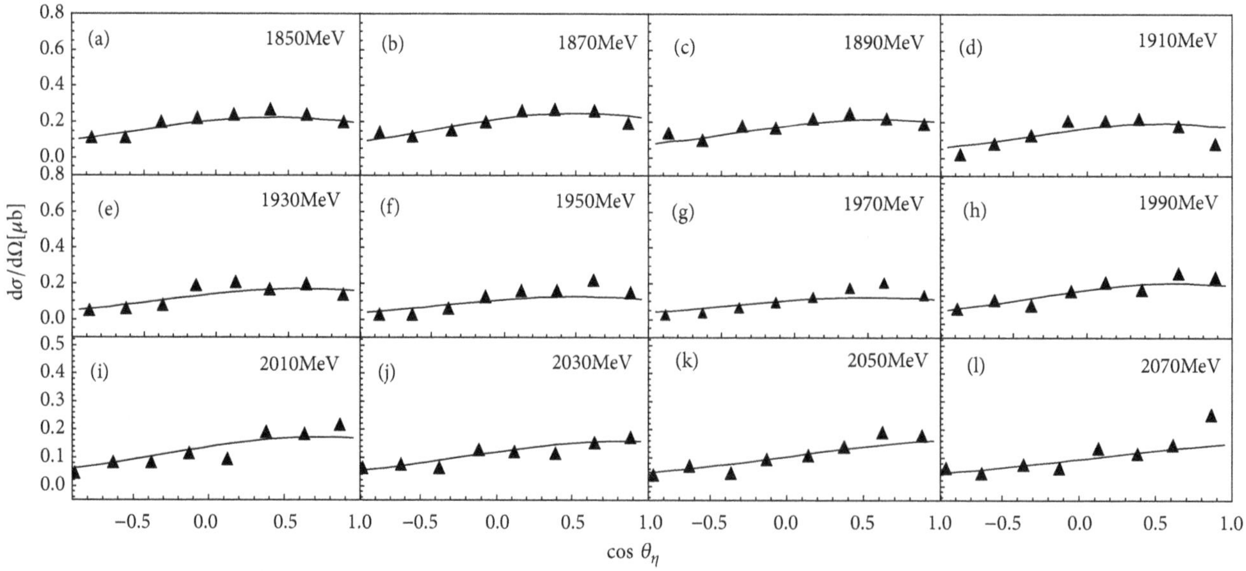

FIGURE 6: Same as Figure 4, but showing angular distributions for the different bins of final state energy 1850 MeV ≤ W ≤ 2070 MeV.

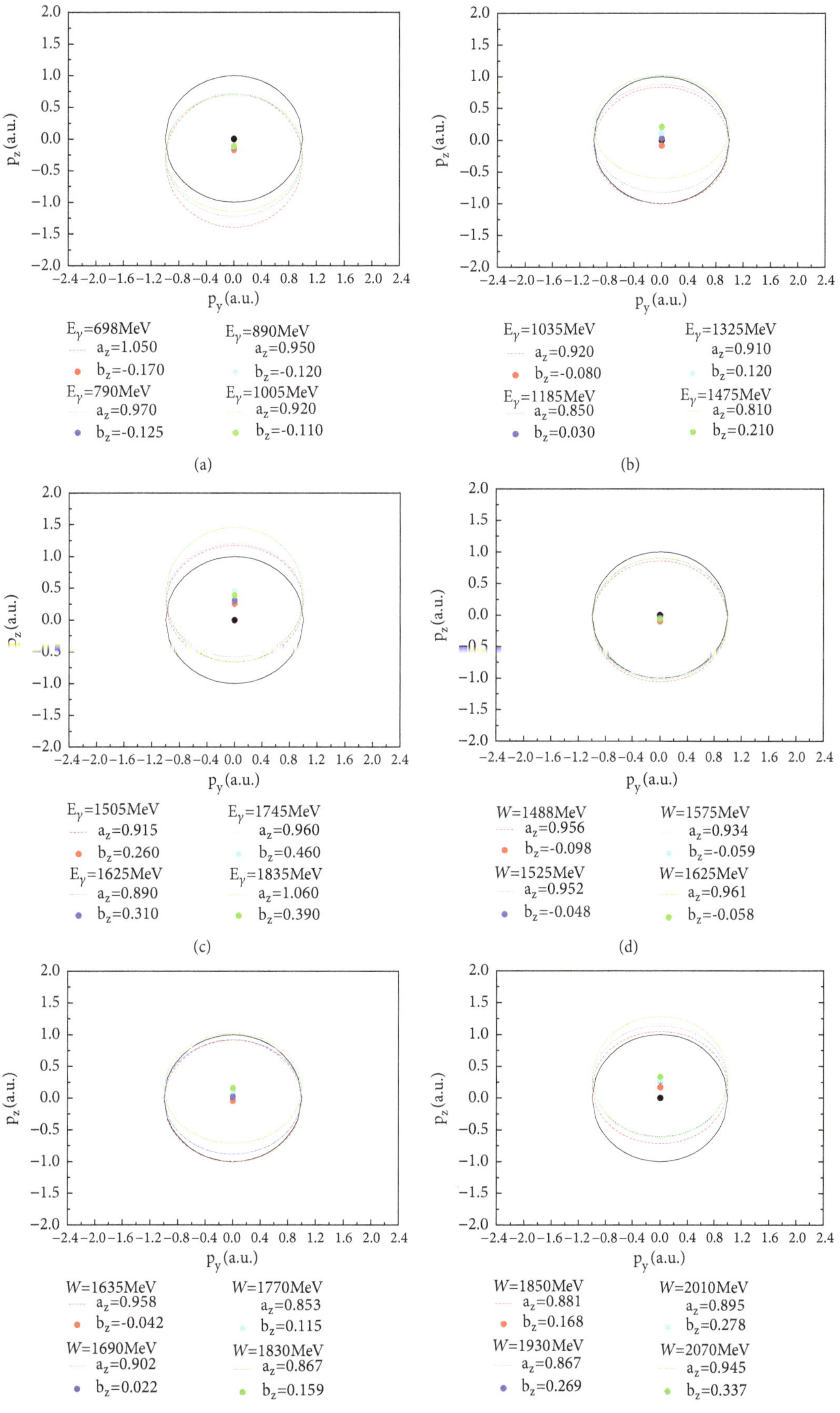

FIGURE 7: The deformable and translational source in the reaction plane for different bins of incident photon energy E_γ or final state energy W.

TABLE 6: Values of a_z and b_z taken in Figure 6 model results.

Figure 6	W (MeV)	a_z	b_z	χ^2/dof
(a)	1850	0.881	0.168	0.127
(b)	1870	0.872	0.221	0.132
(c)	1890	0.883	0.241	0.135
(d)	1910	0.856	0.252	0.153
(e)	1930	0.867	0.269	0.218
(f)	1950	0.853	0.253	0.295
(g)	1970	0.860	0.249	0.321
(h)	1990	0.875	0.261	0.184
(i)	2010	0.895	0.278	0.359
(j)	2030	0.906	0.299	0.337
(k)	2050	0.932	0.348	0.285
(l)	2070	0.945	0.337	0.304

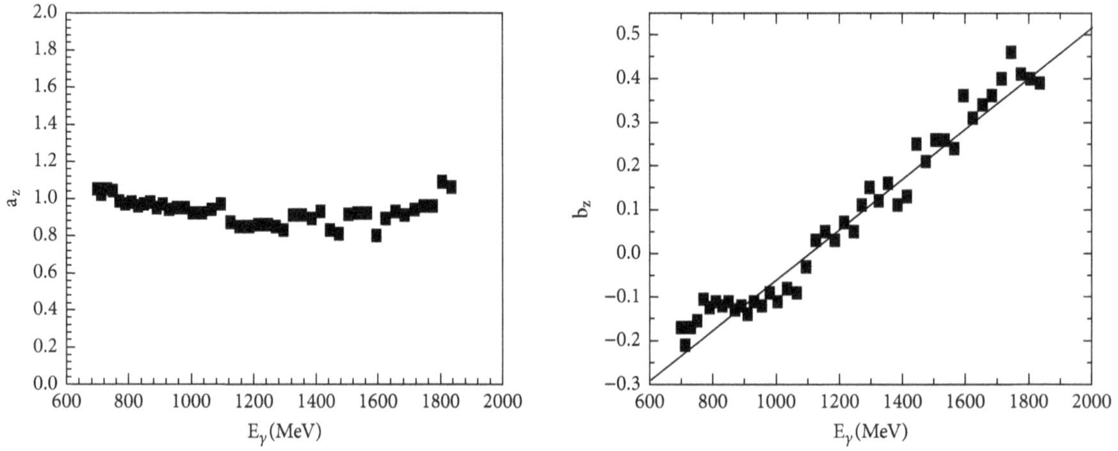

FIGURE 8: a_Z and b_Z for different bins of incident photon energy E_γ. The symbols are the values taken in Figures 1–3. The straight line is a fitted curve.

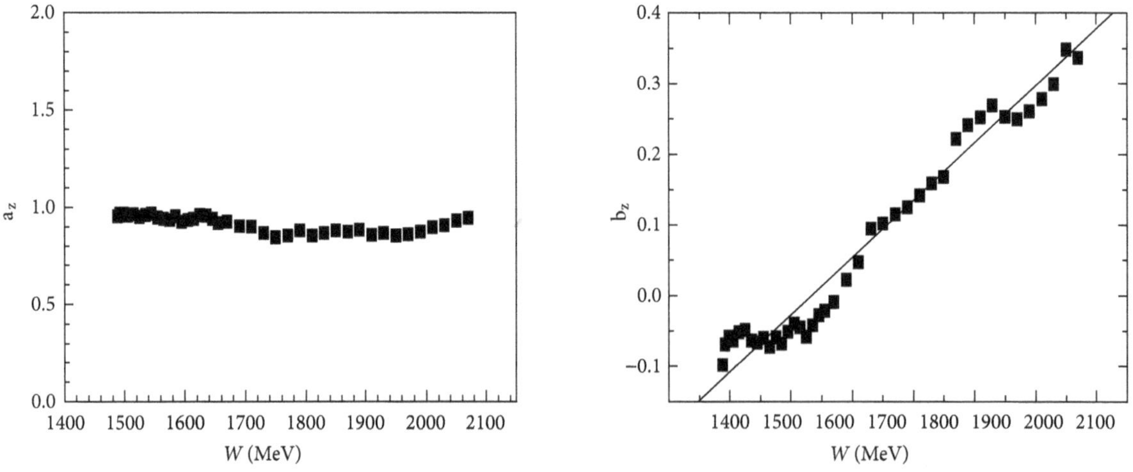

FIGURE 9: The a_Z and b_Z values taken for different bins of final state energy W. The symbols are the values taken in Figures 4–6. The straight line is a fitted curve.

source. Due to the source interaction, the sources emit particles anisotropically. The η mesons are emitted from these sources. In our previous work, the model can successfully describe transverse momentum spectra and pseudorapidity spectra of final-state particles produced in proton-proton (pp) collisions, proton-nucleus (pA) collisions, and nucleus -nucleus (AA) collisions at intermediate energy and at high energy [17–21]. In this work, we extend the multisource thermal model to the statistical investigation of final-state particles produced in the photon-induced reaction. The model is improved to describe the angular dependence of the η photoproduction from quasi-free protons and neutrons. The information of the source deformation and translation is obtained with different beam energies. It is helpful for us to understand the η photoproduction.

Conflicts of Interest

The authors declare that they have no conflicts of interest.

Acknowledgments

This work is supported by National Natural Science Foundation of China under Grants No. 11247250 and No. 11575103, Shanxi Provincial Natural Science Foundation under Grant No. 201701D121005, and Scientific and Technological Innovation Programs of Higher Education Institutions in Shanxi (STIP) Grant No. 201802017.

References

[1] B. Krusche, C. Wilkin, and Prog. Part, "Production of η and η' mesons on nucleons and nuclei," *Progress in Particle and Nuclear Physics*, vol. 80, pp. 43–95, 2014.

[2] M. Dieterle, "First measurement of the polarization observable E and helicity-dependent cross sections in single π^0 photoproduction from quasi-free nucleons," *Phys. Lett. B*, vol. 770, no. 523, 2017.

[3] B. C. Li, Y. Y. Fu, L. L. Wang, and F.-H. Liu, "Dependence of elliptic flows on transverse momentum and number of participants in Au+Au collisions at $\sqrt{s_{NN}}$=200 GeV," *Journal of Physics G: Nuclear and Particle Physics*, vol. 40, no. 2, Article ID 025104, 2013.

[4] T. Sekihara, H. Fujioka, and T. Ishikawa, "Possible η'd bound state and its bound state and its s-channel formation in the γd $\longrightarrow$ ηd reaction," *Physical Review C*, vol. 97, Article ID 045202, 2018.

[5] A. V. Anisovich, "Neutron helicity amplitudes," *Physical Review C*, vol. 96, Article ID 055202, 2017.

[6] V. Kuznetsov, "Observation of Narrow $N^+(1685)$ and N^o Resonances in $_\gamma N \longrightarrow \pi\eta N$ Reactions," *JETP Lett*, vol. 106, no. 11, pp. 693–699, 2017.

[7] J. D. Holt, J. Menéndez, J. Simonis, and A. Schwenk, "Three-nucleon forces and spectroscopy of neutron-rich calcium isotopes," *Physical Review C: Nuclear Physics*, vol. 90, no. 2, Article ID 024312, 2014.

[8] S. A. Voloshin, "Collective phenomena in ultra-relativistic nuclear collisions: anisotropic flow and more," *Progress in Particle and Nuclear Physics*, vol. 67, no. 2, pp. 541–546, 2012.

[9] L. Witthauer, "Photoproduction of η mesons from the neutron: Cross sections and double polarization observable E," *The European Physical Journal A*, vol. 53, no. 58, 2017.

[10] J. Rafelski and J. Letessier, "Testing limits of statistical hadronization," *Nuclear Physics A*, vol. 715, p. 98, 2003.

[11] A. Andronic, P. Braun-Munzinger, and J. Stachel, "Thermal hadron production in relativistic nuclear collisions: The hadron mass spectrum, the horn, and the QCD phase transition," *Physics Letters B*, vol. 673, no. 2, pp. 142–145, 2009.

[12] J. Cleymans, H. Oeschler, K. Redlich, and S. Wheaton, "Comparison of chemical freeze-out criteria in heavy-ion collisions," *Physical Review C: Nuclear Physics*, vol. 73, no. 4, Article ID 034905, 2006.

[13] P. Braun-Munzinger, J. Stachel, and C. Wetterich, "Chemical freeze-out and the QCD phase transition temperature," *Physics Letters B*, vol. 596, no. 1-2, pp. 61–69, 2004.

[14] A. Adare, S. Afanasiev, and C. Aidala, "Measurement of neutral mesons in p+p collisions at $\sqrt{s}$=200 GeV and scaling properties of hadron production," *Physical Review D: Particles, Fields, Gravitation and Cosmology*, vol. 83, Article ID 052004, 2011.

[15] K. Aamodt, N. Abel, and U. Abeysekara, "Transverse momentum spectra of charged particles in proton–proton collisions at $\sqrt{s}$ = 900 GeV with ALICE at the LHC," *Physics Letters B*, vol. 693, no. 2, pp. 53–68, 2010.

[16] C. Y. Wong and G. Wilk, "Tsallis fits to p_T spectra and multiple hard scattering in pp collisions at the LHC," *Physical Review D: Particles, Fields, Gravitation and Cosmology*, vol. 87, Article ID 114007, 2013.

[17] B. C. Li, Y. Y. Fu, E. Q. Wang, L. L. Wang, and F. H. Liu, "Transverse momentum dependence of charged and strange hadron elliptic flows in Cu–Cu collisions," *Journal of Physics G: Nuclear and Particle Physics*, vol. 39, no. 8, Article ID 025009, 2012.

[18] S. Andringa, E. Arushanova, and S. Asahi, "Current status and future prospects of the SNO+ experiment," *Advances in High Energy Physics*, vol. 2016, Article ID 6194250, 21 pages, 2016.

[19] B.-C. Li, Y.-Z. Wang, F.-H. Liu, X.-J. Wen, and Y.-E. Dong, "Particle production in relativistic pp and AA collisions at RHIC and LHC energies with Tsallis statistics using the two-cylindrical multisource thermal model," *Physical Review D: Particles, Fields, Gravitation and Cosmology*, vol. 89, Article ID 054014, 2014.

[20] B. C. Li, Y. Z. Wang, and F. H. Liu, "Formulation of transverse mass distributions in Au–Au collisions at $\sqrt{sNN}$ = 200 GeV/nucleon," *Physics Letters B*, vol. 725, no. 4-5, pp. 352–356, 2013.

[21] B.-C. Li, Z. Zhang, J.-H. Kang, G.-X. Zhang, and F.-H. Liu, "Tsallis statistical interpretation of transverse momentum spectra in high-energy pA collisions," *Advances in High Energy Physics*, vol. 2015, Article ID 741816, 10 pages, 2015.

2

Probing the Effects of New Physics in $\overline{B}^{*} \longrightarrow P\ell\nu_{\ell}$ Decays

Qin Chang,[1] Jie Zhu,[1,2] Na Wang,[1] and Ru-Min Wang[2]

[1]*Institute of Particle and Nuclear Physics, Henan Normal University, Henan 453007, China*
[2]*Institute of Theoretical Physics, Xinyang Normal University, Henan 464000, China*

Correspondence should be addressed to Qin Chang; changqin@htu.edu.cn

Academic Editor: Enrico Lunghi

The significant divergence between the SM predictions and experimental measurements for the ratios, $R_{D^{(*)}} \equiv \mathscr{B}(\overline{B} \longrightarrow D^{(*)}\tau^{-}\overline{\nu}_{\tau})/\mathscr{B}(\overline{B} \longrightarrow D^{(*)}\ell'^{-}\overline{\nu}_{\ell'})$ with $(\ell' = e, \mu)$, implies possible hint of new physics in the flavor sector. In this paper, motivated by the "$R_{D^{(*)}}$ puzzle" and abundant B^{*} data samples at high-luminosity heavy-flavor experiments in the future, we try to probe possible effects of new physics in the semileptonic $\overline{B}^{*}_{u,d,s} \longrightarrow P\ell^{-}\overline{\nu}_{\ell}$ $(P = D, D_s, \pi, K)$ decays induced by $b \longrightarrow (u,c)\ell^{-}\overline{\nu}_{\ell}$ transitions in the model-independent vector and scalar scenarios. Using the spaces of NP parameters obtained by fitting to the data of R_D and R_{D^*}, the NP effects on the observables including branching fraction, ratio R^{*}_{P}, lepton spin asymmetry, and lepton forward-backward asymmetry are studied in detail. We find that the vector type couplings have large effects on the branching fraction and ratio R^{*}_{P}. Meanwhile, the scalar type couplings provide significant contributions to all of the observables. The future measurements of these observables in the $\overline{B}^{*} \longrightarrow P\ell^{-}\overline{\nu}_{\ell}$ decays at the LHCb and Belle-II could provide a way to crosscheck the various NP solutions to the "$R_{D^{(*)}}$ puzzle".

1. Introduction

Thanks to the fruitful running of the B factories and Large Hadron Collider (LHC) in the past years, most of the $B_{u,d}$ mesons decays with branching fractions $\gtrsim \mathcal{O}(10^{-7})$ have been measured. The rare B-meson decays play an important role in testing the standard model (SM) and probing possible hints of new physics (NP). Although most of the experimental measurements are in good agreement with the SM predictions, several indirect hints for NP, the tensions or the so-called puzzles, have been observed in the flavor sector.

The semileptonic $\overline{B} \longrightarrow D^{(*)}\ell\overline{\nu}_{\ell}$ decays are induced by the CKM favored tree-level charged current, and therefore, their physical observables could be rather reliably predicted in the SM and the effects of NP are expected to be tiny. In particular, the ratios defined by $R_{D^{(*)}} \equiv \mathscr{B}(\overline{B} \longrightarrow D^{(*)}\tau^{-}\overline{\nu}_{\tau})/\mathscr{B}(\overline{B} \longrightarrow D^{(*)}\ell'^{-}\overline{\nu}_{\ell'})$ $(\ell' = e, \mu)$ are independent of the CKM matrix elements, and the hadronic uncertainties canceled to a large extent; thus they could be predicted with a rather high accuracy. However, the BaBar [1, 2], Belle [3–5], and LHCb [6] collaborations have recently observed some anomalies in these ratios. The latest experimental average values for $R_{D^{(*)}}$ reported by the Heavy-Flavor Average Group (HFAG) are [7]

$$R_{D}^{\mathrm{Exp}} = 0.403 \pm 0.040 \pm 0.024,$$
$$R_{D^*}^{\mathrm{Exp}} = 0.310 \pm 0.015 \pm 0.008, \quad (1)$$

which deviate from the SM predictions

$$R_{D}^{\mathrm{SM}} = 0.300 \pm 0.008,$$
$$R_{D^*}^{\mathrm{SM}} = 0.252 \pm 0.003, \quad (2)$$

(see [8] in the first equation of (2) and [9] in the second equation of (2)) at the levels of 2.2σ and 3.4σ errors, respectively. Moreover, when the correlations between R_D and R^{*}_{D} are taken into account, the tension would reach up to 3.9σ level [7]. Besides, the ratio $R_{J/\psi} \equiv \mathscr{B}(B_c \longrightarrow J/\psi\tau^{-}\overline{\nu}_{\tau})/\mathscr{B}(B_c \longrightarrow J/\psi\mu^{-}\overline{\nu}_{\mu})$ has recently been measured by the LHCb collaboration [10], which also shows an excess of about 2σ from the central value range of the corresponding SM predictions [0.25, 0.28]. In addition, another mild hint of NP in the $b \longrightarrow u\ell\bar{\nu}$ induced $B \longrightarrow \tau\bar{\nu}$ decay has been observed by the BaBar and Belle collaborations [11–14]; the deviation is at the level of 1.4σ [15].

The large deviations in $R_{D^{(*)}}$ and possible anomalies in the other decay channels mentioned above imply possible hints of NP relevant to the lepton flavor violation (LFV) [15]. The investigations for these anomalies have been made extensively both within model-independent frameworks [16–37] and in some specific NP models where the $b \longrightarrow c\tau\bar{\nu}_\tau$ transition is mediated by leptoquarks [16, 17, 38–46], charged Higgses [16, 47–59], charged vector bosons [16, 60, 61], and sparticles [62–65].

In addition to B mesons, the vector ground states of $b\bar{q}$ system, B^* mesons, with quantum number of $n^{2s+1}L_J = 1^3S_1$ and $J^P = 1^-$ [66–69], also can decay through the $b \longrightarrow (u,c)\ell\bar{\nu}_\ell$ transitions at quark-level. Therefore, in principle, the corresponding NP effects might enter into the semileptonic B^* decays as well. The B^* decay occurs mainly through the electromagnetic process $\bar{B}^* \longrightarrow \bar{B}\gamma$, and the weak decay modes are very rare. Fortunately, thanks to the rapid development of heavy-flavor experiments instruments and techniques, the B^* weak decays are hopeful to be observed by the running LHC and forthcoming SuperKEK/Belle-II experiments [70–72] in the near future. For instance, the annual integrated luminosity of Belle-II is expected to reach up to $\sim 13\ ab^{-1}$ and the B^* weak decays with branching fractions $> \mathcal{O}(10^{-9})$ are hopeful to be observed [70, 73, 74]. Moreover, the LHC experiment also will provide a lot of experimental information for B^* weak decays due to the much larger beauty production cross-section of pp collision relative to e^+e^- collision [75].

Recently, some interesting theoretical studies for the B^* weak decays have been made within the SM in [73, 74, 76–82]. In this paper, motivated by the possible NP explanation for the $R_{D^{(*)}}$ puzzles, the corresponding NP effects on the semileptonic B^* decays will be studied in a model-independent way. In the investigation, the scenarios of vector and scalar NP interactions are studied, respectively; their effects on the branching fraction, differential branching fraction, lepton spin asymmetry, forward-backward asymmetry, and ratio R_P^* $(P = D, \pi, K)$ of semileptonic B^* decays are explored by using the spaces of various NP couplings obtained through the measured $R_{D^{(*)}}$.

Our paper is organized as follows. In Section 2, after a brief description of the effective Lagrangian for the $b \longrightarrow (u,c)\ell\bar{\nu}_\ell$ transitions, the theoretical framework and calculations for the $\bar{B}^* \longrightarrow P\ell\bar{\nu}_\ell$ decays in the presence of various NP couplings are presented. Section 3 is devoted to the numerical results and discussions for the effects of various NP couplings. Finally, we give our conclusions in Section 4.

2. Theoretical Framework and Calculation

2.1. Effective Lagrangian and Amplitudes. We employ the effective field theory approach to compute the amplitudes of $\bar{B}^* \longrightarrow P\ell\bar{\nu}_\ell$ decays in a model-independent scheme. The most general effective Lagrangian at $\mu = \mathcal{O}(m_b)$ for the $b \longrightarrow p\ell^-\bar{\nu}_\ell$ $(p = u, c)$ transition can be written as [19, 21, 40, 46]

$$\begin{aligned}\mathcal{L}_{\text{eff}} = -2\sqrt{2}G_F \sum_{p=u,c} V_{pb} \{&(1+V_L)\,\bar{p}_L\gamma^\mu b_L \bar{\ell}_L\gamma_\mu\nu_L \\ &+ V_R\bar{p}_R\gamma^\mu b_R\bar{\ell}_L\gamma_\mu\nu_L + \widetilde{V}_L\bar{p}_L\gamma^\mu b_L\bar{\ell}_R\gamma_\mu\nu_R \\ &+ \widetilde{V}_R\bar{p}_R\gamma^\mu b_R\bar{\ell}_R\gamma_\mu\nu_R + S_L\bar{p}_R b_L\bar{\ell}_R\nu_L + S_R\bar{p}_L b_R\bar{\ell}_R\nu_L \\ &+ \widetilde{S}_L\bar{p}_R b_L\bar{\ell}_L\nu_R + \widetilde{S}_R\bar{p}_L b_R\bar{\ell}_L\nu_R + T_L\bar{p}_R\sigma^{\mu\nu}b_L\bar{\ell}_R\sigma_{\mu\nu}\nu_L \\ &+ \widetilde{T}_L\bar{p}_L\sigma^{\mu\nu}b_R\bar{\ell}_L\sigma_{\mu\nu}\nu_R\} + \text{h.c.},\end{aligned} \tag{3}$$

where G_F is the Fermi coupling constant, V_{pb} denotes the CKM matrix elements, and $P_{L,R} = (1 \pm \gamma_5)/2$ is the negative/positive projection operator. Assuming the neutrinos are left-handed and neglecting the tensor couplings, the effective Lagrangian can be simplified as

$$\begin{aligned}\mathcal{L}_{\text{eff}} = &-\frac{G_F}{\sqrt{2}} \\ &\cdot \sum_{p=u,c} V_{pb}\{(1+V_L)\,\bar{p}\gamma_\mu(1-\gamma_5)\,b\bar{\ell}\gamma^\mu(1-\gamma_5)\,\nu \\ &+ V_R\bar{p}\gamma_\mu(1+\gamma_5)\,b\bar{\ell}\gamma^\mu(1-\gamma_5)\,\nu \\ &+ S_L\bar{p}(1-\gamma_5)\,b\bar{\ell}(1-\gamma_5)\,\nu \\ &+ S_R\bar{p}(1+\gamma_5)\,b\bar{\ell}(1-\gamma_5)\,\nu\} + \text{h.c.},\end{aligned} \tag{4}$$

where $V_{L,R}$ and $S_{L,R}$ are the effective NP couplings (Wilson coefficients) defined at $\mu = \mathcal{O}(m_b)$. In the SM, all the NP couplings will be zero.

We use the method of [83–87] to calculate the helicity amplitudes. The square of amplitudes for the $\bar{B}^* \longrightarrow P\ell^-\bar{\nu}_\ell$ decay can be written as the product of leptonic ($L_{\mu\nu}$) and hadronic ($H^{\mu\nu}$) tensors,

$$\begin{aligned}\left|\mathcal{M}\left(\bar{B}^* \longrightarrow P\ell^-\bar{\nu}_\ell\right)\right|^2 &= \left|\left\langle P\ell^-\bar{\nu}_\ell \left|\mathcal{L}_{\text{eff}}\right| \bar{B}^*\right\rangle\right|^2 \\ &= \sum_{i,j} L^{ij}_{\mu\nu}H^{ij,\mu\nu},\end{aligned} \tag{5}$$

where the superscripts i and j refer to four operators in the effective Lagrangian given by (4) (the tensors related to the scalar and pseudoscalar operators can be understood through the relations given by (21) and (22)); in the SM, $i = j$ corresponds to the operator $\bar{p}\gamma_\mu(1-\gamma_5)b\bar{\ell}\gamma^\mu(1-\gamma_5)\nu$. For convenience in writing, these superscripts are omitted below. Inserting the completeness relation

$$\sum_{m,n} \bar{\epsilon}_\mu(m)\,\bar{\epsilon}^*_\nu(n)\,g_{mn} = g_{\mu\nu}, \tag{6}$$

the product of $L_{\mu\nu}$ and $H^{\mu\nu}$ can be further expressed as

$$L_{\mu\nu}H^{\mu\nu} = \sum_{m,m',n,n'} L(m,n)\,H(m',n')\,g_{mm'}g_{nn'}. \tag{7}$$

Here, $\bar{\epsilon}_\mu$ is the polarization vector of the virtual intermediate states, which is W^* boson in the SM and named as ω in this paper for convenience of expression. The quantities $L(m,n) \equiv L^{\mu\nu}\bar{\epsilon}_\mu(m)\bar{\epsilon}^*_\nu(n)$ and $H(m,n) \equiv H^{\mu\nu}\bar{\epsilon}^*_\mu(m)\bar{\epsilon}_\nu(n)$ are Lorentz invariant and therefore can be evaluated in different reference frames. In the following evaluation, $H(m,n)$ and $L(m,n)$ will be calculated in the B^*-meson rest frame and the $\ell-\bar{\nu}_\ell$ center-of-mass frame, respectively.

2.2. Kinematics for $\overline{B}^* \longrightarrow P\ell^-\bar{\nu}_\ell$ Decays. In the B^*-meson rest frame with daughter P-meson moving in the positive z-direction, the momenta of particles B^* and P are

$$\begin{aligned} p^\mu_{B^*} &= (m_{B^*}, 0, 0, 0), \\ p^\mu_P &= \left(E_P, 0, 0, \left|\vec{p}\right|\right). \end{aligned} \tag{8}$$

For the four polarization vectors, $\bar{\epsilon}^\mu(\lambda_\omega = t, 0, \pm)$, one can conveniently choose [83, 84]

$$\begin{aligned} \bar{\epsilon}^\mu(t) &= \frac{1}{\sqrt{q^2}}\left(q_0, 0, 0, -\left|\vec{p}\right|\right), \\ \bar{\epsilon}^\mu(0) &= \frac{1}{\sqrt{q^2}}\left(\left|\vec{p}\right|, 0, 0, -q_0\right), \\ \bar{\epsilon}^\mu(\pm) &= \frac{1}{\sqrt{2}}(0, \pm 1, -i, 0), \end{aligned} \tag{9}$$

where $q_0 = (m^2_{B^*} - m^2_P + q^2)/2m_{B^*}$ and $|\vec{p}| = \lambda^{1/2}(m^2_{B^*}, m^2_P, q^2)/2m_{B^*}$, with $\lambda(a,b,c) \equiv a^2+b^2+c^2-2(ab+bc+ca)$ and $q^2 = (p_{B^*} - p_P)^2$ being the momentum transfer squared, are the energy and momentum of the virtual ω. The polarization vectors of the initial B^*-meson can be written as

$$\begin{aligned} \epsilon^\mu(0) &= (0, 0, 0, 1), \\ \epsilon^\mu(\pm) &= \frac{1}{\sqrt{2}}(0, \mp 1, -i, 0). \end{aligned} \tag{10}$$

In the $\ell - \bar{\nu}_\ell$ center-of-mass frame, the four momenta of lepton and antineutrino pair are given as

$$\begin{aligned} p^\mu_\ell &= \left(E_\ell, \left|\vec{p}_\ell\right|\sin\theta, 0, \left|\vec{p}_\ell\right|\cos\theta\right), \\ p^\mu_{\bar{\nu}_\ell} &= \left(\left|\vec{p}_\ell\right|, -\left|\vec{p}_\ell\right|\sin\theta, 0, -\left|\vec{p}_\ell\right|\cos\theta\right), \end{aligned} \tag{11}$$

where $E_\ell = (q^2 + m^2_\ell)/2\sqrt{q^2}$, $|\vec{p}_\ell| = (q^2 - m^2_\ell)/2\sqrt{q^2}$, and θ is the angle between the P and ℓ three-momenta. In this frame, the polarization vector $\bar{\epsilon}^\mu$ takes the form

$$\begin{aligned} \bar{\epsilon}^\mu(t) &= (1, 0, 0, 0), \\ \bar{\epsilon}^\mu(0) &= (0, 0, 0, 1), \\ \bar{\epsilon}^\mu(\pm) &= \frac{1}{\sqrt{2}}(0, \mp 1, -i, 0). \end{aligned} \tag{12}$$

2.3. Hadronic Helicity Amplitudes. For the $\overline{B}^* \longrightarrow P\ell^-\bar{\nu}_\ell$ decay, the hadronic helicity amplitudes $H^{V_{L,R}}_{\lambda_{B^*}\lambda_\omega}$ and $H^{S_{L,R}}_{\lambda_{B^*}\lambda_\omega}$ are defined by

$$\begin{aligned} &H^{V_L}_{\lambda_{B^*}\lambda_\omega}\left(q^2\right) \\ &\quad = \bar{\epsilon}^{*\mu}(\lambda_\omega)\left\langle P(p_P)\left|\bar{p}\gamma_\mu(1-\gamma_5)b\right|\overline{B}^*(p_{B^*}, \lambda_{B^*})\right\rangle, \end{aligned} \tag{13}$$

$$\begin{aligned} &H^{V_R}_{\lambda_{B^*}\lambda_\omega}\left(q^2\right) \\ &\quad = \bar{\epsilon}^{*\mu}(\lambda_\omega)\left\langle P(p_P)\left|\bar{p}\gamma_\mu(1+\gamma_5)b\right|\overline{B}^*(p_{B^*}, \lambda_{B^*})\right\rangle, \end{aligned} \tag{14}$$

$$H^{S_L}_{\lambda_{B^*}\lambda_\omega}\left(q^2\right) = \left\langle P(p_P)\left|\bar{p}(1-\gamma_5)b\right|\overline{B}^*(p_{B^*}, \lambda_{B^*})\right\rangle, \tag{15}$$

$$H^{S_R}_{\lambda_{B^*}\lambda_\omega}\left(q^2\right) = \left\langle P(p_P)\left|\bar{p}(1+\gamma_5)b\right|\overline{B}^*(p_{B^*}, \lambda_{B^*})\right\rangle, \tag{16}$$

which describe the decay of three helicity states of B^* meson into a pseudoscalar P meson and the four helicity states of virtual ω. It should be noted that λ_ω in $H^{S_{L,R}}_{\lambda_{B^*}\lambda_\omega}(q^2)$, (15) and (16), should always be equal to t.

For $B^* \longrightarrow P$ transition, the matrix elements of the vector and axial-vector currents can be written in terms of form factors $V(q^2)$ and $A_{0,1,2}(q^2)$ as

$$\begin{aligned} &\left\langle P(p_P)\left|\bar{p}\gamma_\mu b\right|\overline{B}^*(\epsilon, p_{B^*})\right\rangle = -\frac{2iV\left(q^2\right)}{m_{B^*} + m_P} \\ &\quad \cdot \varepsilon_{\mu\nu\rho\sigma}\epsilon^\nu p^\rho_P p^\sigma_{B^*}, \end{aligned} \tag{17}$$

$$\begin{aligned} &\left\langle P(p_P)\left|\bar{p}\gamma_\mu\gamma_5 b\right|\overline{B}^*(\epsilon, p_{B^*})\right\rangle = 2m_{B^*}A_0\left(q^2\right)\frac{\epsilon\cdot q}{q^2}q_\mu \\ &\quad + (m_P + m_{B^*})A_1\left(q^2\right)\left(\epsilon_\mu - \frac{\epsilon\cdot q}{q^2}q_\mu\right) + A_2\left(q^2\right) \\ &\quad \cdot \frac{\epsilon\cdot q}{m_P + m_{B^*}}\left[(p_{B^*} + p_P)_\mu - \frac{m^2_{B^*} - m^2_P}{q^2}q_\mu\right], \end{aligned} \tag{18}$$

with the sign convention $\epsilon_{0123} = -1$. Furthermore, using the equations of motion,

$$i\partial_\mu\left(\bar{p}\gamma^\mu b\right) = \left[m_b(\mu) - m_p(\mu)\right]\bar{p}b, \tag{19}$$

$$i\partial_\mu\left(\bar{p}\gamma^\mu\gamma_5 b\right) = -\left[m_b(\mu) + m_p(\mu)\right]\bar{p}\gamma_5 b, \tag{20}$$

one can write the matrix elements of scalar and pseudoscalars currents as

$$\begin{aligned} &\left\langle P(p_P)\left|\bar{p}b\right|\overline{B}^*(\epsilon, p_{B^*})\right\rangle = \frac{1}{m_b(\mu) - m_p(\mu)} \\ &\quad \cdot q_\mu\left\langle P(p_P)\left|\bar{p}\gamma^\mu b\right|\overline{B}^*(\epsilon, p_{B^*})\right\rangle = 0, \end{aligned} \tag{21}$$

$$\begin{aligned} &\left\langle P(p_P)\left|\bar{p}\gamma_5 b\right|\overline{B}^*(\epsilon, p_{B^*})\right\rangle = -\frac{1}{m_b(\mu) + m_p(\mu)} \\ &\quad \cdot q_\mu\left\langle P(p_P)\left|\bar{p}\gamma^\mu\gamma_5 b\right|\overline{B}^*(\epsilon, p_{B^*})\right\rangle = -(\epsilon\cdot q) \\ &\quad \cdot \frac{2m_{B^*}}{m_b(\mu) + m_p(\mu)}A_0\left(q^2\right), \end{aligned} \tag{22}$$

in which $m_b(\mu)$ and $m_p(\mu)$ are the running quark masses.

Then, by contracting above hadronic matrix elements with the polarization vectors in the B^*-meson rest frame, we obtain five nonvanishing helicity amplitudes

$$H_{0t}(q^2) = H_{0t}^{V_L}(q^2) = -H_{0t}^{V_R}(q^2) = \frac{2m_{B^*}|\vec{p}|}{\sqrt{q^2}} \cdot A_0(q^2), \tag{23}$$

$$H_{00}(q^2) = H_{00}^{V_L}(q^2) = -H_{00}^{V_R}(q^2) = \frac{1}{2m_{B^*}\sqrt{q^2}}\left[(m_{B^*}+m_P)(m_{B^*}^2-m_P^2+q^2) \cdot A_1(q^2) + \frac{4m_{B^*}^2|\vec{p}|^2}{m_{B^*}+m_P}A_2(q^2)\right], \tag{24}$$

$$H_{\pm\mp}(q^2) = H_{\pm\mp}^{V_L}(q^2) = -H_{\mp\pm}^{V_R}(q^2) = -(m_{B^*}+m_P) \cdot A_1(q^2) \mp \frac{2m_{B^*}|\vec{p}|}{m_{B^*}+m_P}V(q^2), \tag{25}$$

$$H'_{0t}(q^2) = H_{0t}^{S_L}(q^2) = -H_{0t}^{S_R}(q^2) = -\frac{2m_{B^*}|\vec{p}|}{m_b(\mu)+m_c(\mu)}A_0(q^2). \tag{26}$$

It is obvious that only the amplitudes with $\lambda_{B^*} = \lambda_P - \lambda_\omega = -\lambda_\omega$ survive.

2.4. Leptonic Helicity Amplitudes. Expanding the leptonic tensor in terms of a complete set of Wigner's d^J-functions [9, 83, 87], $L_{\mu\nu}H^{\mu\nu}$ can be rewritten as a compact form

$$L_{\mu\nu}H^{\mu\nu} = \frac{1}{8}\sum_{\lambda_\ell,\lambda_{\bar{\nu}_\ell},\lambda_\omega,\lambda'_\omega,J,J'}(-1)^{J+J'} \cdot h^i_{\lambda_\ell,\lambda_{\bar{\nu}_\ell}} h^{j*}_{\lambda_\ell,\lambda_{\bar{\nu}_\ell}} \delta_{\lambda_{B^*},-\lambda_\omega}\delta_{\lambda_{B^*},-\lambda'_\omega} \times d^J_{\lambda_\omega,\lambda_\ell-1/2} d^{J'}_{\lambda'_\omega,\lambda_\ell-1/2} H^i_{\lambda_{B^*}\lambda_\omega} H^{j*}_{\lambda_{B^*}\lambda'_\omega}, \tag{27}$$

in which J and J' run over 1 and 0, $\lambda_\omega^{(\prime)}$ and λ_ℓ run over their components, and massless right-handed antineutrinos with $\lambda_{\bar{\nu}_\ell} = 1/2$. In (27), $h^{i,j}_{\lambda_\ell,\lambda_{\bar{\nu}_\ell}}$ are the leptonic helicity amplitudes defined as

$$h^{V_{L,R}}_{\lambda_\ell,\lambda_{\bar{\nu}_\ell}} = \bar{u}_\ell(\lambda_\ell)\gamma^\mu(1-\gamma_5)\nu_{\bar{\nu}}\left(\frac{1}{2}\right)\bar{\epsilon}_\mu(\lambda_\omega), \tag{28}$$

$$h^{S_{L,R}}_{\lambda_\ell,\lambda_{\bar{\nu}_\ell}} = \bar{u}_\ell(\lambda_\ell)(1-\gamma_5)\nu_{\bar{\nu}}\left(\frac{1}{2}\right). \tag{29}$$

In the $\ell - \bar{\nu}_\ell$ center-of-mass frame, taking the exact forms of the spinors and polarization vectors, we finally obtain four nonvanishing contributions

$$\left|h^{V_{L,R}}_{-1/2,1/2}\right|^2 = 8(q^2 - m_\ell^2), \tag{30}$$

$$\left|h^{V_{L,R}}_{1/2,1/2}\right|^2 = 8\frac{m_\ell^2}{2q^2}(q^2 - m_\ell^2), \tag{31}$$

$$\left|h^{S_{L,R}}_{1/2,1/2}\right|^2 = 8\frac{q^2 - m_\ell^2}{2}, \tag{32}$$

$$\left|h^{V_{L,R}}_{1/2,1/2}\right| \times \left|h^{S_{L,R}}_{1/2,1/2}\right| = 8\frac{m_\ell}{2\sqrt{q^2}}(q^2 - m_\ell^2). \tag{33}$$

2.5. Observables of $\overline{B}^* \longrightarrow P\ell^-\overline{\nu}_\ell$ Decays. With the amplitudes obtained in above subsections, we then present the observables considered in our following evaluations. The double differential decay rate of $\overline{B}^* \longrightarrow P\ell^-\overline{\nu}_\ell$ decay is written as

$$\frac{d\Gamma}{dq^2 d\cos\theta} = \frac{G_F^2|V_{pb}|^2}{(2\pi)^3}\frac{|\vec{p}|}{8m_{B^*}^2}\frac{1}{3}\left(1-\frac{m_\ell^2}{q^2}\right) \cdot \left|\mathcal{M}\left(\overline{B}^* \longrightarrow P\ell^-\overline{\nu}_\ell\right)\right|^2, \tag{34}$$

where the factor 1/3 is caused by averaging over the spins of initial state $\overline{B}^*$. Using the standard convention for d^J-function [88], we finally obtain the double differential decay rates with a given leptonic helicity state ($\lambda_\ell = \pm 1/2$), which are

$$\begin{aligned}\frac{d^2\Gamma[\lambda_\ell = -1/2]}{dq^2 d\cos\theta} &= \frac{G_F^2|V_{pb}|^2|\vec{p}|}{256\pi^3 m_{B^*}^2}\frac{1}{3}q^2\left(1-\frac{m_\ell^2}{q^2}\right)^2 \\ &\times \Big\{|1+V_L|^2\left[(1-\cos\theta)^2 H_{-+}^2 + (1+\cos\theta)^2 H_{-+}^2 + 2\sin^2\theta H_{00}^2\right] + |V_R|^2\left[(1-\cos\theta)^2 H_{+-}^2 + (1+\cos\theta)^2 H_{-+}^2 + 2\sin^2\theta H_{00}^2\right] \\ &- 4\mathcal{Re}\left[(1+V_L)V_R^*\right]\left[(1+\cos\theta^2)H_{+-}H_{-+} + \sin^2\theta H_{00}^2\right]\Big\},\end{aligned} \tag{35}$$

$$\begin{aligned}\frac{d^2\Gamma[\lambda_\ell = 1/2]}{dq^2 d\cos\theta} &= \frac{G_F^2|V_{pb}|^2|\vec{p}|}{256\pi^3 m_{B^*}^2}\frac{1}{3}q^2\left(1-\frac{m_\ell^2}{q^2}\right)^2\frac{m_\ell^2}{q^2} \\ &\times \Bigg\{\left(|1+V_L|^2+|V_R|^2\right)\left[\sin^2\theta\left(H_{-+}^2+H_{+-}^2\right) + 2\left(H_{0t}-\cos\theta H_{00}\right)^2\right] - 4\mathcal{Re}\left[(1+V_L)V_R^*\right] \\ &\cdot \left[\sin^2\theta H_{-+}H_{+-} + \left(H_{0t}-\cos\theta H_{00}\right)^2\right] + 4\mathcal{Re}\left[(1+V_L-V_R)(S_L^*-S_R^*)\right] \\ &\cdot \frac{\sqrt{q^2}}{m_\ell}\left[H'_{0t}\left(H_{0t}-\cos\theta H_{00}\right)\right] + 2|S_L-S_R|^2\frac{q^2}{m_\ell^2} \cdot H'^2_{0t}\Bigg\}.\end{aligned} \tag{36}$$

Using (35) and (36), one can get the explicit forms of various observables of $\overline{B}^* \longrightarrow P\ell^-\overline{\nu}_\ell$ decays as follows:

(i) The differential decay rate is

$$\frac{d\Gamma}{dq^2} = \frac{G_F^2 \left|V_{pb}\right|^2 \left|\overrightarrow{p}\right|}{96\pi^3 m_{B^*}^2} \frac{1}{3} q^2 \left(1 - \frac{m_\ell^2}{q^2}\right)^2 \times \left\{ \left(\left|1 + V_L\right|^2 + \left|V_R\right|^2\right) \cdot \left[\left(H_{-+}^2 + H_{+-}^2 + H_{00}^2\right)\left(1 + \frac{m_\ell^2}{2q^2}\right) + \frac{3m_\ell^2}{2q^2} H_{0t}^2\right] - 2\mathcal{Re}\left[(1 + V_L) V_R^*\right] \cdot \left[\left(2H_{-+}H_{+-} + H_{00}^2\right)\left(1 + \frac{m_\ell^2}{2q^2}\right) + \frac{3m_\ell^2}{2q^2} H_{0t}^2\right] + 3\mathcal{Re}\left[(1 + V_L - V_R)(S_L^* - S_R^*)\right] H_{0t}' H_{0t} \frac{m_\ell}{\sqrt{q^2}} + \frac{3}{2}\left|S_L - S_R\right|^2 H_{0t}'^2 \right\}. \tag{37}$$

(ii) The q^2 dependent ratio is

$$R_P^*\left(q^2\right) \equiv \frac{d\Gamma\left(\overline{B}^* \longrightarrow P\tau^-\overline{\nu}_\tau\right)/dq^2}{d\Gamma\left(\overline{B}^* \longrightarrow P\ell'^-\overline{\nu}_{\ell'}\right)/dq^2}, \tag{38}$$

where ℓ' denotes the light lepton.

(iii) The lepton spin asymmetry is

$$A_\lambda^P\left(q^2\right) = \frac{d\Gamma\left[\lambda_\ell = -1/2\right]/dq^2 - d\Gamma\left[\lambda_\ell = 1/2\right]/dq^2}{d\Gamma\left[\lambda_\ell = -1/2\right]/dq^2 + d\Gamma\left[\lambda_\ell = 1/2\right]/dq^2}. \tag{39}$$

(iv) The forward-backward asymmetry is

$$A_\theta^P\left(q^2\right) = \frac{\int_{-1}^0 d\cos\theta\left(d^2\Gamma/dq^2 d\cos\theta\right) - \int_0^1 d\cos\theta\left(d^2\Gamma/dq^2 d\cos\theta\right)}{d\Gamma/dq^2}. \tag{40}$$

The SM results can be obtained from above formulae by taking $V_L = V_R = S_L = S_R = 0$.

In the following evaluations, in order to fit the NP spaces, we also need the observables of $\overline{B} \longrightarrow D^{(*)}\ell^-\overline{\nu}_\ell$ decays, which have been fully calculated in the past years. In this paper, we adopt the relevant theoretical formulae given in [46].

3. Numerical Results and Discussions

3.1. Input Parameters. Before presenting our numerical results and analyses, we would like to clarify the values of input parameters used in the calculation. For the CKM matrix elements, we use [89]

$$\left|V_{cb}\right| = 4.181_{-0.060}^{+0.028} \times 10^{-2}, \quad \left|V_{ub}\right| = 3.715_{-0.060}^{+0.060} \times 10^{-3}. \tag{41}$$

For the well-measured Fermi coupling constant G_F, the masses of mesons and leptons, and the running masses of quarks at $\mu = m_b$, we take their central values given by PDG [88]. The total decay widths (or lifetimes) of B^* mesons are essential for estimating the branching fraction; however there is no available experimental data until now. According to the fact that the electromagnetic process $B^* \longrightarrow B\gamma$ dominates the decays of B^* meson, we take the approximation $\Gamma_{\text{tot}}(B^*) \simeq \Gamma(B^* \longrightarrow B\gamma)$; the latter has been evaluated within different theoretical models [90–96]. In this paper, we adopt the most recent results [95, 96]

$$\Gamma_{\text{tot}}\left(B^{*+}\right) \simeq \Gamma\left(B^{*+} \longrightarrow B^+\gamma\right) = \left(468_{-75}^{+73}\right) \text{eV}, \tag{42}$$

$$\Gamma_{\text{tot}}\left(B^{*0}\right) \simeq \Gamma\left(B^{*0} \longrightarrow B^0\gamma\right) = (148 \pm 20) \text{eV}, \tag{43}$$

$$\Gamma_{\text{tot}}\left(B_s^{*0}\right) \simeq \Gamma\left(B_s^{*0} \longrightarrow B_s^0\gamma\right) = (68 \pm 17) \text{eV}. \tag{44}$$

Then the residual inputs are the transition form factors, which are crucial for evaluating the observables of $\overline{B}^* \longrightarrow P\ell^-\overline{\nu}_\ell$ and $\overline{B} \longrightarrow D^{(*)}\ell^-\overline{\nu}_\ell$ decays. For the $B \longrightarrow D^{(*)}$ transitions, the scheme of Caprini, Lellouch, and Neubert (CLN) parametrization [97] is widely used, and the CLN parameters can be precisely extracted from the well-measured $\overline{B} \longrightarrow D^{(*)}\ell^-\overline{\nu}_\ell$ decays; numerically, their values read [7]

$$\rho_D^2 = 1.128 \pm 0.033, \quad V_1(1)\left|V_{cb}\right| = (41.30 \pm 0.99) \times 10^{-3}; \tag{45}$$

$$\rho_{D^*}^2 = 1.205 \pm 0.026, \quad h_{A_1}(1)\left|V_{cb}\right| = (35.38 \pm 0.43) \times 10^{-3}, \quad R_1(1) = 1.404 \pm 0.032, \quad R_2(1) = 0.854 \pm 0.020. \tag{46}$$

However, for the $\overline{B}_{u,d,s}^* \longrightarrow P_{u,d,s}$ transition, there is no experimental data and ready-made theoretical results to use at present. Here, we employ the Bauer-Stech-Wirbel (BSW) model [98, 99] to evaluate the form factors for both $\overline{B}^* \longrightarrow P$ and $\overline{B} \longrightarrow D^{(*)}$ transitions. Using the inputs $m_u = m_d = 0.35\,\text{GeV}$, $m_s = 0.55\,\text{GeV}$, $m_c = 1.7\,\text{GeV}$, $m_b = 4.9\,\text{GeV}$, and $\omega = \sqrt{\langle \overrightarrow{p}_\perp^2 \rangle} = 0.4\,\text{GeV}$, we obtain the results at $q^2 = 0$,

$$A_0^{\overline{B}^* \longrightarrow D}(0) = 0.71,$$

$$A_1^{\overline{B}^* \longrightarrow D}(0) = 0.75,$$

$$A_2^{\bar{B}^* \to D}(0) = 0.62,$$
$$V^{\bar{B}^* \to D}(0) = 0.76; \tag{47}$$

$$A_0^{\bar{B}_s^* \to D_s}(0) = 0.66,$$
$$A_1^{\bar{B}_s^* \to D_s}(0) = 0.69,$$
$$A_2^{\bar{B}_s^* \to D_s}(0) = 0.59, \tag{48}$$
$$V^{\bar{B}_s^* \to D_s}(0) = 0.72;$$

$$A_0^{\bar{B}^* \to \pi}(0) = 0.34,$$
$$A_1^{\bar{B}^* \to \pi}(0) = 0.38,$$
$$A_2^{\bar{B}^* \to \pi}(0) = 0.30, \tag{49}$$
$$V^{\bar{B}^* \to \pi}(0) = 0.35;$$

$$A_0^{\bar{B}_s^* \to K}(0) = 0.28,$$
$$A_1^{\bar{B}_s^* \to K}(0) = 0.29,$$
$$A_2^{\bar{B}_s^* \to K}(0) = 0.26, \tag{50}$$
$$V^{\bar{B}_s^* \to K}(0) = 0.30;$$

$$F_0^{\bar{B} \to D}(0) = F_1^{\bar{B} \to D}(0) = 0.70; \tag{51}$$

$$A_0^{\bar{B} \to D^*}(0) = 0.63,$$
$$A_1^{\bar{B} \to D^*}(0) = 0.66,$$
$$A_2^{\bar{B} \to D^*}(0) = 0.69, \tag{52}$$
$$V^{\bar{B} \to D^*}(0) = 0.71.$$

To be conservative, 15% uncertainties are assigned to these values in our following evaluation. Moreover, with the assumption of nearest pole dominance, the dependences of form factors on q^2 read [98, 99]

$$F_0(q^2) \simeq \frac{F_0(0)}{1 - q^2/m^2_{B_q(0^+)}},$$
$$F_1(q^2) \simeq \frac{F_1(0)}{1 - q^2/m^2_{B_q(1^-)}},$$
$$A_0(q^2) \simeq \frac{A_0(0)}{1 - q^2/m^2_{B_q(0^-)}},$$
$$A_1(q^2) \simeq \frac{A_1(0)}{1 - q^2/m^2_{B_q(1^+)}},$$
$$A_2(q^2) \simeq \frac{A_2(0)}{1 - q^2/m^2_{B_q(1^+)}},$$
$$V(q^2) \simeq \frac{V(0)}{1 - q^2/m^2_{B_q(1^-)}}, \tag{53}$$

where $B_q(J^P)$ is the state of B_q with quantum number of J^P (J and P are the quantum numbers of total angular momenta and parity, respectively).

With the theoretical formulae and inputs given above, we then proceed to present our numerical results and discussion, which are divided into two scenarios with different simplification for our attention to the types of NP couplings as follows:

(i) Scenario I: taking $S_L = S_R = 0$, *i.e.*, only considering the NP effects of $V_{L,R}$ couplings

(ii) Scenario II: taking $V_L = V_R = 0$, *i.e.*, only considering the NP effects of $S_{L,R}$ couplings

In these two scenarios, we consider all the NP parameters to be real for our analysis. In addition, we assume that only the third generation leptons get corrections from the NP in the $b \longrightarrow (u,c)\ell\bar{\nu}_\ell$ processes and for $\ell = e, \mu$ the NP is absent. In the following discussion, the allowed spaces of NP couplings are obtained by fitting to R_D and R_{D^*} (1), with the data varying randomly within their 1σ error, while the theoretical uncertainties are also considered and obtained by varying the inputs randomly within their ranges specified above.

3.2. Scenario I: Effects of V_L and V_R Type Couplings. In this subsection, we vary couplings V_L and V_R while keeping all other NP couplings to zero. Under the constraints from the data of R_D and R_D^*, the allowed spaces of new physics parameters, V_L and V_R, are shown in Figure 1. In the fit, the $B \longrightarrow D^{(*)}$ form factors based on CLN parametrization and BSW model are used, respectively; it can be seen from Figure 1 that their corresponding fitting results are consistent with each other, but the constraint with the former is much stronger due to the relatively small theoretical error. Therefore, in the following evaluations and discussions, the results obtained by using CLN parametrization are used. In addition, our fitting result in Figure 1 agrees well with the ones obtained in the previous works, for instance, [26, 35].

From Figure 1, we find that (i) the allowed spaces of (V_L, V_R) are bounded into four separate regions, namely, solutions A-D. (ii) Except for solution A, the other solutions are all far from the zero point $(0,0)$ and result in very large NP contributions. Taking solution C (D) as an example, the SM contribution is completely canceled out by the NP contribution related to V_L, and the V_R coupling presents sizable positive (negative) NP contribution to fit data. The situation of solution B is similar, but only V_L coupling presents sizable NP contribution. Numerically, one can easily conclude that the NP contributions of solutions B-D are about two times larger than the SM, which seriously exceeds our general

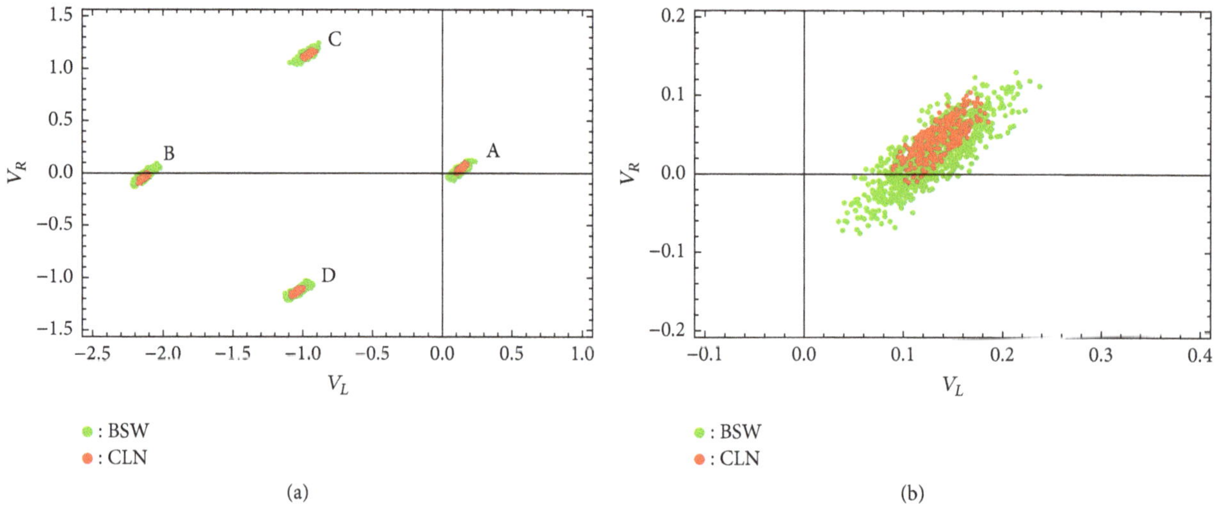

FIGURE 1: The allowed spaces of V_L and V_R obtained by fitting to R_D and R_{D^*}. The red and green regions are obtained by using the form factors of CLN parametrization and BSW model, respectively. (b) shows the minimal result (solution A) of the four solutions shown in (a).

expectation that the amplitudes should be dominated by the SM and the NP only presents minor corrections. In this point of view, the minimal solution (solution A) is much favored than solutions B-D. So, in our following discussions, we pay attention only to solution A, which is replotted in Figure 1(b) and numerical result is

$$V_L = 0.14^{+0.06}_{-0.06},$$
$$V_R = 0.05^{+0.06}_{-0.07}. \quad (54)$$

solution A

Using the values of NP couplings given by (54), we then present our theoretical predictions for $\mathscr{B}(\overline{B}^* \longrightarrow P\tau^-\overline{\nu}_\tau)$ and q^2-integrated R_P^* in Table 1, in which the SM results are also listed for comparison. The q^2-dependence of differential observables $d\Gamma/dq^2$, R_P^*, A_λ^P, and A_θ^P for $B^{*-} \longrightarrow D^0\tau^-\overline{\nu}_\tau$ and $\pi^0\tau^-\overline{\nu}_\tau$ decays is shown in Figure 2; the case of $\overline{B}^{*0} \longrightarrow D^+\tau^-\overline{\nu}_\tau$ and $\overline{B}_s^{*0} \longrightarrow D_s^+\tau^-\overline{\nu}_\tau$ ($\overline{B}^{*0} \longrightarrow \pi^+\tau^-\overline{\nu}_\tau$ and $\overline{B}_s^{*0} \longrightarrow K^+\tau^-\overline{\nu}_\tau$) is similar to the one of $B^{*-} \longrightarrow D^0\tau^-\overline{\nu}_\tau$ ($B^{*-} \longrightarrow \pi^0\tau^-\overline{\nu}_\tau$) decay, not shown here. The following are some discussions and comments:

(1) From Table 1, it can be seen that the branching fractions of $b \longrightarrow c\tau\overline{\nu}_\tau$ induced $\overline{B}^*_{u,d,s}$ decays are at the level of $\mathcal{O}(10^{-8} - 10^{-7})$, while the $b \longrightarrow u\tau\overline{\nu}_\tau$ induced decays are relatively rare due to the suppression caused by the CKM factor. In addition, the difference between the branching fractions of three decay modes induced by $b \longrightarrow c\tau\overline{\nu}_\tau$ (or $b \longrightarrow u\tau\overline{\nu}_\tau$) transition is mainly attributed to the relation of total decay widths, $\Gamma_{\text{tot}}(B^{*-}) : \Gamma_{\text{tot}}(\overline{B}^{*0}) : \Gamma_{\text{tot}}(\overline{B}_s^{*0}) \sim 1 : 2 : 6$, illustrated by (42), (43), and (44).

(2) Comparing with the SM results, one can easily find from Table 1 that $\mathscr{B}(\overline{B}^* \longrightarrow P\tau^-\overline{\nu}_\tau)$ are enhanced about 20% by the NP contributions of V_L and V_R. It can also be clearly seen from Figures 2(a) and 2(b). However, as shown in Figures 2(a) and 2(b), due to the large theoretical uncertainties caused by the form factors, the NP hints are hard to be totally distinguished from the SM results.

(3) The theoretical uncertainties can be well controlled by using the ratio R_P^* instead of decay rate due to the cancelation of nonperturbative errors; therefore R_P^* is much suitable for probing the NP hints. From the last three rows of Table 1, it can be found that the NP prediction for R_P^* significantly deviates from the SM result. In particular, as Figures 2(c) and 2(d) show, the NP effects can be totally distinguished from the SM at $q^2 \gtrsim 7\,\text{GeV}^2$ even though the theoretical errors are considered. So, future measurements on $\overline{B}^* \longrightarrow P\tau^-\overline{\nu}_\tau$ decays can make further test on the NP models which provide possible solutions to the R_D and R_{D^*} problems.

(4) From Figures 2(e)–2(h) it can be found that the NP contribution of solution A has little effect on the observables A_λ^P and A_θ^P in the whole q^2 region, which can be understood from the following analyses. Because the NP contribution of solution A is dominated by the left-handed coupling V_L, we can find that $|\mathscr{M}(\overline{B}^* \longrightarrow P\ell^-\overline{\nu}_\ell)| \propto |(1 + V_L)|^2$ in the limit of $(1 + V_L) \gg V_R$. As a result, the NP contributions (solution A) to the numerator and denominator of A_λ^P and A_θ^P cancel each other out to a large extent. For A_λ^P, the cases of solutions B, C, and D are similar to solution A.

3.3. Scenario II: Effects of S_L and S_R Type Couplings. In this subsection, we only consider the effects of scalar interactions S_L and S_R and take the other NP couplings to be zero. Under the 1σ constraint from the data of R_D and R_D^*, the allowed

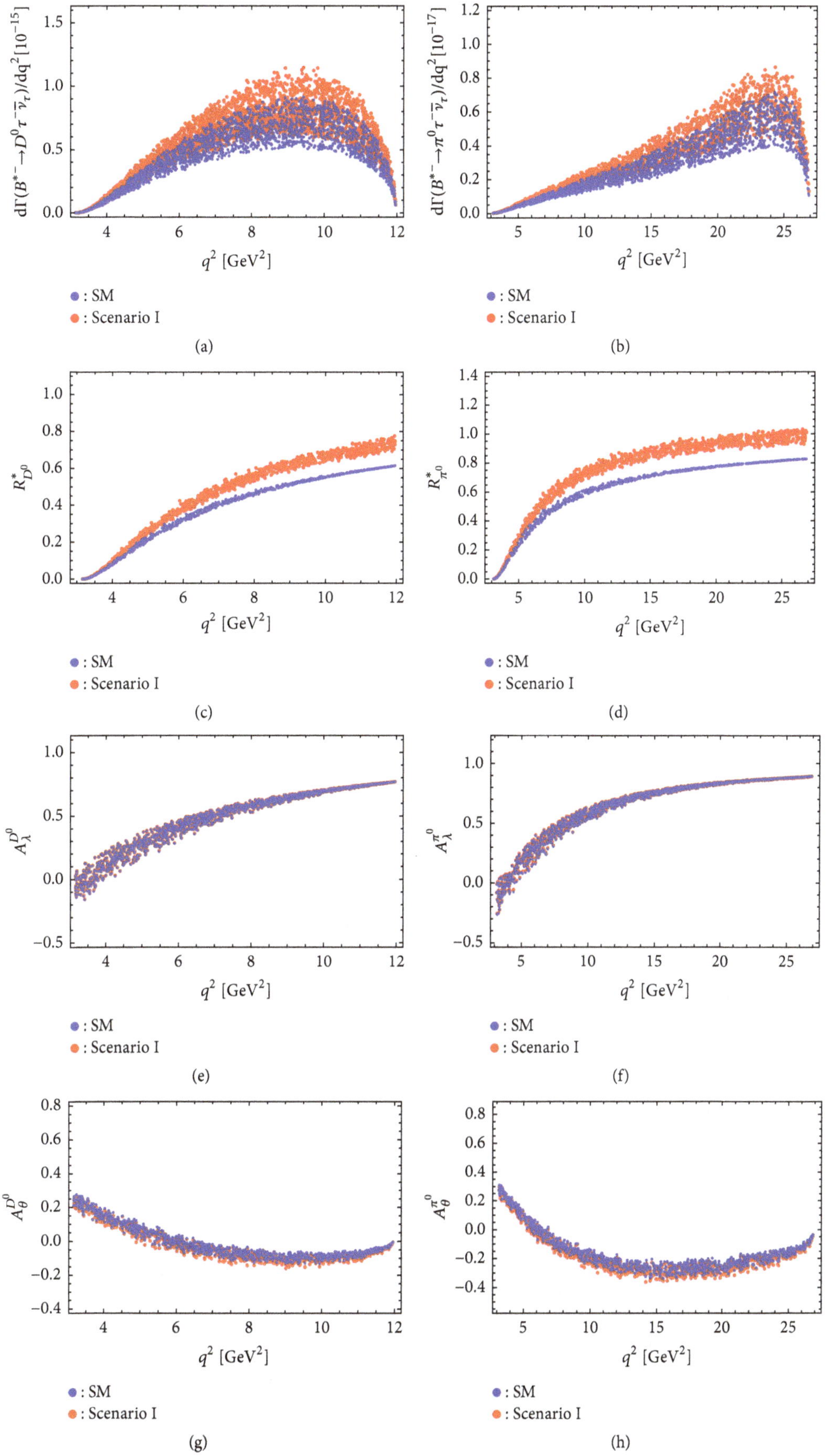

FIGURE 2: The q_2-dependence of the differential observables $d\Gamma/dq_2$, R_{*P}, $A_{P\lambda}$, and $A_{P\theta}$ for $B_{*-} \longrightarrow D_0\tau_{-}\nu_{\tau}$ and $\pi_0\tau_{-}\nu_{\tau}$ decays within the SM and scenario I.

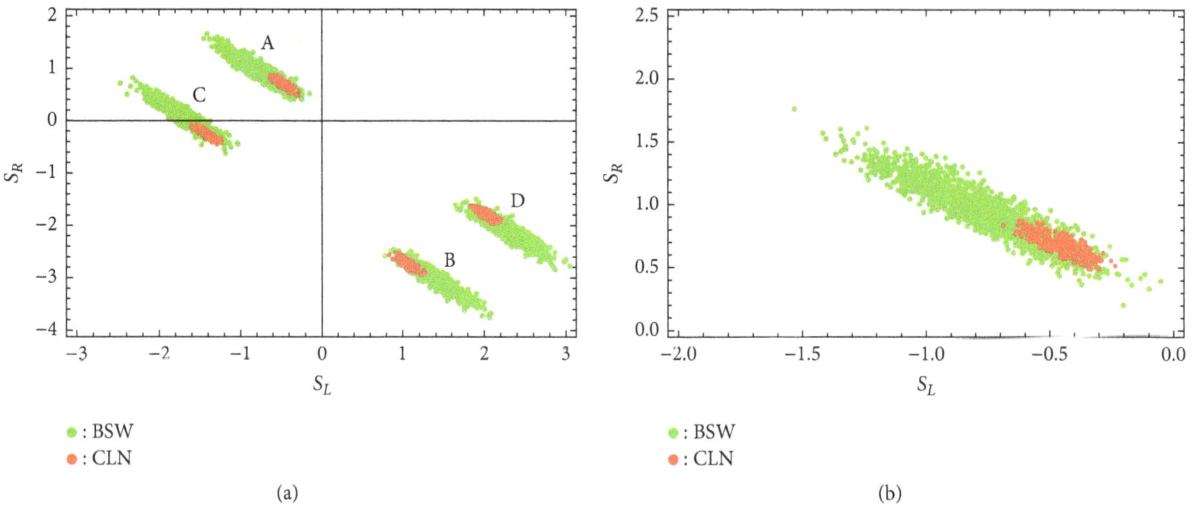

FIGURE 3: The allowed spaces of S_L and S_R obtained by fitting to the data of R_D and R_{D^*}. The other captions are the same as in Figure 1.

TABLE 1: The theoretical predictions for the branching fractions of $\overline{B}^* \longrightarrow P\tau^-\overline{\nu}_\tau$ decays and R_P^* within the SM and the two scenarios. The first error is caused by the uncertainties of form factors, CKM factors, and $\Gamma_{tot}(B^*)$; and the second error given in the last two columns is caused by the NP couplings.

Obs.	SM Prediction	Scenario I	Scenario II
$\mathcal{B}(B^{*-} \longrightarrow D^0\tau^-\overline{\nu}_\tau)$	$0.87^{+0.46}_{-0.32} \times 10^{-8}$	$1.04^{+0.54+0.06}_{-0.38-0.05} \times 10^{-8}$	$1.00^{+0.51+0.03}_{-0.36-0.04} \times 10^{-8}$
$\mathcal{B}(\overline{B}^{*0} \longrightarrow D^+\tau^-\overline{\nu}_\tau)$	$2.74^{+1.29}_{-0.94} \times 10^{-8}$	$3.27^{+1.66+0.19}_{-1.14-0.15} \times 10^{-8}$	$3.13^{+1.52+0.10}_{-1.13-0.11} \times 10^{-8}$
$\mathcal{B}(\overline{B}_s^{*0} \longrightarrow D_s^+\tau^-\overline{\nu}_\tau)$	$5.13^{+3.67}_{-2.13} \times 10^{-7}$	$6.13^{+4.51+0.35}_{-2.48-0.28} \times 10^{-7}$	$5.89^{+3.93+0.20}_{-2.39-0.22} \times 10^{-7}$
$\mathcal{B}(B^{*-} \longrightarrow \pi^0\tau^-\overline{\nu}_\tau)$	$1.42^{+0.79}_{-0.50} \times 10^{-10}$	$1.71^{+0.91+0.09}_{-0.63-0.07} \times 10^{-10}$	$1.74^{+0.94+0.10}_{-0.62-0.10} \times 10^{-10}$
$\mathcal{B}(\overline{B}^{*0} \longrightarrow \pi^+\tau^-\overline{\nu}_\tau)$	$0.99^{+0.38}_{-0.41} \times 10^{-9}$	$1.08^{+0.55+0.06}_{-0.37-0.05} \times 10^{-9}$	$1.09^{+0.52+0.06}_{-0.39-0.06} \times 10^{-9}$
$\mathcal{B}(\overline{B}_s^{*0} \longrightarrow K^+\tau^-\overline{\nu}_\tau)$	$0.95^{+0.65}_{-0.40} \times 10^{-9}$	$1.14^{+0.78+0.06}_{-0.46-0.05} \times 10^{-9}$	$1.20^{+0.87+0.08}_{-0.47-0.08} \times 10^{-9}$
R_D^*	$0.298^{+0.012}_{-0.010}$	$0.355^{+0.015+0.020}_{-0.011-0.016}$	$0.341^{+0.048+0.011}_{-0.026-0.012}$
R_π^*	$0.677^{+0.013}_{-0.014}$	$0.816^{+0.017+0.044}_{-0.012-0.035}$	$0.827^{+0.126+0.046}_{-0.073-0.048}$
R_K^*	$0.638^{+0.017}_{-0.015}$	$0.770^{+0.021+0.042}_{-0.015-0.034}$	$0.810^{+0.144+0.052}_{-0.084-0.054}$

spaces of S_L and S_R are shown in Figure 3. Similar to scenario I, four solutions for S_L and S_R are found in scenario II, which can be seen from Figure 3(a); and the fitting results obtained by using form factors in CLN parametrization and BSW model are consistent with each other. Solutions B-D result in so large NP contributions; therefore, in the following discussion, we pay our attention to solution A, which are replotted in Figure 3(b). The numerical result of solution A is

$$S_L = -0.46^{+0.24}_{-0.24},$$
$$S_R = 0.70^{+0.23}_{-0.24}. \quad (55)$$

Using these values, we present in Table 1 our numerical predictions of scenario II for the observables, $\mathcal{B}(\overline{B}^* \longrightarrow P\tau^-\overline{\nu}_\tau)$ and q^2-integrated R_P^*. Moreover, the q^2 distributions of differential observables $d\Gamma/dq^2$, R_P^*, A_λ^P, and A_θ^P are shown in Figure 4. The following are some discussions for these results:

(1) From Table 1 and Figures 4(a) and 4(b), it can be found that the $\mathcal{B}(\overline{B}^* \longrightarrow P\tau^-\overline{\nu}_\tau)$ and R_P^* can be enhanced about 15% compared with the SM results by the NP contributions. Similar to the situation of scenario I, the NP effect of S_L and S_R on R_P^* is much more significant than the one on branching fraction due to the theoretical uncertainties of R_P^* which can be well controlled. In particular, as Figures 4(a) and 4(b) show, the spectra of the SM and NP for R_P^* can be clearly distinguished at middle q^2 region.

(2) The main difference between the effects of scalar and vector couplings on the $\overline{B}^* \longrightarrow P\tau^-\overline{\nu}_\tau$ decays is that the former only contributes to the longitudinal amplitude, which can be found from (37). As a result, their effects on $\mathcal{B}(\overline{B}^* \longrightarrow P\tau^-\overline{\nu}_\tau)$ and R_P^* are a little different, which can be seen by comparing Figures 2(a)–2(d) with Figures 4(a)–4(d).

(3) Another significant difference between the scalar and vector couplings is that only the leptonic helicity amplitudes of scalar type with $\lambda_\ell = 1/2$ survive, which can be easily found from (35) and (36). Therefore, as Figures 4(e) and 4(f) show, the scalar couplings lead to significant NP effects on A_λ^P, which

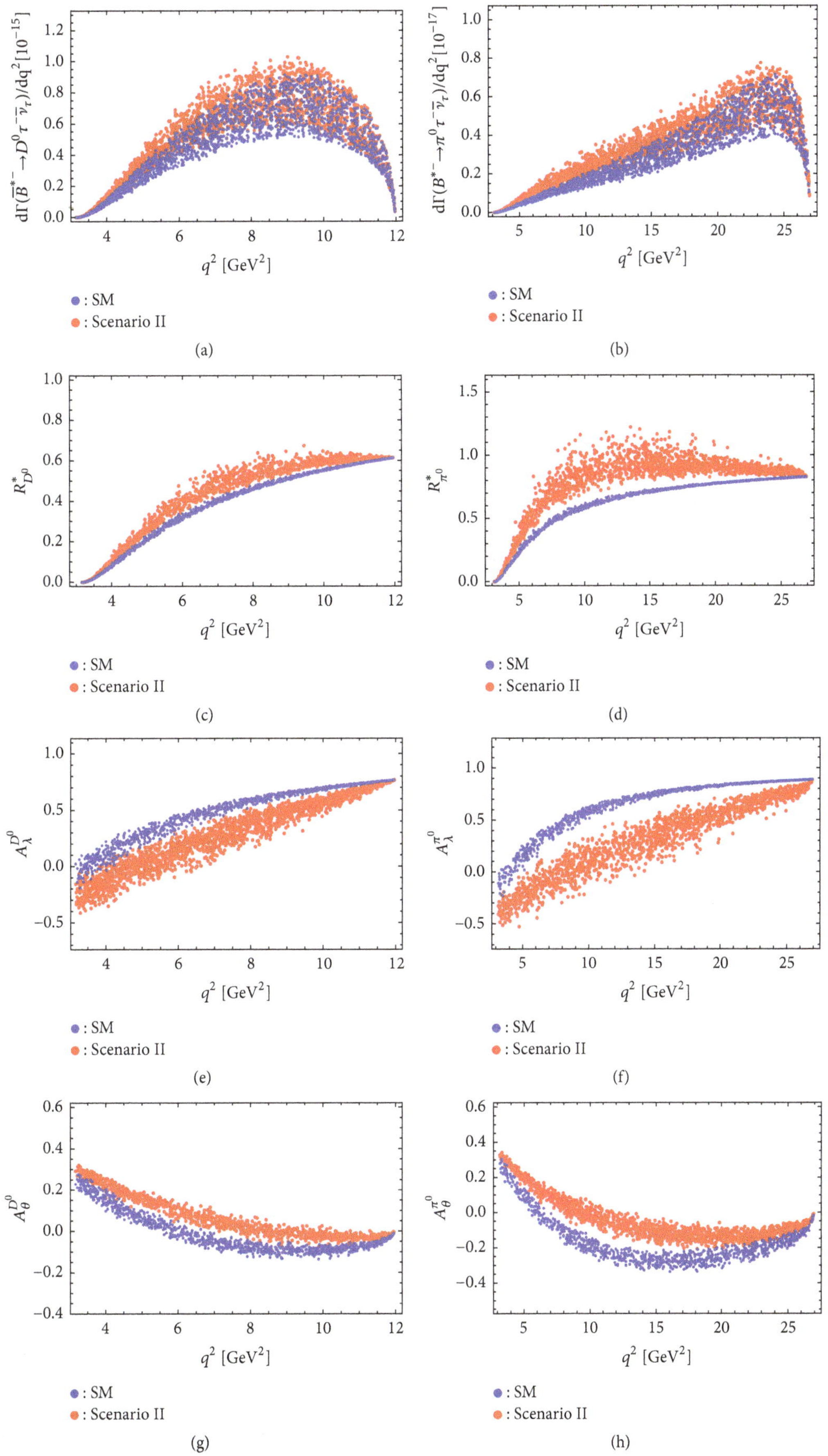

FIGURE 4: The q_2-dependence of the differential observables $d\Gamma/dq_2$, $R_{\underset{P}{*}}$, $A_{\underset{\lambda}{P}}$, and $A_{\underset{\theta}{P}}$ for $B_{*-} \longrightarrow D_0\tau_-\underset{\tau}{\nu}$ and $\pi_0\tau_-\underset{\tau}{\nu}$ decays within the SM and scenario II.

is obviously different from predictions of vector couplings in scenario I (Figures 2(e) and 2(f)). Besides, as Figures 4(e) and 4(f) show, S_L and S_R couplings also have large contributions to the A_θ^P at all q^2 region, which is another difference with the vector couplings (Figures 2(g) and 2(h)). Therefore, the future measurements on these observables will provide strict tests on the SM and various NP models.

4. Summary

In this paper, motivated by the observed "R_{D^*} and R_D puzzles" and its implication of NP, we have studied the NP effects on the $b \longrightarrow (c,u)\ell^-\overline{\nu}_\ell$ induced semileptonic $\overline{B}^*_{u,d,s} \longrightarrow P\ell^-\overline{\nu}_\ell$ ($P = D, D_s, \pi, K$) decays in a model-independent scheme. Using the allowed spaces of vector and scalar couplings obtained by fitting to the data of R_{D^*} and R_D, the NP effects on the decay rate, ratio R_P^*, lepton spin asymmetry, and forward-backward asymmetry are studied in vector and scalar scenarios, respectively. It is found that the vector couplings present large contributions to the decay rate and R_P^*, but their effects on A_λ^P and A_θ^P are very tiny. Different from the vector couplings, the scalar couplings present significant effects not only on the decay rate and R_P^* but also on A_λ^P and A_θ^P. The future measurements on the $\overline{B}^*_{u,d,s} \longrightarrow P\ell^-\overline{\nu}_\ell$ decays will further test the predictions of the SM and NP and confirm or refute possible NP solutions to R_{D^*} and R_D.

Acknowledgments

This work is supported by the National Natural Science Foundation of China (Grant No. 11475055), the Foundation for the Author of National Excellent Doctoral Dissertation of China (Grant No. 201317), and Program for Innovative Research Team in University of Henan Province (Grant No. 19IRTSTHN018).

References

[1] J. P. Lees et al., "Evidence for an excess of $\overline{B} \longrightarrow D^{(*)}\tau^-\overline{\nu}_\tau$ decays," *Physical Review Letters*, vol. 109, no. 10, 2012.

[2] J. P. Lees et al., "Measurement of the mass of the D^0 meson," *Physical Review D*, vol. 88, no. 7, 2013.

[3] M. Huschle et al., "Measurement of the branching ratio of $\overline{B} \longrightarrow D^{(*)}\tau^-\overline{\nu}_\tau$ relative to $\overline{B} \longrightarrow D^{(*)}\ell^-\overline{\nu}_\ell$ decays with hadronic tagging at Belle," *Physical Review D*, vol. 92, no. 7, 2015.

[4] Y. Sato et al., "Measurement of the branching ratio of $\overline{B}^0 \longrightarrow D^{*+}\tau^-\overline{\nu}_\tau$ relative to $\overline{B}^0 \longrightarrow D^{*+}\ell^-\overline{\nu}_\ell$ decays with a semileptonic tagging method," *Physical Review D*, vol. 94, no. 7, 2016.

[5] A. Abdesselam, I. Adachi, K. Adamczyk, H. Aihara et al., "Measurement of the τ lepton polarization in the decay $\overline{B} \longrightarrow D^*\tau^-\overline{\nu}_\tau$," 2016, https://arxiv.org/abs/1608.06391.

[6] R. Aaij et al., "Measurement of the ratio of branching fractions $\mathscr{B}(\overline{B}^0 \longrightarrow D^{*+}\tau^-\overline{\nu}_\tau)/\mathscr{B}(\overline{B}^0 \longrightarrow D^{*+}\mu^-\overline{\nu}_\mu)$," *Physical Review Letters*, vol. 115, no. 11, 2015, Addendum: [Physical Review Letters, vol. 115, no. 15, 2015].

[7] Y. Amhis et al., "Averages of b-hadron, c-hadron, and τ -lepton properties as of summer 2016," *The European Physical Journal C*, vol. 77, no. 12, 2017.

[8] H. Na, M. Chris, G. Bouchard, L. Peter, M. Chris, and S. Junko, "$B \longrightarrow D\ell\nu$ form factors at nonzero recoil and extraction of $|V_{cb}|$," *Physical Review D*, vol. 92, no. 5, 2015.

[9] S. Fajfer, J. F. Kamenik, and I. Nišandžić, "$B \longrightarrow D^*\tau\overline{\nu}_\tau$ sensitivity to new physics," *Physical Review D: Particles, Fields, Gravitation and Cosmology*, vol. 85, no. 9, 2012.

[10] R. Aaij et al., "Measurement of the Ratio of Branching Fractions $\mathscr{B}(B_c^+ \longrightarrow J/\psi\tau^+\nu_\tau)/\mathscr{B}(B_c^+ \longrightarrow J/\psi\mu^+\nu_\mu)$," *Physical Review Letters*, vol. 120, no. 12, 2018.

[11] J. P. Lees et al., "Evidence of $B^+ \longrightarrow \tau^+\nu$ decays with hadronic B tags," *Physical Review D*, vol. 88, no. 3, 2013.

[12] B. Aubert et al., "Search for $B^+ \longrightarrow \ell^+\nu_\ell$ recoiling against $B^- \longrightarrow D^0\ell^-\overline{\nu}X$," *Physical Review D*, vol. 81, no. 5, 2010.

[13] I. Adachi et al., "Evidence for $B^- \longrightarrow \tau^-\overline{\nu}_\tau$ with a hadronic tagging method using the full data sample of belle," *Physical Review Letters*, vol. 110, no. 13, 2013.

[14] B. Kronenbitter et al., "Measurement of the branching fraction of $B^+ \longrightarrow \tau^+\nu_\tau$ decays with the semileptonic tagging method," *Physical Review D*, vol. 92, no. 5, 2015.

[15] G. Ciezarek, M. Franco Sevilla, B. Hamilton et al., "A challenge to lepton universality in B-meson decays," *Nature*, vol. 546, pp. 227–233, 2017.

[16] M. Freytsis, Z. Ligeti, and J. T. Ruderman, "Flavor models for $\overline{B} \longrightarrow D^{(*)}\tau\overline{\nu}$," *Physical Review D: Particles, Fields, Gravitation and Cosmology*, vol. 92, no. 5, 2015.

[17] L. Calibbi, A. Crivellin, and T. Ota, "Effective field theory approach to $b \longrightarrow s\ell\ell^{(\prime)}$, $B \longrightarrow K^{(*)}\nu\overline{\nu}$ and $B \longrightarrow D^{(*)}\tau\nu$ with third generation couplings," *Physical Review Letters*, vol. 115, no. 18, 2015.

[18] R. Alonso, B. Grinstein, and J. M. Camalich, "Lepton universality violation with lepton flavor conservation in B-meson decays," *Journal of High Energy Physics*, vol. 2015, no. 10, 2015.

[19] M. Tanaka and R. Watanabe, "New physics in the weak interaction of $\overline{B} \longrightarrow D^{(*)}\tau\overline{\nu}$," *Physical Review D: Particles, Fields, Gravitation and Cosmology*, vol. 87, no. 3, 2013.

[20] S. Fajfer, J. F. Kamenik, I. Nišandžić, and J. Zupan, "Implications of lepton flavor universality violations in B decays," *Physical Review Letters*, vol. 109, no. 16, 2012.

[21] D. Becirevic, S. Fajfer, I. Nisandzic, and A. Tayduganov, "Angular distributions of $\overline{B} \longrightarrow D^{(*)}\ell\overline{\nu}_\ell$ decays and search of new physics," 2016, https://arxiv.org/abs/1602.03030.

[22] S. Bhattacharya, S. Nandi, and S. K. Patra, "Optimal-observable analysis of possible new physics in $B \longrightarrow D^{(*)}\tau\nu_\tau$," *Physical Review D: Particles, Fields, Gravitation and Cosmology*, vol. 93, no. 3, 2016.

[23] B. Bhattacharya, A. Datta, D. London, and S. Shivashankara, "Simultaneous explanation of the R_K and $R(D^{(*)})$ puzzles," *Physics Letters B*, vol. 742, pp. 370–374, 2015.

[24] M. Duraisamy, P. Sharma, and A. Datta, "Azimuthal $B \longrightarrow D^*\tau^-\bar{\nu}_\tau$ angular distribution with tensor operators," *Physical Review D: Particles, Fields, Gravitation and Cosmology*, vol. 90, no. 7, 2014.

[25] K. Hagiwara, M. M. Nojiri, and Y. Sakaki, "*CP* violation in $B \longrightarrow D\tau\nu_\tau$ using multipion tau decays," *Physical Review D: Particles, Fields, Gravitation and Cosmology*, vol. 89, no. 9, 2014.

[26] R. Dutta, A. Bhol, and A. K. Giri, "Effective theory approach to new physics in $b \longrightarrow u$ and $b \longrightarrow c$ leptonic and semileptonic decays," *Physical Review D: Particles, Fields, Gravitation and Cosmology*, vol. 88, no. 11, 2013.

[27] M. Duraisamy and A. Datta, "The full $B \longrightarrow D^*\tau^-\bar{\nu}_\tau$ angular distribution and CP violating triple products," *Journal of High Energy Physics*, vol. 2013, no. 9, 2013.

[28] P. Biancofiore, P. Colangelo, and F. De Fazio, "Anomalous enhancement observed in $B \longrightarrow D^{(*)}\tau\bar{\nu}_\tau$ decays," *Physical Review D: Particles, Fields, Gravitation and Cosmology*, vol. 87, no. 7, 2013.

[29] S. Faller, T. Mannel, and S. Turczyk, "Limits on new physics from exclusive $B \longrightarrow D^{(*)}\ell\bar{\nu}$ decays," *Physical Review D: Particles, Fields, Gravitation and Cosmology*, vol. 84, no. 1, 2011.

[30] C. H. Chen and C. Q. Geng, "Lepton angular asymmetries in semileptonic charmful *B* decays," *Physical Review D: Particles, Fields, Gravitation and Cosmology*, vol. 71, no. 7, 2005.

[31] A. K. Alok, D. Kumar, S. Kumbhakar, and S. U. Sankar, "D^* polarization as a probe to discriminate new physics in $\bar{B} \longrightarrow D^*\tau\bar{\nu}$," *Physical Review D: Particles, Fields, Gravitation and Cosmology*, vol. 95, no. 11, 2017.

[32] M. A. Ivanov, J. G. Korner, and C. T. Tran, "Analyzing new physics in the decays $\bar{B}^0 \longrightarrow D^{(*)}\tau^-\bar{\nu}_\tau$ with form factors obtained from the covariant quark model," *Physical Review D*, vol. 94, no. 9, 2016.

[33] Y. Y. Fan, Z. J. Xiao, R. M. Wang, and B. Z. Li, "The $B \longrightarrow D^{(*)}\ell\nu_\ell$ decays in the pQCD approach with the Lattice QCD input," *Science Bulletin*, vol. 60, no. 23, pp. 2009–2015, 2015.

[34] Y. Y. Fan, W. F. Wang, S. Cheng, and Z. J. Xiao, "Semileptonic decays $B \longrightarrow D^{(*)}\ell\nu$ in the perturbative QCD factorization approach," *Chinese Science Bulletin*, vol. 59, no. 2, pp. 125–132, 2014.

[35] A. Datta, M. Duraisamy, and D. Ghosh, "Diagnosing new physics in $b \longrightarrow c\tau\nu_\tau$ decays in the light of the recent *BABAR* result," *Physical Review D: Particles, Fields, Gravitation and Cosmology*, vol. 86, no. 3, 2012.

[36] J. A. Bailey et al., "Refining new-physics searches in $B \longrightarrow D\tau\nu$ with lattice QCD," *Physical Review Letters*, vol. 109, no. 7, 2012.

[37] D. Becirevic, N. Kosnik, and A. Tayduganov, "$\bar{B} \longrightarrow D\tau\bar{\nu}_\tau$ vs. $\bar{B} \longrightarrow D\mu\bar{\nu}_\mu$," *Physics Letters B*, vol. 716, no. 1, pp. 208–213, 2012.

[38] F. Deppisch, S. Kulkarni, H. Päs, and E. Schumacher, "Leptoquark patterns unifying neutrino masses, flavor anomalies, and the diphoton excess," *Physical Review D: Particles, Fields, Gravitation and Cosmology*, vol. 94, no. 1, 2016.

[39] B. Dumont, K. Nishiwaki, and R. Watanabe, "LHC constraints and prospects for S_1 scalar leptoquark explaining the $\bar{B} \longrightarrow D^{(*)}\tau\bar{\nu}$ anomaly," *Physical Review D: Particles, Fields, Gravitation and Cosmology*, vol. 94, no. 3, 2016.

[40] I. Doršner, S. Fajfer, A. Greljo, J. F. Kamenik, and N. Košnik, "Physics of leptoquarks in precision experiments and at particle colliders," *Physics Reports*, vol. 641, pp. 1–68, 2016.

[41] Y. Sakaki, M. Tanaka, A. Tayduganov, and R. Watanabe, "Probing new physics with q^2 distributions in $\bar{B} \longrightarrow D^{(*)}\tau\bar{\nu}$," *Physical Review D: Particles, Fields, Gravitation and Cosmology*, vol. 91, no. 11, 2015.

[42] M. Bauer and M. Neubert, "Minimal Leptoquark Explanation for the $R_{D^{(*)}}$, R_K, and $(g-2)_\mu$ anomalies," *Physical Review Letters*, vol. 116, no. 14, 2016.

[43] S. Fajfer and N. Košnik, "Vector leptoquark resolution of R_K and $R_{D^{(*)}}$ puzzles," *Physics Letters B*, vol. 755, pp. 270–274, 2016.

[44] S. Sahoo and R. Mohanta, "Lepton flavor violating *B* meson decays via a scalar leptoquark," *Physical Review D: Particles, Fields, Gravitation and Cosmology*, vol. 93, no. 11, 2016.

[45] R. Barbieri, G. Isidori, A. Pattori, and F. Senia, "Anomalies in *B*-decays and *U*(2) flavor symmetry," *The European Physical Journal C*, vol. 76, no. 2, 2016.

[46] Y. Sakaki, R. Watanabe, M. Tanaka, and A. Tayduganov, "Testing leptoquark models in $\bar{B} \longrightarrow D^{(*)}\tau\bar{\nu}$," *Physical Review D: Particles, Fields, Gravitation and Cosmology*, vol. 88, no. 9, 2013.

[47] A. Celis, M. Jung, X. Li, and A. Pich, "Sensitivity to charged scalars in $B \longrightarrow D^{(*)}\tau\nu_\tau$ and $B \longrightarrow \tau\nu_\tau$ decays," *Journal of High Energy Physics*, vol. 13, no. 1, 2013.

[48] A. Celis, M. Jung, X. Q. Li, and A. Pich, "Scalar contributions to $b \longrightarrow c(u)\tau\nu$ transitions," *Physics Letters B*, vol. 771, pp. 168–179, 2017.

[49] X. Q. Li, "$R(D)$ and $R(D^*)$ anomalies and their phenomenological implications," *Nuclear and Particle Physics Proceedings*, vol. 287-288, pp. 181–184, 2017.

[50] J. M. Cline, "Scalar doublet models confront τ and b anomalies," *Physical Review D*, vol. 93, no. 7, 2016.

[51] C. S. Kim, Y. W. Yoon, and X. B. Yuan, "Exploring top quark FCNC within 2HDM type III in association with flavor physics," *Journal of High Energy Physics*, vol. 2015, no. 12, pp. 1–30, 2015.

[52] A. Crivellin, J. Heeck, and P. Stoffer, "Perturbed lepton-specific two-higgs-doublet model facing experimental hints for physics beyond the standard model," *Physical Review Letters*, vol. 116, no. 8, 2016.

[53] D. S. Hwang, "Transverse spin polarization of τ^- in $\bar{B}^0 \longrightarrow D^+\tau^-\bar{\nu}$ and charged higgs boson," 2015, https://arxiv.org/abs/1504.06933.

[54] A. Crivellin, C. Greub, and A. Kokulu, "Flavor-phenomenology of two-Higgs-doublet models with generic Yukawa structure," *Physical Review D: Particles, Fields, Gravitation and Cosmology*, vol. 87, no. 9, 2013.

[55] U. Nierste, S. Trine, and S. Westhoff, "Charged-Higgs-boson effects in the $B \longrightarrow D\tau\nu_\tau$ differential decay distribution," *Physical Review D: Particles, Fields, Gravitation and Cosmology*, vol. 78, no. 1, 2008.

[56] K. Kiers and A. Soni, "Improving constraints on $\tan\beta/m_H$ using $B \longrightarrow D\tau\bar{\nu}$," *Physical Review D: Particles, Fields, Gravitation and Cosmology*, vol. 56, no. 9, 1997.

[57] M. Tanaka, "Charged Higgs effects on exclusive semi-tauonic *B* decays," *Zeitschrift für Physik C Particles and Fields*, vol. 67, no. 2, pp. 321–326, 1995.

[58] W. S. Hou, "Enhanced charged Higgs boson effects in $B^- \longrightarrow \tau\bar{\nu}$, $\mu\bar{\nu}$ and $b \longrightarrow \tau\bar{\nu} + X$," *Physical Review D: Particles, Fields, Gravitation and Cosmology*, vol. 48, no. 5, 1993.

[59] Y. Sakaki and H. Tanaka, "Constraints on the charged scalar effects using the forward-backward asymmetry on $\bar{B} \longrightarrow D^{(*)}\tau\bar{\nu}_\tau$," *Physical Review D*, vol. 87, no. 5, 2013.

[60] S. M. Boucenna, A. Celis, J. Fuentes-Martin, A. Vicente, and J. Virto, "Non-abelian gauge extensions for B-decay anomalies," *Physics Letters B*, vol. 760, pp. 214–219, 2016.

[61] C. Hati, G. Kumar, and N. Mahajan, "$\overline{B} \longrightarrow D^{(*)}\tau\overline{\nu}$ excesses in ALRSM constrained from B, D decays and $D^0 \longrightarrow \overline{D}^0$ mixing," *Journal of High Energy Physics*, vol. 2016, no. 1, 2016.

[62] D. Das, C. Hati, G. Kumar, and N. Mahajan, "Towards a unified explanation of $R_{D^{(*)}}$, R_K and $(g-2)_\mu$ anomalies in a left-right model with leptoquarks," *Physical Review D: Particles, Fields, Gravitation and Cosmology*, vol. 94, no. 5, 2016.

[63] R.-M. Wang, J. Zhu, H. M. Gan, Y. Y. Fan, Q. Chang, and Y. G. Xu, "Probing R-parity violating supersymmetric effects in the exclusive $b \longrightarrow c\ell^-\overline{\nu}_\ell$ decays," *Physical Review D: Particles, Fields, Gravitation and Cosmology*, vol. 93, no. 9, 2016.

[64] B. Wei, J. Zhu, J. H. Shen, R. M. Wang, and G. R. Lu, "Probing the R-parity violating supersymmetric effects in $B_c \longrightarrow J/\psi\ell^-\overline{\nu}_\ell, \eta_c\ell^-\overline{\nu}_\ell$ and $\Lambda_b \longrightarrow \Lambda_c\ell^-\overline{\nu}_\ell$ decays," 2018, https://arxiv.org/abs/1801.00917.

[65] N. G. Deshpande and A. Menon, "Hints of R-parity violation in B decays into $\tau\nu$," *Journal of High Energy Physics*, vol. 2013, no. 1, 2013.

[66] N. Isgur and M. B. Wise, "Spectroscopy with heavy-quark symmetry," *Physical Review Letters*, vol. 66, no. 9, 1991.

[67] S. Godfrey and R. Kokoski, "Properties of P-wave mesons with one heavy quark," *Physical Review D: Particles, Fields, Gravitation and Cosmology*, vol. 43, no. 5, 1991.

[68] E. J. Eichten, C. T. Hill, and C. Quigg, "Properties of orbitally excited heavy-light (Qq^-) mesons," *Physical Review Letters*, vol. 71, no. 25, 1993.

[69] D. Ebert, V. O. Galkin, and R. N. Faustov, "Mass spectrum of orbitally and radially excited heavy-light mesons in the relativistic quark model," *Physical Review D: Particles, Fields, Gravitation and Cosmology*, vol. 57, no. 9, 1998, Erratum: Physical Review D, vol. 59, no. 1, 1998.

[70] T. Abe et al., "Belle II technical design report," 2010, https://arxiv.org/abs/1011.0352.

[71] R. Aaij et al., "Implications of LHCb measurements and future prospects," *The European Physical Journal C*, vol. 73, no. 4, 2013.

[72] R. Aaij et al., "LHCb detector performance," *International Journal of Modern Physics A*, vol. 30, no. 7, 2015.

[73] Q. Chang, P. P. Li, X. H. Hu, and L. Han, "Study of nonleptonic $B_s^* \longrightarrow M_1M_2$ $(M = D, D_s, \pi, K)$ weak decays with factorization approach," *International Journal of Modern Physics A*, vol. 30, no. 27, 2015.

[74] Q. Chang, X. Hu, J. Sun, X. Wang, and Y. Yang, "Study on nonleptonic $B_q^* \longrightarrow D_qV$ and P_qD^* weak decays," *Advances in High Energy Physics*, vol. 2015, Article ID 767523, 8 pages, 2015.

[75] R. Aaij et al., "Measurement of $\sigma(pp \longrightarrow b\overline{b}X)$ at $\sqrt{s} = 7$TeV in the forward region," *Physics Letters B*, vol. 694, no. 3, pp. 209–216, 2010.

[76] B. Grinstein and J. Martin Camalich, "Weak decays of excited B mesons," *Physical Review Letters*, vol. 116, no. 14, 2016.

[77] Z. G. Wang, "Semileptonic decays $B^*_c \longrightarrow \eta_c\ell\overline{\nu}_\ell$ with QCD sum rules," *Communications in Theoretical Physics*, vol. 61, no. 1, 2014.

[78] K. Zeynali, V. Bashiry, and F. Zolfagharpour, "Form factors and decay rate of $B_c^* \longrightarrow D_s1^+1^-$ decays in the QCD sum rules," *The European Physical Journal A*, vol. 50, 2014.

[79] V. Bashiry, "Investigation of the rare exclusive $B_c^* \longrightarrow D_s\nu\overline{\nu}$ decays in the framework of the QCD sum rules," *Advances in High Energy Physics*, vol. 2014, Article ID 503049, 10 pages, 2014.

[80] G. Z. Xu, Y. Qiu, C. P. Shen, and Y. J. Zhang, "$B^*_{s,d} \longrightarrow \mu^+\mu^-$ and its impact on $B_{s,d} \longrightarrow \mu^+\mu^-$," *The European Physical Journal C*, vol. 76, no. 11, 2016.

[81] Q. Chang, L. X. Chen, Y. Y. Zhang, J. F. Sun, and Y. L. Yang, "$\overline{B}_{d,s} \longrightarrow D^*_{d,s}V$ and $\overline{B}^*_{d,s} \longrightarrow D_{d,s}V$ decays in QCD factorization and possible puzzles," *The European Physical Journal C*, vol. 76, no. 10, 2016.

[82] Q. Chang, J. Zhu, X. L. Wang, J. F. Sun, and Y. L. Yang, "Study of semileptonic $\overline{B}^* \longrightarrow P\ell\overline{\nu}_\ell$ decays," *Nuclear Physics B*, vol. 909, pp. 921–933, 2016.

[83] J. G. Korner and G. A. Schuler, "Exclusive semi-leptonic decays of bottom mesons in the spectator quark model," *Zeitschrift für Physik C Particles and Fields*, vol. 38, no. 3, pp. 511–518, 1988, Erratum: Zeitschrift für Physik C Particles and Fields, vol. 41, no. 4, 1989.

[84] J. G. Körner and G. A. Schuler, "Exclusive semileptonic heavy meson decays including lepton mass effects," *Zeitschrift für Physik C Particles and Fields*, vol. 46, no. 1, pp. 93–109, 1990.

[85] K. Hagiwara, A. D. Martin, and M. F. Wade, "Exclusive semileptonic B-meson decays," *Nuclear Physics B*, vol. 327, no. 3, pp. 569–594, 1989.

[86] K. Hagiwara, A. D. Martin, and M. F. Wade, "Helicity amplitude analysis of $B \longrightarrow D^*l\nu$ decays," *Physics Letters B*, vol. 228, no. 1, pp. 144–148, 1989.

[87] A. Kadeer, J. G. Körner, and U. Moosbrugger, "Helicity analysis of semileptonic hyperon decays including lepton-mass effects," *The European Physical Journal C*, vol. 59, no. 1, pp. 27–47, 2009.

[88] C. Patrignani et al., "Review of particle physics," *Chinese Physics C*, vol. 40, no. 10, 2016.

[89] J. Charles et al., "*CP* violation and the CKM matrix: assessing the impact of the asymmetric B factories," *The European Physical Journal C*, vol. 41, no. 1, pp. 1–131, 2005, Updated results and plots available at: http://ckmfitter.in2p3.fr.

[90] J. L. Goity and W. Roberts, "Radiative transitions in heavy mesons in a relativistic quark model," *Physical Review D: Particles, Fields, Gravitation and Cosmology*, vol. 64, no. 9, 2001.

[91] D. Ebert, R. N. Faustov, and V. O. Galkin, "Radiative M1-decays of heavy-light mesons in the relativistic quark model," *Physics Letters B*, vol. 537, no. 3-4, pp. 241–248, 2002.

[92] S.-L. Zhu, Z.-S. Yang, and W.-Y. P. Hwang, "$D^* \longrightarrow D\gamma$ and $B^* \longrightarrow B\gamma$ as derived from QCD sum rules," *Modern Physics Letters A*, vol. 12, no. 39, pp. 3027–3035, 1997.

[93] T. M. Aliev, D. A. Demir, E. Iltan, and N. K. Pak, "Radiative $B^* \longrightarrow B\gamma$ and $D^* \longrightarrow D\gamma$ decays in light-cone QCD sum rules," *Physical Review D: Particles, Fields, Gravitation and Cosmology*, vol. 54, no. 1, 1996.

[94] P. Colangelo, F. De Fazio, and G. Nardulli, "Radiative heavy meson transitions," *Physics Letters B*, vol. 316, no. 4, pp. 555–560, 1993.

[95] H. M. Choi, "Decay constants and radiative decays of heavy mesons in light-front quark model," *Physical Review D: Particles, Fields, Gravitation and Cosmology*, vol. 75, no. 7, 2007.

[96] C. Y. Cheung and C. W. Hwang, "Strong and radiative decays of heavy mesons in a covariant model," *Journal of High Energy Physics*, vol. 2014, no. 4, 2014.

[97] I. Caprini, L. Lellouch, and M. Neubert, "Dispersive bounds on the shape of $B \longrightarrow D^{(*)}\ell\nu$ form factors," *Nuclear Physics B*, vol. 530, no. 1-2, pp. 153–181, 1998.

[98] M. Wirbel, B. Stech, and M. Bauer, "Exclusive semileptonic decays of heavy mesons," *Zeitschrift für Physik C Particles and Fields*, vol. 29, no. 4, pp. 637–642, 1985.

[99] M. Bauer and M. Wirbel, "Formfactor effects in exclusive D and B decays," *Zeitschrift für Physik C Particles and Fields*, vol. 42, no. 4, pp. 671–678, 1989.

3

Analysis of Top Quark Pair Production Signal from Neutral 2HDM Higgs Bosons at LHC

Majid Hashemi and Mahbobeh Jafarpour

Physics Department, College of Sciences, Shiraz University, Shiraz 71946-84795, Iran

Correspondence should be addressed to Majid Hashemi; majid.hashemi@cern.ch

Academic Editor: Burak Bilki

In this paper, the top quark pair production events are analyzed as a source of neutral Higgs bosons of the two Higgs doublet model type I at LHC. The production mechanism is $pp \longrightarrow H/A \longrightarrow t\bar{t}$ assuming a fully hadronic final state through $t \longrightarrow Wb \longrightarrow jjb$. In order to distinguish the signal from the main background which is the standard model $t\bar{t}$, we benefit from the fact that the top quarks in signal events acquire large Lorentz boost due to the heavy neutral Higgs boson. This feature leads to three collinear jets (a fat jet) which is a discriminating tool for identification of the top quarks from the Higgs boson resonances. Events with two identified top jets are selected and the invariant mass of the top pair is calculated for both signal and background. It is shown that the low $\tan\beta$ region has still some parts which can be covered by this analysis and has not yet been excluded by flavor physics data.

1. Introduction

The standard model (SM) of particle physics has taken a major step forward by observing the Higgs boson at LHC [1, 2] based on a theoretical framework known as the Higgs mechanism [3–8]. The observed particle may belong to a single SU(2) doublet (SM) or a two Higgs doublet model (2HDM) [9–11] whose lightest Higgs boson respects the observed particle properties.

One of the motivations for the two Higgs doublet model is supersymmetry where each particle has a superpartner. The supersymmetry provides an elegant solution to the gauge coupling unification, dark matter candidate, and the Higgs boson mass radiative correction by a natural parameters tuning. In such a model two Higgs doublets are required to give mass to the double space of the particles [12–14].

There are four types of 2HDMs with different scenarios of Higgs-fermion couplings. The ratio of vacuum expectation values of the two Higgs doublets ($\tan\beta = v_2/v_1$) is a measure of the Higgs-fermion coupling in all 2HDM types [15].

In general, 2HDM involves five physical Higgs bosons due to the extended degrees of freedom added to the model by introducing the second Higgs doublet. The lightest Higgs boson, h, is like the SM Higgs boson. The rest are two neutral Higgs bosons, H, A (subjects of this study), and two charged bosons, $H^{\pm}$. A review of the theory and phenomenology of 2HDM can be found in [16].

In addition to direct searches for the 2HDM Higgs bosons at colliders, there are indirect searches based on flavor physics data by investigating sources of deviations from SM when processes containing 2HDM Higgs bosons are introduced [17]. Limits obtained from these types of studies are one of the strongest limits on the mass of the charged and neutral Higgs bosons and $\tan\beta$ and will be referred to when presenting the final results.

The adopted scenario in this analysis is a search for heavy neutral Higgs boson with mass in the range 0.5-1 TeV at LHC operating at $\sqrt{s} = 14$ TeV. All heavy Higgs bosons (CP-even, CP-odd, and the charged Higgs) are assumed to be degenerate, i.e., $m_H = m_A = m_{H^{\pm}}$. The region of interest is low $\tan\beta$ and the final results will be limited to $\tan\beta < 2$. The signal process is $pp \longrightarrow H/A \longrightarrow t\bar{t} \longrightarrow W^+bW^-\bar{b} \longrightarrow jjbjjb$. The fully hadronic final state is expected to result in two fat jets (each consisting of three sub-jets associated with the top quark) which are examined using the updated HEPTopTagger 2 [18, 19]. Events which contain two identified (tagged) top jets are used to fill the top pair invariant mass distribution histogram. The same approach is

TABLE 1: Different types of 2HDM in terms of the Higgs boson couplings with U (up-type quarks), D (down-type quarks), and L (leptons).

	Type I	Type II	Type III	Type IV
ρ^D	$\kappa^D \cot\beta$	$-\kappa^D \tan\beta$	$-\kappa^D \tan\beta$	$\kappa^D \cot\beta$
ρ^U	$\kappa^U \cot\beta$	$\kappa^U \cot\beta$	$\kappa^U \cot\beta$	$\kappa^U \cot\beta$
ρ^L	$\kappa^L \cot\beta$	$-\kappa^L \tan\beta$	$\kappa^L \cot\beta$	$-\kappa^L \tan\beta$

applied on background events and final shape discrimination is performed to evaluate the signal significance. Before going to the details of the analysis, a brief review of the theoretical framework is presented in the next section.

2. The Higgs Sector of 2HDM

The 2HDM Lagrangian for neutral Higgs-fermion couplings as introduced in [20] takes the following form:

$$\begin{aligned}\mathscr{L}_Y &= \frac{1}{\sqrt{2}}\sum_f \bar{f}\left[\kappa^f s_{\beta-\alpha} + \rho^f c_{\beta-\alpha}\right] fh \\ &+ \frac{1}{\sqrt{2}}\sum_f \bar{f}\left[\kappa^f c_{\beta-\alpha} - \rho^f s_{\beta-\alpha}\right] fH \\ &+ \frac{i}{\sqrt{2}}\bar{f}\gamma_5\rho^f fA\end{aligned} \quad (1)$$

where h, H, A are the neutral Higgs boson fields, $\kappa^f = \sqrt{2}(m_f/v)$ for any fermion type f and $s_{\beta-\alpha} = \sin(\beta-\alpha)$, and $c_{\beta-\alpha} = \cos(\beta-\alpha)$. The ρ^f parameters define the model type and are proportional to κ^f as in Table 1 [21]. Therefore the four types of interactions (2HDM types) depend on the values of ρ^f [22].

In this study, we require $s_{\beta-\alpha} = 1$ which has two advantages. The first one is that the $s_{\beta-\alpha}$ factor in the lightest Higgs-gauge coupling is set to unity while the heavier Higgs, H, decouples from gauge bosons [16]. On the other hand, the SM-like Higgs-fermion interactions are $\tan\beta$ independent.

According to Table 1, the type I is interesting for low $\tan\beta$ as all couplings in the neutral Higgs sector are proportional to $\cot\beta$. This feature leads to cancellation of this factor as long as Higgs boson branching ratio of decay to leptons and quarks is concerned. The mass of the fermion thus plays an important role in the decay rate, and as seen from Figures 1 and 2, the Higgs boson decay to $t\bar{t}$ dominates for all relevant Higgs boson masses and $\tan\beta$ values. The decay to a pair of gluons proceeds through a preferably top quark loop and stands as the second channel. The third channel is $H/A \longrightarrow b\bar{b}$ which has been shown to be visible at LHC [23]. The current study focuses on $H/A \longrightarrow t\bar{t}$ with branching ratio being near unity and independent of the Higgs boson mass (Figure 1) and $\tan\beta$ (Figure 2).

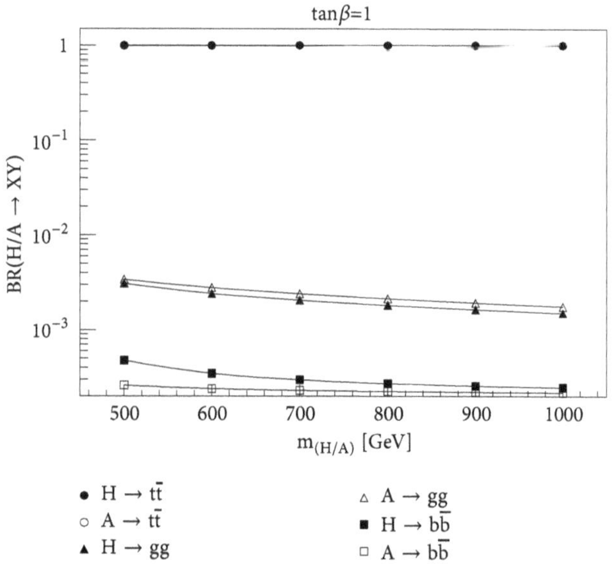

FIGURE 1: The branching ratio of neutral Higgs boson decays as a function of the mass. The $\tan\beta$ is set to 1.

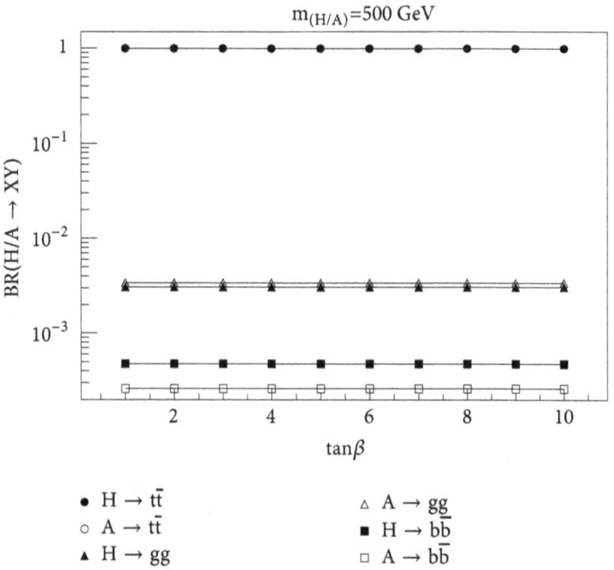

FIGURE 2: The branching ratio of neutral Higgs boson decays as a function of $\tan\beta$. The Higgs boson mass is set to 500 GeV.

3. Signal Identification and the Search Scenario

The signal process under study is a Higgs boson production with the Higgs boson masses in the range 500 – 1000 GeV. The three Higgs bosons masses are set to be equal for minimizing $\Delta\rho$ [24]. All selected points are checked to be consistent with the potential stability, perturbativity, and unitarity requirements and the current experimental limits on Higgs boson masses using 2HDMC 1.6.3 [25, 26].

There has been phenomenological searches for leptophilic Higgs boson within type IV 2HDM at LHC [27] and linear colliders [28, 29]. These searches are based on leptonic decay of the Higgs boson. On the other hand, the type I 2HDM can be considered as a leptophobic model where the Higgs boson decay to quarks plays an important role. At the first glance, decays to all fermions are relevant at low $\tan\beta$ values. However, the fermion mass in the Higgs-fermion vertex enhances the top quark coupling dramatically compared to other channels. This is due to the fact that the common $\cot\beta$ factors cancel out when calculating branching ratio of Higgs decays to fermions. Therefore in this analysis, the Higgs boson decay to $t\bar{t}$ is considered as the signal.

While the neutral Higgs boson searches at LEP [30, 31] lead to $m_A \geq 93.4$ GeV, the LHC results [32, 33] indicate that the neutral Higgs boson mass in the range $m_{H/A} = 200-400$ GeV is excluded for $\tan\beta \geq 5$. This result is based on minimal supersymmetric standard model (MSSM) which has a different Higgs boson spectrum from 2HDM due to supersymmetry constraints. Since our region of interest is Higgs boson masses above 500 GeV, no constraints from LEP or LHC limit the current analysis and the Higgs boson masses under study.

There are also results from flavor physics data. The strongest limit in this category comes from $b \longrightarrow s\gamma$ analysis which imposes lower limit on the charged Higgs mass in types II and III at 600 GeV [34–36]. There are other analyses such as $B \longrightarrow \tau\nu$, $D_s \longrightarrow \tau\nu$, $B_s \longrightarrow \mu\mu$, $B \longrightarrow K^*\gamma$, and meson mixing. Such observables have a smaller impact than $b \longrightarrow s\gamma$. All limits from the above observables as well as the one from $b \longrightarrow s\gamma$ are mainly relevant at types II and III while types I and IV are less affected due to the fact that the charged Higgs-quark coupling in processes which raise deviation from standard model is suppressed in types I and IV with increasing $\tan\beta$.

In order to compare the two categories of types I/IV and II/III, one may notice that types I and IV behave differently from types II and III as far as the charged Higgs coupling to quarks is concerned. In the former, the charged Higgs coupling to all quark types is suppressed at low $\tan\beta$, while, in the latter, coupling with at least one type of the quarks (up type or down type) is enhanced with $\tan\beta$. Therefore charged Higgs limits from flavor physics in types I and IV are very soft and basically relevant at $\tan\beta$ values as low as 2. This is the region of search in this analysis. Although we are dealing with neutral Higgs bosons, since the scenario under study is a degenerate scenario based on $m_H = m_A = m_{H^\pm}$, limits on the charged Higgs are propagated into the final results.

4. Software Setup and Cross Sections

The signal cross section is obtained from PYTHIA 8.2.15 [37] using 2HDM spectrum files in LHA format [38, 39] extracted from 2HDMC 1.6.3 [25, 26]. The LHA files contain information about the parameters of the theoretical model as well as properties of any particle which may not be present in standard model, like 2HDM Higgs bosons. In this case,

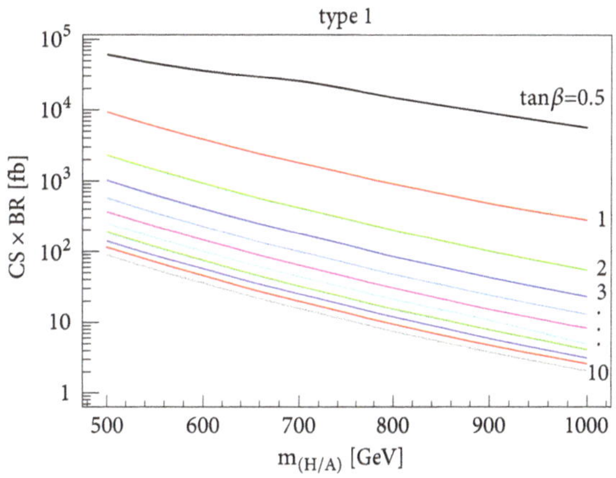

FIGURE 3: The signal cross section times BR($H/A \longrightarrow t\bar{t}$) at $\sqrt{s} = 14$ TeV as a function of the Higgs boson mass.

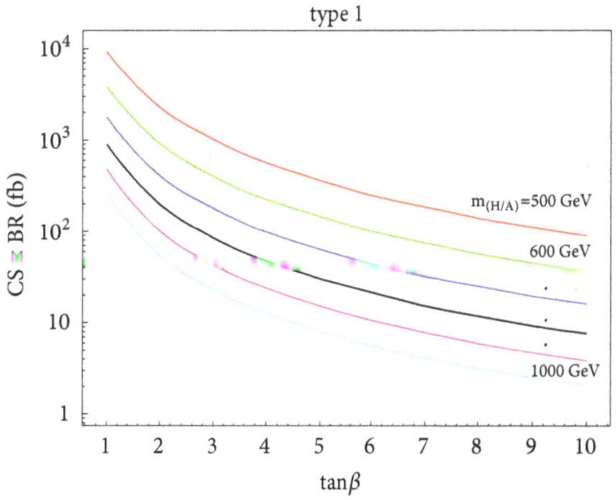

FIGURE 4: The signal cross section times BR($H/A \longrightarrow t\bar{t}$) at $\sqrt{s} = 14$ TeV as a function of $\tan\beta$.

it contains Higgs bosons masses and their branching ratio of decays. For each benchmark point a separated LHA file is generated using 2HDMC and is passed to PYTHIA for event generation and cross section calculation. Results are shown in Figures 3 and 4 which show that the cross section decreases with increasing the Higgs boson mass as well as $\tan\beta$. Therefore the most suitable area for search is where the mass is as low as possible and $\tan\beta$ is also very small. The main SM background processes are $t\bar{t}$, gauge boson pair production WW, WZ, ZZ, s−channel and t−channel single top, single W and single Z/γ^*, and QCD multijet events. These background processes are generated using PYTHIA except for the QCD multijet background for which Alpgen 2.14 [40–42] is used for the hard scattering generation. The output of Alpgen is stored as LHE file [38, 39] and is passed to PYTHIA for multiparticle interaction and final state showering. The cross sections are obtained using PYTHIA except for QCD samples which is obtained from Alpgen and $t\bar{t}$ for which we adopt the NLO (next to leading order) cross section calculated using MCFM [43–46]. The signal (benchmark points) and background cross sections are listed in Table 2. The QCD

TABLE 2: The signal and background cross sections at $\sqrt{s} = 14$ TeV. The "sts" and "stt" denote the s-channel and t-channel single top processes and QCD 6j represents QCD multijet events with 6 jets in the final state.

(a)

	Signal					
$m_{(H/A)}$ [GeV]	500	600	700	800	900	1000
$\sigma \times BR$ [fb]	61.1	36.0	25.4	15.0	9.1	5.6

(b)

	Background								
	tt	WW	WZ	ZZ	sts	stt	W	Z	QCD 6j
$\sigma \times BR$ [pb]	390	32.6	12.1	5.33	5.6	117	1.02×10^5	5.7×10^4	4×10^3

TABLE 3: Signal selection efficiencies and number of events at $\tan\beta = 0.5$ before the mass window cut.

	Signal					
m_H [GeV]	500	600	700	800	900	1000
eff	0.0005	0.00244	0.0055	0.0079	0.096	0.104
Events	9238	25981	41967	35703	26150	17614

multijet has a large cross section. Therefore only events with 6 jets in the final state are generated to produce the same jet multiplicity in the final state. This is based on the assumption that events with more or less number of jets do not contribute to the signal region at the end. Samples with less number of jets have a larger cross section but do not pass the selection cuts (e.g., 6 jet requirement) and those with more number of jets have much smaller cross sections and do not contribute to the signal region sizably.

5. Signal Selection and Analysis

The generation of signal and background events starts with PYTHIA 8 [37] followed by jet reconstruction using FASTJET 2.8 [47, 48].

The analysis uses the top tagging algorithm to identify two top jets in the final state; however, before going to that step, two selection cuts are applied to purify the signal sample. The first requirement is to have at least 2 b-jets in the final state (as there are two b-jets from the top quark decays in signal events). The b-tagging is based on a matching algorithm which uses generator level information of b quarks and compares them with reconstructed jets. If a jet is flying adjacent to a b-quark, it is considered as a b-jet with 70% probability. A 10% fake rate from c-jets is also considered as the mistagging rate.

The second requirement is a lepton veto which requires events to be free of leptons (with a transverse energy threshold of 10 GeV). This is to select fully hadronic events and reject QCD multijet events with possibility of heavy meson leptonic decays.

At this step the top tagging algorithm is applied on signal and background events. The top tagging algorithm uses a different jet reconstruction algorithm from the one used for b-tagging. The b-tagging jet algorithm is anti-kt while the top tagging algorithm is CA (Cambridge-Aachen) as discussed in what follows.

The jet reconstruction algorithms are classified according to their different subjet distance measures which can be written as $d_{j1j2} = \Delta R^2_{j1j2}/D2 \times \min(p^{2n}_{T,j1}, p^{2n}_{T,j2})$ with $n = -1, 0, 1$ for anti-k_T, Cambridge/Aachen (CA), and k_T algorithms, respectively. The k_T algorithm first combines the soft and collinear subjets and is suitable for reconstructing the QCD splitting history in top tagging algorithm. The anti-k_T algorithm, first combines the hardest subjets to obtain a stable jet with clean jet boundary. The CA algorithm always combines the most collinear subjets while not being sensitive to soft splittings and therefore is suitable for top tagging reconstruction. The algorithm adopted by HEPTopTagger is thus CA with a cone size of $\Delta R = 1.5$.

The HEPTopTagger is one of recent algorithms introduced for boosted top quark reconstruction [49]. It is based on a CA jet reconstruction with $\Delta R = 1.5$ and the top jet candidate p_T above 200 GeV. The threshold can be lowered down to 150 GeV without significant loss of efficiency [50, 51]. Having the collection of fat jets in the first step, the top tagging algorithm starts with undoing the last clustering of the top jet candidate j and requiring the mass drop criterion as $\min m_{j_i} < 0.8 m_j$ where j_i is the ith subjet from the jet j. Subjets with $m_j < 30$ GeV are not considered to end the unclustering iteration.

In the second step a filtering is applied to find a three-subjet combination with a jet mass within $m_t \pm 25$ GeV.

In the last step, having sorted jets in p_T, several requirements are applied to find the best combination of subjets with two subjets giving the best W boson invariant mass and the whole three subjets to be consistent with the top quark invariant mass. Details of these criteria are expressed in [50].

Performing the algorithm, selection efficiency for each signal sample is obtained. The same procedure is applied on background samples. Results are shown in Tables 3 and 4 for signal and background samples, respectively. These tables also include the number of events before the mass window cut. An event is required to have two top jets identified.

TABLE 4: Background selection efficiencies and number of events at $\tan\beta = 0.5$ before the mass window cut. The QCD sample is the 6-jet final state and STS (STT) denote the single top s-(t-) channel.

	Background								
Process	$t\bar{t}$	QCD	WW	WZ	ZZ	STS	STT	W	Z
eff	0.0027	3×10^{-5}	6×10^{-7}	2.3×10^{-6}	2.5×10^{-6}	3.8×10^{-5}	1.0×10^{-5}	0	0
Events	353525	40000	5.9	8.3	4.0	63.8	365.7	0	0

TABLE 5: Signal and background analysis results at $\tan\beta = 0.5$ and 1. The S and B denote the final number of signal and background events after the mass window cut.

	$m_{(H/A)}$ [GeV]					
	500	600	700	800	900	1000
Mass window [GeV]	450 − 600	530 − 620	630 − 710	710 − 820	810 − 920	870 − 1020
S	4600	16312	26915	24619	16438	12041
B	59183	63947	75728	82082	55146	52082
$\frac{S}{B}$	0.08	0.25	0.35	0.3	0.3	0.23
$\frac{S}{\sqrt{B}}$ $\tan\beta = 0.5$	18.9	64.5	97.8	85.9	70	52.8
$\frac{S}{\sqrt{B}}$ $\tan\beta = 1$	2.8	6.7	6.8	5.1	3.7	2.6

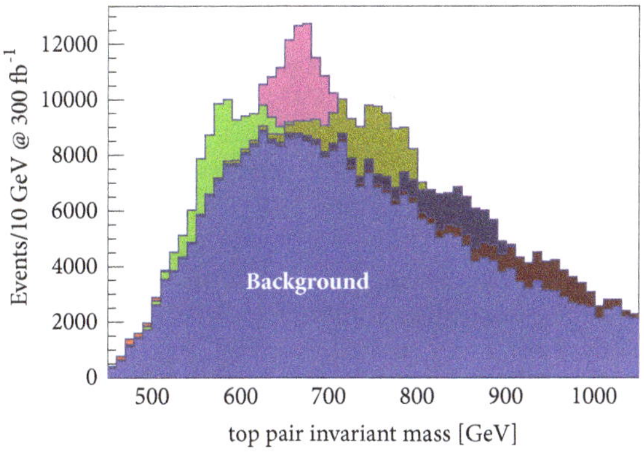

FIGURE 5: The signal distributions on top of the total background (shown in blue). The signal from Higgs boson masses in the range 500-1000 GeV is shown in different colors.

The invariant mass of the two top jets is calculated as the Higgs boson candidate mass. Both signal and background distributions of top quark pair invariant masses are normalized according to the corresponding cross sections. The signal on top of the background is then plotted for each benchmark point as seen in Figure 5.

The invariant mass of the top pair has a large resolution due to uncertainties in the four-momentum reconstruction of the jets as well as the top tagging algorithm. In the top tagging algorithm a χ^2 based method is used to find the correct combination of the jets with their invariant mass in agreement with the W boson and the top quark masses. There can be still ways to improve the top tagging algorithm such as normalizing the light jet four momenta to set their invariant mass equal to the W boson mass before reconstructing the top quark. Since there are two W bosons in the event, this method has some difficulties but can be studied in a detailed analysis related to the performance of the algorithm.

It should be noted that the signal distribution shown in Figure 5 has a single peak for each Higgs boson mass hypothesis due to equal masses of the Higgs bosons. Different masses hypothesis can also be considered. However, the signal distribution can not be distinguished from the equal mass scenario as long as the difference between the Higgs bosons masses is within the invariant mass resolution. Therefore scenarios with $|m_H - m_A| > 100$ GeV might be observable with two distinguishable peaks; however, such a large mass splitting raises the problem of large $\Delta\rho$ which should be avoided.

Since a large number of backgrounds fill the signal region, a mass window cut is applied to select the signal and increase the signal to background ratio. The position of the mass window (both left and right sides) is determined in an automatic search based on requiring the maximum signal significance. This is performed in a loop over bins of the histogram and finding the left and right bins inside which the signal significance is maximum.

Table 5 shows mass window position, total efficiencies for signal and background events, final number of signal and background events passing the mass window cut, their ratio, and the signal significance as $S/\sqrt{B}$ at two values of $\tan\beta = 0.5$ and 1. The integrated luminosity is set to 300 fb^{-1}. Table 5 clearly shows the high sensitivity of the signal significance to $\tan\beta$ parameter. The analysis is thus relevant to $\tan\beta$ values as low as ~ 2.

Figure 6 shows the signal significance as a function of the Higgs boson mass for different $\tan\beta$ values. The dashed horizontal line indicates the 5σ significance. As seen from

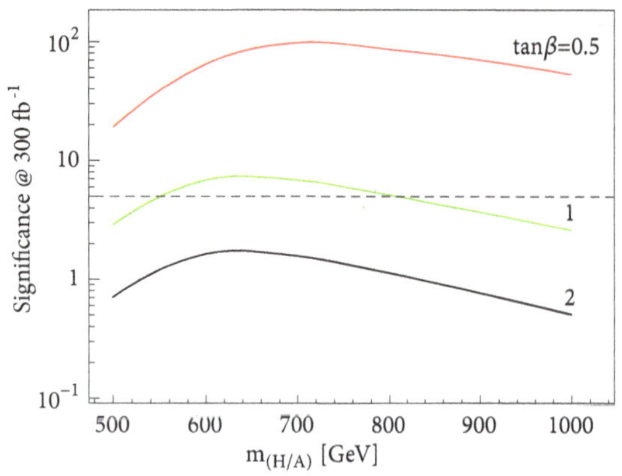

FIGURE 6: The signal significance at 300 fb^{-1} as a function of the Higgs boson mass for different values of $\tan\beta$.

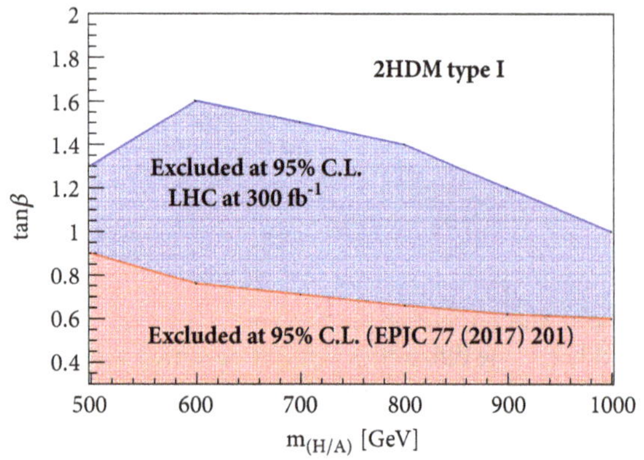

FIGURE 7: The 95% C.L. exclusion region at 300 fb^{-1}.

Figure 3 and Table 3, the signal cross section decreases with increasing Higgs boson mass while selection efficiencies increase. Therefore the product of the cross section times selection efficiency has a peak somewhere near the middle of the Higgs boson mass range where none of the cross section or selection efficiency are too small. This peak happens at m_H = 700 GeV in this analysis. Lower masses suffer from the low selection efficiency while higher masses have the problem of low cross section.

Using the analysis results for Higgs boson masses from 500 GeV to 1000 GeV, one can obtain the 95% C.L. exclusion region and the 5σ discovery contours. Figure 7 shows the exclusion region at 95% C.L. including the recent result from [35] (the result reported in [35] is based on charged Higgs mass as a function of $\tan\beta$; however, it is included in the current work as a limit for all Higgs bosons since the Higgs boson masses are equal in the scenario adopted in this analysis). The 5σ contour is also shown in Figure 8.

As seen from Figures 7 and 8, both exclusion and discovery are possible at regions not yet excluded by LHC. Therefore any sign of extra top pair signals on top of SM background could be regarded as a signal for new physics especially 2HDM. It should be noted that, in this analysis, a full set of background processes was studied. However, all background processes led to very small number of events

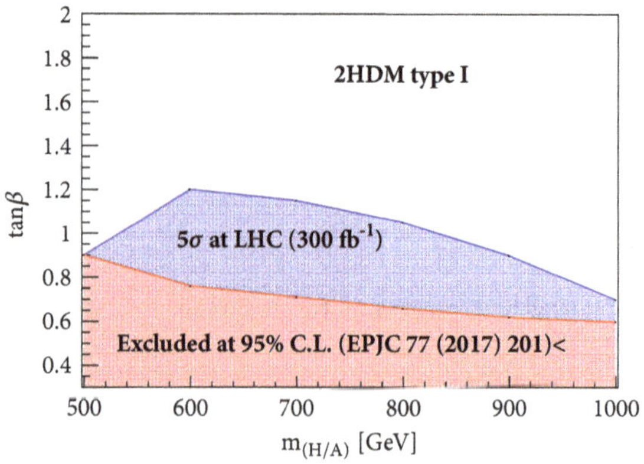

FIGURE 8: The 5σ discovery region at 300 fb^{-1}.

which were negligible compared to the SM $t\bar{t}$. Therefore final plots are based on signal on top of the $t\bar{t}$ distribution without any sizable error.

The LHC sensitivity to the signal studied in this analysis at integrated luminosity of 3000 fb^{-1} can be estimated as follows. If the signal significance grows like $\sqrt{L}$ (L is the integrated luminosity), at $L = 3000\ fb^{-1}$, the signal significance will be roughly three times larger compared to when $L = 300\ fb^{-1}$. The signal cross section, however, decreases from $\tan\beta = 1$ to 2 by a factor of 4 as shown in Figure 4. Therefore the signal significance acquires a factor of 3/4 by increasing $\tan\beta$ and integrated luminosity from 1 and 300 fb^{-1} to 2 and 3000 fb^{-1}. This means that points with m_H = 600 and 700 GeV will be observable at $\tan\beta = 2$ at 3000 fb^{-1}. All above considerations are of course affected by the systematic uncertainties due to the jet energy scale and four-momentum resolution as well as uncertainties in theoretical cross section calculation. A detailed analysis and estimate of such uncertainties are needed before the final assessment.

6. Discussion

The Higgs boson decay to $t\bar{t}$ is already known to dominate at 2HDM type I, and there are thorough studies of 2HDM which cover this area [52]. The aim of this work was to perform an event selection analysis based on LHC data environment and show the signal on top of the background and present exclusion contours. The scenario is taken to be a very limited case (degenerate Higgs bosons masses) to study the best possible cases (benchmark points in the parameter space). The selected benchmark points indeed lead to positive results which make the whole analysis interesting for LHC program. Furthermore, we benefit from top tagging technique which enhances the signal to background ratio and reduces the fake rate. This is in turn a test of the algorithm itself as well as benefiting from its ability in identifying the signal and reducing the background.

7. Conclusions

Extra sources of $t\bar{t}$ events from what we expect from standard model can appear from theories beyond standard model such as two Higgs doublet models. In 2HDM type I, the heavy neutral (CP-even or odd) Higgs decay to $t\bar{t}$ dominates at low $\tan\beta$. In such a scenario a proton-proton collision may create a neutral Higgs decaying to $t\bar{t}$. The signal from such a process can be observed as an excess of top pair events over what is expected from SM. The discriminating tool can be a top pair invariant mass distribution filled with events containing two top jets from both signal and background processes. The analysis performed in this work shows that such a signal is observable at integrated luminosity of 300 fb^{-1} for $\tan\beta$ values which depend on the Higgs boson mass. The exclusion at 95% C.L. is also possible at the same integrated luminosity for $\tan\beta < 2$ with $m_{(H/A)} = 600$ GeV as the best point.

Conflicts of Interest

The authors declare that they have no conflicts of interest.

Acknowledgments

This work was performed using the computing cluster at college of sciences, Shiraz University. We would like to appreciate the personnel involved in the operation and maintenance of the cluster.

References

[1] S. Chatrchyan, V. Khachatryan, A. M. Sirunyan et al., "Observation of a new boson at a mass of 125 GeV with the CMS experiment at the LHC," *Physics Letters B*, vol. 716, p. 30, 2012.

[2] ATLAS Collaboration, G. Aad, T. Abajyan, B. Abbott et al., "Observation of a new particle in the search for the Standard Model Higgs boson with the ATLAS detector at the LHC," *Physics Letters B*, vol. 716, p. 1, 2012.

[3] F. Englert and R. Brout, "Broken symmetry and the mass of gauge vector mesons," *Physical Review Letters*, vol. 13, no. 9, p. 321, 1964.

[4] P. W. Higgs, "Broken symmetries and the masses of gauge bosons," *Physical Review Letters*, vol. 13, pp. 508-509, 1964.

[5] P. Higgs, "Broken symmetries, massless particles and gauge fields," *Physics Letters*, vol. 12, no. 2, pp. 132-133, 1964.

[6] G. S. Guralnik, C. R. Hagen, and T. W. B. Kibble, "Global Conservation Laws and Massless Particles," *Physical Review Letters*, vol. 13, p. 585, 1964.

[7] P. W. Higgs, "Spontaneous symmetry breakdown without massless bosons," *Physical Review A: Atomic, Molecular and Optical Physics*, vol. 145, no. 4, pp. 1156–1163, 1966.

[8] T. W. B. Kibble, "Symmetry breaking in n-Abelian gauge theories," *Physical Review A: Atomic, Molecular and Optical Physics*, vol. 155, no. 5, pp. 1554–1561, 1967.

[9] T. D. Lee, "A Theory of Spontaneous *T* Violation," *Physical Review D: Particles, Fields, Gravitation and Cosmology*, vol. D8, p. 1226, 1973.

[10] S. L. Glashow and S. Weinberg, "Natural conservation laws for neutral currents," *Physical Review D: Particles, Fields, Gravitation and Cosmology*, vol. 15, no. 7, pp. 1958–1965, 1977.

[11] G. C. Branco, " Spontaneous ," *Physical Review D: Particles, Fields, Gravitation and Cosmology*, vol. 22, no. 11, pp. 2901–2905, 1980.

[12] I. J. R. Aitchison, "Supersymmetry and the MSSM: An Elementary Introduction," https://arxiv.org/abs/hep-ph/0505105.

[13] E. Ma and D. Ng, "New supersymmetric option for two Higgs doublets," *Physical Review D*, vol. D49, p. 6164, 1994.

[14] A. Djouadi, "The anatomy of electroweak symmetry breaking Tome II: The Higgs bosons in the Minimal Supersymmetric Model," *Physics Reports*, vol. 459, Article ID 0503173, p. 1, 1016.

[15] H. E. Haber and D. O'Neil, "Erratum: Basis-independent methods for the two-Higgs-doublet model. II. The significance of tanβ," *Physical Review D: Particles, Fields, Gravitation and Cosmology*, vol. 74, p. 059905, 2006.

[16] G. C. Branco, P. M. Ferreira, L. Lavoura, M. N. Rebelo, M. Sher, and J. P. Silva, "Theory and phenomenology of two-Higgs-doublet models," *Physics Reports*, vol. 516, p. 1, 2012.

[17] F. Mahmoudi and O. Stal, "Flavor constraints on two-Higgs-doublet models with general diagonal Yukawa couplings," *Physical Review D*, vol. 81, Article ID 035016, p. 035016, 1103, arXiv:0907.1791.

[18] T. Plehn, M. Spannowsky, M. Takeuchi, and D. Zerwas, "Stop reconstruction with tagged tops," *Journal of High Energy Physics*, vol. 2010, p. 078, 2010.

[19] T. Plehn, M. Spannowsky, and M. Takeuchi, "How to improve top-quark tagging," *Physical Review D: Particles, Fields, Gravitation and Cosmology*, vol. 85, p. 034029, 2012.

[20] S. Davidson and H. E. Haber, "Erratum: Basis-independent methods for the two-Higgs-doublet model," *Physical Review D: Particles, Fields, Gravitation and Cosmology*, vol. 72, no. 9, p. 035004, 2005.

[21] V. Barger, J. L. Hewett, and R. J. Phillips, "New constraints on the charged Higgs sector in two-Higgs-doublet models," *Physical Review D: Particles, Fields, Gravitation and Cosmology*, vol. 41, no. 11, pp. 3421–3441, 1990.

[22] M. Aoki, S. Kanemura, K. Tsumura, and K. Yagyu, "Models of Yukawa interaction in the two Higgs doublet model, and their collider phenomenology," *Physical Review D: Particles, Fields, Gravitation and Cosmology*, vol. 80, p. 015017, 2009.

[23] G. C. Dorsch, S. J. Huber, K. Mimasu, and J. M. No, "Echoes of the Electroweak Phase Transition: Discovering a Second Higgs Doublet through $A_0 \longrightarrow ZH_0$," *Physical Review Letters*, vol. 113, p. 211802, 2014.

[24] W. Grimus, L. Lavoura, O. M. Ogreid, and P. Osland, "A precision constraint on multi-Higgs-doublet models," *Journal of Physics G: Nuclear and Particle Physics*, vol. 35, no. 7, p. 075001, 2008.

[25] D. Eriksson, J. Rathsman, and O. Stal, "2HDMC - two-Higgs-doublet model calculator," *Computer Physics Communications*, vol. 181, no. 1, pp. 189–205, 2010.

[26] D. Eriksson, J. Rathsman, and O. Stal, "2HDMC — two-Higgs-doublet model calculator," *Computer Physics Communications*, vol. 181, 833, no. 4, p. 834, 2010.

[27] S. Su and B. Thomas, "The LHC Discovery Potential of a Leptophilic Higgs," *Physical Review D*, vol. 79, p. 095014, 2009.

[28] M. Hashemi, "Final state interactions in $K \longrightarrow \pi\pi$ decays: ΔI=1/2 rule vs. ε'/ε," *The European Physical Journal C*, vol. 77, p. 302, 2017.

[29] M. Hashemi and G. Haghighat, "Search for heavy neutral CP-even Higgs within lepton-specific 2HDM at a future linear collider," *Physics Letters B*, vol. 772, p. 426, 1016.

[30] R. Barate, D. Decamp, P. Ghez et al., "Search for charged Higgs bosons in e+e- collisions at energies up to sqrt(s) = 189GeV," *Physics Letters B*, vol. 487, pp. 253–263, 2000.

[31] M. Acciarri, P. Achard, O. Adrian et al., "Search for Charged Higgs Bosons in e+e- Collisions at Centre-of-Mass Energies up to 202 GeV," *Physics Letters B*, vol. 496, no. 1–2, pp. 34–42, 2000.

[32] The CMS Collaboration, A. M. Sirunyan, A. Tumasyan et al., "Search for additional neutral MSSM Higgs bosons in the $\tau\tau$ final state in proton-proton collisions at $\sqrt{s} = 13$ TeV," *Journal of High Energy Physics*, vol. 09, p. 7, 2018.

[33] M. Aaboud, G. Aad, B. Abbott et al., "Search for minimal supersymmetric standard model Higgs Bosons *H/A* and for a Z' boson in the $\tau\tau$ final state produced in *pp* collisions at $\sqrt{s} =$ 13 TeV with the ATLAS detector," *The European Physical Journal C*, vol. C76, no. 11, p. 585, 2016.

[34] M. Misiak, H. M. Asatrian, R. Boughezal et al., "Updated NNLO QCD predictions for the weak radiative B-meson decays," *Physical Review Letters*, vol. 114, Article ID 221801, p. 221801, 2015.

[35] M. Misiak and M. Steinhauser, "Weak radiative decays of the B meson and bounds on $M_H \pm$ in the Two-Higgs-Doublet Model," *The European Physical Journal C*, vol. 77, p. 201, 2017.

[36] A. Arbey, F. Mahmoudi, O. Stal, and T. Stefaniak, "Status of the Charged Higgs Boson in Two Higgs Doublet Models," *The European Physical Journal C*, vol. 78, p. 182, 2018.

[37] T. Sjostrand, S. Mrenna, and P. Z. Skands, "A brief introduction to PYTHIA 8.1," *Computer Physics Communications*, vol. 178, no. 11, pp. 852–867, 1016.

[38] J. Alwall, A. Ballestrero, P. Bartalini et al., "Monte Carlos for the LHC," in *Proceedings of the Workshop on the Tools for LHC Event Simulation (MC4LHC)*, Geneva, Switzerland, July, 2006.

[39] J. Alwall, A. Ballestrero, P. Bartalini et al., "A standard format for Les Houches Event Files," *Computer Physics Communications*, vol. 176, no. 4, pp. 300–304, 2007.

[40] F. Caravaglios, M. L. Mangano, M. Moretti, and R. Pittau, "A new approach to multi-jet calculations in hadron collisions," *Nuclear Physics B*, vol. 539, no. 1-2, pp. 215–232, 1999.

[41] M. L. Mangano, M. Moretti, and R. Pittau, "Multijet matrix elements and shower evolution in hadronic collisions: Wbb̄+n-jets as a case study," *Nuclear Physics B*, vol. 632, pp. 343–362, 2002.

[42] M. L. Mangano, F. Piccinini, A. D. Polosa, M. Moretti, and R. Pittau, "ALPGEN, a generator for hard multiparton processes in hadronic collisions," *Journal of High Energy Physics*, vol. 2003, no. 07, p. 001, 2003.

[43] J. M. Campbell, R. K. Ellis, and F. Tramontano, "Single top production and decay at next-to-leading order," *Physical Review D*, vol. D70, p. 094012, 2004.

[44] J. M. Campbell and F. Tramontano, "Next-to-leading order corrections to Wt production and decay," *Nuclear Physics B*, vol. 726, p. 109, 2005.

[45] J. M. Campbell, R. Frederix, F. Maltoni, and F. Tramontano, "Next-to-Leading-Order Predictions for," *Physical Review Letters*, vol. 102, p. 182003, 2009.

[46] J. M. Campbell and R. K. Ellis, "Top-quark processes at NLO in production and decay," https://arxiv.org/abs/1204.1513.

[47] M. Cacciari, "FastJet: a code for fast k_t clustering, and more," in *Proceedings of the 14th International Workshop, DIS*, Tsukuba, Japan, https://arxiv.org/abs/hep-ph/0607071.

[48] M. Cacciari, G. P. Salam, and G. Soyez, "FastJet user manual: (For version 3.0.2)," *The European Physical Journal C*, vol. 72, no. 3, article 1896, pp. 1–54, 2012.

[49] T. Plehn and M. Spannowsky, "Top Tagging," *Journal of Physics G: Nuclear and Particle Physics*, vol. G39, p. 083001, 2012.

[50] S. Yang and Q.-S. Yan, "Searching for Heavy Charged Higgs Boson with Jet Substructure at the LHC," *Journal of High Energy Physics*, vol. 02, p. 074, 2012.

[51] M. Hashemi and G. Haghighat, "Observability of s-channel Heavy Charged Higgs at LHC Using Top Tagging Technique," *Journal of High Energy Physics*, vol. 02, p. 040, 2016.

[52] B. Dumont, J. F. Gunion, Y. Jiang, and S. Kraml, "Constraints on and future prospects for Two-Higgs-Doublet Models in light of the LHC Higgs signal," *Physical Review D*, vol. D90, p. 035021, 2014.

4

Approximate Scattering State Solutions of DKPE and SSE with Hellmann Potential

O. J. Oluwadare[1] **and K. J. Oyewumi**[2]

[1]*Department of Physics, Federal University Oye-Ekiti, Ekiti State, Nigeria*
[2]*Theoretical Physics Section, Department of Physics, University of Ilorin, Ilorin, Nigeria*

Correspondence should be addressed to O. J. Oluwadare; oluwatimilehin.oluwadare@fuoye.edu.ng

Academic Editor: Edward Sarkisyan-Grinbaum

We study the approximate scattering state solutions of the Duffin-Kemmer-Petiau equation (DKPE) and the spinless Salpeter equation (SSE) with the Hellmann potential. The eigensolutions, scattering phase shifts, partial-waves transitions, and the total cross section for all the partial waves are obtained and discussed. The dependence of partial-waves transitions on total angular momentum number, angular momentum number, mass combination, and potential parameters was presented in the figures.

1. Introduction

The relativistic and nonrelativistic quantum mechanical study of Hellmann potential is a long-standing and a well-known problem. The Hellmann potential in this study may be written as [1–5]

$$V(r) = -\frac{a}{r} + \frac{b}{r}e^{-\rho r}, \tag{1}$$

where the first part is the Coulomb potential with the strength parameter a and the second part is the screening Coulomb and/or Yukawa potential with the strength parameter b. The parameter ρ is the potential screening parameter which regulates the shape of the potential. The Coulomb potential has been investigated by some authors in all the limits of quantum mechanics due to its importance in atomic physics [6–8]. Both eigenfunctions and eigenvalues and their structures have been presented in the previous work. The bound and scattering states of the screening Coulomb and/or Yukawa potential have been studied by some researchers in various dimensions [9–12]. The energy levels, wave functions, phase shifts, scattering amplitude, and the effect of the screening parameter on quantum systems have been extensively discussed. The importance of these two parts necessitates the study of Hellmann potential in quantum mechanics.

However, most of the recent studies on the quantum mechanical treatment of Hellmann potential focused on the relativistic and nonrelativistic bound state problems [1–5]. Just recently a good number of researchers have explored the study of the scattering states of Hellmann problems so as to obtain new results that will provide a better understanding of quantum systems. In this regard, Yazarloo et al. extended the study to scattering states of Dirac equation with Hellmann potential under the spin and pseudospin symmetries [4]. The Dirac phase shift and normalized wave function for the spin and pseudospin symmetries were reported. Arda and Sever studied the approximate nonrelativistic bound and scattering states with any l values using the PT-/non-PT symmetry and non-Hermitian Hellmann potential [13]. The phase shift was calculated in terms of the angular momentum quantum number. Also, in one of our previous papers, we studied the scattering state solution of Klein-Gordon equation with Hellmann potential [14]. Again, Arda studied the approximate bound state solution of two-body spinless Salpeter equation for Hellmann potential [15]. He obtained energy levels and eigenfunctions in terms of hypergeometric functions. He also treated Yukawa potential and Coulomb potential as special cases. The Hellmann potential finds its applications in nuclear and high energy physics.

The motivation behind this work is to investigate the approximate scattering state solutions of Duffin-Kemmer-Petiau equation (DKPE) and spinless Salpeter equation (SSE) with Hellmann Potential. The SSE explains in detail the dynamics of semirelativistic of and two-body effects particle [see [16] and the references therein] whereas the DKPE explains explicitly the dynamics of relativistic spin 0 and spin 1 particles [see [17–26] and the references therein].

This work is organized as follows: Section 2 presents scattering state solutions of DKPE with Hellmann potential. The scattering state solution of SSE with Hellmann potential is presented in Section 3. In Section 4, we discuss the results and the conclusion are given in Section 5.

2. Scattering States of the Duffin-Kemmer-Petiau Equation (DKPE) with Hellman Potential

The DKP equation with energy $E_{n,J}$, total angular momentum centrifugal term, and the mass m of the particle is given as follows [17–22]:

$$U''_{n,J}(r) - \frac{J(J+1)}{r^2} + \left[\left(E_{n,J} + V_v^0\right)^2 - m^2\right] U_{n,J}(r) = 0, \quad (2)$$

where $U_{n,J}(r)$ is the radial wave function depending on the principal quantum number n and total angular momentum quantum number J and V_v^0 is the vector potential representing Hellmann potential of (1) in this study. The subscript "v" symbolizes vector while the superscript zero ("0") stands for the spin zero for the particle. $E_{n,J}$ is the energy levels of the spin-zero particle.

The effect of total angular momentum centrifugal term in (2) can be subdued using approximation scheme of the type [1–3, 14, 22, 27]

$$\frac{1}{r^2} \approx \frac{\rho^2}{(1-e^{-\rho r})^2}. \quad (3)$$

The above approximation has been reported to be valid for $\rho r \ll 1$ [14, 22, 27]. The approximate schemes to centrifugal terms have been applied by several authors in several important quantum problems. Also, its development is used in treating centrifugal terms by several authors in several important works; [see [28–32] and the references therein]. Substituting (1) and (3) into (2) and transform using mapping function $z = 1 - e^{-\rho r}$ lead to

$$\begin{aligned} &z^2(1-z)^2 U''_{n,J}(z) - z^2(1-z) U'_{n,J}(z) \\ &+ \left[-\beta_1 z^2 + \beta_2 z - \beta_3\right] U_{n,J}(z) = 0, \end{aligned} \quad (4)$$

where we have employed the following parameters for simplicity:

$$\begin{aligned} -\beta_1 &= a\left(a - \frac{2E_{n,J}}{\rho}\right) + b\left(b - \frac{2E_{n,J}}{\rho}\right) - J(J+1) \\ &- \frac{k^2}{\rho^2}, \end{aligned} \quad (5)$$

$$\beta_2 = -\frac{2E_{n,J}}{\rho}(a+b) + 2b(a-b), \quad (6)$$

$$-\beta_3 = J(J+1) - (a-b)^2, \quad (7)$$

and $k = \sqrt{(E_{n,J}^2 - m^2) + a^2\rho^2 - 2a\rho E_{n,J} - J(J+1)\rho^2}$ is the wave propagation constant.

Choosing the trial wave function of the type,

$$U_{n,J}(z) = z^{\gamma}(1-z)^{-i(k/\rho)} u_{n,J}(z), \quad (8)$$

and substituting it into (4), we obtain the hypergeometric type equation [33]

$$\begin{aligned} &z(1-z) u''_{n,J}(z) + \left[2\gamma - \left(2\gamma - 2i\frac{k}{\rho} + 1\right) z\right] u'_{n,J}(z) \\ &+ \left[\left(\gamma - i\frac{k}{\rho}\right)^2 + \beta_1\right] u_{n,J}(z) = 0, \end{aligned} \quad (9)$$

where

$$\gamma = \frac{1}{2} + \sqrt{\left(J + \frac{1}{2}\right)^2 - (a-b)^2}, \quad (10)$$

$$\begin{aligned} \tau_1 &= \gamma - i\frac{k}{\rho} \\ &- \sqrt{a\left(a - \frac{2E_{n,J}}{\rho}\right) + b\left(b - \frac{2E_{n,J}}{\rho}\right) - J(J+1) - \frac{k^2}{\rho^2}}, \end{aligned} \quad (11)$$

$$\begin{aligned} \tau_2 &= \gamma - i\frac{k}{\rho} \\ &+ \sqrt{a\left(a - \frac{2E_{n,J}}{\rho}\right) + b\left(b - \frac{2E_{n,J}}{\rho}\right) - J(J+1) - \frac{k^2}{\rho^2}}, \end{aligned} \quad (12)$$

$$\tau_3 = 2\gamma. \quad (13)$$

Therefore, the DKP radial wave functions for any arbitrary $J-$ states may be written as

$$\begin{aligned} &U_{n,J}(r) \\ &= N_{n,J}(1-e^{-\rho r})^{\gamma} e^{ikr} {}_2F_1(\tau_1, \tau_2, \tau_3; 1 - e^{-\rho r}), \end{aligned} \quad (14)$$

where $N_{n,J}$ is the normalization factor.

The phase shifts δ_J and normalization factor $N_{n,J}$ can be obtained by applying the analytic-continuation formula [33].

$$\begin{aligned} {}_2F_1(\tau_1, \tau_2, \tau_3; z) &= \frac{\Gamma(\tau_3)\Gamma(\tau_3 - \tau_1 - \tau_2)}{\Gamma(\tau_3 - \tau_1)\Gamma(\tau_3 - \tau_2)} \\ &\cdot {}_2F_1(\tau_1; \tau_2; 1 + \tau_1 + \tau_2 - \tau_3; 1 - z) \\ &+ (1-z)^{\tau_3 - \tau_1 - \tau_2} \frac{\Gamma(\tau_3)\Gamma(\tau_1 + \tau_2 - \tau_3)}{\Gamma(\tau_1)\Gamma(\tau_2)} \\ &\cdot {}_2F_1(\tau_3 - \tau_1; \tau_3 - \tau_2; \tau_3 - \tau_1 - \tau_2 + 1; 1 - z). \end{aligned} \quad (15)$$

Considering (15) with the property ${}_2F_1(\tau_1, \tau_2, \tau_3; 0) = 1$, when $r \longrightarrow \infty$, yields

$$ {}_2F_1\left(\tau_1,\tau_2,\tau_3;1-e^{-\rho r}\right)\ \overrightarrow{r\longrightarrow\infty}\ \Gamma(\tau_3)\left|\frac{\Gamma(\tau_3-\tau_1-\tau_2)}{\Gamma(\tau_3-\tau_1)\Gamma(\tau_3-\tau_2)}+e^{-2ikr}\left|\frac{\Gamma(\tau_3-\tau_1-\tau_2)}{\Gamma(\tau_3-\tau_1)\Gamma(\tau_3-\tau_2)}\right|^*\right|. \tag{16} $$

The following relations have been introduced in the process of derivation

$$ \tau_3-\tau_1-\tau_2=(\tau_1+\tau_2-\tau_3)^*=2i\left(\frac{k}{\rho}\right), \tag{17} $$

$$ \begin{aligned} \tau_3-\tau_2&=\gamma+i\frac{k}{\rho}\\ &\quad-\sqrt{a\left(a-\frac{2E_{n,J}}{\rho}\right)+b\left(b-\frac{2E_{n,J}}{\rho}\right)-J(J+1)-\frac{k^2}{\rho^2}}\\ &=\tau_1^*, \end{aligned} \tag{18} $$

$$ \begin{aligned} \tau_3-\tau_1&=\gamma+i\frac{k}{\rho}\\ &\quad+\sqrt{a\left(a-\frac{2E_{n,J}}{\rho}\right)+b\left(b-\frac{2E_{n,J}}{\rho}\right)-J(J+1)-\frac{k^2}{\rho^2}}\\ &=\tau_2^*. \end{aligned} \tag{19} $$

Now, defining a relation,

$$ \frac{\Gamma(\tau_3-\tau_1-\tau_2)}{\Gamma(\tau_3-\tau_1)\Gamma(\tau_3-\tau_2)}=\left|\frac{\Gamma(\tau_3-\tau_1-\tau_2)}{\Gamma(\tau_3-\tau_1)\Gamma(\tau_3-\tau_2)}\right|e^{i\delta}, \tag{20} $$

and inserting it into (16) yield

$$ {}_2F_1\left(\tau_1,\tau_2,\tau_3;1-e^{-\rho r}\right)\ \overrightarrow{r\longrightarrow\infty}\ \Gamma(\tau_3)\left[\frac{\Gamma(\tau_3-\tau_1-\tau_2)}{\Gamma(\tau_3-\tau_1)\Gamma(\tau_3-\tau_2)}\right]e^{-ikr}\left[e^{i(kr-\delta)}+e^{-i(kr-\delta)}\right] \tag{21} $$

Thus, we have the asymptotic form of (14) when $r\longrightarrow\infty$ as

$$ U_{n,J}(r)\ \overrightarrow{r\longrightarrow\infty}\ 2N_{n,J}\Gamma(\tau_3)\left[\frac{\Gamma(\tau_3-\tau_1-\tau_2)}{\Gamma(\tau_3-\tau_1)\Gamma(\tau_3-\tau_2)}\right]\sin\left(kr+\frac{\pi}{2}+\delta\right). \tag{22} $$

Accordingly, with the appropriate boundary condition imposed by [34], (22) yields

$$ U_{n,J}(\infty)\longrightarrow 2\sin\left(kr-\frac{l\pi}{2}+\delta_J\right). \tag{23} $$

Comparing (22) and (23), the DKP phase shift and the corresponding normalization factor can be found, respectively, as follows:

$$ \begin{aligned} \delta_J&=\frac{\pi}{2}(J+1)+\arg\Gamma(\tau_3-\tau_1-\tau_2)-\arg\Gamma(\tau_3-\tau_1)\\ &\quad-\arg\Gamma(\tau_3-\tau_2)\\ &=\frac{\pi}{2}(J+1)+\arg\Gamma\left(\frac{2ik}{\rho}\right)-\arg\Gamma(\tau_2^*)\\ &\quad-\arg\Gamma(\tau_1^*) \end{aligned} \tag{24} $$

and

$$ \begin{aligned} N_{n,J}&=\frac{1}{\sqrt{\tau_3}}\left|\frac{\Gamma(\tau_3-\tau_2)\Gamma(\tau_3-\tau_1)}{\Gamma(\tau_3-\tau_1-\tau_2)}\right|\\ &=\frac{1}{\sqrt{\tau_3}}\left|\frac{\Gamma(\tau_1^*)\Gamma(\tau_2^*)}{\Gamma(2i(k/\rho))}\right|. \end{aligned} \tag{25} $$

The DKP total cross section for the sum of partial-wave cross sections σ_J is defined as [27]

$$ \sigma_{total}=\sum_{l=0}^{\infty}\sigma_J=\frac{\pi}{k^2}\sum_{l=0}^{\infty}(2l+1)T_J, \tag{26} $$

where $T_J=4\sin^2\delta_J$ defines the DKP partial-wave transitions.

A straightforward substitution of phase shift formula in (24) into (26) yields the total cross section

$$ \begin{aligned} \sigma_{total}&=\frac{4\pi}{k^2}\sum_{l=0}^{\infty}(2l+1)\sin^2\left[\frac{\pi}{2}(J+1)+\arg\Gamma\left(\frac{2ik}{\rho}\right)\right.\\ &\quad\left.-\arg\Gamma(\tau_2^*)-\arg\Gamma(\tau_1^*)\right] \end{aligned} \tag{27} $$

Also, we need to analyze the gamma function $\Gamma(\tau_3-\tau_1)$ [34] from the S-matrix as

$$ \begin{aligned} \tau_3-\tau_1&=\gamma+i\frac{k}{\rho}\\ &\quad+\sqrt{a\left(a-\frac{2E_{n,J}}{\rho}\right)+b\left(b-\frac{2E_{n,J}}{\rho}\right)-J(J+1)-\frac{k^2}{\rho^2}}. \end{aligned} \tag{28} $$

The first-order poles of $\Gamma(\gamma+i(k/\rho)+\sqrt{a(a-2E_{n,J}/\rho)+b(b-2E_{n,J}/\rho)-J(J+1)-k^2/\rho^2})$ are situated at

$$\Gamma\left(\gamma + i\frac{k}{\rho} + \sqrt{a\left(a - \frac{2E_{n,J}}{\rho}\right) + b\left(b - \frac{2E_{n,J}}{\rho}\right) - J(J+1) - \frac{k^2}{\rho^2}}\right) + n = 0 \quad (n = 0, 1, 2, \ldots). \tag{29}$$

By applying algebraic means to (29), we obtain the DKP bound state energy levels equation for the Hellmann potential as follows:

$$k^2 = -\rho^2 \left[\frac{(n+\gamma)^2 + a(2E_{n,J}/\rho - a) + b(2E_{n,J}/\rho - b) - J(J+1)}{2(n+\gamma)}\right]^2. \tag{30}$$

3. Scattering States Solutions of the Spinless Salpeter Equation (SSE) with Hellmann Potential

The spinless Salpeter equation for two different particles interacting in a spherically symmetric potential in the center of mass system is given by [see [35–38] and the references therein]

$$\left[\sum_{i=1,2}\left(\sqrt{-\Delta + m_i^2} - m_i\right) + (V(r) - E_{n,l})\right]\chi(r) = 0, \tag{31}$$

where $\chi(r) = R_{nl}(r)Y_{lm}(\theta, \varphi)$. Also, using appropriate transformation equation $R_{nl}(r) = \psi_{nl}(r)/r$, the radial component of SSE in the case of heavy interacting particles may be written as [see details in [35–38]]

$$\psi_{nl}''(r) + \left[-\frac{l(l+1)}{r^2} + 2\mu(E_{n,l} - V(r)) + \left(\frac{\mu}{\eta}\right)^3 (E_{n,l} - V(r))^2\right]\psi_{nl}(r) = 0, \tag{32}$$

where

$$\mu = \frac{m_1 m_2}{(m_1 + m_2)}, \tag{33}$$

$$\left(\frac{\eta}{\mu}\right)^3 = \frac{m_1 m_2}{(m_1 m_2 - 3\mu^2)}. \tag{34}$$

The units $\hbar = c = 1$ have been employed in the process of derivation and $E_{n,l}$ is the semirelativistic energy of the two particles having arbitrary masses m_1 and m_2. μ and η are the reduced mass and mass index, respectively. The solution to (32) becomes nonrelativistic as the term with the mass index tends to zero.

Substituting the potential in (1) and approximation in (3) into (32) and applying the same procedure in Section 2, the radial wave functions for the spinless Salpeter equation with Hellmann potential are obtained as follows:

$$\psi_{nl}(r) = N_{n,l}(1 - e^{-\rho r})^{\upsilon} e^{ikr} {}_2F_1(\xi_1, \xi_2, \xi_3; 1 - e^{-\rho r}), \tag{35}$$

having the following useful parameters:

$$k = \sqrt{2\mu(E_{n,l} + a\rho) + \left(\frac{\mu}{\eta}\right)^3 (E_{n,l} + a)^2 - l(l+1)\rho^2}, \tag{36}$$

$$\upsilon = \frac{1}{2} + \sqrt{\left(l + \frac{1}{2}\right)^2 - \left(\frac{\mu}{\eta}\right)^3 \left(\frac{a}{\rho} - b\right)^2}, \tag{37}$$

$$\xi_1 = \upsilon - i\frac{k}{\rho} - \sqrt{\frac{2\mu a}{\rho} + \left(\frac{\mu}{\eta}\right)^3 \left[\frac{2E_{n,l}}{\rho}\left(\frac{a}{\rho} - b\right) + \left(\frac{a}{\rho} - b\right)\left(\frac{a}{\rho} + b\right)\right] - l(l+1) - \frac{k^2}{\rho^2}}, \tag{38}$$

$$\xi_2 = \upsilon - i\frac{k}{\rho} + \sqrt{\frac{2\mu a}{\rho} + \left(\frac{\mu}{\eta}\right)^3 \left[\frac{2E_{n,l}}{\rho}\left(\frac{a}{\rho} - b\right) + \left(\frac{a}{\rho} - b\right)\left(\frac{a}{\rho} + b\right)\right] - l(l+1) - \frac{k^2}{\rho^2}}, \tag{39}$$

$$\xi_3 = 2\upsilon. \tag{40}$$

The corresponding phase shift for the spinless Salpeter equation containing Hellmann potential is obtained as

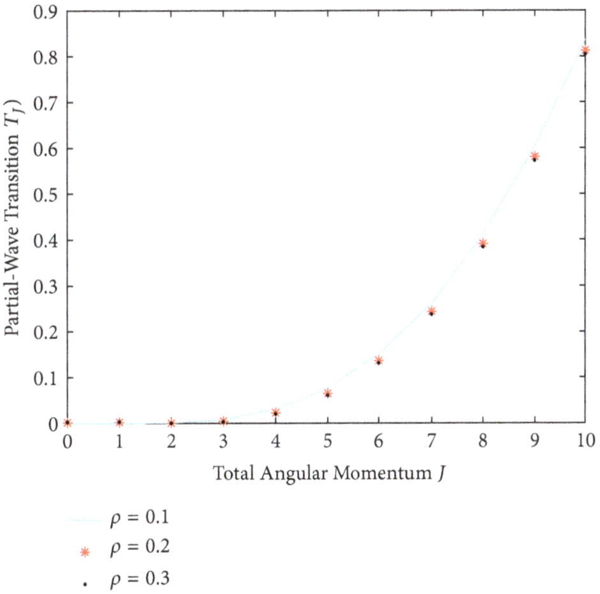

FIGURE 1: DKP partial-wave transition for the Hellman potential as a function of total angular momentum J with $a = b = 0.15$ and $E_{n,J} = m = 1$.

$$\delta_l = \frac{\pi}{2}(l+1) + \arg\Gamma\left(2i\left(\frac{k}{\rho}\right)\right) - \arg\Gamma\left(\xi_2{}^*\right) - \arg\Gamma\left(\xi_1{}^*\right), \tag{41}$$

$$\xi_1{}^* = \xi_3 - \xi_2 = \upsilon + i\frac{k}{\rho} - \sqrt{\frac{2\mu a}{\rho} + \left(\frac{\mu}{\eta}\right)^3\left[\frac{2E_{n,l}}{\rho}\left(\frac{a}{\rho} - b\right) + \left(\frac{a}{\rho} - b\right)\left(\frac{a}{\rho} + b\right)\right] - l(l+1) - \frac{k^2}{\rho^2}}, \tag{42}$$

$$\xi_2{}^* = \xi_3 - \xi_1 = \upsilon + i\frac{k}{\rho} + \sqrt{\frac{2\mu a}{\rho} + \left(\frac{\mu}{\eta}\right)^3\left[\frac{2E_{n,l}}{\rho}\left(\frac{a}{\rho} - b\right) + \left(\frac{a}{\rho} - b\right)\left(\frac{a}{\rho} + b\right)\right] - l(l+1) - \frac{k^2}{\rho^2}}. \tag{43}$$

and the normalization constant

$$N_{n,l} = \frac{1}{\Gamma(\xi_3)}\left|\frac{\Gamma(\xi_1{}^*)\Gamma(\xi_2{}^*)}{\Gamma(2i(k/\rho))}\right|. \tag{44}$$

The energy eigenvalue equation for the spinless Salpeter equation with Hellmann potential is

$$k^2 = -\rho^2\left[\frac{(n+\upsilon)^2 - 2\mu a/\rho + (\mu/\eta)^3\left(2bE_{n,l}/\rho - 2aE_{n,l}/\rho^2 - a^2/\rho^2 + b^2\right) + l(l+1)}{2(n+\upsilon)}\right]^2. \tag{45}$$

The total scattering cross section for the sum of partial-wave cross sections σ_l is given as

$$\sigma_{tot.} = \sum_{l=0}^{\infty}\sigma_l = \frac{\pi}{k^2}\sum_{l=0}^{\infty}(2l+1)T_l, \tag{46}$$

where

$$T_l = 4\sin^2\delta_l \tag{47}$$

which defines the partial-wave transitions for the SSE with Hellmann potential in this present study.

4. Discussion

We have used the units $\hbar = c = 1$ in partial-wave transition illustrations. For equal mass cases, we used $(\mu/\eta)^3 = 1/4$ and $\mu = m_1/2$ while $(\mu/\eta)^3 = 1$ and $\mu = m_1/100$ were used for unequal masses cases. In all the cases, we consider $m_2 = E_{n,l} = 1$ and $m_1 = 1$ for the equal masses case only. For the screening parameters $\rho = 0.1$, $\rho = 0.2$, and $\rho = 0.3$, the DKP partial-waves transitions increase exponentially (see Figure 1). The two-body effect here appears as a shift of the phases of the partial waves. For lower partial-waves, say $l < 5$, the

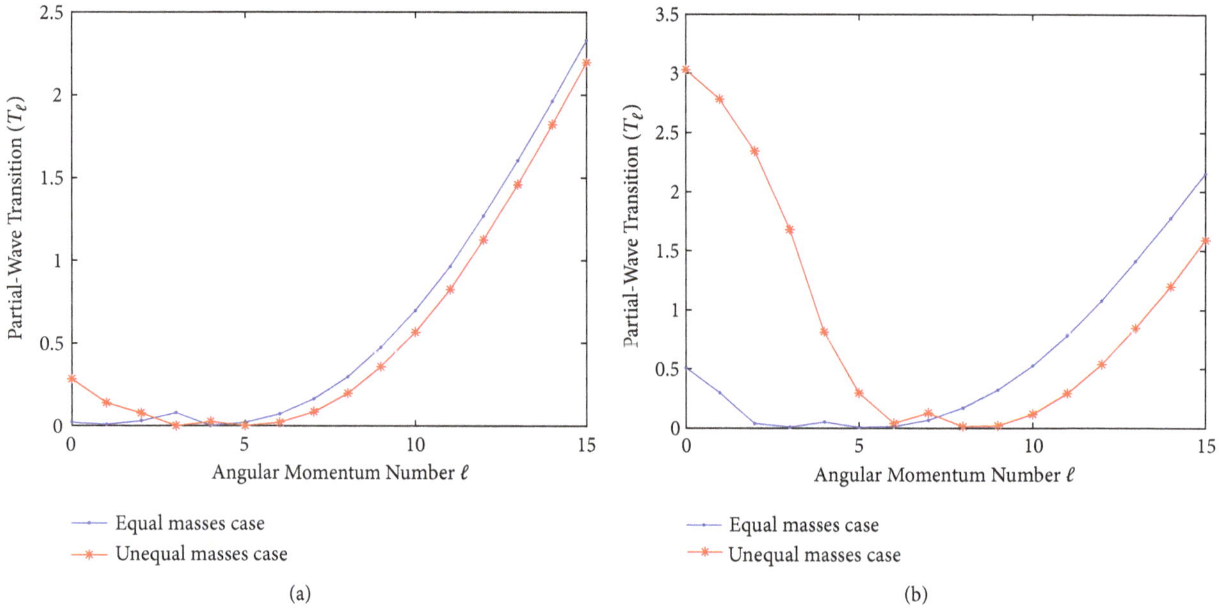

FIGURE 2: (a) Partial-wave transition for the spinless Salpeter equation with the Hellmann potential as a function of angular momentum quantum number l with $a = 0.2, b = -1, E_{n,l} = 1, \rho = 0.5$. (b) Partial-wave transition for the spinless Salpeter equation with the Hellmann potential as a function of angular momentum quantum number l with $a = 2, b = -1, E_{n,l} = 1, \rho = 0.5$.

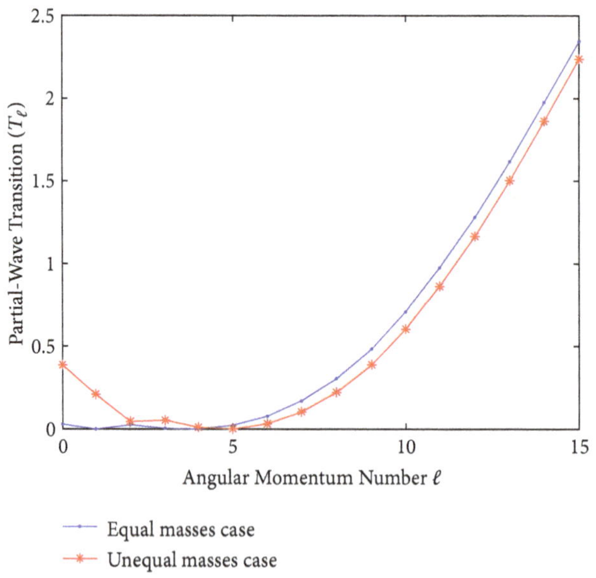

FIGURE 3: Partial-wave transition for the spinless Salpeter equation with the Hellmann potential as a function of angular momentum quantum number l with $a = 0, b = -3, E_{n,l} = 1, \rho = 0.5$.

partial-waves transition decays exponentially whereas, for higher partial waves, say $l > 5$, the partial-waves transition rises exponentially (see Figures 2–4). Also alteration of potential parameters has a serious effect on the partial-wave transition illustrations. Compare Figure 2(a) with Figures 2(b) and 3 and Figure 4(a) with Figure 4(b).

5. Conclusion

We have investigated the approximate scattering state solutions of DKPE and SSE with Hellman potential via analytical method. The approximate DKP and semirelativistic scattering phase shifts, partial-wave transitions, eigenvalues, and

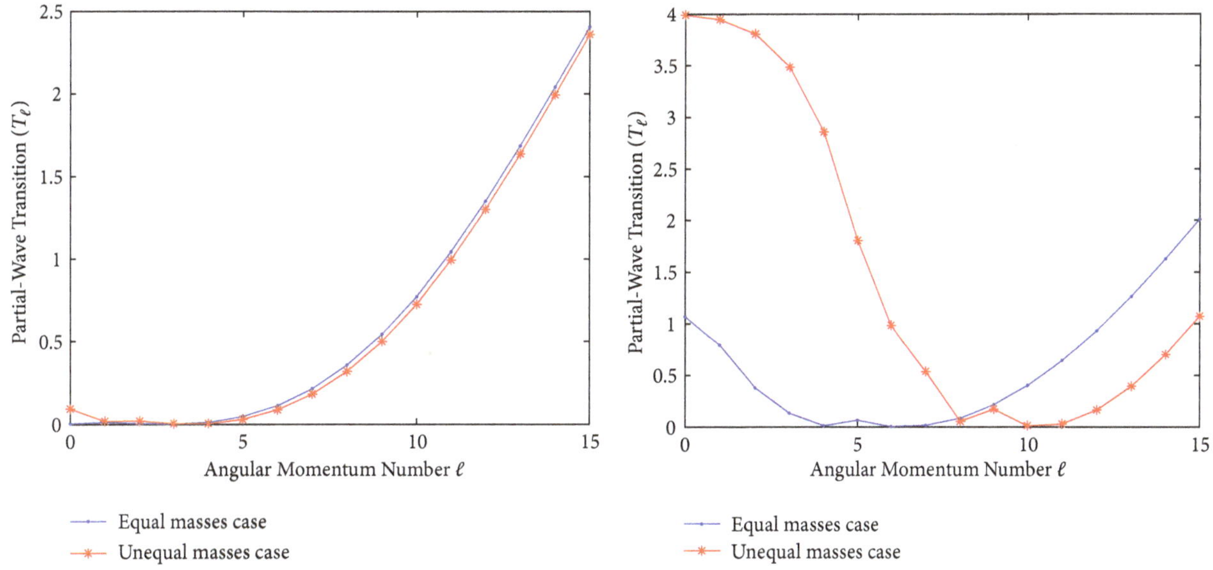

(a) (b)

FIGURE 4: (a) Partial-wave transition for the spinless Salpeter equation with the Hellmann potential as a function of angular momentum quantum number l with $a = -2, b = 0, E_{n,l} = 1, \rho = 0.5$. (b) Partial-wave transition for the spinless Salpeter equation with the Hellmann potential as a function of angular momentum quantum number l with $a = 3, b = 0, E_{n,l} = 1, \rho = 0.5$.

normalized eigenfunctions have been obtained. The DKP and semirelativistic partial-wave transition calculations for the Hellmann potential were shown in Figures 1–4, respectively.

It is clearly shown that the total angular momentum number, angular momentum number, and potential parameters contribute significantly to the partial-wave transition and that the two-body effects modify the phases of the partial waves and are usually noticeable for lower partial waves.

Conflicts of Interest

The authors declare that there are no conflicts of interest regarding this paper.

References

[1] A. Arda and R. Sever, "Pseudospin and spin symmetric solutions of the Dirac equation: Hellmann potential, Wei–Hua potential, Varshni potential," *Zeitschrift für Naturforschung A*, vol. 69, pp. 163–172, 2014.

[2] M. Hamzavi, K.-E. Thylwe, and A. A. Rajabi, "Approximate bound states solution of the Hellmann potential," *Communications in Theoretical Physics*, vol. 60, no. 1, 2013.

[3] C. A. Onate, J. O, A. Ojonubah, E. J. Eweh, and M. Ugboja, "Approximate eigen solutions of D.K.P. and Klein-Gordon equations with Hellmann potential," *The African Review of Physics*, vol. 9, p. 0062, 2014.

[4] B. H. Yazarloo, H. Mehraban, and H. Hassanabadi, "Relativistic scattering states of the Hellmann potential," *Acta Physica Polonica A*, vol. 127, p. 684, 2015.

[5] S. Hassanabadi, M. Ghominejad, B. H. Yazarloo, M. Solaimani, and H. Hassanabadi, "Approximate solution of scattering states of the spinless Salpeter equation with the Yukawa potential," *Chinese journal of physics*, vol. 52, p. 1194, 2014.

[6] H. Tezuka, "Positive energy bound state solutions in relativistic Coulomb potential," *Japan Journal of Industrial and Applied Mathematics*, vol. 14, no. 1, pp. 39–50, 1997.

[7] S.-H. Dong, "On the bound states of the Dirac equation with a Coulomb potential in 2 + 1 dimensions," *Physica Scripta. An International Journal for Experimental and Theoretical Physics*, vol. 67, no. 2, pp. 89–92, 2003.

[8] S.-H. Dong, "The Dirac equation with a Coulomb potential in D dimensions," *Journal of Physics A: Mathematical and General*, vol. 36, no. 18, pp. 4977–4986, 2003.

[9] R. Dutt, A. Ray, and P. Ray, "Bound states of screened Coulomb potentials," *Physics Letters A*, vol. 83, pp. 65–68, 1981.

[10] X. W. Cao, W. L. Chen, Y. Y. Li, and G. F. Wei, "The scattering of the screened Coulomb potential," *Physica Scripta*, vol. 89, no. 8, Article ID 085001, 2014.

[11] N. Salehi and H. Hassanabadi, "Scattering amplitude of the Duffin-Kemmer-Petiau equation for the Yukawa potential for J = 0," *The European Physical Journal A*, vol. 51, no. 100, 2015.

[12] F. Pakdel, A. A. Rajabi, and M. Hamzavi, "Scattering and bound state solutions of the Yukawa potential within the Dirac equation," *Advances in High Energy Physics*, vol. 2014, Article ID 867483, 7 pages, 2014.

[13] A. Arda and R. Sever, "PT-/non-PT-symmetric and non-Hermitian Hellmann potential: approximate bound and scattering states with any ℓ-values," *Physica Scripta*, vol. 89, no. 10, Article ID 105204, 2014.

[14] O. J. Oluwadare and K. J. Oyewumi, "Scattering states solutions of Klein-Gordon equation with three physically solvable potential models," *Chinese Journal of Physics*, vol. 55, no. 6, pp. 2422–2435, 2017.

[15] A. Arda, "Analytical solution of two-body spinless salpeter equation for Hellman potential," *Indian Journal of Physics*, vol. 91, p. 903, 2017.

[16] W. Lucha and F. F. Schoberl, "The spinless relativistic Woods-Saxon problem," *International Journal of Modern Physics A*, vol. 29, no. 10, Article ID 1450057, 15 pages, 2014.

[17] S. Zarrinkamar, A. A. Rajabi, B. H. Yazarloo, and H. Hassanabadi, "An approximate solution of the DKP equation under the Hulthén vector potential," *Chinese Physics C*, vol. 37, Article ID 023101, 2013.

[18] S. Sargolzaeipor, H. Hassanabadi, and A. Boumali, "Morse potential of the q-deformed in the Duffin–Kemmer–Petiau equation," *International Journal of Geometric Methods in Modern Physics*, vol. 14, Article ID 1750112, 2017.

[19] S. Zarrinkamar, S. F. Forouhandeh, B. H. Yazarloo, and H. Hassanabadi, "Scattering states of the Duffin-Kemmer-Petiau equation for the Hulthén potential," *The European Physical Journal Plus*, vol. 128, p. 109, 2013.

[20] A. N. Ikot, H. Hassanabadi, S. Zarrinkamar, and B. H. Yazarloo, "Solutions of the duffin-kemmer-petiau equation under a vector hellman potential," *Few-Body Systems*, vol. 55, no. 3, pp. 211–218, 2014.

[21] A. N. Ikot, H. P. Obong, J. D. Olisa, and H. Hassanabadi, "Scattering state of coupled Hulthen–Woods–Saxon potentials for the Duffin–Kemmer–Petiau equation with pekeris approximation for the centrifugal term," *Zeitschrift für Naturforschung A*, vol. 70, p. 185, 2015.

[22] K. J. Oyewumi and O. J. Oluwadare, "Scattering state solutions of the Duffin-Kemmer-Petiau equation with the Varshni potential model," *The European Physical Journal A*, vol. 53, p. 29, 2017.

[23] G. Petiau, *[Ph.D. thesis]*, Academie Royale de Belgique. Classe des Sciences. Memoires. Collection. University of Paris, 1936.

[24] N. Kemmer, "Quantum theory of einstein-bose particles and nuclear interaction," *Proceedings of the Royal Society A Mathematical, Physical and Engineering Sciences*, vol. 166, no. 924, pp. 127–153, 1938.

[25] R. J. Duffin, "On the characteristic matrices of covariant systems," *Physical Review A: Atomic, Molecular and Optical Physics*, vol. 54, no. 12, p. 1114, 1938.

[26] N. Kemmer, "The particle aspect of meson theory," in *Proceedings of the Royal Society A*, vol. 173, pp. 91–116, 1938.

[27] O. J. Oluwadare and K. J. Oyewumi, "The semi-relativistic scattering states of the two-body spinless Salpeter equation with the Varshni potential model," *The European Physical Journal Plus*, vol. 132, p. 277, 2017.

[28] B. Gonul, O. Ozer, Y. Cancelik, and M. KocaK, "Hamiltonian hierarchy and the Hulthen potential," *Physics Letters A*, vol. 275, no. 4, pp. 238–243, 2000.

[29] G. F Wei, C. Y. Long, and S. H. Dong, "The scattering of the Manning–Rosen potential with centrifugal term," *Physics Letters A*, vol. 372, p. 2592, 2008.

[30] G. F. Wei and S. H. Dong, "Algebraic approach to pseudospin symmetry for the Dirac equation with scalar and vector modified Pöschl-Teller potentials," *Europhysics Letters Association*, vol. 87, no. 4, p. 40004, 2009.

[31] G. F. Wei and S. H. Dong, "Pseudospin symmetry in the relativistic Manning–Rosen potential including a Pekeris-type approximation to the pseudo-centrifugal term," *Physics Letters B*, vol. 686, pp. 288–292, 2010.

[32] G. F. Wei, W. L. Chen, and S. H. Dong, "The arbitrary l continuum states of the hyperbolic molecular potential," *Physics Letters A*, vol. 378, p. 2367, 2014.

[33] M. Abramowitz and I. A. Stegun, *Handbook of Mathematical Functions with Formulas, Graphs, and Mathematical Tables*, vol. 55 of *Applied Mathematics Series*, National Bureau of Standards, Washington, DC, USA, 1964.

[34] L. D. Landau and E. M. Lifshitz, *Quantum mechanics: Non-Relativistic Theory*, Pergamon Press Ltd., New York, NY, USA, 3rd edition, 1977.

[35] H. Hassanabadi and B. H. Yazarloo, "Exact solutions of the Spinless-Salpeter equation under Kink-Like potential," *Chinese Physics C*, vol. 37, no. 12, 2013.

[36] H. Hassanabadi, B. H. Yazarloo, S. Hassanabadi, S. Zarrinkamar, and N. Salehi, "The semi-relativistic scattering states of the Hulthén and Hyperbolic-type potentials," *Acta Physica Polonica A*, vol. 124, p. 20, 2013.

[37] A. A. Rajabi, E. Momtazi, B. H. Yazarloo, and H. Hassanabadi, "Solutions of the two-body Salpeter equation under the Coulomb and exponential potential for any l state with Laplace approach," *Chinese Physics C*, vol. 38, Article ID 023101, 2014.

[38] A. N. Ikot, C. N. Isonguyo, Y. E. Chad-Umoren, and H. Hassanabadi, "Solution of spinless salpeter equation with generalized hulthén potential using SUSYQM," *Acta Physica Polonica A*, vol. 127, 2015.

5

Analysis of $D_s^{*}D^{*}K^{*}$ and $D_{s1}D_1K^{*}$ Vertices in Three-Point Sum Rules

M. Janbazi ,[1] R. Khosravi ,[2] and E. Noori[3]

[1] *Young Researchers and Elites Club, Shiraz Branch, Islamic Azad University, Shiraz, Iran*
[2] *Department of Physics, Isfahan University of Technology, Isfahan 84156-83111, Iran*
[3] *Young Researchers and Elite Club, Karaj Branch, Islamic Azad University, Karaj, Iran*

Correspondence should be addressed to M. Janbazi; mehdijanbazi@yahoo.com

Academic Editor: Ricardo G. Felipe

In this study, the coupling constants of $D_s^* D^* K^*$ and $D_{s1} D_1 K^*$ vertices were determined within the three-point Quantum chromodynamics sum rules method with and without consideration of the $SU_f(3)$ symmetry. The coupling constants were calculated for off-shell charm and K* cases. Considering the nonperturbative effect of the correlation function, as the most important contribution, the quark-quark, quark-gluon, and gluon-gluon condensate corrections were estimated and were compared with other predictive methods.

1. Introduction

Considerable attention has been focused on the strong form factors and coupling constants of meson vertices in the context of quantum chromodynamics (QCD) since the last decade. In high energy physics, understanding the functional form of the strong form factors plays very important role in explaining of the meson interactions. Therefore, accurate determination of the strong form factors and coupling constants associated with the vertices involving mesons has attracted great interest in recent studies of the high energy physics.

Quantum chromodynamics sum rules (QCDSR) formalism has been used extensively to study about the "exotic" mesons made of quark- gluon hybrid ($q\bar{q}g$), tetraquark states ($q\bar{q}q\bar{q}$), molecular states of two ordinary mesons, glueballs [1], and vertices involving charmed mesons such as $D^*D^*\rho$ [2, 3], $D^*D\pi$ [2, 4], $DD\rho$ [5], $D^*D\rho$ [6], DDJ/ψ [7], D^*DJ/ψ [8], D^*D_sK, D_s^*DK, $D_0^*D_sK$, D_{s0}^*DK [9], D^*D^*P, D^*DV, DDV [10], $D^*D^*\pi$ [11], D_sD^*K, D_s^*DK [12], $DD\omega$ [13], $D_{s0}DK$ and D_0D_sK [14], $D_{s1}D^*K$, $D_{s1}D^*K_0^*$ [15, 16], D_sD_sV, $D_s^*D_s^*V$ [17, 18], and $D_1D^*\pi$, $D_1D_0\pi$, $D_1D_1\pi$ [19].

In this study, 3-point sum rules (3PSR) method is used to calculate the strong form factors and coupling constants of the $D_s^*D^*K^*$ and $D_{s1}D_1K^*$ vertices. The 3PSR correlation function is investigated from the phenomenological and the theoretical points of view. Regarding the phenomenological (physical) approach, the representation can be expressed in terms of hadronic degrees of freedom which can be considered as responsible for the introduction of the form factors, decay constant, and masses. The theoretical (QCD) approach usually can be divided into two main contributions as perturbative and nonperturbative. In this approach, the quark-gluon language and Wilson operator product expansion (OPE) are usually used to evaluate the correlation function in terms of the QCD degrees of freedom such as quark condensate, gluon condensate, etc. Equating the two sides and applying the double Borel transformations with respect to the momentum of the initial and final states to suppress the contribution of the higher states, and continuum, the strong form factors can be estimated.

The effective Lagrangian of the interaction for the $D_s^*D^*K^*$ and $D_{s1}D_1K^*$ vertices can be written as [20]

$$\begin{aligned}\mathcal{L}_{D_s^*D^*K^*} = {} & i g_{D_s^*D^*K^*}\left[D_s^{*\mu}\left(\partial_\mu K^{*\nu}\overline{D}_\nu^{*} - K^{*\nu}\partial_\mu\overline{D}_\nu^{*}\right)\right. \\ & + \left(\partial_\mu D_s^{*\nu}K_\nu^{*} - D_s^{*\nu}\partial_\mu K_\nu^{*}\right)\overline{D}^{*\mu} \\ & \left. + K^{*\mu}\left(D_s^{*\nu}\partial_\mu\overline{D}_\nu^{*} - \partial_\mu D_s^{*\nu}\overline{D}_\nu^{*}\right)\right],\end{aligned} \tag{1}$$

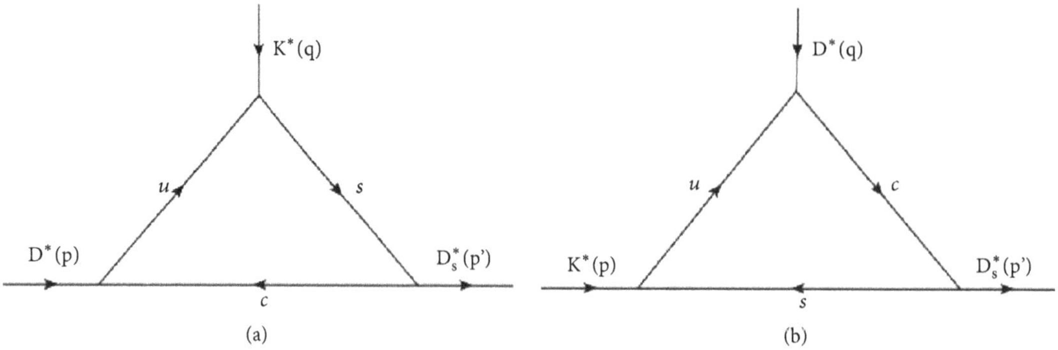

FIGURE 1: Perturbative diagrams for off-shell K^* (a) and off-shell D^* (b).

$$\begin{aligned}\mathscr{L}_{D_{s1}D_1K^*} &= ig_{D_{s1}D_1K^*}\left[D_{s1}{}^{\mu}\left(\partial_{\mu}K^{*\nu}\overline{D}_{1\nu} - K^{*\nu}\partial_{\mu}\overline{D}_{1\nu}\right)\right.\\ &\quad + \left(\partial_{\mu}D_{s1}{}^{\nu}K^*_{\nu} - D_{s1}{}^{\nu}\partial_{\mu}K^*_{\nu}\right)\overline{D}_1^{\mu}\\ &\quad \left. + K^{*\mu}\left(D_{s1}{}^{\nu}\partial_{\mu}\overline{D}_{1\nu} - \partial_{\mu}D_{s1}{}^{\nu}\overline{D}_{1\nu}\right)\right],\end{aligned} \tag{2}$$

where $g_{D_s^*D^*K^*}$ and $g_{D_{s1}D_1K^*}$ are the strong form factor. Using the introduced form of the Lagrangian, the elements related to the $D_s^*D^*K^*$ and $D_{s1}D_1K^*$ vertices can be derived in terms of the strong form factor as

$$\begin{aligned}&\left\langle D^*(p,\varepsilon)\,D_s^*(p',\varepsilon')\mid K^*(q,\varepsilon'')\right\rangle\\ &\quad = ig_{D_s^*D^*K^*}(q^2)\\ &\qquad \times\left[\left(q^{\alpha}+p'^{\alpha}\right)g^{\mu\nu} - \left(q^{\mu}+p^{\mu}\right)g^{\nu\alpha} + q^{\nu}g^{\alpha\mu}\right]\\ &\qquad \times \varepsilon_{\alpha}(p)\,\varepsilon'_{\mu}(p')\,\varepsilon''_{\nu}(q),\end{aligned} \tag{3}$$

$$\begin{aligned}&\left\langle D_1(p,\varepsilon)\,D_{s1}(p',\varepsilon')\mid K^*(q,\varepsilon'')\right\rangle\\ &\quad = ig_{D_{s1}D_1K^*}(q^2)\\ &\qquad \times\left[\left(q^{\alpha}+p'^{\alpha}\right)g^{\mu\nu} - \left(q^{\mu}+p^{\mu}\right)g^{\nu\alpha} + q^{\nu}g^{\alpha\mu}\right]\\ &\qquad \times \varepsilon_{\alpha}(p)\,\varepsilon'_{\mu}(p')\,\varepsilon''_{\nu}(q),\end{aligned} \tag{4}$$

where $q = p - p'$.

The organization of the paper is as follows: In Section 2, the quark-quark, quark-gluon, and gluon-gluon condensate contributions, considering the nonperturbative effects of the Borel transform scheme, are discussed in order to calculate the strong form factors of the $D_s^*D^*K^*$ and $D_{s1}D_1K^*$ vertices in the framework of the 3PSR. The numerical analysis of the strong form factors estimation as well as the coupling constants, with and without consideration of the $SU_f(3)$ symmetry, is described in Section 3 and the conclusion is made in Section 4.

2. The Strong Form Factor of $D_s^*D^*K^*$ and $D_{s1}D_1K^*$ Vertices

To compute the strong form factor of the $D_s^*D^*K^*$ and $D_{s1}D_1K^*$ vertices via the 3PSR, we start with the correlation function. When the K^* meson is off-shell, the correlation function can be written in the following form:

$$\begin{aligned}\Pi^{K^*}_{\mu\nu\alpha}(p,p') &= i^2\int d^4x\,d^4y\,e^{i(p'x-py)}\\ &\quad \cdot\left\langle 0\mid\mathscr{T}\left\{j_{\mu}^{D_s^*}(x)\,j_{\nu}^{K^*\dagger}(0)\,j_{\alpha}^{D^*\dagger}(y)\right\}\mid 0\right\rangle,\end{aligned} \tag{5}$$

$$\begin{aligned}\Pi^{K^*}_{\mu\nu\alpha}(p,p') &= i^2\int d^4x\,d^4y\,e^{i(p'x-py)}\\ &\quad \cdot\left\langle 0\mid\mathscr{T}\left\{j_{\mu}^{D_{s1}}(x)\,j_{\nu}^{K^*\dagger}(0)\,j_{\alpha}^{D_1\dagger}(y)\right\}\mid 0\right\rangle.\end{aligned} \tag{6}$$

For off-shell charm meson, the correlation function can be written as

$$\begin{aligned}\Pi^{D^*}_{\mu\alpha\nu}(p,p') &= i^2\int d^4x\,d^4y\,e^{i(p'x-py)}\\ &\quad \cdot\left\langle 0\mid\mathscr{T}\left\{j_{\mu}^{D_s^*}(x)\,j_{\alpha}^{D^*\dagger}(0)\,j_{\nu}^{K^*\dagger}(y)\right\}\mid 0\right\rangle,\end{aligned} \tag{7}$$

$$\begin{aligned}\Pi^{D_1}_{\mu\alpha\nu}(p,p') &= i^2\int d^4x\,d^4y\,e^{i(p'x-py)}\\ &\quad \cdot\left\langle 0\mid\mathscr{T}\left\{j_{\mu}^{D_{s1}}(x)\,j_{\alpha}^{D_1\dagger}(0)\,j_{\nu}^{K^*\dagger}(y)\right\}\mid 0\right\rangle,\end{aligned} \tag{8}$$

where $j_{\mu}^{D_s^*} = \bar{c}\gamma_{\mu}s$, $j_{\alpha}^{D^*} = \bar{c}\gamma_{\alpha}u$, $j_{\mu}^{D_{s1}} = \bar{c}\gamma_{\mu}\gamma_5 s$, $j_{\alpha}^{D_1} = \bar{c}\gamma_{\alpha}\gamma_5 u$, and $j_{\nu}^{K^*} = \bar{u}\gamma_{\nu}s$ are interpolating currents with the same quantum numbers of D_s^*, D^*, D_{s1}, D_1, and K^* mesons. As described in Figure 1, $\mathscr{T}$, p, and p' are time ordering product and four momentum instances of the initial and final mesons, respectively.

Considering the OPE scheme in the phenomenological approach, the correlation functions (see (5) to (8)) can be written in terms of several tensor structures and their coefficients are found using the sum rules. It is clear from (3) and (4) that the form factor $g_{D_s^*D^*K^*}$ is used for the fourth Lorentz structure which can be extracted from the sum rules. We choose the Lorentz structure because of its fewer ambiguities in the 3PSR approach, i.e., less influence of higher dimension of the condensates and better stability as function of the Borel mass parameter [2]. For these reasons, the $g_{\mu\alpha}q_{\nu}$

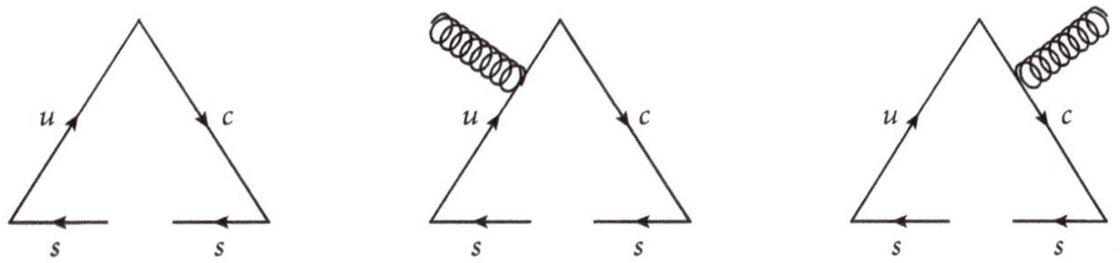

FIGURE 2: Nonperturbative diagrams for the off-shell D^* meson.

structure is chosen which is assumed to better formulate the problem.

In order to calculate the phenomenological part of the correlation functions in (5) to (8), three complete sets of intermediate states with the same quantum number as the currents $j_{\mu}^{D_s^*}$, $j_{\alpha}^{D^*}$, $j_{\mu}^{D_{s1}}$, $j_{\alpha}^{D_1}$, and $j_{\nu}^{K^*}$ are selected. The matrix elements $\langle 0 \mid j_{\mu}^{D_s^*} \mid D_s^*(p,\varepsilon)\rangle$, $\langle 0 \mid j_{\alpha}^{D^*} \mid D^*(p)\rangle$, $\langle 0 \mid j_{\mu}^{D_{s1}} \mid D_{s1}(p,\varepsilon)\rangle$, $\langle 0 \mid j_{\alpha}^{D_1} \mid D_1(p)\rangle$, and $\langle 0 \mid j_{\nu}^{K^*} \mid K^*(q,\varepsilon)\rangle$ are defined as

$$\left\langle 0 \mid j_{\mu}^{V} \mid V(p,\varepsilon)\right\rangle = m_V f_V \varepsilon_{\mu}(p), \tag{9}$$

where m_V and f_V are the masses and decay constants of mesons $V(D_s^*, D^*, D_{s1}, D_1, K^*)$ and ε_{μ} is introduced as the polarization vector of the vector meson $V(D_s^*, D^*, D_{s1}, D_1, K^*)$.

The phenomenological part of the $g_{\mu\alpha}q_{\nu}$ structure associated with the $D_s^* D^* K^*$ vertex for off-shell D^* and K^* mesons can be expressed as

$$\begin{aligned}
\Pi_{\mu\nu\alpha}^{D^*} &= -g_{D_s^* D^* K^*}^{D^*}\left(q^2\right) \\
&\cdot \frac{m_{K^*} f_{K^*} f_{D^*} f_{D_s^*}\left(3m_{D^*}^2 + m_{K^*}^2 - q^2\right)}{2m_{D_s^*}\left(q^2 - m_{D^*}^2\right)\left(p^2 - m_{K^*}^2\right)\left(p'^2 - m_{D_s^*}^2\right)} + \cdots, \\
\Pi_{\mu\nu\alpha}^{K^*} &= g_{D_s^* D^* K^*}^{K^*}\left(q^2\right) \\
&\cdot \frac{m_{K^*} f_{K^*} f_{D^*} f_{D_s^*}\left(3m_{D_s^*}^2 + m_{D^*}^2 - q^2\right)}{2m_{D_s^*}\left(q^2 - m_{K^*}^2\right)\left(p^2 - m_{D^*}^2\right)\left(p'^2 - m_{D_s^*}^2\right)} + \cdots,
\end{aligned} \tag{10}$$

The phenomenological part of the $g_{\mu\alpha}q_{\nu}$ structure associated with the $D_{s1}D_1K^*$ vertex for off-shell D_1 and K^* mesons can be expressed as

$$\begin{aligned}
\Pi_{\mu\nu\alpha}^{D_1} &= -g_{D_{s1} D_1 K^*}^{D_1}\left(q^2\right) \\
&\cdot \frac{m_{K^*} f_{K^*} f_{D_1} f_{D_{s1}}\left(3m_{D_1}^2 + m_{K^*}^2 - q^2\right)}{2m_{D_{s1}}\left(q^2 - m_{D_1}^2\right)\left(p^2 - m_{K^*}^2\right)\left(p'^2 - m_{D_{s1}}^2\right)} + \cdots, \\
\Pi_{\mu\nu\alpha}^{K^*} &= g_{D_{s1} D_1 K^*}^{K^*}\left(q^2\right) \\
&\cdot \frac{m_{K^*} f_{K^*} f_{D_1} f_{D_{s1}}\left(3m_{D_{s1}}^2 + m_{D_1}^2 - q^2\right)}{2m_{D_{s1}}\left(q^2 - m_{K^*}^2\right)\left(p^2 - m_{D^*}^2\right)\left(p'^2 - m_{D_{s1}}^2\right)} + \cdots.
\end{aligned} \tag{11}$$

Using the operator product expansion in Euclidean region and assuming $p^2, p'^2 \longrightarrow -\infty$, one can calculate the QCD side of the correlation function (see (5) to (8)) which contains perturbative and nonperturbative terms. Using the double dispersion relation for the coefficient of the Lorentz structure $g_{\mu\alpha}q_{\nu}$ appearing in the correlation function (see (3) and (4)), we get

$$\begin{aligned}
&\Pi_{per}^{M}\left(p^2, p'^2, q^2\right) \\
&= -\frac{1}{4\pi^2}\int ds \int ds' \frac{\rho^M\left(s, s', q^2\right)}{\left(s - p^2\right)\left(s' - p'^2\right)} \\
&\quad + \text{subtraction terms},
\end{aligned} \tag{12}$$

where $\rho^M(s, s', q^2)$ is spectral density and M stands for off-shell charm and K^* mesons. The Cutkoskys rule allows us to obtain the spectral densities of the correlation function for the Lorentz structure appearing in the correlation function. As shown in Figure 1, the leading contribution comes from the perturbative term. As a result, the spectral densities are obtained in the case of the double discontinuity in (12) for the vertices; see Appendix A.

In order to consider the nonperturbative part of the correlation functions for the case of spectator light quark (for off-shell charm meson), we proceed to calculate the nonperturbative contributions in the QCD approach which contain the quark-quark and quark-gluon condensates [23]. Figure 2 describes the important quark-quark and quark-gluon condensates from the nonperturbative contribution of the off-shell charm mesons [23].

In the 3PSR frame work, when the heavy quark is a spectator (for off-shell K^* meson), the gluon-gluon contribution can be considered. Figure 3 shows related diagrams of the gluon-gluon condensate. More details about the nonperturbative contributions $C_{D_s^* D^* K^*}^{D^*}$ and $C_{D_{s1} D_1 K^*}^{D_1}$ (sum contributions of quark-quark and quark-gluon condensates) and $C_{D_s^* D^* K^*}^{K^*}$ and $C_{D_{s1} D_1 K^*}^{K^*}$ (for gluon-gluon condensates) corresponding to Figures 2 and 3 are given in Appendix B, respectively.

Considering the perturbative and nonperturbative parts of the correlation function in order to suppress the contributions of the higher states, the strong form factors can be calculated in the phenomenological side by equating the two representations of the correlation function and applying the Borel transformations with respect to $p^2 (p^2 \longrightarrow M_1^2)$ and

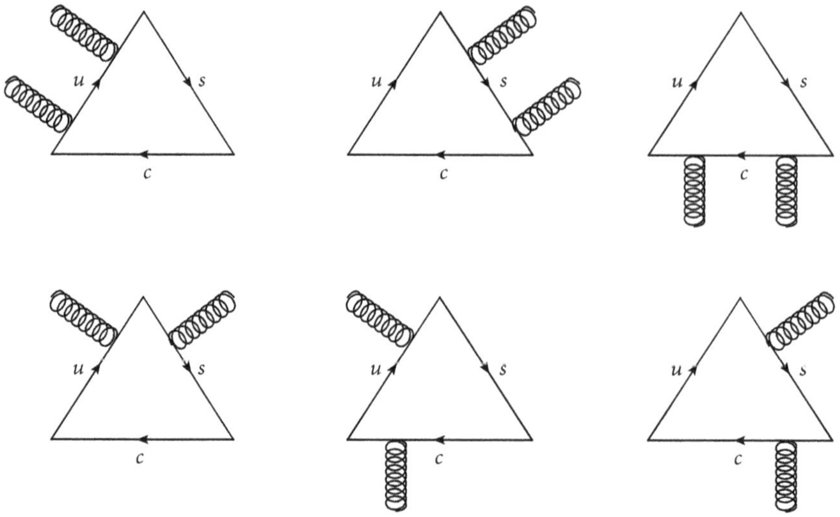

FIGURE 3: Nonperturbative diagrams for the off-shell K^* meson.

$p'^2 (p'^2 \longrightarrow M_2^2)$. The equations for the strong form factors $g^{K^*}_{D_s^* D^* K^*}$ and $g^{D^*}_{D_s^* D^* K^*}$ are obtained as

$$g^{K^*}_{D_s^* D^* K^*}\left(q^2\right) = \frac{2 m_{D_s^*}\left(q^2 - m_{K^*}^2\right)}{m_{K^*} f_{K^*} f_{D^*} f_{D_s^*}\left(3 m_{D_s^*}^2 + m_{D^*}^2 - q^2\right)} \cdot e^{m_{D^*}^2/M_1^2} e^{m_{D_s^*}^2/M_2^2} \left\{ -\frac{1}{4\pi^2} \int_{(m_c+m_s)^2}^{s_0^{D_s^*}} ds' \cdot \int_{(m_u+m_c)^2}^{s_0^{D^*}} ds \rho^{K^*}_{D_s^* D^* K^*}\left(s, s', q^2\right) e^{-s/M_1^2} e^{-s'/M_2^2} - i M_1^2 M_2^2 \left\langle \frac{\alpha_s}{\pi} G^2 \right\rangle \times C^{K^*}_{D_s^* D^* K^*} \right\},$$

$$g^{D^*}_{D_s^* D^* K^*}\left(q^2\right) = \frac{2 m_{D_s^*}\left(q^2 - m_{D^*}^2\right)}{m_{K^*} f_{K^*} f_{D^*} f_{D_s^*}\left(3 m_{D^*}^2 + m_{K^*}^2 - q^2\right)} \cdot e^{m_{K^*}^2/M_1^2} e^{m_{D_s^*}^2/M_2^2} \left\{ -\frac{1}{4\pi^2} \int_{(m_c+m_s)^2}^{s_0^{D_s^*}} ds' \cdot \int_{(m_u+m_s)^2}^{s_0^{K^*}} ds \rho^{D^*}_{D_s^* D^* K^*}\left(s, s', q^2\right) e^{-s/M_1^2} e^{-s'/M_2^2} + M_1^2 M_2^2 \langle s\bar{s} \rangle \times C^{D^*}_{D_s^* D^* K^*} \right\}. \quad (13)$$

The equations describing the strong form factors $g^{K^*}_{D_{s1} D_1 K^*}$ and $g^{D_1}_{D_{s1} D_1 K^*}$ can be written as

$$g^{K^*}_{D_{s1} D_1 K^*}\left(q^2\right) = \frac{2 m_{D_{s1}}\left(q^2 - m_{K^*}^2\right)}{m_{K^*} f_{K^*} f_{D_1} f_{D_{s1}}\left(3 m_{D_{s1}}^2 + m_{D_1}^2 - q^2\right)} \cdot e^{m_{D_1}^2/M_1^2} e^{m_{D_{s1}}^2/M_2^2} \left\{ -\frac{1}{4\pi^2} \int_{(m_c+m_s)^2}^{s_0^{D_{s1}}} ds' \cdot \int_{(m_u+m_c)^2}^{s_0^{D_1}} ds \rho^{K^*}_{D_{s1} D_1 K^*}\left(s, s', q^2\right) e^{-s/M_1^2} e^{-s'/M_2^2} - i M_1^2 M_2^2 \left\langle \frac{\alpha_s}{\pi} G^2 \right\rangle \times C^{K^*}_{D_{s1} D_1 K^*} \right\},$$

$$g^{D_1}_{D_{s1} D_1 K^*}\left(q^2\right) = \frac{2 m_{D_{s1}}\left(q^2 - m_{D_1}^2\right)}{m_{K^*} f_{K^*} f_{D_1} f_{D_{s1}}\left(3 m_{D_1}^2 + m_{K^*}^2 - q^2\right)} \cdot e^{m_{K^*}^2/M_1^2} e^{m_{D_{s1}}^2/M_2^2} \left\{ -\frac{1}{4\pi^2} \int_{(m_c+m_s)^2}^{s_0^{D_{s1}}} ds' \cdot \int_{(m_u+m_s)^2}^{s_0^{K^*}} ds \rho^{D_1}_{D_{s1} D_1 K^*}\left(s, s', q^2\right) e^{-s/M_1^2} e^{-s'/M_2^2} + M_1^2 M_2^2 \langle s\bar{s} \rangle \times C^{D_1}_{D_{s1} D_1 K^*} \right\}. \quad (14)$$

where the quantities $s_0^{D_s^*}$, $s_0^{D^*}$ $s_0^{D_{s1}}$, $s_0^{D_1}$, and $s_0^{K^*}$ are introduced as the continuum thresholds in D_s^*, D^* D_{s1}, D_1, and K^* mesons, respectively, and $\rho^{K^*}_{D_s^* D^* K^*}$, $\rho^{D^*}_{D_s^* D^* K^*}$, $C^{K^*}_{D_s^* D^* K^*}$, $C^{D^*}_{D_s^* D^* K^*}$, $\rho^{K^*}_{D_{s1} D_1 K^*}$, $\rho^{D_1}_{D_{s1} D_1 K^*}$, $C^{K^*}_{D_{s1} D_1 K^*}$, and $C^{D_1}_{D_{s1} D_1 K^*}$ are defined in Appendices A and B.

3. Numerical Analysis

In order to numerically estimate the strong form factors and coupling constants of the vertices $D_s^* D^* K^*$ and $D_{s1} D_1 K^*$, the values of the quark and meson masses are chosen as $m_s = 0.14 \pm 0.01$ GeV, $m_{K^*} = 0.89$ GeV, $m_{D_s^*} = 2.11$ GeV, $m_{D_{s1}} = 2.46$ GeV, and $m_{D_1} = 2.42$ GeV [24]. Moreover, the leptonic decay constants of the vertices are $f_{K^*} = 220 \pm 5$ [24], $f_{D_s^*} = 314 \pm 19$ [25], $f_{D^*} = 242 \pm 12$ [25], $f_{D_{s1}} = 225 \pm 20$ [26], and $f_{D_1} = 219 \pm 11$ [27]

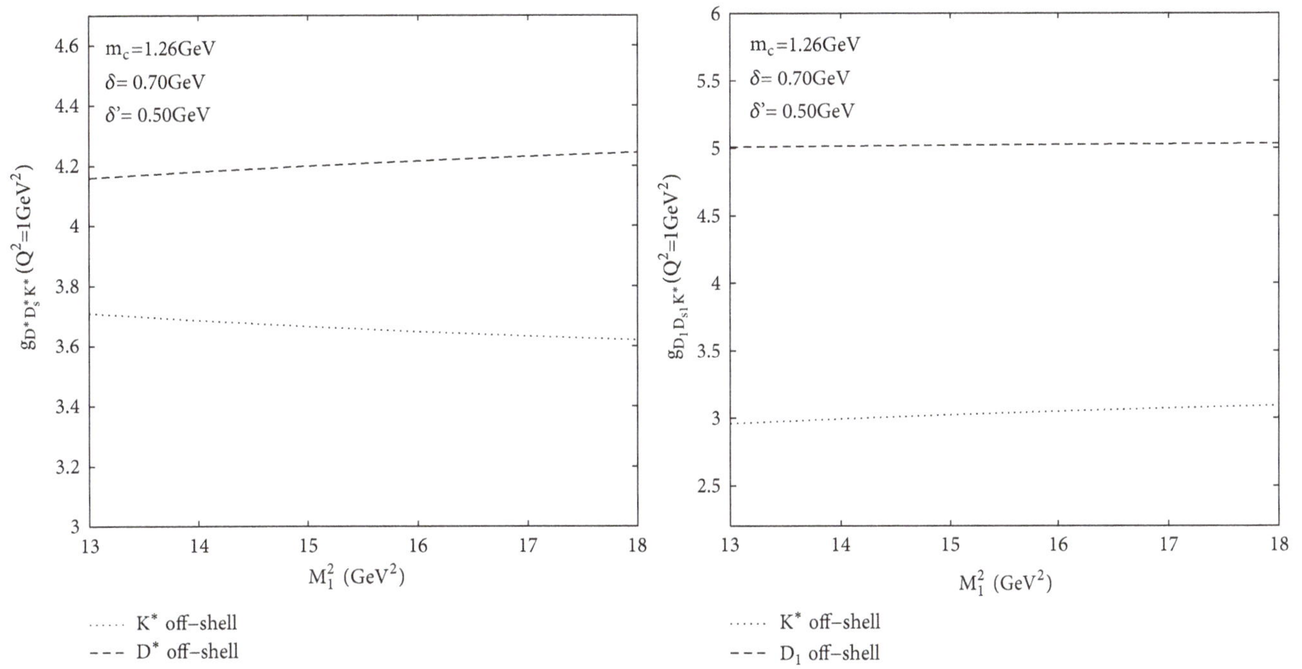

FIGURE 4: The strong form factors $g_{D_s^* D^* K^*}$ (left) and $g_{D_{s1} D_1 K^*}$ (right) as functions of the Borel mass parameter M_1^2 for off-shell charm and K* mesons.

TABLE 1: Parameters appearing in the fit functions for the $D_s^* D^* K^*$ and $D_{s1} D_1 K^*$ vertices for various m_c and (δ, δ'), where $(\delta_1, \delta_1') = (0.50, 0.30)$, $(\delta_2, \delta_2') = (0.70, 0.50)$, and $(\delta_3, \delta_3') = (0.90, 0.70)$ GeV.

	Set I						Set II	
Form factor	$A(\delta_1, \delta_1')$	$B(\delta_1, \delta_1')$	$A(\delta_2, \delta_2')$	$B(\delta_2, \delta_2')$	$A(\delta_3, \delta_3')$	$B(\delta_3, \delta_3')$	$A(\delta_2, \delta_2')$	$B(\delta_2, \delta_2')$
$g^{K^*}_{D_s^* D^* K^*}(Q^2)$	4.04	183.10	4.95	197.49	5.93	215.62	4.43	266.52
$g^{D^*}_{D_s^* D^* K^*}(Q^2)$	3.67	61.46	4.42	52.33	5.56	47.96	4.18	57.95
$g^{K^*}_{D_{s1} D_1 K^*}(Q^2)$	4.19	15.24	4.39	21.18	4.68	29.21	4.22	7.69
$g^{D_1}_{D_{s1} D_1 K^*}(Q^2)$	2.62	13.27	3.13	19.47	4.02	25.32	3.03	14.87

There are four auxiliary parameters containing the Borel mass parameters M_1 and M_2 and continuum thresholds $s_0^{K^*}$, $s_0^{D^*(D_1)}$, and $s_0^{D_s^*(D_{s1})}$ in ((13) and (14)). The strong form factors and coupling constants are physical quantities which are independent of the mass parameters and continuum thresholds. However, the continuum thresholds are not completely arbitrary and can be related to the energy of the first exited state. The values of the continuum thresholds are taken to be $s_0^{K^*} = (m_{K^*} + \delta)^2$, $s_0^{D^*(D_1)} = (m_{D^*(D_1)} + \delta')^2$, and $s_0^{D_s^*(D_{s1})} = (m_{D_s^*(D_{s1})} + \delta')^2$ where $0.50 \text{ GeV} \le \delta \le 0.90$ GeV and $0.30 \text{ GeV} \le \delta' \le 0.70$ GeV [2–4].

Our results should be almost insensitive to the intervals of the Borel parameters. In this work, the Borel masses are related as $M_1^2/M_2^2 = (m_{K^*}^2 + m_c^2)/m_{D_s^*(D_{s1})}^2$ and $M_1^2 = M_2^2$ for off-shell charm mesons and K^*, respectively [5, 6]. The form factors for the $D_s^* D^* K^*$ and $D_{s1} D_1 K^*$ vertices with respect to the Borel parameters M_1^2 are shown in Figure 4. It is found from the figure that the stability of the form factors, as function of Borel parameters, is good in the region of $13 \text{ GeV}^2 < M_1^2 < 18 \text{ GeV}^2$ for off-shell K^* and charm mesons. We get $M_1^2 = 15 \text{ GeV}^2$ and calculate the strong form factors $g_{D_s^* D^* K^*}$ in some points of Q^2 via the 3PSR formalism.

To extract the coupling constants from the form factors, it is needed to extend the Q^2 dependency of the strong form factors to the ranges that the sum rule results are not valid. Therefore, we fitted two sets of points (boxes and circles) imposing the condition that the two resulting parameterizations lead to the same result for $Q^2 = -m_m^2$, where m_m is the mass of the off-shell mesons. This procedure is sufficient to reduce the uncertainties. It is found that the sum rule predictions of the form factors in ((13) and (14)) are well fitted to the function

$$g(Q^2) = Ae^{-Q^2/B}. \tag{15}$$

The values of the parameters A and B are given in Table 1.

Variations of the strong form factors $g^{K^*}_{D_s^* D^* K^*}$ and $g^{D^*}_{D_s^* D^* K^*}$ for $D_s^* D^* K^*$ vertex and $g^{K^*}_{D_{s1} D_1 K^*}$ and $g^{D_1}_{s1 D_1 K^*}$ for $D_{s1} D_1 K^*$ vertex with respect to the Q^2 parameter are shown in Figure 5. The boxes and circles show the results of the

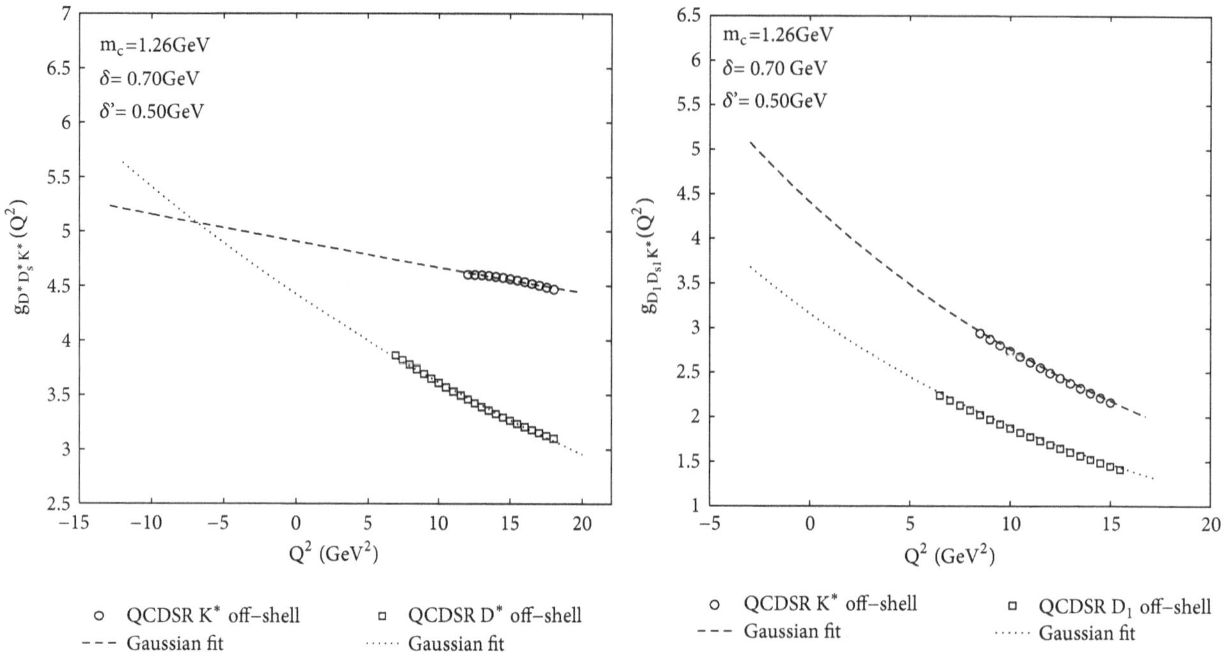

FIGURE 5: The strong form factors $g_{D_s^*D^*K^*}$ (left) and $g_{D_{s1}D_1K^*}$ (right) as functions of Q^2 for off-shell charm and K* mesons.

TABLE 2: The coupling constant of the vertices $D_s^*D^*K^*$ and $D_{s1}D_1K^*$.

g	Set I off-shell charmed	off-shell K^*	Set II off-shell charmed	off-shell K^*
$g_{D_s^*D^*K^*}$	4.77 ± 0.63	4.96 ± 0.64	4.48 ± 0.58	4.45 ± 0.58
$g_{D_{s1}D_1K^*}$	4.22 ± 0.55	4.56 ± 0.59	4.48 ± 0.58	4.67 ± 0.62

TABLE 3: Parameters appearing in the fit functions for the $g_{D_s^*D^*K^*}$ and $g_{D_{s1}D_1K^*}$ form factors in $SU_f(3)$ symmetry with $m_c = 1.26$ GeV and $(\delta, \delta') = (0.70, 0.50)$ GeV.

Form factor	A	B	Form factor	A	B
$g^{K^*}_{D_s^*D^*K^*}(Q^2)$	5.01	218.91	$g^{K^*}_{D_{s1}D_1K^*}(Q^2)$	4.56	20.63
$g^{D^*}_{D_s^*D^*K^*}(Q^2)$	4.54	52.08	$g^{D_1}_{D_{s1}D_1K^*}(Q^2)$	3.66	20.89

numerical evaluation of the form factors via the 3PSR. It is clear from the figure that the form factors are in good agreement with the fitted function.

So-called harder is used. In the present analysis we find that the form factor is harder when the lighter meson is off-shell. This is in line with the results of our previous work [17], whereas this is in contrast with other previous calculations quoted by the authors [28, 29].

The value of the strong form factors at $Q^2 = -m_m^2$ is defined as coupling constant. Calculation results of the coupling constant of the vertices $D_s^*D^*K^*$ and $D_{s1}D_1K^*$ are summarized in Table 2. It should be noted that the coupling constants $g_{D_s^*D^*K^*}$ and $g_{D_{s1}D_1K^*}$ are in the unit of GeV^{-1}.

In order to estimate the error of the calculated parameters, variations of the Borel parameter, continuum thresholds, and leptonic decay constants, as the most significant reasons of the uncertainties, are considered.

To investigate the value of the strong coupling constant via the $SU_f(3)$ symmetry, the mass of the s quark is ignored in all equations. Calculated parameters A and B for the $g_{D_s^*D^*K^*}$ and $g_{D_{s1}D_1K^*}$ vertices, considering $(\delta, \delta') = (0.70, 0.50)$ GeV, are given in Table 3.

Estimated coupling constants of the vertices $D_s^*D^*K^*$, and $D_{s1}D_1K^*$, considering the $SU_f(3)$ symmetry, are summarized in Table 4. The comparisons of the coupling constants $g_{D_s^*D^*K^*}$ with $g_{D^*D^*\rho}$, considering other methods described in [2, 3], are given in Table 5. It is found that the results of the calculated parameters are in reasonable agreement with that of [2, 3] and a factor of two orders of magnitude larger in comparison with [21, 22].

4. Conclusion

Strong form factors and coupling constants of $D_s^*D^*K^*$ and $D_{s1}D_1K^*$ vertices were calculated in the frame work of

TABLE 4: The coupling constant of the vertices $D_s^* D^* K^*$ and $D_{s1} D_1 K^*$, in $SU_f(3)$ symmetry.

g	off-shell charmed	off-shell K^*	g	off-shell charmed	off-shell K^*
$g_{D_s^* D^* K^*}$	4.88 ± 0.64	5.03 ± 0.65	$g_{D_{s1} D_1 K^*}$	4.85 ± 0.63	4.75 ± 0.62

TABLE 5: Values of the strong coupling constant reporting different reference of the coupling constant $g_{D^* D^* \rho}$ [2, 3, 21, 22].

g	Ours	Reference [2, 3]	Reference [21, 22]
$g_{D_s^* D^* K^*}$	4.95 ± 0.64	6.60 ± 0.30	2.52

3-point sum rules of quantum chromodynamics with and without consideration of the $SU_f(3)$ symmetry. Considering nonperturbative contributions of the correlation functions, the quark-quark, quark-gluon, and gluon-gluon condensate corrections were estimated as the most effective terms. It was found from the numerical results that the obtained coupling constants are in good agreement with the other prediction methods described in [2, 3].

Appendix

A. Explicit Expressions of Spectral Densities

In this appendix, the explicit expressions of spectral densities are given as

$$\begin{aligned}
\rho^{D^*}_{D_s^* D^* K^*} &= 3I_0 \left[3m_s^2 - 2m_c m_s - s - \Delta + 4A \right. \\
&\quad \left. + (C_1 - C_2)(2u + 2m_c m_s - 2s) - 8(E_1 - E_2)\right] \\
\rho^{K^*}_{D_s^* D^* K^*} &= 3I_0 \left[3m_c^2 - 2m_c m_s - s - \Delta' + 4A' \right. \\
&\quad \left. + 2(C_1' - C_2')(u + 2m_c m_s - 2s) - 8(E_1' - E_2')\right] \\
\rho^{D_1}_{D_{s1} D_1 K^*} &= 3I_0 \left[3m_s^2 + 2m_c m_s - s - \Delta + 4A \right. \\
&\quad \left. + (C_1 - C_2)(2u - 2m_c m_s - 2s) - 8(E_1 - E_2)\right] \\
\rho^{K^*}_{D_{s1} D_1 K^*} &= 3I_0 \left[3m_c^2 + 2m_c m_s - s - \Delta' + 4A' \right. \\
&\quad \left. + 2(C_1' - C_2')(u - 2m_c m_s - 2s) - 8(E_1' - E_2')\right]
\end{aligned} \tag{A.1}$$

where coefficients in the spectral densities are given as

$$\begin{aligned}
I_0(s, s', q^2) &= \frac{1}{4\lambda^{1/2}(s, s', q^2)}, \\
\lambda(a, b, c) &= a^2 + b^2 + c^2 - 2ac - 2bc - 2ac, \\
\Delta &= s' + m_s^2 - m_c^2, \\
\Delta' &= s' + m_c^2 - m_s^2, \\
\Delta'' &= s + m_s^2, \\
u &= s + s' - q^2, \\
C_1 &= \frac{1}{\lambda(s, s', q^2)}\left[2s'\Delta'' - u\Delta\right], \\
C_2 &= \frac{1}{\lambda(s, s', q^2)}\left[2s\Delta - u\Delta''\right], \\
A &= -\frac{1}{2\lambda(s, s', q^2)}\left[4ss'm_s^2 - s\Delta^2 - s'\Delta''^2 - m_s^2 u^2 \right. \\
&\quad \left. + u\Delta\Delta''\right], \\
E_1 &= \frac{1}{2\lambda^2(s, s', q^2)}\left[8ss'^2 m_s^2 \Delta'' - 2s' m_s^2 u^2 \Delta'' \right. \\
&\quad - 4ss' m_s^2 u\Delta + m_s^2 u^3 \Delta - 2s'^2 \Delta''^3 + 3s' u\Delta\Delta''^2 \\
&\quad \left. - 2ss'\Delta^2\Delta'' - u^2\Delta^2\Delta'' + su\Delta^3\right], \\
E_2 &= \frac{1}{2\lambda^2(s, s', q^2)}\left[8s^2 s' m_s^2 \Delta - 2s m_s^2 u^2 \Delta'' \right. \\
&\quad - 4ss' m_s^2 u\Delta'' + m_s^2 u^3 \Delta'' - 2s^2\Delta^3 + 3su\Delta^2\Delta'' \\
&\quad \left. - 2ss'\Delta\Delta''^2 - u^2\Delta\Delta''^2 + s'u\Delta''^3\right],
\end{aligned} \tag{A.2}$$

and also $A' = A|_{m_c \longleftrightarrow m_s}$, $C_1' = C_1|_{m_c \longleftrightarrow m_s}$, $C_2' = C_2|_{m_c \longleftrightarrow m_s}$, $E_1' = E_1|_{m_c \longleftrightarrow m_s}$, and $E_2' = E_2|_{m_c \longleftrightarrow m_s}$.

B. Explicit Expressions of the Coefficients

In this appendix, the explicit expressions of the coefficients of the quark and gluon condensate contributions of the strong form factors in the Borel transform scheme for all the vertices are presented.

$$\begin{aligned}
C^{D^*}_{D_s^* D^* K^*} &= \left(-6\frac{m_s m_c{}^2}{M_2{}^2} - 2\frac{m_0{}^2 m_c}{M_2{}^2} + 6\frac{m_c m_s{}^2}{M_2{}^2} \right. \\
&\quad + 3\frac{m_s q^2}{M_2{}^2} + 3\frac{m_c q^2 m_s{}^2}{M_1{}^2 M_2{}^2} - \frac{m_0{}^2 m_c q^2}{M_1{}^2 M_2{}^2} - 3\frac{m_c{}^3 m_s{}^2}{M_1{}^2 M_2{}^2} \\
&\quad \left. + \frac{m_0{}^2 m_c{}^3}{M_1{}^2 M_2{}^2} - 3\frac{m_c{}^3 m_s{}^2}{M_2{}^4} + \frac{3}{2}\frac{m_0{}^2 m_c{}^3}{M_2{}^4} + 3m_s\right) \\
&\quad \times e^{-m_c^2/M_2^2}, \\
C^{D_1}_{D_{s1} D_1 K^*} &= \left(2\frac{m_0{}^2 m_c}{M_2{}^2} - 6\frac{m_s m_c{}^2}{M_2{}^2} + 3\frac{m_s q^2}{M_2{}^2}\right.
\end{aligned}$$

$$- 6\frac{m_c m_s^2}{M_2^2} - \frac{m_0^2 m_c^3}{M_1^2 M_2^2} - 3\frac{m_c q^2 m_s^2}{M_1^2 M_2^2} + \frac{m_0^2 m_c q^2}{M_1^2 M_2^2}$$

$$+ 3\frac{m_c^3 m_s^2}{M_1^2 M_2^2} + 3\frac{m_c^3 m_s^2}{M_2^4} - \frac{3}{2}\frac{m_0^2 m_c^3}{M_2^4} + 3m_s\Bigg)$$

$$\times e^{-m_c^2/M_2^2}, \tag{B.1}$$

$$C^{K^*}_{D_s^* D^* K^*} = \widehat{I}_0(3,2,2)\, m_c^6 - \widehat{I}_0(3,2,2)\, m_c^5 m_s$$

$$+ \widehat{I}_0(3,2,2)\, m_c^3 m_s^3 + \widehat{I}_1^{[1,0]}(3,2,2)\, m_c^4$$

$$+ 3\widehat{I}_0(4,1,1)\, m_c^4 - \widehat{I}_2^{[1,0]}(3,2,2)\, m_c^4 + 2\widehat{I}_6(3,2,2)$$

$$\cdot m_c^4 + 2\widehat{I}_2(2,1,3)\, m_c^3 m_s + 2\widehat{I}_0(2,1,3)\, m_c^3 m_s$$

$$- 2\widehat{I}_1(2,2,2)\, m_c^3 m_s - 2\widehat{I}_1(2,1,3)\, m_c^3 m_s$$

$$- \widehat{I}_1(3,2,1)\, m_c^3 m_s + \widehat{I}_2(3,2,1)\, m_c^3 m_s$$

$$- 3\widehat{I}_0(4,1,1)\, m_c^3 m_s + 2\widehat{I}_2(2,2,2)\, m_c^3 m_s$$

$$+ \widehat{I}_0(2,2,2)\, m_c^2 m_s^2 - 2\widehat{I}_6(3,2,2)\, m_c^2 m_s^2$$

$$+ \widehat{I}_0^{[0,1]}(3,2,2)\, m_c^2 m_s^2 + \widehat{I}_2^{[1,0]}(3,2,2)\, m_c^2 m_s^2$$

$$- \widehat{I}_1^{[1,0]}(3,2,2)\, m_c^2 m_s^2 + 2\widehat{I}_2(3,1,2)\, m_c m_s^3$$

$$- \widehat{I}_2^{[1,0]}(3,2,2)\, m_c m_s^3 - 2\widehat{I}_1(3,1,2)\, m_c m_s^3$$

$$+ 6\widehat{I}_0(1,1,4)\, m_c m_s^3 + \widehat{I}_1^{[1,0]}(3,2,2)\, m_c m_s^3$$

$$+ \widehat{I}_0(3,1,2)\, m_s^4 + 2\widehat{I}_6^{[1,0]}(3,2,2)\, m_c^2 + \widehat{I}_0(2,1,2)$$

$$\cdot m_c^2 + 8\widehat{I}_8(3,2,1)\, m_c^2 + 4\widehat{I}_8^{[0,1]}(3,2,2)\, m_c^2$$

$$+ 2\widehat{I}_6^{[0,1]}(3,2,2)\, m_c^2 - 4\widehat{I}_6(3,2,1)\, m_c^2$$

$$- 4\widehat{I}_7^{[0,1]}(3,2,2)\, m_c^2 + 3\widehat{I}_1^{[1,0]}(4,1,1)\, m_c^2$$

$$+ 6\widehat{I}_6(4,1,1)\, m_c^2 + 6\widehat{I}_6(3,1,2)\, m_c^2 - 8\widehat{I}_7(3,2,1)$$

$$\cdot m_c^2 - 12\widehat{I}_7(4,1,1)\, m_c^2 + 2\widehat{I}_0(2,2,1)\, m_c^2$$

$$+ 12\widehat{I}_8(4,1,1)\, m_c^2 - 3\widehat{I}_2^{[1,0]}(4,1,1)\, m_c^2$$

$$+ 8\widehat{I}_6(2,1,3)\, m_c m_s + 4\widehat{I}_6(2,2,2)\, m_c m_s$$

$$- \widehat{I}_0^{[1,1]}(3,2,2)\, m_c m_s - 3\widehat{I}_1(1,3,1)\, m_c m_s$$

$$- 2\widehat{I}_1^{[0,1]}(3,1,2)\, m_c m_s + 3\widehat{I}_1^{[1,0]}(3,2,1)\, m_c m_s$$

$$- 2\widehat{I}_6(3,2,1)\, m_c m_s + 3\widehat{I}_2(1,3,1)\, m_c m_s$$

$$+ 2\widehat{I}_2^{[0,1]}(3,1,2)\, m_c m_s - \widehat{I}_0(2,1,2)\, m_c m_s$$

$$- 2\widehat{I}_6(3,1,2)\, m_c m_s - 3\widehat{I}_2^{[1,0]}(3,2,1)\, m_c m_s$$

$$- 2\widehat{I}_6^{[1,0]}(3,2,2)\, m_s^2 - 12\widehat{I}_6(1,1,4)\, m_s^2$$

$$- \widehat{I}_0(2,1,2)\, m_s^2 - 3\widehat{I}_0(1,1,3)\, m_s^2 - 2\widehat{I}_0^{[0,1]}(3,1,2)$$

$$\cdot m_s^2 + 3\widehat{I}_0^{[1,0]}(1,1,4)\, m_s^2 + \widehat{I}_0^{[1,1]}(3,2,2)\, m_s^2$$

$$+ 6\widehat{I}_2^{[1,0]}(1,1,4)\, m_s^2 + \widehat{I}_0(3,1,1)\, m_s^2$$

$$+ 24\widehat{I}_7(1,1,4)\, m_s^2 - 24\widehat{I}_8(1,1,4)\, m_s^2 + 4\widehat{I}_6(2,2,2)$$

$$\cdot m_s^2 - 6\widehat{I}_1^{[1,0]}(1,1,4)\, m_s^2 - 2\widehat{I}_0^{[1,0]}(2,2,2)\, m_s^2$$

$$+ 3\widehat{I}_0^{[0,1]}(1,1,4)\, m_s^2 + 2\widehat{I}_6(3,1,2)\, m_s^2$$

$$+ 3\widehat{I}_1^{[0,1]}(2,1,2) - 2\widehat{I}_2(2,1,1) + 3\widehat{I}_6(1,3,1)$$

$$- 3\widehat{I}_2^{[0,1]}(2,1,2) - 4S_1, 1\,(3,2,2) + 4\widehat{I}_7^{[1,0]}(3,2,1)$$

$$- 8\widehat{I}_8(2,2,1) - 2\widehat{I}_6^{[0,1]}(2,2,2) - 2\widehat{I}_6^{[0,1]}(3,1,2)$$

$$+ 8\widehat{I}_7(2,2,1) - 4\widehat{I}_8^{[1,0]}(3,2,1) - 4\widehat{I}_6^{[1,0]}(3,2,1)$$

$$+ 2\widehat{I}_6^{[1,1]}(3,2,2) - 12\widehat{I}_8(3,1,1) + 3\widehat{I}_1^{[1,0]}(3,1,2)$$

$$+ 2\widehat{I}_1(2,1,1) - 2\widehat{I}_2^{[1,0]}(1,2,2) + 2\widehat{I}_1^{[1,0]}(1,2,2)$$

$$+ 4\widehat{I}_7^{[0,1]}(2,2,2) - 4\widehat{I}_8^{[0,1]}(2,2,2) + 4\widehat{I}_6(2,2,1)$$

$$- \widehat{I}_0^{[0,1]}(3,1,1) - 3\widehat{I}_2^{[1,1]}(3,1,2) + \widehat{I}_2^{[2,0]}(2,2,2)$$

$$- \widehat{I}_1^{[2,0]}(2,2,2) + 12\widehat{I}_7(3,1,1) - 8\widehat{I}_6(2,1,2)$$

$$+ 4\widehat{I}_8^{[1,1]}(3,2,2) + \widehat{I}_0^{[2,0]}(3,2,1) - 2\widehat{I}_0^{[1,0]}(1,2,2)$$

$$- \widehat{I}_0^{[0,1]}(2,1,2) - 3\widehat{I}_0(2,1,1) - 4\widehat{I}_6(3,1,1)$$

$$- 2\widehat{I}_6^{[1,0]}(2,2,2) + 4\widehat{I}_7^{[1,0]}(2,2,2) - 4\widehat{I}_8^{[1,0]}(2,2,2)$$

$$+ 2\widehat{I}_0(1,1,2) - \widehat{I}_0^{[1,0]}(3,1,1) - 6\widehat{I}_7(1,3,1)$$

$$+ 6\widehat{I}_8(1,3,1),$$

$$C^{K^*}_{D_{s1} D_1 K^*} = \widehat{I}_0(3,2,2)\, m_c^6 + \widehat{I}_2(3,2,2)\, m_c^3 m_s^3$$

$$- \widehat{I}_1(3,2,2)\, m_c^3 m_s^3 - \widehat{I}_0(3,2,2)\, m_c^3 m_s^3$$

$$- 4\widehat{I}_7(3,2,2)\, m_c^4 + 3\widehat{I}_0(2,2,2)\, m_c^4 + 2\widehat{I}_6(3,2,2)$$

$$\cdot m_c^4 + 4\widehat{I}_8(3,2,2)\, m_c^4 + 3\widehat{I}_0(4,1,1)\, m_c^4$$

$$- 2\widehat{I}_2(2,2,2)\, m_c^3 m_s + 2\widehat{I}_1(2,2,2)\, m_c^3 m_s$$

$$- 2\widehat{I}_2(2,1,3)\, m_c^3 m_s + 2\widehat{I}_1(2,1,3)\, m_c^3 m_s$$

$$+ \widehat{I}_1(3,2,1)\, m_c^3 m_s + \widehat{I}_0^{[0,1]}(3,2,2)\, m_c^3 m_s$$

$$- \widehat{I}_2(3,2,1)\, m_c^3 m_s - \widehat{I}_0(3,1,2)\, m_c^3 m_s$$

$$- 2\widehat{I}_6(3,2,2)\, m_c^2 m_s^2 - \widehat{I}_1^{[1,0]}(3,2,2)\, m_c^2 m_s^2$$

$$- 6\widehat{I}_0(1,1,4)\, m_c^2 m_s^2 + \widehat{I}_2^{[1,0]}(3,2,2)\, m_c^2 m_s^2$$

$$+ 6\widehat{I}_2(1,1,4)\, m_c m_s^3 - 6\widehat{I}_1(1,1,4)\, m_c m_s^3$$

$$
\begin{aligned}
&+ \widehat{I}_0(3,1,2)\, m_s^4 + \widehat{I}_1(2,1,2)\, m_c^2 - \widehat{I}_1(3,1,1)\, m_c^2 \\
&+ 2\widehat{I}_0(2,2,1)\, m_c^2 + \widehat{I}_1^{[2,0]}(3,2,2)\, m_c^2 + 8\widehat{I}_8(3,2,1) \\
&\cdot m_c^2 + 4\widehat{I}_8^{[0,1]}(3,2,2)\, m_c^2 + 2\widehat{I}_6^{[0,1]}(3,2,2)\, m_c^2 \\
&- 12\widehat{I}_7(4,1,1)\, m_c^2 - 4\widehat{I}_7^{[0,1]}(3,2,2)\, m_c^2 \\
&+ 4\widehat{I}_8^{[1,0]}(3,2,2)\, m_c^2 + \widehat{I}_2(3,1,1)\, m_c^2 \\
&+ 2\widehat{I}_8^{[1,0]}(3,2,2)\, m_c^2 - \widehat{I}_2^{[2,0]}(3,2,2)\, m_c^2 \\
&- 4\widehat{I}_6(3,2,1)\, m_c^2 - 8\widehat{I}_7(3,2,1)\, m_c^2 - \widehat{I}_0^{[0,1]}(2,2,2) \\
&\cdot m_c^2 + \widehat{I}_0(2,1,2)\, m_c^2 + 6\widehat{I}_6(4,1,1)\, m_c^2 \\
&- \widehat{I}_2(2,1,2)\, m_c^2 - 4\widehat{I}_7^{[1,0]}(3,2,2)\, m_c^2 + 6\widehat{I}_6(3,1,2) \\
&\cdot m_c^2 + 12\widehat{I}_8(4,1,1)\, m_c^2 - 4\widehat{I}_1(1,1,3)\, m_c m_s \\
&+ 4\widehat{I}_2(1,1,3)\, m_c m_s - 4\widehat{I}_6(2,2,2)\, m_c m_s \\
&- 3\widehat{I}_2(1,3,1)\, m_c m_s + 2\widehat{I}_0(1,2,2)\, m_c m_s \\
&+ 2\widehat{I}_6(3,1,2)\, m_c m_s + \widehat{I}_0(2,1,2)\, m_c m_s \\
&- 2\widehat{I}_2^{[0,1]}(3,1,2)\, m_c m_s - 8\widehat{I}_6(2,1,3)\, m_c m_s \\
&+ 2\widehat{I}_0^{[1,0]}(2,1,3)\, m_c m_s + 2\widehat{I}_6(3,2,1)\, m_c m_s \\
&+ \widehat{I}_2(2,1,2)\, m_c m_s + 2\widehat{I}_1^{[0,1]}(3,1,2)\, m_c m_s \\
&- \widehat{I}_1(2,1,2)\, m_c m_s + 3\widehat{I}_1(1,3,1)\, m_c m_s \\
&+ 3\widehat{I}_0(2,2,1)\, m_c m_s - 3\widehat{I}_0(1,1,3)\, m_s^2 + 4\widehat{I}_6(2,2,2) \\
&\cdot m_s^2 - 6\widehat{I}_1^{[1,0]}(1,1,4)\, m_s^2 + 3\widehat{I}_0^{[0,1]}(1,1,4)\, m_s^2 \\
&- 12\widehat{I}_6(1,1,4)\, m_s^2 - 2\widehat{I}_0^{[0,1]}(3,1,2)\, m_s^2 \\
&- \widehat{I}_0(2,1,2)\, m_s^2 + 6\widehat{I}_2^{[1,0]}(1,1,4)\, m_s^2 + 2\widehat{I}_6(3,1,2) \\
&\cdot m_s^2 - 2\widehat{I}_6^{[1,0]}(3,2,2)\, m_s^2 - 2\widehat{I}_2^{[1,0]}(1,2,2) \\
&+ \widehat{I}_2^{[1,0]}(3,1,1) - 4\widehat{I}_6^{[1,0]}(3,2,1) - \widehat{I}_0^{[1,0]}(3,1,1) \\
&- 8\widehat{I}_8(2,2,1) - \widehat{I}_0^{[0,1]}(2,1,2) - 12\widehat{I}_8(3,1,1) \\
&- 2\widehat{I}_6^{[1,0]}(2,2,2) + 12\widehat{I}_7(3,1,1) - 8\widehat{I}_8(2,1,2) \\
&- 8\widehat{I}_6(2,1,2) - \widehat{I}_1^{[1,0]}(3,1,1) + 8\widehat{I}_7(2,1,2) \\
&+ 4\widehat{I}_7^{[1,0]}(3,2,1) - 4\widehat{I}_8^{[1,0]}(3,2,1) + 4\widehat{I}_7^{[0,1]}(3,1,2) \\
&- 4\widehat{I}_8^{[0,1]}(3,1,2) + 2\widehat{I}_1^{[0,1]}(1,2,2) + 2\widehat{I}_1(2,1,1) \\
&- 4\widehat{I}_7^{[1,1]}(3,2,2) + 4\widehat{I}_8^{[1,1]}(3,2,2) + 3\widehat{I}_6(1,3,1) \\
&- 2\widehat{I}_6^{[0,1]}(3,1,2) - 2\widehat{I}_6^{[0,1]}(2,2,2) - 4\widehat{I}_6(3,1,1) \\
&+ \widehat{I}_0^{[0,2]}(3,1,2) - \widehat{I}_0^{[0,1]}(3,1,1) - 3\widehat{I}_0(2,1,1) \\
&+ 4\widehat{I}_6(2,2,1) + 2\widehat{I}_6^{[1,1]}(3,2,2) - 2\widehat{I}_2(2,1,1) \\
&+ 8\widehat{I}_7(2,2,1) - 2\widehat{I}_0^{[0,1]}(2,2,1),
\end{aligned}
\tag{B.2}
$$

where

$$
\begin{aligned}
\widehat{I}_\mu^{[\alpha,\beta]}(a,b,c) &= \left[M_1^2\right]^\alpha \\
&\cdot \left[M_2^2\right]^\beta \frac{d^\alpha}{d\left(M_1^2\right)^\alpha} \frac{d^\beta}{d\left(M_2^2\right)^\beta} \left[M_1^2\right]^\alpha \left[M_2^2\right]^\beta \widehat{I}_\mu(a,b,c),
\end{aligned}
$$

$$
\begin{aligned}
\widehat{I}_k(a,b,c) &= i\frac{(-1)^{a+b+c+1}}{16\pi^2\Gamma(a)\Gamma(b)\Gamma(c)}\left(M_1^2\right)^{1-a-b+k} \\
&\cdot \left(M_2^2\right)^{4-a-c-k} U_0(a+b+c-5, 1-c-b), \\
\widehat{I}_m(a,b,c) &= i\frac{(-1)^{a+b+c+1}}{16\pi^2\Gamma(a)\Gamma(b)\Gamma(c)}\left(M_1^2\right)^{-a-b-1+m} \\
&\cdot \left(M_2^2\right)^{7-a-c-m} U_0(a+b+c-5, 1-c-b), \\
\widehat{I}_6(a,b,c) &= i\frac{(-1)^{a+b+c+1}}{32\pi^2\Gamma(a)\Gamma(b)\Gamma(c)}\left(M_1^2\right)^{3-a-b}\left(M_2^2\right)^{3-a-c} \\
&\cdot U_0(a+b+c-6, 2-c-b), \\
\widehat{I}_n(a,b,c) &= i\frac{(-1)^{a+b+c}}{32\pi^2\Gamma(a)\Gamma(b)\Gamma(c)}\left(M_1^2\right)^{-4-a-b+n} \\
&\cdot \left(M_2^2\right)^{11-a-c-n} U_0(a+b+c-7, 2-c-b),
\end{aligned}
\tag{B.3}
$$

where $k = 1, 2$, $m = 3, 4, 5$ and $n = 7, 8$. We can define the function $U_0(a,b)$ as

$$
\begin{aligned}
U_0(a,b) &= \int_0^\infty dy\left(y + M_1^2 + M_2^2\right)^a \\
&\cdot y^b \exp\left[-\frac{B_{-1}}{y} - B_0 - B_1 y\right],
\end{aligned}
\tag{B.4}
$$

where

$$
\begin{aligned}
B_{-1} &= \frac{1}{M_2^2 M_1^2}\left(m_s^2\left(M_1^2 + M_2^2\right)^2 - M_2^2 M_1^2 Q^2\right), \\
B_0 &= \frac{1}{M_1^2 M_2^2}\left(m_s^2 + m_c^2\right)\left(M_1^2 + M_2^2\right), \\
B_1 &= \frac{m_c^2}{M_1^2 M_2^2}.
\end{aligned}
\tag{B.5}
$$

Conflicts of Interest

The authors of the manuscript declare that there are no conflicts of interest regarding publication of this article.

References

[1] M. Du, W. Chen, X. Chen, and S. Zhu, "Exotic $QQ\bar{q}\bar{q}$, $QQ\bar{q}\bar{s}$, and $QQ\bar{s}\bar{s}$ states," *Physical Review D*, vol. 87, Article ID 014003, 2013.

[2] F. Navarra, M. Nielsen, M. Bracco, M. Chiapparini, and C. Schat, "$D^* D\pi$ and $B^* B\pi$ form factors from QCD sum rules," *Physics Letters B*, vol. 489, no. 3, pp. 319–328, 2000.

[3] M. E. Bracco, M. Chiapparini, F. S. Navarra, and M. Nielsen, "$\rho D^* D^*$ vertex from QCD sum rules," *Physics Letters B*, vol. 659, no. 3, pp. 559–564, 2008.

[4] F. S. Navarra, M. Nielsen, and M. E. Bracco, *Physical Review D: Particles, Fields, Gravitation and Cosmology*, vol. 65, no. 3, 2002.

[5] M. Bracco, M. Chiapparini, A. Lozéa, F. Navarra, and M. Nielsen, "D and ρ mesons: who resolves whom?" *Physics Letters B*, vol. 521, no. 1-2, pp. 1–6, 2001.

[6] B. O. Rodrigues, M. E. Bracco, M. Nielsen, and F. S. Navarra, "$D^* D\rho$ vertex from QCD sum rules," *Nuclear Physics A*, vol. 852, p. 127, 2011.

[7] R. Matheus, F. Navarra, M. Nielsen, and R. Rodrigues da Silva, "The J/ψDD vertex in QCD sum rules," *Physics Letters B*, vol. 541, no. 3-4, pp. 265–272, 2002.

[8] R. R. da Silva, R. D. Matheus, F. S. Navarra, and M. Nielsen, "The $J/\psi DD^*$ vertex," https://arxiv.org/abs/hep-ph/0310074.

[9] Z. G. Wang and S. L. Wan, "Analysis of the vertices $D^* D_s K$, $D_s^* DK$, $D_0 D_s K$, and $D_{s0} DK$ with the light-cone QCD sum rules," *Physical Review D*, vol. 74, 2006.

[10] Z. G. Wang, "Analysis of the vertices $D^* D^* P, D^* DV$ DVand DDV with light-cone QCD sum rules," *Nuclear Physics A*, vol. 796, p. 61, 2007.

[11] F. Carvalho, F. O. Durães, F. S. Navarra, and M. Nielsen, " Hadronic form factors and the ," *Physical Review C: Nuclear Physics*, vol. 72, no. 2, 2005.

[12] M. E. Bracco, A. J. Cerqueira, M. Chiapparini, A. Lozea, and M. Nielsen, "$D^* D_s K$ and $D_s^* DK$ vertices in a QCD sum rule approach," *Physics Letters B*, vol. 641, no. 3, pp. 286–293, 2006.

[13] L. Holanda, R. Marques de Carvalho, and A. Mihara, "The ωDD vertex in a sum rule approach," *Physics Letters B*, vol. 644, no. 4, pp. 232–236, 2007.

[14] Z. G. Wang and S. L. Wan, "Analysis of the vertices $D^* D_s K$, $D_s^* DK$, $D_0 D_s K$, and $D_{s0} DK$ with the light-cone QCD sum rules," *Physical Review D*, vol. 74, Article ID 014017, 2006.

[15] Z. G. Wang, "Radiative decay of the dynamically generated open and hidden charm scalar mesonresonances $D_{s0}^*(2317)$ and X(3700)," https://arxiv.org/pdf/0709.2339.pdf.

[16] N. Ghahramany, R. Khosravi, and M. Janbazi, "The strong coupling constants gDslD* K and in QCD sum rules," in *Proceedings of the International Journal of Modern Physics A*, vol. 27, World Scientific Publishing Company, 2012.

[17] R. Khosravi and M. Janbazi, "Vertices of the vector mesons with the strange charmed mesons in QCD," *Physical Review D: Particles, Fields, Gravitation and Cosmology*, vol. 87, no. 1, 2013.

[18] R. Khosravi and M. Janbazi, "Analysis of $VD_{s0}^* D_{s1}$ and $VD_s D_s^*$ vertices," *Physical Review D*, vol. 89, Article ID 016001, 2014.

[19] M. Janbazi, N. Ghahramany, and E. Pourjafarabadi, "Coupling constants of bottom (charmed) mesons with the pion from three-point QCD sum rules," *The European Physical Journal C*, vol. 74, no. 2, 2014.

[20] M. E. Bracco, M. Chiapparini, F. S. Navarra, and M. Nielsen, "Charm couplings and form factors in QCD sum rules," *Progress in Particle and Nuclear Physics*, vol. 67, pp. 1019–1052, 2011.

[21] Z. W. Lin, C. M. Ko, and B. Zhang, "Hadronic scattering of charmed mesons," *Physical Review C: Nuclear Physics*, vol. 61, no. 2, Article ID 024904, 7 pages, 2000.

[22] Y. Oh, T. Song, and S. H. Lee, "J/ψ absorption by π and ρ mesons in a meson exchange model with anomalous parity interactions," *Physical Review C*, vol. 63, Article ID 034901, 2001.

[23] P. Colangelo and A. Khodjamirian, "QCD sum rules, a modern perspective," in *The Frontier of Particle Physics*, pp. 1495–1576, 2001.

[24] J. Beringer, J.-F. Arguin, and R. M. Barnett, "Review of particle physics," *Physical Review D: Particles, Fields, Gravitation and Cosmology*, vol. 86, Article ID 010001, 2012.

[25] Gelhausen., Patrick. et al., "Decay constants of heavy-light vector mesons from QCD sum rules," *Physical Review D*, vol. 88, Article ID 014015, 2013.

[26] C. E. Thomas, "Nonleptonic weak decays of B to D_s and D mesons," *Physical Review D*, vol. 73, Article ID 033005, 2006.

[27] A. Bazavov, T. Bhattacharya, and M. Cheng, "Chiral and deconfinement aspects of the QCD transition," *Physical Review D: Particles, Fields, Gravitation and Cosmology*, vol. 85, Article ID 054503, 2012.

[28] M. E. Bracco et al., *Progress in Particle and Nuclear Physics*, Charm couplings and form factors in QCD sum rules, Ed., vol. 67 of *1052*, 1019, 4 edition, 2012.

[29] B. Rodrigues, M. E. Osório, and A. Cerqueira Jr., "The $gD_s D_{s\phi}$ strong coupling constant from QCD Sum Rules," *Nuclear Physics A*, vol. 957, pp. 109–122, 2017.

6

Degeneracy Resolution Capabilities of NOνA and DUNE in the Presence of Light Sterile Neutrino

Akshay Chatla[1], **Sahithi Rudrabhatla**[2] **and Bindu A. Bambah**[1]

[1]*School of Physics, University of Hyderabad, Hyderabad 500046, India*
[2]*Department of Physics, University of Illinois at Chicago, Chicago, IL 60607, USA*

Correspondence should be addressed to Akshay Chatla; chatlaakshay@gmail.com

Academic Editor: Hiroyasu Ejiri

We investigate the implications of a sterile neutrino on the physics potential of the proposed experiment DUNE and future runs of NOνA using latest NOνA results. Using combined analysis of the disappearance and appearance data, NOνA reported preferred solutions at normal hierarchy (NH) with two degenerate best-fit points: one in the lower octant (LO) and $\delta_{13} = 1.48\pi$ and the other in higher octant (HO) and $\delta_{13} = 0.74\pi$. Another solution of inverted hierarchy (IH), which is 0.46σ away from best fit, was also reported. We discuss chances of resolving these degeneracies in the presence of sterile neutrino.

1. Introduction

Sterile neutrinos are hypothetical particles that do not interact via any of the fundamental interactions other than gravity. The term sterile is used to distinguish them from active neutrinos, which are charged under weak interaction. The theoretical motivation for sterile neutrino explains the active neutrino mass after spontaneous symmetry breaking, by adding a gauge singlet term (sterile neutrino) to the Lagrangian under $SU(3)_c \otimes SU(2)_L \otimes U(1)_r$ where the Dirac term appears through the Higgs mechanism, and Majorana mass term is a gauge singlet and hence appears as a bare mass term [1]. The diagonalization of the mass matrix gives masses to all neutrinos due to the See-Saw mechanism.

Some experimental anomalies also point towards the existence of sterile neutrinos. Liquid Scintillator Neutrino Detector (LSND) detected $\overline{\nu}_\mu \longrightarrow \overline{\nu}_e$ transitions indicating $\Delta m^2 \approx 1eV^2$ which is inconsistent with $\Delta m^2_{32}, \Delta m^2_{21}$ (LSND anomaly) [2]. Measurement of the width of Z boson by LEP gave number of active neutrinos to be 2.984 ± 0.008 [3]. Thus the new neutrino introduced to explain the anomaly has to be a sterile neutrino. MiniBooNE, designed to verify the LSND anomaly, observed an unexplained excess of events in low-energy region of $\overline{\nu}_e, \nu_e$ spectra, consistent with LSND [4]. SAGE and GALLEX observed lower event rate than expected, explained by the oscillations of ν_e due to $\Delta m^2 \geq 1eV^2$(Gallium anomaly) [5–7]. Recent precise predictions of reactor antineutrino flux have increased the expected flux by 3% over old predictions. With the new flux evaluation, the ratio of observed and predicted flux deviates at 98.6% C.L (Confidence level) from unity; this is called "reactor antineutrino anomaly" [8]. This anomaly can also be explained using sterile neutrino model.

Short-baseline (SBL) experiments are running to search for sterile neutrinos. SBL experiments are the best place to look for sterile neutrino, as they are sensitive to new expected mass-squared splitting $\Delta \mathrm{m}^2 \simeq 1\mathrm{eV}^2$. However, SBL experiments cannot study all the properties of sterile neutrinos, mainly new CP phases introduced by sterile neutrino models. These new CP phases need long distances to become measurable [9, 10] and thus can be measured using long baseline (LBL) experiments. With the discovery of relatively large value for θ_{13} by Daya Bay [11], the sensitivity of LBL experiments towards neutrino mass hierarchy and CP phases increased significantly. In this context, some phenomenological studies regarding the sensitivity of LBL experiments can be found in recent works [12–16]. Using recent global fits of oscillation parameters in the 3+1 scenario

[17], current LBL experiments can extract two out of three CP phases (one of them being standard δ_{13}) [10]. The phenomenological studies of LBL experiments in presence of sterile neutrino is studied by several groups [18–23]. Now, the sensitivity of LBL experiments towards their original goals decreases due to sterile neutrinos. It is seen in case of the CPV measurement; new CP phases will decrease the sensitivity towards standard CP phase (δ_{13}). This will reduce degeneracy resolution capacities of LBL experiments. In this paper, we study hierarchy-θ_{23}-δ_{13} degeneracies using contours in θ_{23}-δ_{13} plane and how they are affected by the introduction of sterile neutrinos. We attempt to find the extent to which these degeneracies can be resolved in future runs of NOνA and DUNE.

The outline of the paper is as follows. In Section 2, we present the experimental specifications of NOνA and DUNE used in our simulation. We introduce the effect of sterile neutrino on parameter degeneracies resolution in Section 3. Section 4 contains the discussion about the degeneracy resolving capacities of future runs of NOνA and DUNE assuming latest NOνA results—NH- (normal hierarchy-) LO (lower octant); NH-HO (higher octant); and IH- (inverted hierarchy-) HO—as true solutions for both 3 and 3+1 models. Finally, Section 5 contains concluding comments on our results.

2. Experiment Specifications

We used GLoBES (General Long Baseline Experiment Simulator) [24, 25] to simulate the data for different LBL experiments including NOνA and DUNE. The neutrino oscillation probabilities for the 3+1 model are calculated using the new physics engine available from [26].

NOνA [27, 28] is an LBL experiment which started its full operation from October 2014. NOνA has two detectors: the near detector is located at Fermilab (300 ton, 1 km from NuMI beam target) while the far detector (14 Kt) is located at Northern Minnesota 14.6 mrad off the NuMI beam axis at 810 km from NuMI beam target, justifying "off-axis" in the name. This off-axis orientation gives us a narrow beam of flux, peak at 2 GeV [29]. For simulations, we used NOνA setup from [30]. We used the full projected exposure of 3.6×10^{21} p.o.t (protons on target) expected after six years of runtime at 700 kW beam power. Assuming the same runtime for neutrino and antineutrino modes, we get 1.8×10^{21} p.o.t for each mode. Following [31] we considered 5% normalization error for the signal and 10% error for the background for appearance and disappearance channels.

DUNE (Deep Underground Neutrino Experiment) [32, 33] is the next generation LBL experiment. Long Base Neutrino Facility (LBNF) of Fermilab is the source for DUNE. Near detector of DUNE will be at Fermilab. Liquid Argon detector of 40 kt to be constructed at Sanford Underground Research Facility, situated 1300 km from the beam target, will act as the far detector. DUNE uses the same source as of NOνA; we will observe beam flux peak at 2.5GeV. We used DUNE setup give in [34] for our simulations. Since DUNE is still in its early stages, we used simplified systematic treatment, i.e., 5% normalization error on signal and 10% error on the background for both appearance and disappearance spectra. We give experimental details described above in tabular form in Tables 1 and 2.

Oscillation parameters are estimated from the data by comparing observed and predicted ν_e and ν_μ interaction rates and energy spectra. GLoBES calculates event rates of neutrinos for energy bins taking systematic errors, detector resolutions, MSW effect due to earth's crust, etc. into account. The event rates generated for true and test values are used to plot χ^2 contours. GLoBES uses its inbuilt algorithm to calculate χ^2 values numerically considering parameter correlations as well as systematic errors. In our calculations we used χ^2 as

$$\chi^2 = \sum_{i=1}^{\#\text{ofbins}} \sum_{E_n=E_1,E_2\ldots} \frac{\left(O_{E_n,i} - \left(1 + a_F + a_{E_n}\right) T_{E,i}\right)^2}{O_{E_n,i}} + \frac{a_F^2}{\sigma_F^2} + \frac{a_{E_n}^2}{\sigma_{E_n}^2} \qquad (1)$$

where $O_{E_1,i}, O_{E_2,i}\ldots$ are the event rates for the ith bin in the detectors of different experiments, calculated for true values of oscillation parameters; $T_{E_n,i}$ are the expected event rates for the ith bin in the detectors of different experiments for the test parameter values; a_F, a_{E_n} are the uncertainties associated with the flux and detector mass; and σ_F, σ_{E_n} are the respective associated standard deviations. The calculated χ^2 function gives the confidence level in which tested oscillation parameter values can be ruled out with referenced data. It provides an excellent preliminary evaluation model to estimate the experiment performance.

3. Theory

In a 3+1 sterile neutrino model, the flavour and mass eigenstates are connected through a 4×4 mixing matrix. A convenient parametrization of the mixing matrix is [36]

$$U = R_{34}\widetilde{R_{24}}\widetilde{R_{14}}R_{23}\widetilde{R_{13}}R_{12}. \qquad (2)$$

Here R_{ij} and $\widetilde{R_{ij}}$ represent real and complex 4×4 rotation in the plane containing the 2×2 subblock in (i, j) subblock

$$R_{ij}^{2\times2} = \begin{pmatrix} c_{ij} & s_{ij} \\ -s_{ij} & c_{ij} \end{pmatrix} \quad \widetilde{R_{ij}}^{2\times2} = \begin{pmatrix} c_{ij} & \widetilde{s_{ij}} \\ -\widetilde{s_{ij}}^* & c_{ij} \end{pmatrix} \qquad (3)$$

where, $c_{ij} = \cos\theta_{ij}$, $s_{ij} = \sin\theta_{ij}$, $\widetilde{s_{ij}} = s_{ij}e^{-i\delta_{ij}}$, and δ_{ij} are the CP phases.

There are three mass-squared difference terms in 3+1 model: Δm_{21}^2(solar)$\simeq 7.5 \times 10^{-5}\text{eV}^2$, Δm_{31}^2 (atmospheric)$\simeq 2.4 \times 10^{-3}\text{eV}^2$, and Δm_{41}^2(sterile)$\simeq 1\text{eV}^2$. The mass-squared difference term towards which the experiment is sensitive depends on L/E of the experiment. Since SBL experiments have a very small L/E, $\sin^2(\Delta m_{ij}^2 L/4E) \simeq 0$ for Δm_{21}^2 and Δm_{31}^2. Δm_{41}^2 term survives. Hence, SBL experiments

Table 1: Details of experiments.

Name of Exp	NOνA	DUNE
Location	Minnesota	South Dakota
POT(yr^{-1})	6.0x10^{20}	1.1x10^{21}
Baseline(Far/Near)	812 km/1km	1300 km/500 m
Target mass(Far/Near)	14 kt/290 t	40 kt/8 t
Exposure(years)	6	10
Detector type	Tracking Calorimeters	LArTPCs

Table 2: Systematic errors associated with NOνA and DUNE.

Name of Exp	Rule	Normalization error signal(%)	Normalization error background(%)
NOνA	ν_e appearance	5	10
	ν_μ disappearance	2	10
	$\bar{\nu}_e$ appearance	5	10
	$\bar{\nu}_\mu$ disappearance	2	10
DUNE	ν_e appearance	5	10
	ν_μ disappearance	5	10
	$\bar{\nu}_e$ appearance	5	10
	$\bar{\nu}_\mu$ disappearance	5	10

depend only on sterile mixing angles and are insensitive to the CP phases. The oscillation probability, $P_{\mu e}$ for LBL experiments in 3+1 model, after averaging Δm^2_{41} oscillations and neglecting MSW effects, [37] is expressed as sum of the four terms

$$P^{4\nu}_{\mu e} \simeq P_1 + P_2(\delta_{13}) + P_3(\delta_{14}-\delta_{24}) + P_4(\delta_{13}-(\delta_{14}-\delta_{24})). \tag{4}$$

These terms can be approximately expressed as follows:

$$P_1 = \frac{1}{2}\sin^2 2\theta^{4\nu}_{\mu e} + \left[a^2\sin^2 2\theta^{3\nu}_{\mu e} - \frac{1}{4}\sin^2 2\theta_{13}\sin^2 2\theta^{4\nu}_{\mu e}\right] \cdot \sin^2\Delta_{31} + \left[a^2b^2 - \frac{1}{4}\sin^2 2\theta_{12} \cdot \left(\cos^4\theta_{13}\sin^2 2\theta^{4\nu}_{\mu e} + a^2\sin^2 2\theta^{3\nu}_{\mu e}\right)\right]\sin^2\Delta_{21}, \tag{5}$$

$$P_2(\delta_{13}) = a^2 b \sin 2\theta^{3\nu}_{\mu e}\left(\cos 2\theta_{12}\cos\delta_{13}\sin^2\Delta_{21} - \frac{1}{2} \cdot \sin\delta_{13}\sin 2\Delta_{21}\right), \tag{6}$$

$$P_3(\delta_{14}-\delta_{24}) = ab\sin 2\theta^{4\nu}_{\mu e}\cos^2\theta_{13}\left[\cos 2\theta_{12} \cdot \cos(\delta_{14}-\delta_{24})\sin^2\Delta_{21} - \frac{1}{2}\sin(\delta_{14}-\delta_{24}) \cdot \sin 2\Delta_{21}\right], \tag{7}$$

$$P_4(\delta_{13}-(\delta_{14}-\delta_{24})) = a\sin 2\theta^{3\nu}_{\mu e}\sin 2\theta^{4\nu}_{\mu e}\left[\cos 2\theta_{13} \cdot \cos(\delta_{13}-(\delta_{14}-\delta_{24}))\sin^2\Delta_{31} + \frac{1}{2} \cdot \sin(\delta_{13}-(\delta_{14}-\delta_{24}))\sin 2\Delta_{31} - \frac{1}{4}\sin^2 2\theta_{12} \cdot \cos^2\theta_{13}\cos(\delta_{13}-(\delta_{14}-\delta_{24}))\sin^2\Delta_{21}\right], \tag{8}$$

with the parameters defined as

$$\begin{aligned} \Delta_{ij} &\equiv \frac{\Delta m^2_{ij}L}{4E}, \text{a function of baseline (L) and neutrino energy (E)} \\ a &= \cos\theta_{14}\cos\theta_{24}, \\ b &= \cos\theta_{13}\cos\theta_{23}\sin 2\theta_{12}, \\ \sin 2\theta^{3\nu}_{\mu e} &= \sin 2\theta_{13}\sin\theta_{23}, \\ \sin 2\theta^{4\nu}_{\mu e} &= \sin 2\theta_{14}\sin\theta_{24}. \end{aligned} \tag{9}$$

The CP phases introduced due to sterile neutrinos persist in the $P_{\mu e}$ even after averaging out Δm^2_{41} lead oscillations. Last two terms of (4) give the sterile CP phase dependence terms. $P_3(\delta_{14}-\delta_{24})$ depends on the sterile CP phases δ_{14} and δ_{24}, while P_4 depends on a combination of δ_{13} and $\delta_{14}-\delta_{24}$. Thus, we expect LBL experiments to be sensitive to sterile phases. We note that the probability $P_{\mu e}$ is independent θ_{34}. One can see that θ_{34} will effect $P_{\mu e}$ if we consider earth mass effects. Since matter effects are relatively small for

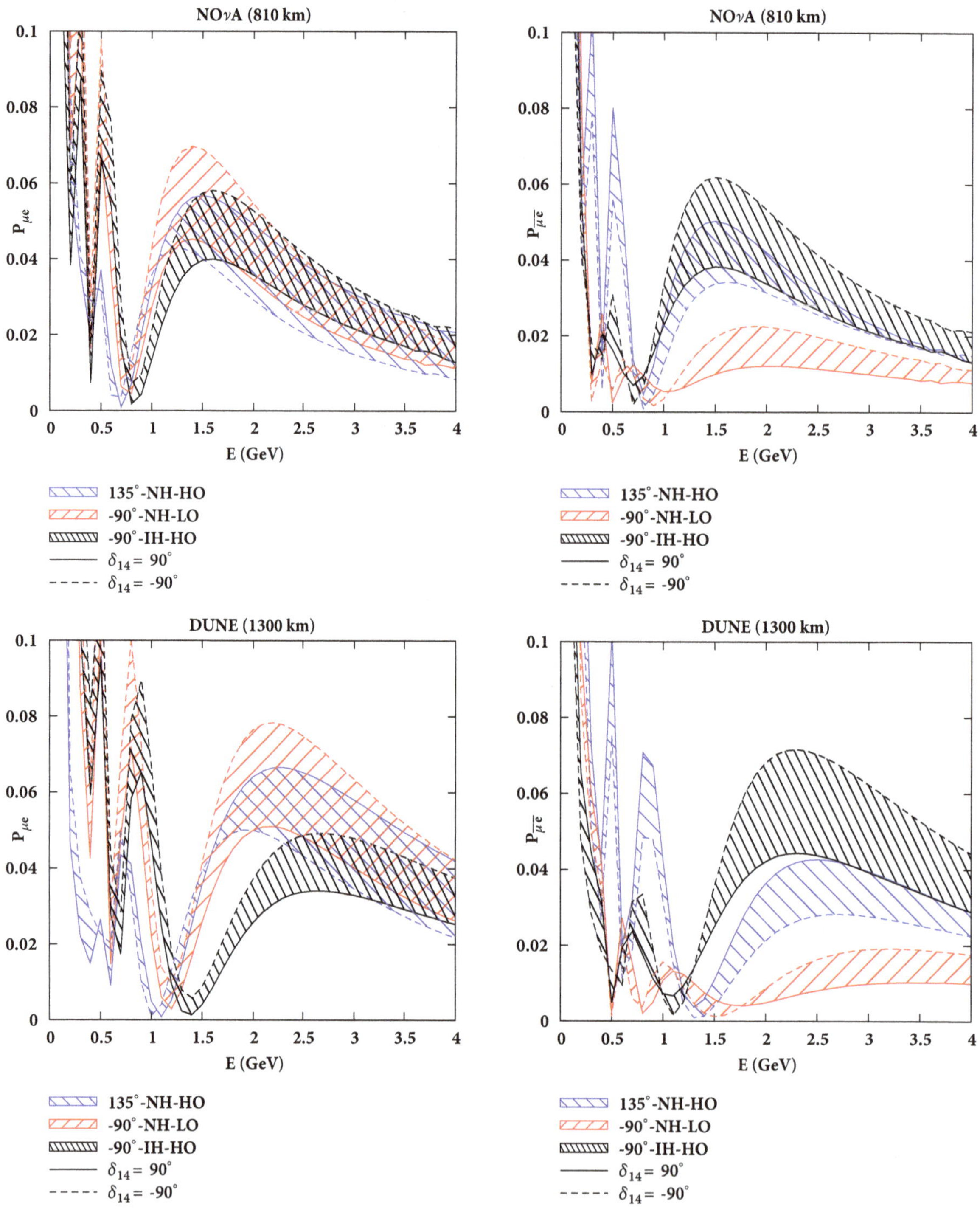

FIGURE 1: The oscillation probability $P_{\mu e}$ as a function of energy. The top (bottom) panel is NOνA (DUNE). The bands correspond to different values of δ_{14}, ranging from -180° to 180° when δ_{24} = 0°. Inside each band, the probability for δ_{14} = 90° (δ_{14} = -90°) case is shown as the solid (dashed) line. The left (right) panel corresponds to neutrinos (antineutrinos).

NOνA and DUNE, their sensitivity towards θ_{34} is negligible. The amplitudes of atmospheric-sterile interference term (8) and solar-atmospheric interference term (6) are of the same order. This new interference term reduces the sensitivity of experiments to the standard CP phase (δ_{13}).

In Figure 1, we plot the oscillation probability ($P_{\mu e}$) as a function of energy while varying δ_{14} (-180° to 180°) and keeping δ_{24} = 0 for the three best-fit values of latest NOνA results [35], i.e., NH-LO-1.48π[δ_{13}], NH-HO-0.74π, and IH-HO-1.48π, where HO implies $\sin^2\theta_{23}$ = 0.62 and LO implies

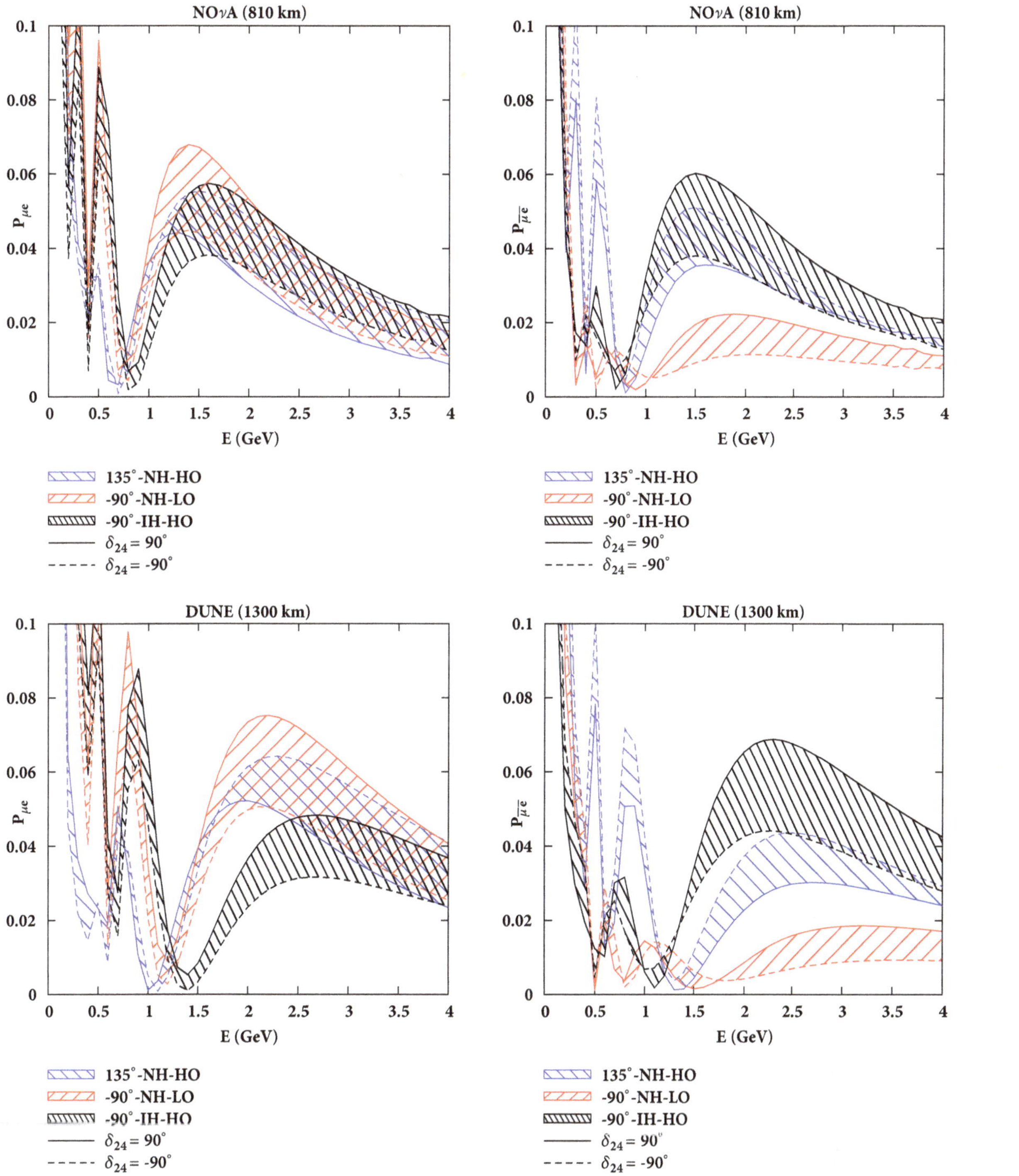

FIGURE 2: The oscillation probability $P_{\mu e}$ as a function of energy. The top (bottom) panel is NOνA (DUNE). The bands correspond to different values of δ_{24}, ranging from -180° to 180° when δ_{14} = 0°. Inside each band, the probability for δ_{24} = 90° (δ_{24} = -90°) case is shown as solid (dashed) line. The left (right) panel is for neutrinos (antineutrinos).

$\sin^2\theta_{23}$ = 0.40. For the flux peak of NOνA, E ≈ 2GeV, we observe a degeneracy between all best-fit values due to the presence of δ_{14} band for neutrino case, while only NH-HO and IH-HO bands overlap in antineutrino case. We see that δ_{14} phase decreases both octant and hierarchy resolution capacity for neutrino case and only mass hierarchy resolution capacity for antineutrino case. The second row plots $P_{\mu e}$ for DUNE at baseline 1300 km. We observe smaller overlap between bands compared to NOνA. Thus, the decrease of degeneracy resolution capacity for DUNE is less than NOνA. Similarly we plot $P_{\mu e}$ while varying δ_{24}(-180° to 180°) in Figure 2 and keeping δ_{14} = 0°. We see that δ_{24} has similar

TABLE 3: Oscillation parameters considered in numerical analysis. The $\sin^2\theta_{23}$ and δ_{13} are taken from latest NOνA results [35].

Parameter	True value	Marginalization Range
$\sin^2\theta_{12}$	0.304	Not Marginalized
$\sin^2 2\theta_{13}$	0.085	[0.075,0.095]
$\sin^2\theta_{23}$	0.623(HO),0.404(LO)	[0.32,0.67]
$\sin^2\theta_{14}$	0.025	Not Marginalized
$\sin^2\theta_{24}$	0.025	Not Marginalized
$\sin^2\theta_{34}$	0.025	Not Marginalized
δ_{13}	135(NH-HO),-90(NH-LO,IH)	[-180,180]
δ_{14}	[-180,180]	[-180,180]
δ_{24}	[-180,180]	[-180,180]
Δm^2_{21}	7.50×10^{-5} eV^2	Not Marginalized
Δm^2_{31}(NH)	2.40×10^{-3} eV^2	Not Marginalized
Δm^2_{31}(IH)	-2.33×10^{-3} eV^2	Not Marginalized
Δm^2_{41}	1 eV^2	Not Marginalized

effect to that of δ_{14}; the only change is reversal of δ_{24} band extrema; i.e., $\delta_{24} = -90^\circ$ gives the same result as $\delta_{14} = 90^\circ$ and vice versa. This can be explained using (4) in which we see δ_{14} and δ_{24} are always together with opposite signs. Overall from the probability plots, we observe that the addition of new CP phases decreases octant and mass hierarchy resolution capacities.

In the next section, we explore how parameter degeneracies are affected in the 3+1 model and the extent to which these degeneracies can be resolved in future runs of NOνA and DUNE.

4. Results for NOνA and DUNE

We explore allowed regions in $\sin^2\theta_{23}$-δ_{cp} plane from NOνA and DUNE simulation data with different runtimes, considering latest NOνA results as true values. Using combined analysis of the disappearance and appearance data, NOνA reported preferred solutions [35] at normal hierarchy (NH) with two degenerate best-fit points: one in the lower octant (LO) and $\delta_{cp} = 1.48\pi$ and the other in higher octant (HO) and $\delta_{cp} = 0.74\pi$. Another solution of inverted hierarchy (IH), 0.46σ away from best fit, is also reported. Table 3 shows true values of oscillation parameters and their marginalization ranges we used in our simulation. By studying the allowed regions, we understand the extent to which future runs of NOνA and DUNE will resolve these degeneracies, if the best-fit values are true values.

In the first row of Figure 3, we show allowed areas for NOνA[3+$\bar{0}$]. In first plot of first row, we show 90% C.L allowed regions for true values of $\delta_{13} = 135^\circ$ and $\theta_{23} = 52^\circ$ and normal hierarchy. We plot test values for both NH and IH, of 3 and 3+1 neutrino models. We observe that introducing sterile neutrino largely decreases the precision of θ_{23}. The WO-RH region, for 3ν case confined between 45° and -180° of δ_{13}, confines the whole δ_{13} region for 4ν case. The WH-RO region of 3ν case doubles, covering the entire region of δ_{13} for 4ν case. The 3+1 model also introduces a small WH-WO region, which was absent in 3ν model. In the second plot of first row (true value $\delta_{13} = -90^\circ$, $\theta_{23} = 40^\circ$ and normal hierarchy), for the 3ν case, we see RH-RO region excluding 45° to 150° of δ_{13}, while RH-WO region covers the whole of the δ_{13} region. In 3+1 model, both RH-RO and RH-WO regions cover the whole of the δ_{13} region. WH-RO solution occupies a small region for 3ν case, covering half of δ_{13} region for 4ν case. WH-WO region covers the whole of the δ_{13} region for 4ν case. In the third plot of first row, true values are taken as $\delta_{13} = -90^\circ$, $\theta_{23} = 52^\circ$ and inverted hierarchy. The RH-RO region covers the entire range of δ_{13} for both 3ν and 4ν case, whereas RH-WO region almost doubles from 3ν case to 4ν case. A small range of δ_{13} excluded from WH-RO for 3ν case is covered in 4ν case. WH-WO region of 3ν case excludes 60° to 150° of δ_{13} while full δ_{13} range is covered for 4ν case.

In the second row of the figure, we plot allowed regions for NOνA[3+$\bar{1}$]. We take true values as best-fit points obtained by NOνA. We observe an increase in precision of parameter measurement, due to an increase in statistics, from added 1 yr of antineutrino run. In the first plot of the second row, the RH-RO octant region covers entire δ_{13} range for both 3ν and 4ν case. RH-WO region includes -180° to 45° of δ_{13} for 3ν case, while the whole range of δ_{13} is covered in 4ν case. A slight increase in the area of WH-RO is observed form 3ν to 4ν case. 4ν introduces WH-WO region which was resolved for 3ν case. In the second plot, RH-RO region allows full range of δ_{13} for 4ν case, while it was restricted to lower half of CP range in 3ν case. We see that WH-RO solution, which was resolved in 3ν case, is reintroduced in 4ν case. We also see a slight increase in the size of WH-WO solution from 3ν to 4ν. In third plot, RH-RO region covers the whole CP range for 4ν while 35° to 125° of δ_{13} are excluded in 3ν case. The almost resolved RH-WO solution for 3ν doubles for 4ν case. WH-RO and WH-WO cover the entire region of δ_{13} for 4ν case.

In the third row, we show allowed regions for NOνA[3+$\bar{3}$]. In the first plot, it can be seen that small area of RH-WO in case of 3ν now covers the whole of δ_{13} region for 4ν case. While the 3ν case has WH-Wδ_{13} degeneracy, 4ν case introduces equal sized WH-WO-Wδ_{13} degeneracy. In second plot, for 3ν case most of δ_{13} values above 0° are excluded, but

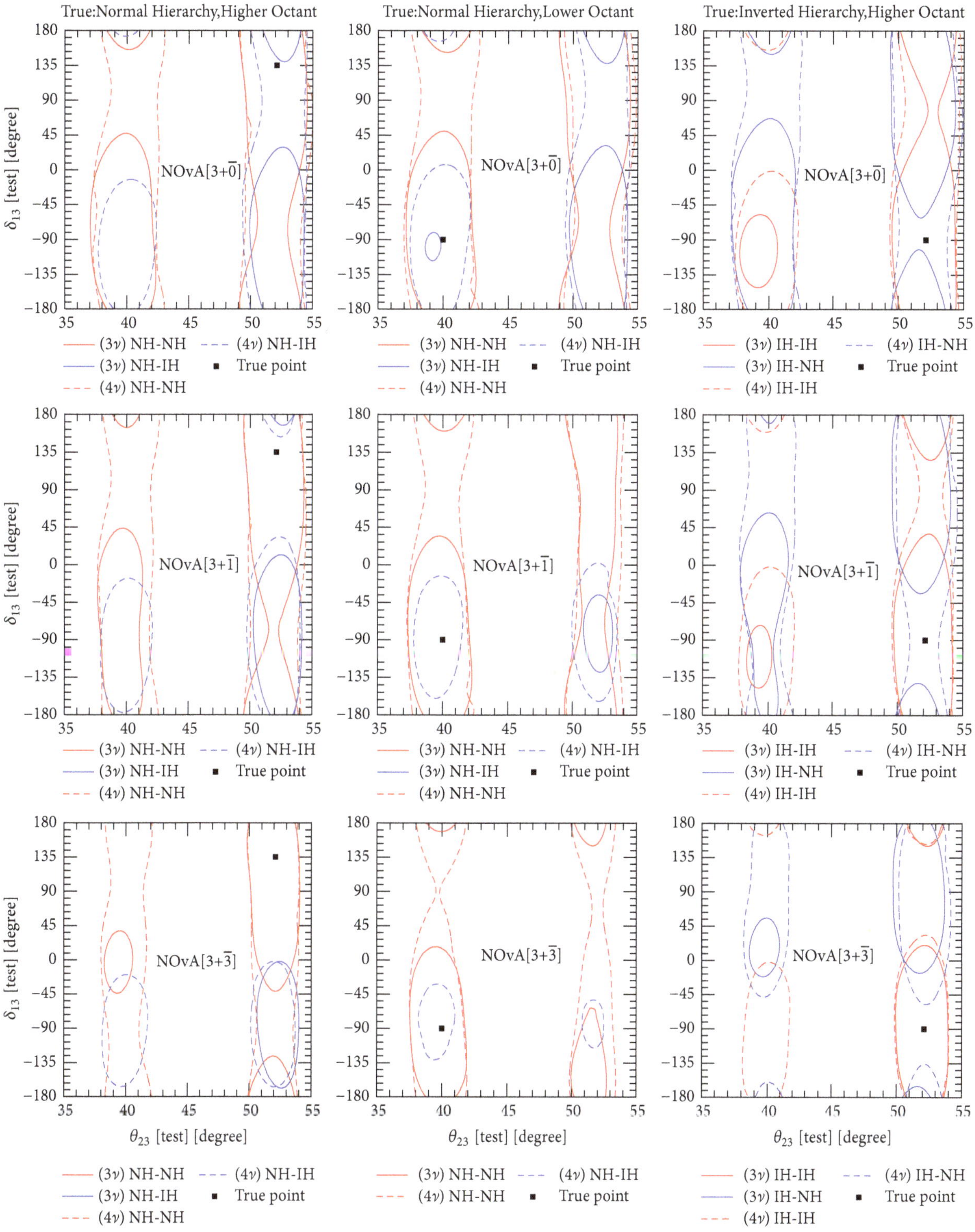

FIGURE 3: Contour plots of allowed regions in the test plane, θ_{23} versus δ_{13}, at 90% C.I with top, middle, and bottom rows for NOνA runs of $3+\bar{0}$, $3+\bar{1}$, and $3+\bar{3}$ years, respectively.

for 4ν case we see that contour covers the whole of δ_{13} range. Already present small area of RH-WO of 3ν is also increased for 4ν case. 4ν case also introduces a small region of WH solutions which were not present in 3ν case. In the third plot, we see that 4ν introduces RH-WO region of the almost equal size of RH-RO region of 3ν case. We observed a slight increase in WH-RO region for 4ν over 3ν case, while the WH-WO region almost triples for 4ν case.

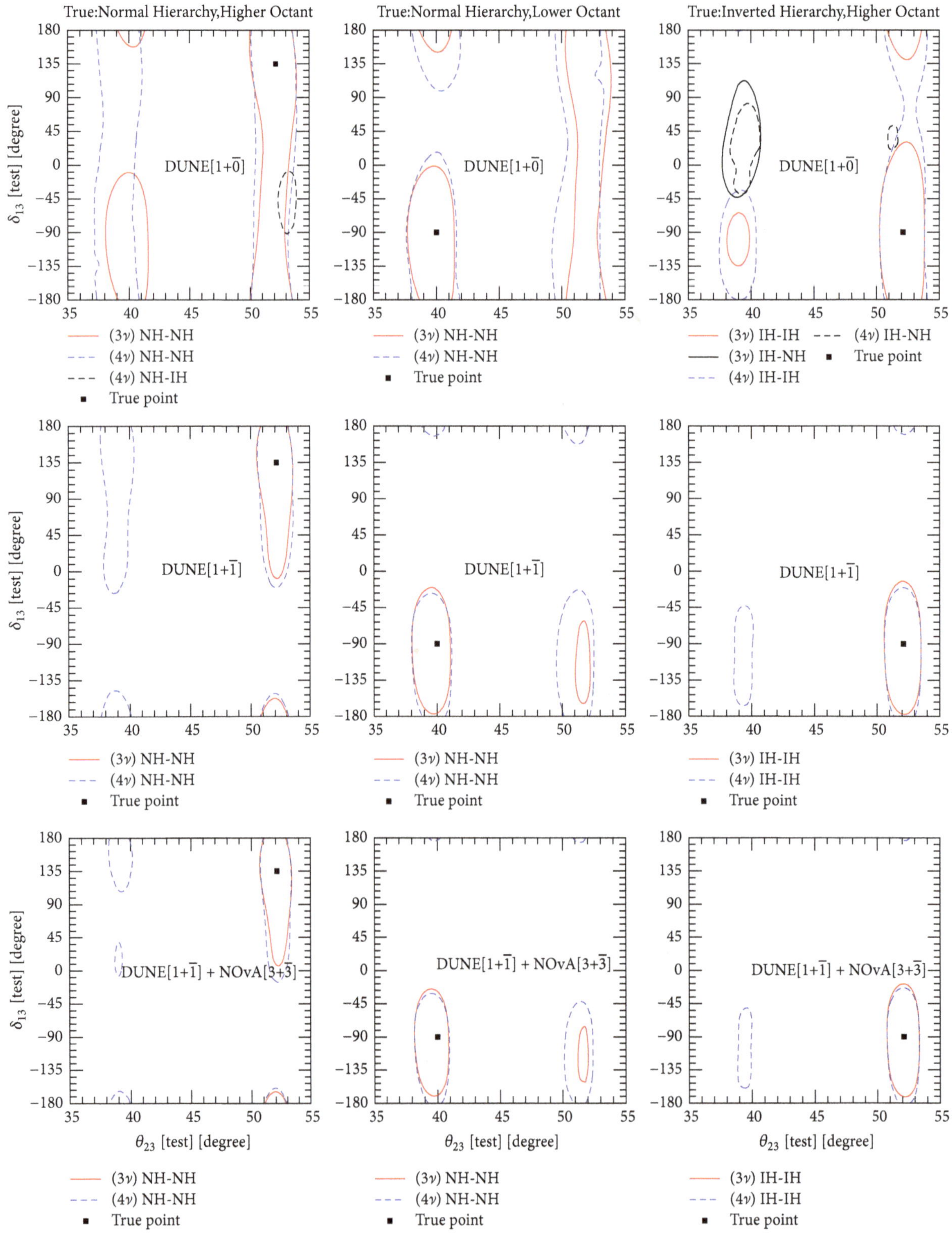

FIGURE 4: Contour plots of allowed regions in the test plane θ_{23} versus δ_{13} at 99% C.L with top, middle, and bottom rows for DUNE runs of $1+\bar{0}$, $1+\bar{1}$ years and DUNE$[1+\bar{1}]$+NOνA$[3+\bar{3}]$, respectively.

In Figure 4, we show allowed parameter regions for DUNE experiment for different runtimes. DUNE, being the next generation LBL experiment, is expected to have excellent statistics. Hence, we plot 99% C.L regions for DUNE. In the first row of Figure 4, we show 99% C.L for DUNE$[1+\bar{0}]$. In the first plot, RH-RO region covers the entire δ_{13} range for both 3ν and 4ν case. The RH-WO region which covers only lower half of δ_{13} region for 3ν case covers the whole range

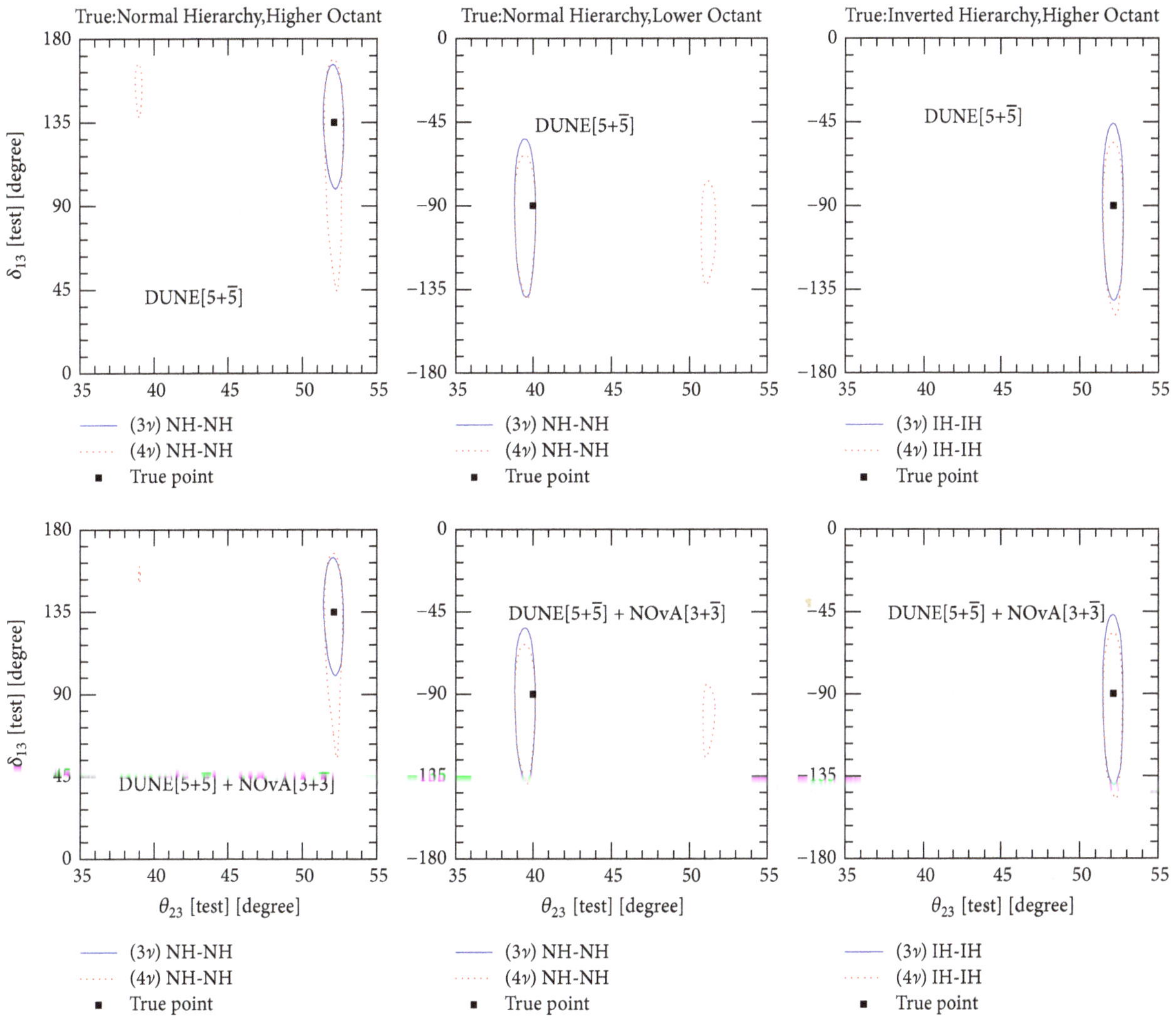

FIGURE 5: Contour plots of allowed regions in the test plane θ_{23} versus δ_{13} at 99% C.L with top and bottom rows for DUNE[5 + $\bar{5}$] and NOνA[3 + $\bar{3}$] + DUNE[5 + $\bar{5}$], respectively.

for 4ν case. A small region of WH is also observed. In the second plot we see that all WH solutions are resolved. RH-WO covers the whole range of δ_{13} for both 3ν and 4ν case. RH-RO solutions exclude 0° to 155° of δ_{13} for 3ν case, while 20° to 100° of δ_{13} are excluded for 4ν case. In third plot, we see that 4ν case extends RH-RO to the whole range of δ_{13} while 30° to 140° of δ_{13} were excluded for 3ν case. We can see that DUNE clearly has better precision than NOνA experiment. In the second row, we show allowed regions for DUNE[1+$\bar{1}$]. We see the WH solutions are resolved for both 3ν and 4ν cases for all the best-fit values. In the first plot, 4ν case introduces RH-WO solution of similar size as RH-RO region of 3ν case. In the second plot, there is no considerable change in 4ν, compared to 3ν case for RH-RO region, while RH-WO octant is approximately doubled for 4ν case compared to 3ν case. In the third plot, 4ν case introduces small region of RH-WO which covers −45° to −170° of δ_{13}. In the third row, we combine statistics of DUNE[1+$\bar{1}$] and NOνA[3+$\bar{3}$]. There is a small improvement in precision from the combined result over the result from DUNE[1+$\bar{1}$] alone. In the first plot, we see that a small RH-WO region is introduced by 4ν case. In the second plot, there is no considerable change between 3ν and 4ν case for RH-RO region, while RH-WO octant almost doubles over 3ν case for 4ν case. In the third plot, 4ν case introduces small region of RH-WO which covers −35° to −160° of δ_{13}.

In Figure 5, we show allowed parameter regions for DUNE experiment, at 99% C.L for DUNE[5+$\bar{5}$]. We see that WH regions completely disappear for all the true value assumptions. In the first plot, RH-RO region covers a small δ_{13} range for both 3ν and 4ν case indicating high precision measurement capacity of DUNE. We see that δ_{13} range for 4ν case is approximately doubled as compared to the 3ν case. A small region of RH-WO is observed for 4ν case. In the second plot, RH-RO region covers small δ_{13} range of equal area for both 3ν and 4ν case. A small region of RH-WO is observed for 4ν case. In the third plot, the RH-WO solution is resolved. There is an increase in precision due to an increase in statistics. DUNE[5+$\bar{5}$] clearly has a better precision compared to the NOνA[3+$\bar{3}$] experiment. In the second row, we combine full run of NOνA and DUNE to check their degeneracy resolution capacity. The WH solutions are resolved for both 3ν and 4ν cases for all the best-fit values. In the first plot, RH-WO solution is almost resolved for 4ν case. In the second plot, RH-RO region covers small δ_{13} range of equal area for both 3ν and 4ν case. A small region of RH-WO is observed for 4ν case. We observe a slight improvement in degeneracy resolution, on consideration of combined statistics of full run DUNE and NOνA, over DUNE[5+$\bar{5}$].

5. Conclusions

We have discussed how the presence of a sterile neutrino will affect the physics potential of the proposed experiment DUNE and future runs of NOνA, in the light of latest NOνA results [35]. The best-fit parameters reported by NOνA still contain degenerate solutions. We attempt to see the extent to which these degeneracies could be resolved in future runs for the 3+1 model. Latest NOνA best-fit values are taken as our true values. First, we show the degeneracy resolution capacity, for future runs of NOνA. We conclude that NOνA[3+$\bar{3}$] could resolve WH-WO solutions for first two true value cases, at 90% C.L for 3ν case, but not for 4ν case. DUNE[1+$\bar{1}$] could resolve WH and RH-Wδ_{cp} solutions for both 3ν and 4ν case. WO degeneracy is resolved for 3ν case at 99% C.L except for small RH-WO region for the second case of true values. DUNE[1+$\bar{1}$] combined with NOνA[3+$\bar{3}$] shows increased sensitivity towards degeneracy resolution. Finally, for the full planned run of DUNE[5+$\bar{5}$], all the degeneracies are resolved at 99% C.L for 3ν case while a tiny region of WO lingers on for 4ν case. For combined statistics of DUNE[5+$\bar{5}$] and NOνA[3+$\bar{3}$], we observe that all the degeneracies are resolved at 99% C.L for both 3ν and 4ν case except for the NH-LO case. Thus, we conclude that NOνA and DUNE experiments together can resolve all the degeneracies at 99% C.L even in the presence of sterile neutrino, if one of the current best-fit values of NOνA is the true value.

Conflicts of Interest

The authors declare that they have no conflicts of interest.

Acknowledgments

Akshay Chatla would like to thank *the Council of Scientific & Industrial Research, Government of India*, for financial support. The work of Sahithi Rudrabhatla was supported by *theDepartment of Science & Technology, Government of India*. We would like to thank Dr. Monojit Ghosh, Dr. C Soumya, and K Siva Prasad for their valuable help.

References

[1] R. R. Volkas, "Introduction to sterile neutrinos," *Progress in Particle and Nuclear Physics*, vol. 48, no. 1, pp. 161–174, 2002.

[2] C. Athanassopoulos et al., "Candidate events in a search for $\bar{\nu}_\mu \longrightarrow \bar{\nu}_e$ oscillations," *Physical Review Letters*, vol. 75, p. 2650, 1995.

[3] S. Schael et al., "ALEPH and DELPHI and L3 and OPAL and SLD collaborations and LEP electroweak working group and SLD electroweak group and SLD heavy flavour group," *Physics Reports*, vol. 427, p. 257, 2006.

[4] A. A. Aguilar-Arevalo et al., "MiniBooNE collaboration," *Physical Review Letters*, vol. 102, Article ID 101802, 2009, arXiv:0812.2243.

[5] J. N. Abdurashitov et al., "Measurement of the response of a Ga solar neutrino experiment to neutrinos from an 37Ar source," *Physical Review C: Nuclear Physics*, vol. 73, Article ID 045805, 2006.

[6] M. A. Acero, C. Giunti, and M. Laveder, "Limits on υ_e and $\bar{\upsilon}_e$ disappearance from Gallium and reactor experiments," *Physical Review D: Particles, Fields, Gravitation and Cosmology*, vol. 78, no. 7, 2008.

[7] C. Giunti and M. Laveder, "Statistical significance of the gallium anomaly," *Physical Review C: Covering Nuclear Physics*, vol. 83, no. 6, p. 065504, 2011.

[8] G. Mention, M. Fechner, T. Lasserre et al., "Reactor antineutrino anomaly," *Physical Review D: Particles, Fields, Gravitation and Cosmology*, vol. 83, no. 7, Article ID 073006, 2011.

[9] N. Klop and A. Palazzo, "Imprints of *CP* violation induced by sterile neutrinos in T2K data," *Physical Review D: Particles, Fields, Gravitation and Cosmology*, vol. 91, no. 7, 2015.

[10] A. Palazzo, "3-flavor and 4-flavor implications of the latest T2K and NOνA electron (anti-)neutrino appearance results," *Physics Letters B*, vol. 757, p. 142, 2016.

[11] F. P. An et al., "Independent measurement of the neutrino mixing angle θ_{13} via neutron capture on hydrogen at Daya Bay," *Physical Review Letters*, vol. 90, no. 7, 2014.

[12] C. Soumya and R. Mohanta, "Towards extracting the best possible results from NOvA," *European Physical Journal C*, vol. 76, no. 6, p. 302, 2016, arXiv:1605.00523.

[13] C. Soumya, K. N. Deepthi, and R. Mohanta, "A Comprehensive Study of the Discovery Potential of NOνA, T2K, and T2HK Experiments," *Advances in High Energy Physics*, vol. 2016, Article ID 9139402, 15 pages, 2016.

[14] K. N. Deepthi, C. Soumya, and R. Mohanta, "Revisiting the sensitivity studies for leptonic CPviolation and mass hierarchy with T2K, NOνA and LBNE experiments," *New Journal of Physics*, vol. 17, no. 2, Article ID 023035, 2015, arXiv:1409.2343.

[15] M. Ghosh, P. Ghoshal, S. Goswami, N. Nath, and S. K. Raut, "New look at the degeneracies in the neutrino oscillation parameters, and their resolution by T2K, NOνA and ICAL," *Physical Review D: Particles, Fields, Gravitation and Cosmology*, vol. 93, no. 1, Article ID 013013, 2016.

[16] S. Goswami and N. Nath, "Implications of the latest NOνA results," *High Energy Physics - Phenomenology*, 2017, arXiv:1705.01274.

[17] J. Kopp, P. A. Machado, M. Maltoni, and T. Schwetz, "Sterile neutrino oscillations: the global picture," *High Energy Physics - Phenomenology*, vol. 1305, p. 050009, 2013.

[18] M. Ghosh, S. Gupta, Z. M. Matthews, P. Sharma, and A. G. Williams, "Study of parameter degeneracy and hierarchy

sensitivity of NOνA in presence of sterile neutrino," *Physical Review D: Particles, Fields, Gravitation and Cosmology*, vol. 96, no. 7, 2017.

[19] S. K. Agarwalla, S. S. Chatterjee, and A. Palazzo, "Physics reach of DUNE with a light sterile neutrino," *High Energy Physics - Phenomenology*, vol. 016, 2016, arXiv:1603.03759.

[20] D. Dutta, R. Gandhi, B. Kayser, M. Masud, and S. Prakash, "Capabilities of long-baseline experiments in the presence of a sterile neutrino," *Journal of High Energy Physics*, vol. 122, 2016.

[21] S. K. Agarwalla, S. S. Chatterjee, and A. Palazzo, "Signatures of a light sterile neutrino in T2HK," *High Energy Physics - Phenomenology*, vol. 1804, no. 091, 2018, arXiv:1801.04855.

[22] S. K. Agarwalla, S. S. Chatterjee, and A. Palazzo, "Octant of θ_{23} in danger with a light sterile neutrino," *Physical Review Letters*, vol. 118, no. 3, 2017.

[23] S. Choubey, D. Dutta, and D. Pramanik, "Measuring the sterile neutrino CP phase at DUNE and T2HK," *The European Physical Journal C*, vol. 78, no. 4, 2018.

[24] P. Huber, M. Lindner, and W. Winter, "Simulation of long-baseline neutrino oscillation experiments with GLoBES: (General Long Baseline Experiment Simulator)," *Computer Physics Communications*, vol. 167, no. 3, pp. 195–202, 2005.

[25] P. Huber, J. Kopp, M. Lindner, M. Rolinec, and W. Winter, "New features in the simulation of neutrino oscillation experiments with GLoBES 3.0. (General Long Baseline Experiment Simulator)," *Computer Physics Communications*, vol. 177, p. 432, 2007.

[26] J. Kopp, "New physics engine for the inclusion sterile neutrinos and non standard interactions in GLoBES," https://www.mpi-hd.mpg.de/personalhomes/globes/tools.html, 2011.

[27] P. Adamson et al., "First measurement of muon-neutrino disappearance in NOvA," *Physical Review Letters D*, vol. 93, no. 5, 2016.

[28] P. Adamson, "First Measurement of Electron Neutrino Appearance in NOvA," *Physical Review Letters*, vol. 116, no. 5, 2016.

[29] D. S. Ayres et al., "The NOvA technical design report," *INIS*, vol. 39, no. 38, 2007, [NOvA Collaboration], doi:10.2172/, doi.

[30] S. K. Agarwalla, S. Prakash, S. K. Raut, and S. U. Sankar, "Potential of optimized NOvA for large theta(13) & combined performance with a LArTPC & T2K," *High Energy Physics - Phenomenology*, vol. 1212, 2012, arXiv:1208.3644.

[31] D. S. Ayres et al., "NOvA proposal to build a 30 kiloton Off-axis detector to study neutrino oscillations in the fermilab NuMI beamline," *High Energy Physics - Experiment*, 2005.

[32] R. Acciarri, "Long-baseline neutrino facility (LBNF) and deep underground neutrino experiment (DUNE) conceptual design report volume 2: The physics program for DUNE at LBNF," *High Energy Physics - Experiment*, 2016, arXiv:1512.06148.

[33] R. Acciarri et al., "Long-baseline neutrino facility (LBNF) and deep underground neutrino experiment (DUNE) conceptual design report volume 1: The LBNF and DUNE projects," *High Energy Physics - Experiment*, 2016, arXiv:1601.05471.

[34] T. Alion et al., "Experiment simulation configurations used in DUNE CDR," *High Energy Physics - Phenomenology*, 2016, arXiv:1606.09550.

[35] P. Adamson et al., "Constraints on Oscillation Parameters from υ_e appearance and υ_μ disappearance in NOvA," *Physical Review Letters*, vol. 118, 2017.

[36] P. Adamson et al., "Search for active-sterile neutrino mixing using neutral-current interactions in NOvA," *Physical Review Letters*, vol. 96, no. 7, Article ID 072006, 2017.

[37] A. Y. Smirnov, "The MSW effect and matter effects in neutrino oscillations," *Physica Scripta*, vol. T121, pp. 57–64, 2005.

7

Bound State of Heavy Quarks using a General Polynomial Potential

Hesham Mansour and Ahmed Gamal

Physics Department, Faculty of Science, Cairo University, Giza, Egypt

Correspondence should be addressed to Ahmed Gamal; alnabolci2010@gmail.com

Academic Editor: Sally Seidel

In the present work, the mass spectra of the bound states of heavy quarks $c\bar{c}$, $b\bar{b}$, and B_c meson are studied within the framework of the nonrelativistic Schrödinger's equation. First, we solve Schrödinger's equation with a general polynomial potential by Nikiforov-Uvarov (NU) method. The energy eigenvalues for any L- value is presented for a special case of the potential. The results obtained are in good agreement with the experimental data and are better than previous theoretical studies.

1. Introduction

The study of quarkonium systems provides a good understating of the quantitative description of quantum chromodynamics (QCD) theory, the standard model and particle physics [1–7]. The quarkonia with a heavy quark and antiquark and their interaction are well described by Schrödinger's equation. The solution of this equation with a spherically symmetric potentials is one of the most important problems in quarkonia systems [8–11]. These potentials should take into account the two important features of the strong interaction, namely, asymptotic freedom and quark confinement [2–6].

In the present work, an interaction potential in the quark-antiquark bound system is taken as a general polynomial to get the general eigenvalue solution. In the next step, we chose a specific potential according to the physical properties of the system. Several methods are used to solve Schrödinger's equation. One of them is the Nikiforov-Uvarov (NU) method [12–14], which gives asymptotic expressions for the eigenfunctions and eigenvalues of the Schrödinger's equation. Hence one can calculate the energy eigenstates for the spectrum of the quarkonia systems [12–15].

The paper is organized as follows: In Section 2, the Nikiforov-Uvarov (NU) method is briefly explained. In Section 3, the Schrödinger equation with a general polynomial potential is solved by the Nikiforov-Uvarov (NU) method. In Section 4, results and discussion are presented. In Section 5, the conclusion is given.

2. The Nikiforov-Uvarov (NU) Method [12–15]

The Nikiforov-Uvarov (NU) method is based on solving the hypergeometric-type second-order differential equation.

$$\ddot{\Psi}(s) + \frac{\tilde{\tau}(s)}{\sigma(s)}\dot{\Psi}(s) + \frac{\tilde{\sigma}(s)}{\sigma^2(s)}\Psi(s) = 0. \quad (1)$$

Here $\sigma(s)$ and $\tilde{\sigma}(s)$ are second-degree polynomials, $\tilde{\tau}(s)$ is a first-degree polynomial, and $\psi(s)$ is a function of the hypergeometric-type.

By taking $\Psi(s) = \varphi(s)\Upsilon(s)$ and substituting in equation (1), we get the following equation

$$\ddot{\Upsilon}(s) + \left[2\frac{\dot{\varphi}(s)}{\varphi(s)} + \frac{\tilde{\tau}(s)}{\sigma(s)}\right]\Upsilon(s) + \left[\frac{\ddot{\varphi}(s)}{\varphi(s)} + \frac{\dot{\varphi}(s)}{\varphi(s)}\frac{\tilde{\tau}(s)}{\sigma(s)} + \frac{\tilde{\sigma}(s)}{\sigma^2(s)}\right]\Upsilon(s) = 0. \quad (2)$$

By taking

$$2\frac{\dot{\varphi}(s)}{\varphi(s)}+\frac{\tilde{\tau}(s)}{\sigma(s)}=\frac{\tau(s)}{\sigma(s)},\qquad \frac{\dot{\varphi}(s)}{\varphi(s)}=\frac{\pi(s)}{\sigma(s)} \tag{3}$$

we get

$$\tau(s)=\tilde{\tau}(s)+2\pi(s), \tag{4}$$

where both $\pi(s)$ and $\tau(s)$ are polynomials of degree at most one.

Also we one can take

$$\frac{\ddot{\varphi}(s)}{\varphi(s)}+\frac{\dot{\varphi}(s)}{\varphi(s)}\frac{\tilde{\tau}(s)}{\sigma(s)}+\frac{\tilde{\sigma}(s)}{\sigma^2(s)}=\frac{\overline{\sigma}(s)}{\sigma^2(s)} \tag{5}$$

where

$$\frac{\ddot{\varphi}(s)}{\varphi(s)}=\left[\frac{\dot{\varphi}(s)}{\varphi(s)}\right]^{\cdot}+\left[\frac{\dot{\varphi}(s)}{\varphi(s)}\right]^2=\left[\frac{\pi(s)}{\sigma(s)}\right]^{\cdot}+\left[\frac{\pi(s)}{\sigma(s)}\right]^2 \tag{6}$$

And

$$\overline{\sigma}(s)=\tilde{\sigma}(s)+\pi^2(s)+\pi(s)\left[\tilde{\tau}(s)-\dot{\sigma}(s)\right]+\dot{\pi}(s)\sigma(s). \tag{7}$$

So equation (2) becomes

$$\ddot{\Upsilon}(s)+\frac{\tau(s)}{\sigma(s)}\Upsilon(s)+\frac{\overline{\sigma}(s)}{\sigma^2(s)}\Upsilon(s)=0. \tag{8}$$

An algebraic transformation from equation (1) to equation (8) is systematic. Hence one can divide $\overline{\sigma}(s)$ by $\sigma(s)$ to obtain a constant λ; i.e.,

$$\overline{\sigma}(s)=\lambda\,\sigma(s). \tag{9}$$

Equation (8) can be reduced to a hypergeometric equation in the form

$$\sigma(s)\ddot{\Upsilon}(s)+\tau(s)\Upsilon(s)+\lambda\Upsilon(s)=0. \tag{10}$$

Substituting from equation (9) in equation (7) and solving the quadratic equation for $\pi(s)$, we obtain

$$\pi^2(s)+\pi(s)\left[\tilde{\tau}(s)-\dot{\sigma}(s)\right]+\tilde{\sigma}(s)-k\sigma(s)=0, \tag{11}$$

where

$$k=\lambda-\dot{\pi}(s). \tag{12}$$

$$\pi(s)=\frac{\dot{\sigma}(s)-\tilde{\tau}(s)}{2}\pm\sqrt{\left(\frac{\dot{\sigma}(s)-\tilde{\tau}(s)}{2}\right)^2-\tilde{\sigma}(s)+k\sigma(s)}. \tag{13}$$

The possible solutions for $\pi(s)$depend on the parameter k according to the plus and minus signs of $\pi(s)$ [13]. Since $\pi(s)$ is a polynomial of degree at most one, the expression under the square root has to be the square of a polynomial. In this case, an equation of the quadratic form is available for the constant k. To determine the parameter k, one must set the discriminant of this quadratic expression to be equal to zero. After determining the values of k one can find the values of $\pi(s)$, λ *and* $\tau(s)$.

Applying the same systematic way for equation (10), we get

$$\lambda_n=-n\dot{\tau}(s)-\frac{n(n-1)}{2}\ddot{\sigma}(s), \tag{14}$$

where n is the principle quantum number.

By comparing equations (12) and (14), we get an equation for the energy eigenvalues.

3. The Schrödinger Equation with a General Polynomial Potential

The radial Schrödinger equation of a quark and antiquark system is

$$\frac{d^2Q}{dr^2}+\left[\frac{2\mu}{h^2}(E-V)-\frac{l(l+1)}{r^2}\right]Q=0. \tag{15}$$

We will use a generalized polynomial potential

$$V(r)=\sum_{m=0}^{m}A_{m-2}r^{m-2},\quad m=0,1,2,3,4,\ldots \tag{16}$$

By substituting in equation (15), we get

$$\frac{d^2Q}{dr^2}+\left[\frac{2\mu}{h^2}E-\frac{2\mu}{h^2}\sum_{m=0}^{m}A_{m-2}r^{m-2}-\frac{l(l+1)}{r^2}\right]Q=0. \tag{17}$$

Let

$$\frac{2\mu}{h^2}A_{m-2}=a_{m-2},\qquad b_{-2}=l(l+1)+a_{-2},\qquad b_0=a_0-\frac{2\mu}{h^2}E=a_0-\epsilon_0 \tag{18}$$

and hence,

$$\frac{d^2Q}{dr^2}+\left[\sum_{m=0}^{m}A(a,b)_{m-2}r^{m-2}\right]Q=0, \tag{19}$$

where

$$\sum_{m=0}^{m}A(a,b)_{m-2}r^{m-2}=-\left[b_{-2}r^{-2}+a_{-1}r^{-1}+b_0+a_1r+a_2r^2+a_3r^3+\cdots\right]. \tag{20}$$

Let r = 1/x; hence

$$\frac{d^2Q}{dr^2} = \frac{d}{dr}\left(\frac{dQ}{dr}\right) = \frac{d}{dx}\frac{dx}{dr}\left(\frac{dQ}{dr}\right) = \frac{d}{dx}\left(x^4\frac{dQ}{dx}\right) = 4x^3\frac{dQ}{dx} + x^4\frac{d^2Q}{dx^2} \tag{21}$$

And

$$\sum_{m=0}^{m} A(a,b)_{m-2}\left(\frac{1}{x}\right)^{m-2} = -\left[b_{-2}x^2 + a_{-1}x + b_0 + \sum_{m=1}^{m} a_m\left(\frac{1}{x}\right)^m\right]. \tag{22}$$

By substituting in equation (19), we get

$$x^4\frac{d^2Q}{dx^2} + 4x^3\frac{dQ}{dx} + \left[-b_{-2}x^2 - a_{-1}x - b_0 - \sum_{m=1}^{m} a_m\left(\frac{1}{x}\right)^m\right]Q = 0. \tag{23}$$

We propose the following approximation scheme on the term $a_m(1/x)^m$. Let us assume that there is a characteristic radius (residual radius) r_0 of the quark and antiquark system (which is the smallest distance between the two quarks where they cannot collide with each other). This scheme is based on the expansion of $a_m(1/x)^m$ in a power series around r_0, i.e., around $\delta = 1/r_0$ in the x-space, up to the second order, so that the a_m, dependent term, preserves the original form of equation (23). This is similar to Pekeris approximation [14, 15], which helps to deform the centrifugal potential such that the modified potential can be solved by the Nikiforov-Uvarov (NU) method. Setting $y = (x - \delta)$ around $y = 0$ (the singularity), one can expand into a power series as follows

$$\sum_{m=1}^{m} a_m\left(\frac{1}{x}\right)^m = \sum_{m=1}^{m}\frac{a_m}{(y+\delta)^m} = \sum_{m=1}^{m}\frac{a_m}{\delta^m}\left[1+\frac{y}{\delta}\right]^{-m}. \tag{24}$$

$$\sum_{m=1}^{m} a_m\left(\frac{1}{x}\right)^m \approx \sum_{m=1}^{m}\left[\frac{a_m}{\delta^m} - \frac{ma_m}{\delta^{m+1}}x + \frac{m(m+1)a_m}{2\delta^{m+2}}x^2\right]. \tag{25}$$

By substituting from equation (25) in equation (23), dividing by x^4 where $x \neq 0$, and rearranging this equation, we get

$$\frac{d^2Q}{dx^2} + \frac{4x}{x^2}\frac{dQ}{dx} + \frac{1}{x^4}\left[-\left(b_0 + \sum_{m=1}^{m}\frac{a_m}{\delta^m}\right) + \left(\sum_{m=1}^{m}\frac{ma_m}{\delta^{m+1}} - a_{-1}\right)x - \left(b_{-2} + \sum_{m=1}^{m}\frac{m(m+1)a_m}{2\delta^{m+2}}\right)x^2\right]Q = 0. \tag{26}$$

We define

$$\left(b_0 + \sum_{m=1}^{m}\frac{a_m}{\delta^m}\right) = q,\quad \left(\sum_{m=1}^{m}\frac{ma_m}{\delta^{m+1}} - a_{-1}\right) = w,\quad \left(b_{-2} + \sum_{m=1}^{m}\frac{m(m+1)a_m}{2\delta^{m+2}}\right) = z \tag{27}$$

And, hence, equation (26) becomes

$$\frac{d^2Q}{dx^2} + \left(\frac{4x}{x^2}\right)\frac{dQ}{dx} + \frac{1}{x^4}\left[-q + wx - zx^2\right]Q = 0. \tag{28}$$

Comparing with equation (1), we get

$$\tilde{\tau} = 4x,\quad \sigma = x^2,\quad \tilde{\sigma} = -q + wx - zx^2 \tag{29}$$

And, by substituting in equation (13), we get

$$\pi(x) = -x \pm \sqrt{(1+k+z)x^2 - wx + q}. \tag{30}$$

Now one can obtain the value of the parameter k, by knowing that $\pi(x)$ is a polynomial of degree at most one and by putting the discriminant of this expression under the square root equal to zero.

$$w^2 - 4(1+k+z)q = 0 \longrightarrow k = \frac{w^2}{4q} - z - 1 \tag{31}$$

By substituting in equation (30) and taking the negative value of $\pi(x)$, for bound state solutions, one finds that the solution is in agreement with the free hydrogen atom spectrum, because of the Coulomb term

$$\pi(x) = -x - \frac{w}{2\sqrt{q}}x + \sqrt{q}. \tag{32}$$

Hence, by substituting in equation (4), we get the following.

$$\tau(x) = \left[2 - \frac{w}{\sqrt{q}}\right]x + 2\sqrt{q}.,\quad \text{where } \left(2 - \frac{w}{\sqrt{q}}\right) < 0 \tag{33}$$

Substituting in equation (12), we obtain

$$\lambda = k + \dot{\pi}(x) = \frac{w^2}{4q} - \frac{w}{2\sqrt{q}} - z - 2. \tag{34}$$

Using equation (14), we obtain

$$\lambda_n = -n\left[2 - \frac{w}{\sqrt{q}}\right] - n(n-1) = -2n + \frac{w}{\sqrt{q}}n - n^2 + n.[5pt] \tag{35}$$

Equalizing equations (34) and (35), we get

$$q = \frac{w^2}{4\left[\sqrt{z + 9/4} + (n + 1/2)\right]^2}. \tag{36}$$

By substituting the values of (q, w, z and δ) in equation (36), we get

$$\epsilon_0 = a_0 + \sum_{m=1}^{m} a_m r_0^{\ m} - \frac{\left(\sum_{m=1}^{m} m a_m r_0^{\ m+1} - a_{-1}\right)^2}{4\left[\sqrt{\left(l(l+1) + a_{-2} + (1/2)\sum_{m=1}^{m} m(m+1) a_m r_0^{\ m+2}\right) + 9/4} + (n + 1/2)\right]^2}. \tag{37}$$

Equation (37) is the desired equation of the energy eigenvalues in spherical symmetric coordinates with a general polynomial radial potential using the Nikiforov-Uvarov (NU) method.

A special case of the above potential was chosen to describe the $q\bar{q}$ interaction, namely,

$$V(r) = \frac{b}{r} + ar + dr^2 + pr^4 \tag{38}$$

where the first term is the Coulomb potential because the two quarks are charged and the second term is the linear term in r which means that $V(r)$ continues growing as $r \longrightarrow \infty$. It is this linear term that leads to quark confinement. One of the striking properties of QCD asymptotic freedom is that the interaction strength between quarks becomes smaller as the distance between them gets shorter.

The third term is a harmonic term and the fourth is an anharmonic term and they are responsible also for quark confinement.

For the above chosen potential, we put $a_{-2} = a_0 = 0$ and $m = 1, 2, 4$, and hence

$$\epsilon_0 = a_1 r_0 + a_2 r_0^{\ 2} + a_4 r_0^{\ 4} - \frac{\left(a_1 r_0^{\ 2} + 2a_2 r_0^{\ 3} + 4a_4 r_0^{\ 5} - a_{-1}\right)^2}{4\left[\sqrt{l(l+1) + a_1 r_0^3 + 3a_2 r_0^4 + 10a_4 r_0^6 + 9/4} + (n + 1/2)\right]^2}. \tag{39}$$

Now, we can rewrite equation (39) in a different form which depends on the parameters of the potential as follows

$$E = ar_0 + dr_0^{\ 2} + pr_0^{\ 4} - \frac{\left(2\mu/h^2\right)\left(ar_0^{\ 2} + 2dr_0^{\ 3} + 4pr_0^{\ 5} - b\right)^2}{4\left[\sqrt{(2\mu a/h^2)\, r_0^3 + (6\mu d/h^2)\, r_0^4 + (20\mu p/h^2)\, r_0^6 + l(l+1) + 9/4} + (n + 1/2)\right]^2}. \tag{40}$$

4. Results and Discussion

In this section, we will calculate the spectra for the bound states of heavy quarks such as charmonium, bottomonium, and B_C meson. To determine the mass spectra in three dimensions, we use the following relation.

$$M = m_q + m_{\bar{q}} + E \tag{41}$$

By substituting in equation (40), we get

$$M = 2m_q + ar_0 + dr_0^{\ 2} + pr_0^{\ 4} - \frac{\left(2\mu/h^2\right)\left(ar_0^{\ 2} + 2dr_0^{\ 3} + 4pr_0^{\ 5} - b\right)^2}{4\left[\sqrt{(2\mu a/h^2)\, r_0^3 + (6\mu d/h^2)\, r_0^4 + (20\mu p/h^2)\, r_0^6 + l(l+1) + 9/4} + (n + 1/2)\right]^2}. \tag{42}$$

It is clear that equation (42) depends on the potential parameters $(a, b, d, p\ and\ r_0)$ which will be obtained from the experimental data.

In the case of charmonium $[\Psi = c\bar{c}]$, the rest mass equation is as follows.

$$M = 2m_c + ar_0 + dr_0^2 + pr_0^4 - \frac{32.743 * \left(ar_0^2 + 2dr_0^3 + 4pr_0^5 - b\right)^2}{4\left[\sqrt{32.743ar_0^3 + 98.229dr_0^4 + 327.43pr_0^6 + l(l+1) + 9/4} + (n + 1/2)\right]^2} \tag{43}$$

In the case of bottomonium $[\Upsilon = b\bar{b}]$, the rest mass equation is as follows.

$$M = 2m_b + ar_0 + dr_0^2 + pr_0^4 - \frac{107.86 * \left(ar_0^2 + 2dr_0^3 + 4pr_0^5 - b\right)^2}{4\left[\sqrt{107.86 * ar_0^3 + 323.58 * dr_0^4 + 1078.6 * pr_0^6 + l(l+1) + 9/4} + (n + 1/2)\right]^2} \tag{44}$$

And in the case of the meson $[B_c = b\bar{c}]$, the rest mass equation is as follows.

$$M = m_c + m_b + ar_0 + dr_0^2 + pr_0^4 - \frac{50.24 * \left(ar_0^2 + 2dr_0^3 + 4pr_0^5 - b\right)^2}{4\left[\sqrt{50.24 * ar_0^3 + 150.72 * dr_0^4 + 502.4 * pr_0^6 + l(l+1) + 9/4} + (n + 1/2)\right]^2} \tag{45}$$

Comparing our theoretical work with the experimental data, we found that the maximum errors are 0.229% for the charmonium, 0.0742% for the bottomonium, and 0.00123% for the B_C meson. These may be due to errors in the measurements of the device. The spin can also be taken into account if one uses relativistic corrections and the appropriate relativistic Schrödinger's equation. Our results are shown in Tables 1, 2, and 3, with a comparison between our results and those obtained in previous calculations in the literature. In the charmonium system, the maximum distance where a quark and antiquark can approach each other is $r_0 = 0.8043\ fm$. Similarly the maximum distances in the cases of the bottomonium system and B_C meson system are $r_0 = 0.47\ fm$ and $r_0 = 0.4256\ fm$, respectively. The positive and negative signs of the coefficient of the harmonic potential refer to the direction of motion of the oscillation. The negative sign of the coefficient of the Coulomb potential refers to the charges of the two quarks, but the positive sign refers to the existence of another negative contribution. The positive and negative signs of the coefficient of the anharmonic potential give a correction to the linear potential.

5. Conclusion

The mass spectra of the quarkonia (charmonium, bottomonium, and B_C meson) using our potential were studied within the framework of the nonrelativistic Schrödinger's equation by using the Nikiforov-Uvarov (NU) method. In [16, 17], the authors used an iteration method and the same potential when $p = 0$. It is found that adding the anharmonic potential gives a good accuracy and our work is comparable with them. In [18, 19], the authors used also the same potential when $a = p = 0$. We found that our work gives better results in comparison with experimental data. In [20, 21, 28, 29], the authors used the same method and the same potential when $p = 0$. We noticed that our work is in better agreement with the experimental data. In [22, 23], the authors used the Cornell potential only and used the same method. It is found that their results are comparable with our work. In [24, 25], the authors used the same potential when $p = 0$ and the same method used in the present work. We found that our work is comparable with them. In [22, 23, 30–37] for the mass spectra of *the* B_C meson, there is

TABLE 1: Mass spectra of charmonium in comparison with other works [r_0 = 0.8043 fm, a = 3.03857 GeV/fm, d = −0.7054 $GeV/(fm)^2$, b = −0.49842 $GeV.fm$, P = −0.2379 $GeV/(fm)^4$].

Type	present work	[16, 17]	[18, 19]	[20, 21]	[22, 23]	[24, 25]	EXP [26, 27]
1s	3.0969	3.078	3.096	3.096	3.096	3.078	3.0969
1p	-	3.415	3.433	3.433	3.255	3.415	-
2s	-	4.187	3.686	3.686	3.686	3.581	-
1d	3.6861	3.752	3.676	3.770	3.504	3.749	3.6861
2p	3.7702	4.143	3.910	4.023	3.779	3.917	3.773
3s	-	5.297	3.984	4.040	4.040	4.085	-
4s	-	6.407	4.150	4.358	4.269	4.589	-
2d	-	-	-	4.096	-	3.078	-
4d	4.160	-	-	-	-	-	4.159
1g	4.039	-	-	-	-	-	4.039
6d	4.263	-	-	-	-	-	4.263

TABLE 2: Mass spectra of bottomonium in comparison with other works [r_0 = 0.47 fm, a = 10.7 GeV/fm, d = −4.95 $GeV/(fm)^2$, b = 6.39286 $GeV.fm$, P = 7.1 $GeV/(fm)^4$].

Type	present work	[16, 17]	[18, 19]	[28, 29]	[22, 23]	[24, 25]	EXP [26, 27]
1s	9.4600	9.510	9.460	9.460	9.460	9.510	9.4601
1p	-	9.612	9.840	9.811	9.916	9.862	-
2s	-	10.627	10.023	10.023	10.023	10.038	-
1d	-	10.211	10.140	10.161	9.864	10.214	-
2p	-	10.944	10.160	10.374	10.114	10.390	-
3s	-	11.726	10.280	10.355	10.355	10.655	-
2d	10.02	-	-	-	-	-	10.023
4s	10.571	12.834	10.420	10.655	10.567	11.094	10.579
3d	10.358	-	-	-	-	-	10.355
6S	11.0198	-	-	-	-	-	11.019
5d	10.873	-	-	-	-	-	10.876

TABLE 3: Mass spectra of B_C *meson* in comparison with other works [r_0 = 0.4256 fm, a = 4.196 GeV/fm, d = −2.6064 $GeV/(fm)^2$, b = 9.5675 $GeV.fm$, P = 6.0631 $GeV/(fm)^4$].

Type	present work	[30–32]	[33, 34]	[35–37]	[22, 23]	EXP [26, 27]
1s	6.227	6.349	6.264	6.270	6.278	6.277
1p	6.287	6.715	6.700	6.699	6.486	-
2s	6.714	6.821	6.856	6.853	6.866	-
1d	6.398	-	-	-	6.772	-
2p	6.759	7.102	7.108	7.091	6.973	-
3s	7.09	7.175	7.244	7.193	7.181	-
4s	7.386	-	-	-	7.369	-
2d	6.85	-	-	-	7.128	-
4d	7.469	-	-	-	-	-
1g	6.728	-	-	-	-	-

no enough experimental data to compare with. In conclusion, comparing with the experimental data, we found that our results are better than those given by previous theoretical estimates.

Disclosure

Hesham Mansour is a Fellow of the Institute of Physics (FInstP).

Conflicts of Interest

The authors declare that they have no conflicts of interest.

References

[1] F. Halzem and A. Martin, *Quarks and Leptons an Introductory Course in Modern Particle Physics*, Wiley Publication, 1984.

[2] D. Griffiths, *Introduction to Elementary Particles*, John Wiley & Sons, New York, NY, USA, 1987.

[3] D. H. Perkins, *Introduction to High Energy Physics*, Cambridge University Press, Cambridge, UK, 2000.

[4] B. R. Martin and G. G. Shaw, *Particle Physics*, The Manchester Physics Series, 2nd edition, 1997.

[5] A. Bettini, *Introduction to Elementary Particle Physics*, Cambridge University Press, 2018.

[6] J. Rathsman, *Quark and Lepton Interactions*, Acta Universitatis Upsalensis Uppsala, 1996.

[7] N. Mistry, *Brief Introduction of Particle Physics*, Cornell University, 2011.

[8] A. T. Paul, *Ralph Llewellyn Modern Physics*, W.H.Freeman, 2012.

[9] B. H. Bransden and C. J. Joachain, *Quantum Mechanics*, Prentice Hall, 2000.

[10] L. D. Landau and E. M. Lifshitz, *Quantum Mechanics*, Butterworth Heinemann, 1981.

[11] D. Griffiths, *Introduction to Elementary Particles*, John Wiley & Sons, New York, NY, USA, 2008.

[12] N. Zettili, *Quantum Mechanics Concepts and Applications*, John Wiley, 2009.

[13] C. Berkdemir, *Theoretical Concept of Quantum Mechanics*, In Tech, 2012.

[14] A. F. Nikiforov and V. B. Uvarov, *Special Functions of Mathematical Physics*, Birkhäauser, Basel, Switzerland, 1988.

[15] H. Karayer, D. Demirhan, and F. Büyükkılıç, "Extension of Nikiforov-Uvarov method for the solution of Heun equation," *Journal of Mathematical Physics*, vol. 56, no. 6, Article ID 063504, 14 pages, 2015.

[16] R. Kumar and F. Chand, "Asymptotic study to the N-dimensional radial schrödinger equation for the quark-antiquark system," *Communications in Theoretical Physics*, vol. 59, no. 5, pp. 528–532, 2013.

[17] M. Abu-Shady, "Heavy Quarkonia and Mesons in the Cornell Potential with Harmonic Oscillator Potential in the N-dimensional Schrödinger Equation," *International Journal of Applied Mathematics and Theoretical Physics*, vol. 2, no. 2, p. 16, 2016.

[18] A. Al-Oun, A. Al-Jamel, and H. Widyan, "Various Properties of Heavy Quarkonia from Flavor-Independent Coulomb Plus Quadratic Potential," *Jordan Journal of Physics*, vol. 8, no. 4, pp. 199–203, 2015.

[19] H. Ke, G. Wang, X. Li, and C. Chang, "The magnetic dipole transitions in the $(c\bar{b})$ binding system," *Science China Physics, Mechanics and Astronomy*, vol. 53, no. 11, pp. 2025–2030, 2010.

[20] N. V. Masksimenko and S. M. Kuchin, "Determination of the mass spectrum of quarkonia by the Nikiforov–Uvarov method," *Russian Physics Journal*, vol. 54, no. 57, 2011.

[21] E. Eichten, K. Gottfried, T. Kinoshita, J. Kogut, K. D. Lane, and T. M. Yan, "Spectrum of charmed quark-antiquark bound states," *Physical Review Letters*, vol. 34, no. 6, pp. 369–372, 1975.

[22] S. M. Kuchin and N. V. Maksimenko, "Theoretical Estimations of the Spin – Averaged Mass Spectra of Heavy Quarkonia and Bc Mesons," *Universal Journal of Physics and Application*, vol. 7, pp. 295–298, 2013.

[23] R. Faccini, BABAR-CONF-07/035 SLAC-PUB-13080.

[24] R. Kumar and F. Chand, "Energy Spectra of the Coulomb Perturbed Potential in N-Dimensional Hilbert Space," *Phys. Scr*, vol. 85, Article ID 055008, 2012.

[25] Y. Li, P. Maris, and J. P. Vary, "Quarkonium as a relativistic bound state on the light front," *Physical Review D: Particles, Fields, Gravitation and Cosmology*, vol. 96, no. 1, 2017.

[26] J. Beringer, J. F. Arguin, and R. M. Barnett, "Review of particle physics," *Physical Review D: Particles, Fields, Gravitation and Cosmology*, vol. 86, Article ID 010001, 2012.

[27] B. Hong, T. Barillari, S. Bethke, S. Kluth, and S. Menke, "Overview of quarkonium production in heavy-ion collisions at LHC," *EPJ Web of Conferences*, vol. 120, p. 06002, 2016.

[28] Z. Ghalenovi, A. A. Rajabi, and A. Tavakolinezhad, "Masses of Charm and Beauty Baryons in the Constituent Quark Model," *International Journal of Modern Physics E*, vol. 21, no. 6, Article ID 1250057, 2012.

[29] J. Niu, L. Guo, H. Ma, and S. Wang, "Heavy quarkonium production through the top quark rare decays via the channels involving flavor changing neutral currents," *The European Physical Journal C*, vol. 78, no. 8, 2018.

[30] A. K. Ray and P. C. Vinodkumar, "Properties of B_c meson," *Pramana*, vol. 66, no. 5, pp. 953–958, 2006.

[31] S. Laachir and A. Laaribi, "Exact Solutions of the Helmholtz equation via the Nikiforov-Uvarov Method," *International Journal of Mathematical, Computational, Physical, Electrical and Computer Engineering*, vol. 7, no. 1, 2013.

[32] H. Ciftci, R. L. Hall, and N. Saad, "Iterative solutions to the Dirac equation," *Physical Review A: Atomic, Molecular and Optical Physics*, vol. 72, no. 2, part A, 022101, 7 pages, 2005.

[33] E. J. Eichten and C. Quigg, "Mesons with beauty and charm: Spectroscopy," *Physical Review D: Particles, Fields, Gravitation and Cosmology*, vol. 49, no. 11, pp. 5845–5856, 1994.

[34] M. Alberg and L. Wilets, "Exact solutions to the Schrödinger equation for potentials with Coulomb and harmonic oscillator terms," *Physics Letters A*, vol. 286, no. 1, pp. 7–14, 2001.

[35] D. Ebert, R. N. Faustov, and V. O. Galkin, "Properties of heavy quarkonia and," *Physical Review D: Particles, Fields, Gravitation and Cosmology*, vol. 67, no. 1, 2003.

[36] B. Ita, P. Tchoua, E. Siryabe, and G. E. Ntamack, "Solutions of the Klein-Gordon Equation with the Hulthen Potential Using the Frobenius Method," *International Journal of Theoretical and Mathematical Physics*, vol. 4, no. 5, pp. 173–177, 2014.

[37] H. Goudarzi and V. Vahidi, "Supersymmetric approach for Eckart potential using the NU method," *Advanced Studies in Theoretical Physics*, vol. 5, no. 9-12, pp. 469–476, 2011.

8

S-Wave Heavy Quarkonium Spectra: Mass, Decays, and Transitions

Halil Mutuk

Physics Department, Faculty of Arts and Sciences, Ondokuz Mayis University, 55139 Samsun, Turkey

Correspondence should be addressed to Halil Mutuk; halilmutuk@gmail.com

Guest Editor: Xian-Wei Kang

In this paper we revisited phenomenological potentials. We studied S-wave heavy quarkonium spectra by two potential models. The first one is power potential and the second one is logarithmic potential. We calculated spin averaged masses, hyperfine splittings, Regge trajectories of pseudoscalar and vector mesons, decay constants, leptonic decay widths, two-photon and two-gluon decay widths, and some allowed M1 transitions. We studied ground and 4 radially excited S-wave charmonium and bottomonium states via solving nonrelativistic Schrödinger equation. Although the potentials which were studied in this paper are not directly QCD motivated potential, obtained results agree well with experimental data and other theoretical studies

1. Introduction

Heavy quarkonium is the bound state of $b\bar{b}$ and $c\bar{c}$ and one of the most important playgrounds for our understanding of the strong interactions of quarks and gluons. Quantum chromodynamics (QCD) is thought to be the *true* theory of these strong interactions. QCD is a nonabelian local gauge field theory with the symmetry group $SU(3)$. In principle, one should be able to calculate hadronic properties such as mass spectrum and transitions by using QCD principles. But QCD does not readily supply us these hadronic properties. This challenge can be attributed to the several features that are not present in other local gauge field theories.

Foremost, being a nonabelian gauge theory, gluons which are gauge bosons, have color charge and interact among themselves. Unlike from quantum electrodynamics (QED), where a photon does not interact with other photon, in QCD one must consider interactions among gluons. This nonabelian nature of the theory makes some calculations complicated, for example, loops in propagators.

There are three other important features of QCD: *asymptotic freedom*, *confinement*, and *dynamical breaking of chiral symmetry*. Asymptotic freedom says that strong interaction coupling constant, α_s, is a function of momentum transfer. When the momentum transfer in a quark-quark collision increases (at short distances), the coupling constant becomes weaker whereas it becomes larger when momentum transfer decreases (at large distances). The idea behind confinement is that, there are no free quarks outside of a hadron; i.e., color charged particles (quarks and gluons) cannot be isolated out of hadrons. Flux tube model gives a reasonable explanation of confinement. When the distance between quark-antiquark (or quarks) pair increases, the gluon field between a pair of color charges forms a flux tube (or string) between them resulting a potential energy which depends linearly on the distance, $V(r) \;=\!\sim\; \sigma r$ where σ is the string constant. As distance increases between quarks, the potential energy can create new quark-antiquark pairs in colorless forms instead of a free quark. Up to now, nobody has been able to prove that confinement from QCD. Lattice QCD calculations simulate this confinement well and give a value for the string tension [1]. The last feature of QCD is the dynamical breaking of chiral symmetry. The QCD Lagrangian with N quark flavor has an exact chiral $SU(N) \times SU(N)$ symmetry but breaks down to $SU(N)$ symmetry because of the nonvanishing expectation value of QCD vacuum [2, 3]. The Goldstone bosons corresponding to this symmetry breaking are the pseudoscalar mesons.

The present aspects of the QCD caused other approaches to deal with these challenges. QCD sum rules, Lattice QCD, and potential models (quark models) are examples of these approaches. These approaches are nonperturbative since the strong interaction coupling constant, which should be the perturbation parameter of QCD is of the order one in low energies, hence the truncation of the perturbative expansion cannot be carried out. Since perturbation theory is not applicable, a nonperturbative approach has to be used to study systems that involve strong interactions. QCD sum rules and lattice QCD are based on QCD itself whereas in potential models, one assumes an interquark potential and solves a Schrödinger-like equation. The advantage of potential model is that, excited states can be studied in the framework of potential models whereas in QCD sum rules and lattice QCD, only the ground state or in some exceptional cases excited states can be studied.

After the discovery of charmonium ($c\bar{c}$) states, potential models have played a key role in understanding of heavy quarkonium spectroscopy [4, 5]. These potentials were in type of Coulomb plus linear confining potential with spin dependent interactions. The discovery of bottomonium ($b\bar{b}$) states were well described by the potential model picture which was used in the charmonium case. Heavy quarkonium spectroscopy was studied since that era with fruitful works [6–18]. A general review about potential models can be found in [19, 20] and references therein.

In the potential models, many features such as mass spectra and decay properties of heavy quarkonium could be described by an interquark potential in two-body Schrödinger equation. Interquark potentials are obtained both from phenomenology and theory. In the phenomenological method, it is assumed that a potential exist with some parameters to be determined by fits to the data. In the theory side, one can use perturbative QCD to determine the potential form at short distances and use lattice QCD at long distances [19]. These potentials can be classified as QCD motivated potentials [21–25] and phenomenological potentials [26–31]. The most commonly used phenomenological potentials are power-law potentials, for example [26] and logarithmic potentials, for example [30]. The detailed properties of these type potentials are studied extensively in [29]. All the potentials which are mentioned here have almost similar behaviour in the range of $0.1\ \text{fm} \leq r \leq 1\ \text{fm}$ which is characteristic region of charmonium and bottomonium systems [32, 33]. Outside the range, the behaviour of potentials differ. Up to now, no one was able to obtain a potential which is compatible at the whole range of distances by using QCD principles.

The potential model calculations have been quite successful in describing the hadron spectrum. Most of the phenomenological potentials must satisfy the following conditions:

$$\frac{dV}{dr} > 0, \qquad \frac{d^2V}{dr^2} \leq 0. \tag{1}$$

It means that static potential is a monotone nondecreasing and concave function of r which is a general property of gauge theories [34].

The great success of quarkonium phenomenology was somehow cracked at 2003 after the observation of $X(3872)$ [35]. The properties of this exotic particle are not compatible with the conventional quark model, the reason why it is named *exotic*. For example in [36], the authors studied $X(3872)$ near threshold zero in the $D^0\overline{D}^{*0}$ S-wave. There are other exotic states, XYZ, and the exotic particle zoo is growing. In this paper we will present some exotic states in the framework of quark model.

Energy spectra of heavy quarkonium are a rich source of the information on the nature of interquark forces and decay mechanisms. The prediction of mass spectrum in accordance with the experimental data does not verify the validity of a model for explaining hadronic interactions. Different potentials can produce reliable spectra with the experimental data. Thus other physical properties such as decay constants, leptonic decay widths, radiative decay widths, etc. need to be calculated.

A specific form of the QCD potential in the whole range of distances is not known. Therefore one needs to use potential models. In this work we revisited a power-law potential [26] and a logarithmic potential [30] to study S-wave heavy quarkonium. These potentials satisfy Eqn. (1), i.e. having nonsingular behaviour for $r \longrightarrow 0$. For our purposes, it must be mentioned that power-law and logarithmic potentials have nice scaling properties when used with a nonrelativistic Schrödinger equation [19]. We generated S-wave charmonium and bottomonium mass spectrum with the decays and M1 transitions. At Section 2 we give out theoretical model. In Sections 3 and 4, we generate S-wave heavy quarkonium spectrum, decays and transitions. In Section 5 we discuss our results and in Section 6 we conclude our results.

2. Formulation of the Model

When quark model was proposed, many authors treated baryons in detail with the harmonic oscillator quark model by using harmonic oscillator wave functions [37–39]. Mesons comparing to baryons are simpler objects since they are composites of two quarks. The reason for using harmonic oscillator wave function is that they form a complete set for a confining potential [40].

In order to obtain mass spectra, we solved Schrödinger equation by variational method. The variational method by using harmonic oscillator wave function gave successful results for heavy and light meson spectrum [15, 41, 42]. The procedure for this method is calculating expectation value of the Hamiltonian via the trial wave function:

$$E = \frac{\langle \Psi | H | \Psi \rangle}{\langle \Psi | \Psi \rangle}. \tag{2}$$

The mass of the meson is found by adding two times the mass of quark to the eigenenergy

$$M = 2m_q + E. \tag{3}$$

The Hamiltonian we consider is

$$H = M + \frac{p^2}{2\mu} + V(r) \tag{4}$$

Table 1: Spin-averaged mass spectrum of charmonium (in MeV).

State	Power	Logarithmic	[15]	[12]
1S	3067	3067	3067	3117
2S	3701	3655	3667	3684
3S	4054	3980	4121	4078
4S	4306	4204	4513	4407
5S	4504	4376	4866	

where $M = m_q + m_{\bar{q}}$, p is the relative momentum, μ is the reduced mass, and $V(r)$ is the potential between quarks. The spectrum can be obtained via solving Schrödinger equation

$$H|\Psi_n\rangle = E_n|\Psi_n\rangle \quad (5)$$

with the harmonic oscillator wave function defined as

$$\Psi_{nlm}(r,\theta,\phi) = R_{nl}(r)\,Y_{lm}(\theta,\phi). \quad (6)$$

Here R_{nl} is the radial wave function given as

$$R_{nl} = N_{nl} r^l e^{-\nu r^2} L_{(n-l)/2}^{l+1/2}\left(2\nu r^2\right) \quad (7)$$

with the associated Laguerre polynomials $L_{(n-l)/2}^{l+1/2}$ and the normalization constant

$$N_{nl} = \sqrt{\sqrt{\frac{2\nu^3}{\pi}}\frac{2((n-l)/2)!\nu^l}{((n+l)/2+1)!!}}. \quad (8)$$

$Y_{lm}(\theta,\phi)$ is the well-known spherical harmonics.

Armed with these, the expectation value of the given Hamiltonian can be calculated. In the variational method, one chooses a trial wave function depending on one or more parameters and then finds the values of these parameters by minimizing the expectation value of the Hamiltonian. It is a good tool for finding ground state energies but as well as energies of excited states. The condition for obtaining excited states energies is that the trial wave function should be orthogonal to all the energy eigenfunctions corresponding to states having a lower energy than the energy level considered. In (7), ν is treated as a variational parameter and it is determined for each state by minimizing the expectation value of the Hamiltonian.

In the following sections we study power-law and logarithmic potentials in order to obtain full spectrum.

3. Mass Spectra of Power-Law and Logarithmic Potentials

Power-law potential is given by [26]

$$V(r) = -8.064\text{ GeV} + 6.898\text{ GeV}\; r^{0.1}. \quad (9)$$

They showed that upsilon and charmonium spectra can be fitted with that potential. The small power of r refers to a situation in which the spacing of energy levels is independent of the quark masses. This situation is also valid for the purely logarithmic potential [30]

$$V(r) = -0.6635\text{ GeV} + 0.733\text{ GeV}\;\ln(r\times 1\text{ GeV}). \quad (10)$$

At first step we obtained spin averaged mass spectrum for $c\bar{c}$ and $b\bar{b}$ systems, respectively. The constituent quark masses are $m_c = 1.8$ GeV and $m_b = 5.174$ GeV for power-law potential and $m_c = 1.5$ GeV and $m_b = 4.906$ GeV for logarithmic potential. Table 1 shows the charmonium spectrum and Table 2 shows the bottomonium spectrum.

Since the interquark potential does not contain the spin dependent part, (2) gives the spin averaged mass for the corresponding states. The calculated masses agree well with the available experimental data and with the values obtained from other theoretical studies. A general potential usually includes spin-spin interaction, spin-orbit interaction, and tensor force terms. To obtain whole picture, it is necessary to consider spin dependent terms within the potential. For $l \geq 1$, there are spin-orbit and tensor force terms which contribute to the fine structure. For equal mass m, the spin-orbit interaction is given by

$$V_{SO} = 2\frac{\alpha_s}{m_q^2 r^3}\left(3\left(\mathbf{S}_1+\mathbf{S}_2\right)\cdot\mathbf{L}\right) \quad (11)$$

and is responsible for the P wave splittings. Again for equal mass m, the tensor potential is given by

$$V_T = \frac{4}{3}\frac{\alpha_s}{m_q^2 r^3}\left(\frac{3\left(\mathbf{S}_1\cdot\mathbf{r}\right)\left(\mathbf{S}_2\cdot\mathbf{r}\right)}{r^2} - \mathbf{S}_1\cdot\mathbf{S}_2\right). \quad (12)$$

For $l = 0$, there is spin-spin term which we will consider in the present work. In the model of the spin averaged mass spectra discussion, all the spin dependent effects are ignored and hence it fails to take into account the splittings due to spin. For example, such splitting exist between the $\eta_c(1S)$ and J/ψ mesons by $\Delta m \simeq 110$ MeV. These mesons occupy the $l = 0$ level. The $c\bar{c}$ in the $\eta_c(1S)$ have $s = 0$, while in the J/ψ, $s = 1$. As a result of this, the mass difference should be related to spin dependent interaction.

3.1. Spin-Spin Interaction. Mass splitting is closely connected with the Lorentz-structure of the quark potential [45]. The origin of the spin-spin interaction term lies in the one-gluon exchange term which is related to $1/r$. Spin is proportional of the magnetic moment of a particle. Magnetic moments generate short range fields $\sim 1/r^3$. In the case of heavy quarkonium systems which are nonrelativistic, wave functions of two particles overlap in a significant amount. This means that particles are very close to each other. So spin-spin

TABLE 2: Spin-averaged mass spectrum of Bottomonium (in MeV).

State	Power	Logarithmic	[15]	[12]
1S	9473	9444	9443	9523
2S	10049	10033	9729	10035
3S	10384	10357	10312	10373
4S	10624	10581	10593	
5S	10813	10753	10840	
6S	10986	10964	11065	

TABLE 3: Charmonium mass spectrum (in MeV). In [18] LP denotes linear potential and SP denotes screened potential.

State	Exp. [43]	Power	Logarithmic	[13]	[11]	[18] LP	[18] SP
$\eta_c(1S)$	2984	2980	2954	2979	2982	2983	2984
$\eta_c(2S)$	3639	3624	3555	3623	3630	3635	3637
$\eta_c(3S)$		3983	3887	3991	4043	4048	4004
$\eta_c(4S)$		4240	4117	4250	4384	4388	4264
$\eta_c(5S)$		4441	4294	4446		4690	4459
J/ψ	3097	3096	3104	3097	3090	3097	3097
$\psi(2S)$	3686	3727	3689	3673	3672	3679	3679
$\psi(3S)$	4040	4078	4011	4022	4072	4078	4030
$\psi(4S)$		4328	4233	4273	4406	4412	4281
$\psi(5S)$		4525	4403	4463		4711	4472

interactions play a significant role in the dynamics. The *spin-spin* interaction term of two particles can be written as

$$V_{SS}(r) = \frac{32\pi\alpha_s}{9m_q m_{\bar{q}}}\mathbf{S_q}\cdot\mathbf{S_{\bar{q}}}\delta(\mathbf{r}). \quad (13)$$

This term can explain *s* wave splittings and has no contribution to $l \neq 0$ states. Putting this term into Schrödinger equation we get

$$E_{HF} = \frac{32\pi\alpha_s}{9m_q m_{\bar{q}}}\int d^3r\Psi^*(\mathbf{r})\Psi(\mathbf{r})\delta(\mathbf{r})\left\langle\mathbf{S_q}\cdot\mathbf{S_{\bar{q}}}\right\rangle. \quad (14)$$

Implementing Dirac-delta function property

$$\int f(x)\delta(x)\,dx = f(0), \quad (15)$$

we obtain

$$E_{HF} = \frac{32\pi\alpha_s}{9m_q m_{\bar{q}}}|\Psi(0)|^2\left\langle\mathbf{S_q}\cdot\mathbf{S_{\bar{q}}}\right\rangle. \quad (16)$$

The matrix element of spin products can be obtained via

$$\mathbf{S}_1\cdot\mathbf{S}_2 = \frac{1}{2}\left(\mathbf{S}^2-\mathbf{S}_1^2-\mathbf{S}_2^2\right) = \frac{1}{2}\left(S(S+1)-\frac{3}{2}\right) \quad (17)$$

so that

$$\left\langle\mathbf{S_q}\cdot\mathbf{S_{\bar{q}}}\right\rangle = \begin{cases}\frac{1}{4}, & \text{for } \vec{S}=1\\ -\frac{3}{4}, & \text{for } \vec{S}=0.\end{cases} \quad (18)$$

Therefore we obtain hyperfine splittings energy as

$$E_{HF} = \begin{cases}\frac{8\pi\alpha_s}{9m_q m_{\bar{q}}}|\Psi(0)|^2, & \text{for } \vec{S}=1\\ -\frac{8\pi\alpha_s}{3m_q m_{\bar{q}}}|\Psi(0)|^2, & \text{for } \vec{S}=0.\end{cases} \quad (19)$$

Here $\Psi(0)$ is the wave function at the origin and can be obtained by the following relation:

$$|\Psi(0)|^2 = \frac{\mu}{2\pi\hbar}\left\langle\frac{dV(r)}{dr}\right\rangle. \quad (20)$$

Expectation value is obtained by the wave function given in (6). S-wave charmonium and bottomonium masses can be seen in Tables 3 and 4. In this calculation, α_s is taken to be 0.37 for charmonium and 0.26 for bottomonium [15].

The mass differences are shown in Tables 5 and 6 for charmonium and bottomonium, respectively.

As can be seen from Tables 3, 4, 5, and 6 our results are compatible with both experimental and theoretical results.

The Regge trajectories for pseudoscalar and vector mesons are shown in Figures 1 and 2 for charmonium and in Figures 3 and 4 for bottomonium.

As can be seen from figures, Regge trajectories show nonlinear behaviour.

4. Dynamical Properties

4.1. Decay Constants. Leptonic decay constants give information about short distance structure of hadrons. In the experiments this regime is testable since the momentum transfer is very large. The pseudoscalar (f_p) and the vector

TABLE 4: Bottomonium mass spectrum (in MeV).

State	Exp. [43]	Power	Logarithmic	[14]	[18]	[44]	[16]
$\eta_b(1S)$	9399	9452	9420	9389	9390	9402	9455
$\eta_b(2S)$	9999	10030	10011	9987	9990	9976	9990
$\eta_b(3S)$		10367	10338	10330	10326	10336	10330
$\eta_b(4S)$		10608	10562	10595	10584	10623	
$\eta_b(5S)$		10798	10735	10817	10800	10869	
$\eta_b(6S)$		11005	10990	11011	10988	11097	
$\Upsilon(1S)$	9460	9480	9452	9460	9460	9465	9502
$\Upsilon(2S)$	10023	10055	10040	10016	10015	10003	10015
$\Upsilon(3S)$	10355	10393	10364	10351	10343	10354	10349
$\Upsilon(4S)$	10579	10629	10588	10611	10597	10635	10607
$\Upsilon(5S)$	10865	10818	10759	10831	10811	10878	10818
$\Upsilon(6S)$	11019	11019	11006	11023	10997	11102	10995

TABLE 5: Mass differences of S-wave charmonium states (in MeV).

State	Exp. [43]	Power	Logarithmic	[13]	[11]	[18] LP	[18] SP
J/ψ-$\eta_c(1S)$	113	116	150	118	108	114	113
$\psi(2S)$-$\eta_c(2S)$	47	103	134	50	42	44	42
$\psi(3S)$-$\eta_c(3S)$		95	124	31	29	30	26
$\psi(4S)$-$\eta_c(4S)$		88	116	23	22	24	17
$\psi(5S)$ - $\eta_c(5S)$		84	109	17		21	13

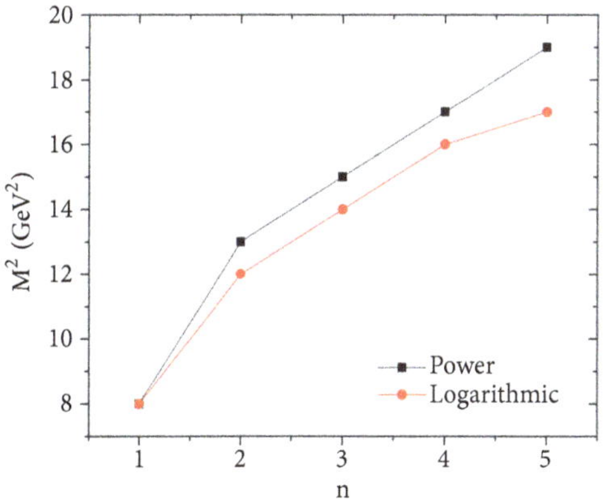

FIGURE 1: Regge trajectories of pseudoscalar charmonium in (n, M^2) plane. The polynomial fit is $M^2 = -0.397857\ n^2 + 5.04014\ n + 4.356$ (GeV^2) for power potential and $M^2 = -0.382143\ n^2 + 4.65786\ n + 4.55$ (GeV^2) for logarithmic potential.

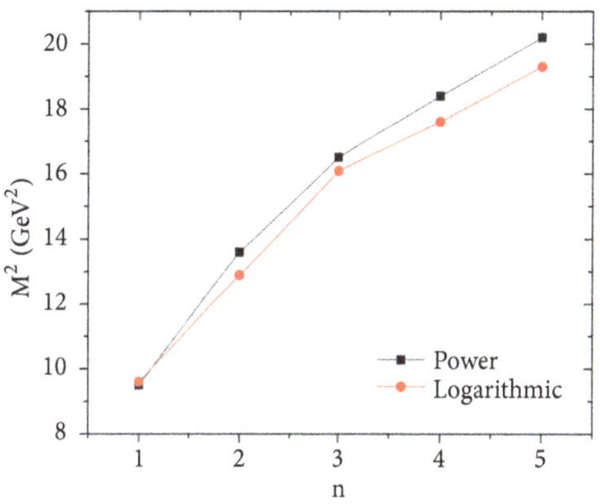

FIGURE 2: Regge trajectories of vector charmonium in (n, M^2) plane. The polynomial fit is $M^2 = -0.405\ n^2 + 5.091\ n + 4.986$ (GeV^2) for power potential and $M^2 = -0.403571\ n^2 + 4.80643\ n + 5.316$ (GeV^2) for logarithmic potential.

(f_v) decay constants are defined, respectively, through the matrix elements [12]

$$p^\mu f_p = i \left\langle 0 \left| \overline{\Psi} \gamma^\mu \gamma^5 \Psi \right| p \right\rangle \quad (21)$$

and

$$m_v f_v \epsilon^\mu = \left\langle 0 \left| \overline{\Psi} \gamma^\mu \Psi \right| v \right\rangle . \quad (22)$$

In the first relation, p^μ is meson momentum and $|p\rangle$ is pseudoscalar meson state. In the second relation, m_v is mass, ϵ^μ is the polarization vector, and $|v\rangle$ is the state vector of meson.

The matrix elements can be calculated by quark model wave function in the momentum space. The result is

$$f_p = \sqrt{\frac{3}{m_p}} \int \frac{d^3k}{(2\pi)^3} \sqrt{1 + \frac{m_q}{E_k}} \sqrt{1 + \frac{m_{\bar{q}}}{E_{\bar{k}}}} \left(1 - \frac{k^2}{\left(E_k + m_q\right)\left(E_{\bar{k}} + m_{\bar{q}}\right)} \right) \phi\left(\vec{k}\right) \quad (23)$$

TABLE 6: Mass differences of S-wave bottomonium states (in MeV).

State	Exp. [43]	Power	Log	[14]	[18]	[44]	[16]
$\Upsilon(1S)$-$\eta_b(1S)$	61	28	32	71	70	63	47
$\Upsilon(2S)$-$\eta_b(2S)$	24	25	29	29	25	27	25
$\Upsilon(3S)$-$\eta_b(3S)$		26	26	21	17	18	19
$\Upsilon(4S)$-$\eta_b(4S)$		21	26	16	13	12	
$\Upsilon(5S)$-$\eta_b(5S)$		20	24	14	11	9	
$\Upsilon(6S)$-$\eta_b(6S)$	14	16	12	9			

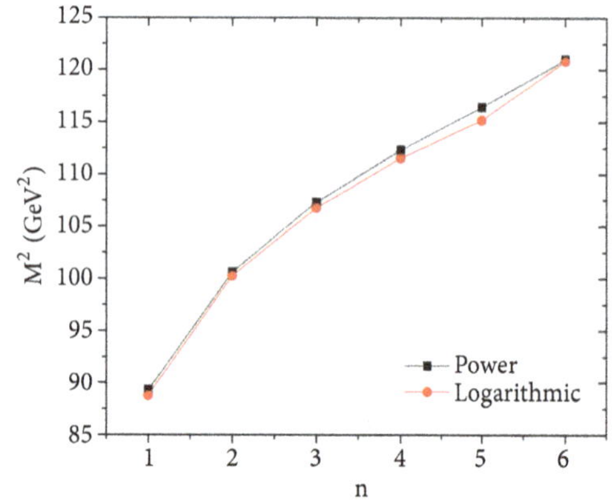

FIGURE 3: Regge trajectories of pseudoscalar bottomonium in (n, M^2) plane. The polynomial fit is $M^2 = -0.79\ n^2 + 11.5586\ n + 79.36$ (GeV^2) for power potential and $M^2 = -0.636071\ n^2 + 10.3054\ n + 80.92$ (GeV^2) for logarithmic potential.

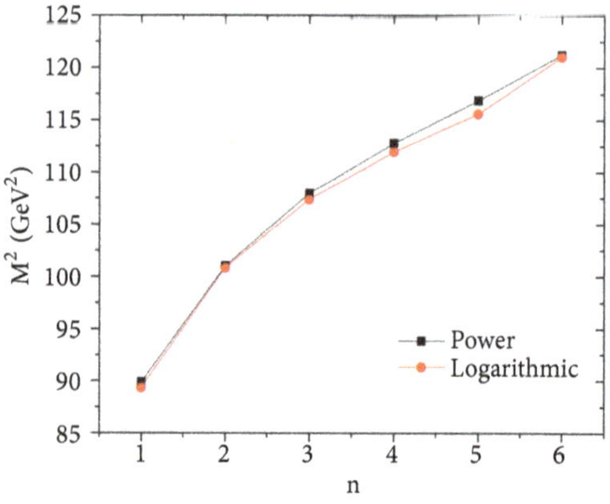

FIGURE 4: Regge trajectories of vector bottomonium in (n, M^2) plane. The polynomial fit is $M^2 = -0.809286\ n^2 + 11.6401\ n + 79.812$ (GeV^2) for power potential and $M^2 = -0.7475\ n^2 + 11.1579\ n + 79.936$ (GeV^2) for logarithmic potential.

for pseudoscalar meson and

$$f_v = \sqrt{\frac{3}{m_v}} \int \frac{d^3k}{(2\pi)^3} \sqrt{1 + \frac{m_q}{E_k}} \sqrt{1 + \frac{m_{\bar{q}}}{E_{\bar{k}}}} \left(1 + \frac{k^2}{3\left(E_k + m_q\right)\left(E_{\bar{k}} + m_{\bar{q}}\right)}\right) \phi\left(\vec{k}\right) \tag{24}$$

for the vector meson [12].

In the nonrelativistic limit, these two equations take a simple form which is known to be Van Royen and Weisskopf relation [46] for the meson decay constants

$$f_{p/v}^2 = \frac{12\left|\Psi_{p/v}(0)\right|^2}{m_{p/v}}. \tag{25}$$

The first-order correction which is also known as QCD correction factor is given by

$$\bar{f}_{p/v}^2 = \frac{12\left|\Psi_{p/v}(0)\right|^2}{m_{p/v}} C^2(\alpha_s) \tag{26}$$

where $C(\alpha_s)$ is given by [47]

$$C(\alpha_s) = 1 - \frac{\alpha_s}{\pi}\left(\Delta_{p/v} - \frac{m_q - m_{\bar{q}}}{m_q + m_{\bar{q}}} \ln \frac{m_q}{m_{\bar{q}}}\right). \tag{27}$$

Here $\Delta_p = 2$ for pseudoscalar mesons and $\Delta_v = 8/3$ for vector mesons. Decay constants are given in Tables 7 and 8 for pseudoscalar and vector mesons, respectively.

4.2. Leptonic Decay Widths. Leptonic decay of a vector meson with $J^{PC} = 1^{--}$ quantum numbers can be pictured by the following annihilation via a virtual photon

$$V(q\bar{q}) \longrightarrow \gamma \longrightarrow e^+e^-. \tag{28}$$

This state is neutral and in principle can decay into a different lepton pair rather than electron-positron pair. The above amplitude can be calculated by the Van Royen and Weisskopf relation [46]

$$\Gamma\left(n^3S_1 \longrightarrow e^+e^-\right) = \frac{16\pi\alpha^2 e_q^2 |\Psi(0)|^2}{m_n^2} \times \left(1 - \frac{16\alpha_s}{3\pi} + \cdots\right), \tag{29}$$

where $\alpha = 1/137$ is the fine structure constant, e_q is the quark charge, m_n is the mass of n^3S_1 state, and $|\Psi_{p/v}(0)|$ is the wave function at the origin of initial state. The term in the parenthesis is the first-order QCD correction factor while $\cdots$ represents higher corrections. The obtained values for leptonic decay widths can be found in Tables 9 and 10 for charmonium and bottomonium, respectively.

TABLE 7: Pseudoscalar decay constants (in MeV).

State	Exp. [43]	Power f_P	Power $\overline{f_P}$	Logarithmic f_P	Logarithmic $\overline{f_P}$	[15] f_P	[15] $\overline{f_P}$	[12]
$\eta_c(1S)$	335 ± 75	543	415	578	442	471	360	402
$\eta_c(2S)$		473	362	497	380	374	286	240
$\eta_c(3S)$		330	252	442	338	332	254	193
$\eta_c(4S)$		325	248	412	315	312	239	
$\eta_c(5S)$		253	193	387	304			
$\eta_b(1S)$		517	431	585	488	834	694	599
$\eta_b(2S)$		479	400	535	447	567	472	411
$\eta_b(3S)$		345	288	504	421	508	422	354
$\eta_b(4S)$		313	261	482	402	481	401	
$\eta_b(5S)$		283	236	465	388			
$\eta_b(6S)$		208	186	434	374			

TABLE 8: Vector decay constants (in MeV).

State	Exp. [43]	Power f_v	Power $\overline{f_v}$	Logarithmic f_v	Logarithmic $\overline{f_v}$	[15] f_P	[15] $\overline{f_P}$	[12]
J/ψ	335 ± 75	529	363	563	386	462	317	393
$\psi(2S)$	279 ± 8	463	318	487	334	369	253	293
$\psi(3S)$	174 ± 18	324	222	436	299	329	226	258
$\psi(4S)$		319	219	406	279	310	212	
$\psi(5S)$		248	170	382	262	290	199	
$\Upsilon(1S)$	708 ± 8	516	402	584	455	831	645	665
$\Upsilon(2S)$	482 ± 10	482	373	535	416	566	439	475
$\Upsilon(3S)$	346 ± 50	350	269	504	393	507	393	418
$\Upsilon(4S)$	325 ± 60	316	243	482	375	481	373	388
$\Upsilon(5S)$	369 ± 93	285	222	464	362	458	356	367
$\Upsilon(6S)$		241	203	442	354	439	341	

TABLE 9: Charmonium leptonic decay widths (in keV). The widths calculated with and without QCD corrections are denoted by $\Gamma_{l^+l^-}$ and $\Gamma^0_{l^+l^-}$.

	Power		Logarithmic		[13]		[15]		Exp. [43]
State	$\Gamma^0_{l^+l^-}$	$\Gamma_{l^+l^-}$	$\Gamma^0_{l^+l^-}$	$\Gamma_{l^+l^-}$	$\Gamma^0_{l^+l^-}$	$\Gamma_{l^+l^-}$	$\Gamma^0_{l^+l^-}$	$\Gamma_{l^+l^-}$	
J/ψ	3.435	1.277	3.154	1.173	11.8	6.60	6.847	2.536	5.55 ± 0.14 ± 0.02
$\psi(2S)$	2.880	1.071	2.362	0.878	4.29	2.40	3.666	1.358	2.33 ± 0.07
$\psi(3S)$	2.153	0.800	1.888	0.702	2.53	1.42	2.597	0.962	0.86 ± 0.07
$\psi(4S)$	1.839	0.684	1.642	0.610	1.73	0.97	2.101	0.778	0.58 ± 0.07
$\psi(5S)$	1.590	0.591	1.551	0.576	1.25	0.70	1.710	0.633	

TABLE 10: Bottomonium leptonic decay widths (in keV). The widths calculated with and without QCD corrections are denoted by $\Gamma_{l^+l^-}$ and $\Gamma^0_{l^+l^-}$.

	Power		Logarithmic		[25]		[15]		Exp. [43]
State	$\Gamma^0_{l^+l^-}$	$\Gamma_{l^+l^-}$	$\Gamma^0_{l^+l^-}$	$\Gamma_{l^+l^-}$	$\Gamma^0_{l^+l^-}$	$\Gamma_{l^+l^-}$	$\Gamma^0_{l^+l^-}$	$\Gamma_{l^+l^-}$	
$\Upsilon(1S)$	0.817	0.456	0.847	0.473	2.31	1.60	1.809	0.998	1.340 ± 0.018
$\Upsilon(2S)$	0.686	0.383	0.709	0.396	0.92	0.64	0.797	0.439	0.612 ± 0.011
$\Upsilon(3S)$	0.610	0.340	0.630	0.352	0.64	0.44	0.618	0.341	0.443 ± 0.008
$\Upsilon(4S)$	0.557	0.311	0.576	0.322	0.51	0.35	0.541	0.298	0.272 ± 0.029
$\Upsilon(5S)$	0.526	0.294	0.535	0.299	0.42	0.29	0.481	0.265	0.31 ± 0.07
$\Upsilon(6S)$	0.492	0.278	0.501	0.282	0.31	0.22	0.432	0.238	0.130 ± 0.030

4.3. Two-Photon Decay Width. $^1S_0$ states with $J^{PC} = 0^{-+}$ quantum number of charmonium and bottomonium can decay into two photons. In the nonrelativistic limit, the decay width for $^1S_0$ state can be written as [48]

$$\Gamma\left(^1S_0 \longrightarrow \gamma\gamma\right) = \frac{12\pi\alpha^2 e_q^4 |\Psi(0)|^2}{m_q^2} \times \left(1 - \frac{3.4\alpha_s}{\pi}\right). \quad (30)$$

The term in the parenthesis is the first-order QCD radiative correction. The results are listed in Table 11.

4.4. Two-Gluon Decay Width. Two-gluon decay width is given by [48]

$$\Gamma\left(^1S_0 \longrightarrow gg\right) = \frac{8\pi\alpha_s^2 |\Psi(0)|^2}{3m_q^2} \times \begin{cases} \left(1 + 4.8\alpha_s\pi\right) & \text{for } \eta_c \\ \left(1 + 4.4\alpha_s\pi\right) & \text{for } \eta_b. \end{cases} \quad (31)$$

The terms in the parenthesis refer to QCD corrections. The obtained results are given in Table 12.

4.5. M1 Transitions. M1 (magnetic dipole transition) decay widths can give more information about spin-singlet states. Moreover M1 transition rates show the validity of theory against experiment [11]. Magnetic transitions conserve both parity and orbital angular momentum of the initial and final states but in the M1 transitions the spin of the state changes. M1 width between two S-wave states is given by [51]

$$\Gamma\left(n^3S_1 \longrightarrow n'^1S_0 + \gamma\right) = \frac{4\alpha e_q^2 E_\gamma^3}{3m_q^2}\left(2J_f + 1\right)\left|\left\langle f \left| j_0\left(\frac{kr}{2}\right)\right| i\right\rangle\right|^2, \quad (32)$$

where $E_\gamma = (M_i^2 - M_f^2)/2M_i$ is the photon energy and $j_0(x)$ is the zeroth-order spherical Bessel function. In the case of small E_γ, spherical Bessel function $j_0(kr/2)$ tends to 1, $j_0(kr/2) \longrightarrow 1$. Thus transitions between the same principal quantum numbers, $n' = n$, are favored and usually known to be *allowed*. In the other case, when $n' \neq n$ the overlap integral between initial and final state is 0 and generally designated as *forbidden* transitions. The obtained transition rates for the allowed ones of S-wave charmonium and bottomonium states are given in Tables 13 and 14, respectively.

5. Results and Discussion

In the present paper we studied S-wave heavy quarkonium spectra with two phenomenological potentials. We have computed spin averaged masses, hyperfine splittings, Regge trajectories for pseudoscalar and vector mesons, decay constants, leptonic decay widths, two-photon and gluon decay widths, and allowed M1 partial widths of S-wave heavy quarkonium states.

In general, most of the quark model potentials tend to be similar, having a Coulomb term and a linear term. For example, in [11] they used standard color Coulomb plus linear scalar form, and also included a Gaussian smeared contact hyperfine interaction in the zeroth-order potential. In [13], the authors used a nonrelativistic potential model with screening effect. In [18] nonrelativistic linear potential and screened potential, in [14, 16, 44] a modified of nonrelativistic potential models and in [15] Hulthen potential are used. Potential models give reliable results with the appropriate parameters in the model. Therefore, the shape of the potential at the limits $r \longrightarrow 0$ and $r \longrightarrow \infty$ have similar behaviours.

Spin averaged mass spectra give idea about the formulation of model since the results are close to experimental values due to contributions from spin dependent interactions are small compared to contribution from potential part. If one ignores all spin dependent interactions, obtained results under this assumption are thought to be averages over related spin states for principal quantum number. Including hyperfine interaction, we obtained the mass spectra for pseudoscalar and vector mesons. The obtained spectra for both charmonium and bottomonium are in good agreement with the experimentally observed spectra and other theoretical studies.

Both power and logarithmic potentials produced approximately same mass differences and are in agreement with experiment for the lowest state in charmonium sector. But for the highest states, the shift is not compatible with the references. The reason for this should be the behaviour of linear part of the potential. In the case of bottomonium sector, mass differences of both power and logarithmic potentials are in accord with the given studies except the lowest state.

The fundamental point in the Regge trajectories is that they can predict masses of unobserved states. For the hadrons constituting of light quarks, the Regge trajectories are approximately linear but for the heavy quarkonium case Regge trajectories can be nonlinear. In the present work, we found that all Regge trajectories show nonlinear properties.

The decay constants are calculated for pseudoscalar and vector mesons by equating their field theoretical definition with the analogous quark model potential definition. This is valid in the nonrelativistic and weak binding limits where quark model state vectors form good representations of the Lorentz group [52, 53]. For pseudoscalar mesons, the corrected value of power and logarithmic potentials are a few MeV above than the available experimental data. For the rest of the pseudoscalar mesons, obtained results are compatible with other studies. In the case of vector mesons, logarithmic potential gave higher values than power potential. In the Υ meson, when the radially states are excited, both two potential gave similar results within the error of experimental value. Computations of the vector decay constant beyond the weak binding limit can be important in the quark potential model frame and need more elaboration [12].

Obtained leptonic decay widths are comparable with the experimental values and other theoretical studies. The QCD corrected factors are more close to experimental values for power and logarithmic potential and this can be referred as the importance of the QCD correction factor in calculating

TABLE 11: Two-photon decay widths (in keV). The widths calculated with and without QCD corrections are denoted by $\Gamma_{\gamma\gamma}$ and $\Gamma^0_{\gamma\gamma}$.

State	Power		Logarithmic		[15]		[8]	[12]	Exp. [43]
	$\Gamma^0_{\gamma\gamma}$	$\Gamma_{\gamma\gamma}$	$\Gamma^0_{\gamma\gamma}$	$\Gamma_{\gamma\gamma}$	$\Gamma^0_{\gamma\gamma}$	$\Gamma_{\gamma\gamma}$	$\Gamma_{\gamma\gamma}$	$\Gamma_{\gamma\gamma}$	
$\eta_c(1S)$	1.10	0.664	1.450	0.869	11.17	6.668	3.69	7.18	$7.2 \pm 0.7 \pm 0.2$
$\eta_c(2S)$	0.987	0.592	1.291	0.774	8.48	5.08	1.4	1.71	
$\eta_c(3S)$	0.907	0.543	1.184	0.710	7.57	4.53	0.930	1.21	
$\eta_c(4S)$	0.847	0.508	1.105	0.662			0.720		
$\eta_c(5S)$	0.801	0.480	1.044	0.620					
$\eta_b(1S)$	0.277	0.199	0.277	0.199	0.58	0.42	0.214	0.45	
$\eta_b(2S)$	0.212	0.153	0.246	0.177	0.29	0.20	0.121	0.11	
$\eta_b(3S)$	0.195	0.142	0.226	0.162	0.24	0.17	0.09	0.063	
$\eta_b(4S)$	0.188	0.136	0.211	0.151			0.07		
$\eta_b(5S)$	0.176	0.129	0.199	0.143					
$\eta_b(6S)$	0.164	0.116	0.182	0.134					

TABLE 12: Two-gluon decay widths (in MeV). The widths calculated with and without QCD corrections are denoted by Γ_{gg} and Γ^0_{gg}.

State	Power		Logarithmic		[15]		[49]	Exp. [43]
	Γ^0_{gg}	Γ_{gg}	Γ^0_{gg}	Γ_{gg}	Γ^0_{gg}	Γ_{gg}	Γ^0_{gg}	Γ_{gg}
$\eta_c(1S)$	32.04	50.15	41.93	32.44	50.82	66.68	15.70	26.7 ± 3.0
$\eta_c(2S)$	28.55	44.70	37.32	24.64	38.61	5.08	8.10	14 ± 7
$\eta_c(3S)$	26.22	41.04	34.23	53.59	21.99			
$\eta_c(4S)$	24.50	38.35	31.96	50.03				
$\eta_c(5S)$	23.15	36.24	30.18	47.24				
$\eta_b(1S)$	5.50	7.50	12.82	17.49	13.72	18.80	11.49	
$\eta_b(2S)$	4.90	6.69	11.41	15.56	6.73	9.22	5.16	
$\eta_b(3S)$	4.50	6.14	10.46	14.28	5.58	7.64	3.80	
$\eta_b(4S)$	4.20	5.74	9.77	13.33				
$\eta_b(5S)$	3.97	5.42	9.22	12.58				
$\eta_b(6S)$	3.62	5.18	8.68	10.86				

TABLE 13: Radiative M1 decay widths of charmonium. In [18] LP stands for linear potential and SP stands for screened potential.

Initial	Final	Power		Logarithmic		[15]	[18]		Exp. [43]
		E_γ (MeV)	Γ (keV)	E_γ (MeV)	Γ (keV)	Γ (keV)	Γ_{LP} (keV)	Γ_{SP} (keV)	Γ (keV)
J/ψ	$\eta_c(1S)$	114.9	1.96	113.8	2.83	3.28	2.39	2.44	1.13 ± 0.35
$\psi(2S)$	$\eta_c(2S)$	111.5	1.39	101.5	2.01	1.45	0.19	0.19	
$\psi(3S)$	$\eta_c(3S)$	93.8	1.10	93.8	1.59		0.051	0.088	
$\psi(4S)$	$\eta_c(4S)$	87.1	0.88	87.1	1.27				
$\psi(5S)$	$\eta_c(5S)$	83.2	0.74	83.2	1.10				

TABLE 14: Radiative M1 decay widths of bottomonium.

Initial	Final	Power		Logarithmic		[10]	[44]	[16]	Exp. [43]
		E_γ (MeV)	Γ (eV)	E_γ (MeV)	Γ (eV)	Γ (eV)	Γ (eV)	Γ (eV)	Γ (eV)
$\Upsilon(1S)$	$\eta_b(1S)$	27.9	0.88	31.9	1.46	5.8	10	9.34	
$\Upsilon(2S)$	$\eta_b(2S)$	24.9	0.62	28.9	1.09	1.4	0.59	0.58	
$\Upsilon(3S)$	$\eta_b(3S)$	25.9	0.54	25.9	0.78	0.8	0.25	0.66	
$\Upsilon(4S)$	$\eta_b(4S)$	20.9	0.37	20.9	0.41				
$\Upsilon(5S)$	$\eta_b(5S)$	19.9	0.32	19.9	0.35				
$\Upsilon(6S)$	$\eta_b(6S)$	14.3	0.29	14.4	0.27				

TABLE 15: Exotic states. Experimental data are taken from [43] unless stated. The units for mass and strong decays are in MeV and two-photon decay is in keV.

	Mass			Strong decay			Two-photon decay		
	Power	Logarithmic	Experiment	Power	Logarithmic	Experiment	Power	Logarithmic	Experiment
$X(3940)$			$3942^{+7}_{-6} \pm 6$			$37^{+26}_{-15} \pm 8$			
$\eta_c(3S)$	3983	3887		41.04	53.59		0.543	0.710	
$X(4160)$			4191 ± 5			70 ± 10			0.48 ± 0.22 [50]
$\eta_c(4S)$	4240	4117		38.35	50.03		0.508	0.612	
$\psi(4415)$			4421 ± 4			62 ± 20			0.58 ± 0.07
$\eta_c(5S)$	4441	4297		36.24	47.24		0.480	0.620	

the decay constants and other short range phenomena using potential models.

$^1S_0$ levels of charmonium and bottomonium states can decay into two photons or gluons. Especially two- photon decays of these levels are important for understanding the accuracy of theoretical models. Obtained results are smaller than the nonrelativistic widths including the one-loop QCD correction factor. For example, results of power and logarithmic potentials in $\eta_c(1S)$ are not in accord with experimental data. The reason of these differences can be due to the static potential between quarks that we used in the solution of two-body Schrödinger equation. For higher states, power and logarithmic potentials results are comparable with other studies. Two-photon decays are complicated processes such as pseudoscalar meson decay to two photons is governed by an intermediate vector meson followed by a meson dominance transition to a photon [12]. These schematic diagrams must be added to calculations to obtain a whole picture. For two-gluon decay widths, two phenomenological potentials gave comparable results with the available experimental data. Notice that in some cases QCD corrected factor is in accord with the experimental data whereas in some cases it is not. The reason for this can be that, there are significant radiative corrections from three-gluon decays so computing only two-gluon decay width could not explain the mechanism in all details.

Finally M1 transitions are calculated. The M1 radiative decay rates are very sensitive to relativistic effects. Even for allowed transitions relativistic and nonrelativistic results differ significantly. An important example is the decay of $J/\psi \longrightarrow \eta_c\gamma$. The nonrelativistic predictions for its rate are more than two times larger than the experimental data [10]. In the charmonium sector, the available experimental data for $J/\psi \longrightarrow \eta_c(1S)$ is comparable with the power potential result, while logarithmic potential result is 1 eV higher. In the bottomonium sector, there is no experimental data available on M1 transitions. Since photon energies and transition rates are very small, the detection of these transitions is an objection. And this can be a reason why no spin-singlet S-wave levels $\eta_b(n^1S_0)$ have been observed yet [10]. The obtained values for M1 transitions are comparable with the references.

Some states in the charmonium and bottomonium sector show properties different from the conventional quarkonium state. Some examples are $X(3940)$, $X(4160)$, and $\psi(4415)$. For $X(3940)$, there is not much available experimental data and more is needed. Wang et al. studied two-body open charm OZI-allowed strong decays of $X(3940)$ and $X(4160)$ considered as $\eta_c(3S)$ and $\eta_c(4S)$, respectively, by the improved Bethe-Salpeter method combined with the $^3P_0$ [54]. They calculated strong decay width of $X(3940)$ as $\Gamma = (33.5^{+18.4}_{-15.3})$ MeV and $X(4160)$ as $\Gamma = (69.9^{+22.4}_{-21.1})$ MeV where the experimental values are $\Gamma = (37^{+26}_{-15} \pm 8)$ MeV for $X(3940)$ and $\Gamma = (70 \pm 10)$ MeV for $X(4160)$ [43]. They concluded that $\eta_c(3S)$ is a good candidate of $X(3940)$ and $\eta_c(4S)$ is a not good candidate of $X(4160)$ due to larger decay width of $\Gamma(D\overline{D}^*)/\Gamma(D^*\overline{D}^*)$ comparing to experimental data. We give our results comparing to these exotic states in Table 15.

Looking at Table 15, we can deduce that, according to our model and results, we can assign $X(3940)$ as $\eta_c(3S)$, $X(4160)$ as $\eta_c(4S)$, and $\psi(4415)$ as $\eta_c(5S)$. To be more accurate, more data is needed to corroborate whether these states are conventional quarkonium or not.

6. Conclusions

Quark potential models have been very successful to study on various properties of mesons. The short distance behaviour of interquark potential appears to be similar where QCD perturbation theory can be applied where at large distance the potential is linear in r where nonperturbative methods are need to be used. The improvements on the potentials can be made and spin-spin, spin-orbit type interactions can be added to model to arrive high accuracy. The potential model approach is a valuable task, which has given to us many insights into the nature of both heavy and light quarkonium physics. Using a relativistic approach together with a model in which $B\overline{B}$ and $D\overline{D}$ thresholds are taken into account, detailed analysis can be made on various aspects of heavy quarkonium.

Conflicts of Interest

The author declares that they have no conflicts of interest.

References

[1] T. Kawanai and S. Sasaki, "Charmonium potential from full lattice QCD," *Physical Review D: Particles, Fields, Gravitation and Cosmology*, vol. 85, no. 9, 2012.

[2] A. Le Yaouanc, L. Oliver, O. Pène, and J. C. Raynal, "Spontaneous breaking of chiral symmetry for confining potentials," *Physical Review D: Particles, Fields, Gravitation and Cosmology*, vol. 29, no. 6, pp. 1233–1257, 1984.

[3] S. Bolognesi, K. Konishi, M. Shifman, and D. Phys. Rev, "Patterns of symmetry breaking in chiral QCD," *Physical Review D, Covering Particles, Fields, Gravitation, and Cosmology*, vol. 97, no. 9, Article ID 094007, 2018.

[4] T. Appelquist and H. D. Politzer, " Heavy Quarks and ," *Physical Review Letters*, vol. 34, no. 1, pp. 43–45, 1975.

[5] A. De Rújula and S. L. Glashow, "Is bound charm found?" *Physical Review Letters*, vol. 34, no. 1, pp. 46–49, 1975.

[6] E. Eichten, K. Gottfried, T. Kinoshita, K. D. Lane, and T. M. Yan, "Interplay of confinement and decay in the spectrum of charmonium," *Physical Review Letters*, vol. 36, p. 500, 1976.

[7] D. P. Stanley and D. Robson, "Nonperturbative potential model for light and heavy quark-antiquark systems," *Physical Review D: Particles, Fields, Gravitation and Cosmology*, vol. 21, no. 11, pp. 3180–3196, 1980.

[8] S. Godfrey and N. Isgur, "Mesons in a relativized quark model with chromodynamics," *Physical Review D: Particles, Fields, Gravitation and Cosmology*, vol. 32, no. 1, pp. 189–231, 1985.

[9] L. P. Fulcher, "Matrix representation of the nonlocal kinetic energy operator, the spinless Salpeter equation and the Cornell potential," *Physical Review D*, vol. 50, p. 447, 1994.

[10] D. Ebert, R. N. Faustov, and V. O. Galkin, " Properties of heavy quarkonia and ," *Physical Review D: Particles, Fields, Gravitation and Cosmology*, vol. 67, no. 1, 2003.

[11] T. Barnes, S. Godfrey, and E. S. Swanson, "Higher charmonia," *Physical Review D: Particles, Fields, Gravitation and Cosmology*, vol. 72, article 054026, 2005.

[12] O. Lakhina and E. S. Swanson, "Dynamic properties of charmonium," *Physical Review D*, vol. 74, 2006.

[13] B. Q. Li and K. T. Chao, "Higher charmonia and X, Y, Z states with screened potential," *Physical Review D: Particles, Fields, Gravitation and Cosmology*, vol. 79, Article ID 013011, 2009.

[14] B. Q. Li and K. T. Chao, "Bottomonium Spectrum with Screened Potential," *Communications in Theoretical Physics*, vol. 52, pp. 653–661, 2009.

[15] K. B. V. K. Bhaghyesh and A. P. Monteiro, "Heavy quarkonium spectra and its decays in a nonrelativistic model with Hulthen potential," *Journal of Physics G: Nuclear and Particle Physics*, vol. 38, Article ID 085001, 2011.

[16] J. Segovia, P. G. Ortega, D. R. Entem, and F. Fernandez, "Bottomonium spectrum revisited," *Physical Review D*, vol. 93, 2016.

[17] S. Godfrey, K. Moats, and E. S. Swanson, "B and B_s meson spectroscopy," *Physical Review D: Particles, Fields, Gravitation and Cosmology*, vol. 94, no. 5, 2016.

[18] W.-J. Deng, H. Liu, L.-C. Gui, and X.-H. Zhong, "Spectrum and electromagnetic transitions of bottomonium," *Physical Review D: Particles, Fields, Gravitation and Cosmology*, vol. 95, Article ID 074002, 2017.

[19] D. B. Lichtenberg, "Excited quark production at Hadron colliders," *International Journal of Modern Physics A*, vol. 2, no. 6, pp. 1669–1705, 1987.

[20] N. Brambilla, *Quarkonium Working Group*, CERN Yellow Report, 2005, arXiv:hep-ph/0412158.

[21] E. Eichten, K. Gottfried, T. Kinoshita, J. Kogut, K. D. Lane, and T.-M. Yan, "Spectrum of charmed quark-antiquark bound states," *Physical Review Letters*, vol. 34, no. 6, pp. 369–372, 1975.

[22] J. Richardson, "The heavy quark potential and the Υ, J/ψ systems," *Physics Letters B*, vol. 82B, p. 272, 1979.

[23] Y.-Q. Chen and Y.-P. Kuang, "Improved QCD-motivated heavy-quark potentials with explicit Λ_{MS} dependence," *Physical Review D*, vol. 46, p. 1165, 1992.

[24] Y.-Q. Chen and Y.-P. Kuang, "Erratum: "Improved QCD-motivated heavy-quark potentials with explicit Λ_{MS} dependence"," *Physical Review D: Particles, Fields, Gravitation and Cosmology*, vol. 47, p. 35, 1993.

[25] V. Khruschev, V. Savrin, and S. Semenov, "On the parameters of the QCD-motivated potential in the relativistic independent quark model," *Physics Letters B*, vol. 525, no. 3-4, pp. 283–288, 2002.

[26] A. Martin, "A simultaneous fit of bb, cc, ss (bcs Pairs) and cs spectra," *Physics Letters B*, vol. 100, p. 511, 1981.

[27] M. Machacek and Y. Tomozawa, "ψ Phenomenology and the nature of quark confinement," *Annals of Physics*, vol. 110, p. 407, 1978.

[28] G. Fogleman, D. B. Lichtenberg, and J. G. Wills, "Heavy-meson spectra calculated with a one-parameter potential," *Lettere al Nuovo Cimento*, vol. 26, p. 369, 1979.

[29] C. Quigg and J. L. Rosner, "Quantum mechanics with applications to quarkonium," *Physics Reports*, vol. 56, no. 4, pp. 167–235, 1979.

[30] C. Quigg and J. L. Rosner, "Quarkonium level spacings," *Physics Letters B*, vol. 71B, pp. 153–157, 1977.

[31] S. Xiaotong and L. Hefen, "A new phenomenological potential for heavy quarkonium," *Zeitschrift für Physik C Particles and Fields*, vol. 34, no. 2, pp. 223–231, 1987.

[32] C. Quigg, H. B. Thacker, and J. L. Rosner, "Constructive evidence for flavor independence of the quark-antiquark potential," *Physical Review D: Particles, Fields, Gravitation and Cosmology*, vol. 21, Article ID 3393, p. 234, 1980.

[33] W. Buchmüller, Y. Jack Ng, and S.-H. H. Tye, "Hyperfine splittings in heavy-quark systems," *Physical Review D: Particles, Fields, Gravitation and Cosmology*, vol. 24, no. 12, pp. 3312–3314, 1981.

[34] C. Bachas, "Concavity of the quarkonium potential," *Physical Review D: Particles, Fields, Gravitation and Cosmology*, vol. 33, no. 9, pp. 2723–2725, 1986.

[35] S. K. Choi, "Belle collaboration," *Physical Review Letters*, vol. 91, Article ID 262001, 2003.

[36] X.-W. Kang and J. A. Oller, "Different pole structures in line shapes of the X(3872)," *The European Physical Journal C*, vol. 77, no. 6, 2017.

[37] R. R. Horgan, "The construction and classification of wavefunctions for the harmonic oscillator model of three quarks," *Journal of Physics G: Nuclear Physics*, vol. 2, p. 625, 1976.

[38] N. Isgur and G. Karl, "*P*-wave baryons in the quark model," *Physical Review D: Particles, Fields, Gravitation and Cosmology*, vol. 18, no. 11, pp. 4187–4205, 1978.

[39] A. W. Hendry, "Decays of high spin Δ^* and N^* resonances in the quark model," *Annals of Physics*, vol. 140, no. 65, 1982.

[40] A. W. Hendry and D. B. Lichtenberg, "Properties of hadrons in the quark model," *Fortschritte der Physik/Progress of Physics banner*, vol. 33, no. 3, pp. 139–231, 1985.

[41] K. B. V. Kumar, B. Hanumaiah, and S. Pepin, "Meson spectrum in a relativistic harmonic model with instanton-induced interaction," *The European Physical Journal A - Hadrons and Nuclei*, vol. 19, no. 9, p. 247, 2004.

[42] K. B. V. Kumar, M. Y.-L. Bhavyashri, and A. P. Monteiro, "P-wave meson spectrum in a relativistic model with instanton induced interaction," *International Journal of Modern Physics A*, vol. 24, 2009.

[43] M. Tanabashi, "Particle Data Group," *Physical Review D: Particles, Fields, Gravitation and Cosmology*, vol. 98, Article ID 030001, 2018.

[44] S. Godfrey and K. Moats, " Erratum: ," *Physical Review D: Particles, Fields, Gravitation and Cosmology*, vol. 92, no. 11, 2015.

[45] V. Lengyel, Y. Fekete, I. Haysak, and A. Shpenik, "Calculation of hyperfine splitting in mesons using configuration interaction approach," *The European Physical Journal C*, vol. 21, no. 2, pp. 355–359, 2001.

[46] R. Van Royen and V. F. Weisskopf, "Protsyessy raspada adronov i modyel cyrillic small soft sign kvarkov," *Il Nuovo Cimento A*, vol. 50, no. 3, pp. 617–645, 1967.

[47] E. Braaten and S. Fleming, "QCD radiative corrections to the leptonic decay rate of the B_c meson," *Physical Review D*, vol. 52, p. 181, 1995.

[48] W. Kwong, P. B. Mackenzie, R. Rosenfeld, and J. L. Rosner, "Quarkonium annihilation rates," *Physical Review D: Particles, Fields, Gravitation and Cosmology*, vol. 37, no. 11, pp. 3210–3215, 1988.

[49] J. T. Laverty, S. F. Radford, and W. W. Repko, "\ga\ga and g g decay rates for equal mass heavy quarkonia," *High Energy Physics - Phenomenology*, 2009, arXiv:hep-ph/0901.3917v3.

[50] M. Ablikim, J. Z. Bai, Y. Ban, X. Cai, H. F. Chen, and H. S. Chen, "Determination of the $\psi(3770)$, $\psi(4040)$, $\psi(4160)$ and $\psi(4415)$ resonance parameters," *Physics Letters B*, vol. 660, no. 3, pp. 315–319, 2008.

[51] E. Eichten, K. Gottfried, T. Kinoshita, K. D. Lane, and T. -. Yan, "Charmonium: The model," *Physical Review D: Particles, Fields, Gravitation and Cosmology*, vol. 17, no. 11, pp. 3090–3117, 1978.

[52] C. Hayne and N. Isgur, "Beyond the wave function at the origin: Some momentum-dependent effects in the nonrelativistic quark model," *Physical Review D: Particles, Fields, Gravitation and Cosmology*, vol. 25, no. 7, pp. 1944–1950, 1982.

[53] N. Isgur, D. Scora, B. Grinstein, and M. B. Wise, "Semileptonic *B* and *D* decays in the quark model," *Physical Review D: Particles, Fields, Gravitation and Cosmology*, vol. 39, no. 3, pp. 799–818, 1989.

[54] Z.-H. Wang, Y. Zhang, L.-B. Jiang, T.-H. Wang, Y. Jiang, and G.-L. Wang, "The strong decays of X(3940) and X(4160)," *European Physical Journal C*, vol. 77, no. 1, article 43, 2017.

BPS Equations of Monopole and Dyon in $SU(2)$ Yang-Mills-Higgs Model, Nakamula-Shiraishi Models, and their Generalized Versions from the BPS Lagrangian Method

Ardian Nata Atmaja[1] **and Ilham Prasetyo**[1,2]

[1]*Research Center for Physics, Indonesian Institute of Sciences (LIPI), Kompleks Puspiptek Serpong, Tangerang 15310, Indonesia*
[2]*Departemen Fisika, FMIPA, Universitas Indonesia, Depok 16424, Indonesia*

Correspondence should be addressed to Ardian Nata Atmaja; ardi002@lipi.go.id and Ilham Prasetyo; ilham.prasetyo@sci.ui.ac.id

Academic Editor: Diego Saez-Chillon Gomez

We apply the BPS Lagrangian method to derive BPS equations of monopole and dyon in the $SU(2)$ Yang-Mills-Higgs model, Nakamula-Shiraishi models, and their generalized versions. We argue that, by identifying the effective fields of scalar field, f, and of time-component gauge field, j, explicitly by $j = \beta f$ with β being a real constant, the usual BPS equations for dyon can be obtained naturally. We validate this identification by showing that both Euler-Lagrange equations for f and j are identical in the BPS limit. The value of β is bounded to $|\beta| < 1$ due to reality condition on the resulting BPS equations. In the Born-Infeld type of actions, namely, Nakamula-Shiraishi models and their generalized versions, we find a new feature that, by adding infinitesimally the energy density up to a constant $4b^2$, with b being the Born-Infeld parameter, it might turn monopole (dyon) to antimonopole (antidyon) and vice versa. In all generalized versions there are additional constraint equations that relate the scalar-dependent couplings of scalar and of gauge kinetic terms or G and w, respectively. For monopole the constraint equation is $G = w^{-1}$, while for dyon it is $w(G-\beta^2 w) = 1-\beta^2$ which further gives lower bound to G as such $G \geq |2\beta\sqrt{1-\beta^2}|$. We also write down the complete square-forms of all effective Lagrangians.

1. Introduction

Monopole has been known to exist in nonabelian gauge theory. One of the main developments was given by 't Hooft in [1] and in parallel with a work by Polyakov in [2], in which he showed that monopole could arise as soliton in a Yang-Mills-Higgs theory, without introducing Dirac's string [3], by spontaneously breaking the symmetry of $SO(3)$ gauge group into $U(1)$ gauge group. Later on, Julia and Zee showed that a more general configuration of soliton called dyon may exist as well within the same model [4]. Furthermore, the exact solutions were given by Prasad and Sommerfiled in [5] by taking some limit where $V \longrightarrow 0$. These solutions were proved by Bogomolnyi in [6] to be solutions of the first-order differential equations which turn out to be closely related to the study of supersymmetric system [7] (in this article, we shall call the limit $V \longrightarrow 0$ as BPS limit and the first-order differential equations as BPS equations).

At high energy the Yang-Mills theory may receive contributions from higher derivative terms. This can be realized in string theory in which the effective action of open string theory may be described by the Born-Infeld type of actions [8]. However, there are several ways in writing the Born-Infeld action for nonabelian gauge theory because of the ordering of matrix-valued field strength [8–13]. Further complications appear when we add Higgs field into the action. One of examples has been given by Nakamula and Shiraishi in which the action exhibits the usual BPS monopole and dyon [14]. Unfortunately, the resulting BPS equations obviously do not capture essential feature of the Born-Infeld action; namely, there is no dependency over the Born-Infeld parameter. In other examples such as in [15], the monopole's profile depends on the Born-Infeld parameter, but the BPS equations are not known so far.

In this article, we would like to derive the well-known BPS equations of monopole and dyon in the $SU(2)$ Yang-Mills-Higgs model and their Born-Infeld type extensions, which we shall call them Nakamula-Shiraishi models, using a procedure called BPS Lagrangian method developed in [16]. We then extend those models to their generalized versions by adding scalar-dependent couplings to each of the kinetic terms and derive the BPS equations for monopole and dyon. In Section 2, we will first discuss in detail the BPS Lagrangian method. In Section 3, we describe how to get the BPS equations for monopole (dyon) from energy density of the $SU(2)$ Yang-Mills-Higgs model using Bogomolny's trick. We write explicitly its effective action and effective actions of the Nakamula-Shiraishi models by taking the 't Hooft-Polyakov (Julia-Zee) ansatz for monopole(dyon). In Section 4, we use the BPS Lagrangian approach to reproduce the BPS equations for monopole and dyon in the $SU(2)$ Yang-Mills-Higgs model and Nakamula-Shiraishi models. Later, in Section 5, we generalize the $SU(2)$ Yang-Mills-Higgs model by adding scalar-dependent couplings to scalar and gauge kinetic terms and derive the corresponding BPS equations. We also generalize Nakamula-Shiraishi models in Section 6 and derive their corresponding BPS equations. We end with discussion in Section 7.

2. BPS Lagrangian Method

In deriving BPS equations of a model, we normally use the so-called Bogomolny's trick by writing the energy density into a complete square form [6]. However, there are several rigorous methods which have been developed in doing so. The first one is based on Bogomolny's trick by assuming the existence of a homotopy invariant term in the energy density that does not contribute to Euler-Lagrange equations [17]. The second method called first-order formalism which works by solving a first integral of the model, together with stressless condition [18–20]. The third method called On-Shell method which works by adding and solving auxiliary fields into the Euler-Lagrange equations and assuming the existence of BPS equations within the Euler-Lagrange equations [16, 21]. The forth method called First-Order Euler-Lagrange (FOEL) formalism, which is generalization of Bogomolnyi decomposition using a concept of strong necessary condition developed in [22], which works by adding and solving a total derivative term into the Lagrangian [23] (in our opinion the procedure looks similar to the On-Shell method by means that adding total derivative terms into the Lagrangian is equivalent to introducing auxiliary fields in the Euler-Lagrange equations. However, we admit that the procedure is written in a more covariant way). The last method, which we shall call BPS Lagrangian method, works by identifying the (effective) Lagrangian with a BPS Lagrangian such that its solutions of the first-derivative fields give out the desired BPS equations [16]. This method was developed based on the On-Shell method by one of the authors of this article and it is much easier to execute compared to the On-Shell method. We chose to use the BPS Lagrangian method to find BPS equations of all models considered in this article. The method is explained in the following paragraphs.

In general the total static energy of N-fields system, $\vec{\phi} = (\phi_1, \dots, \phi_N)$, with Lagrangian density $\mathscr{L}$, is defined by $E_{\text{static}} = -\int d^d x \mathscr{L}$. Bogomolny's trick explains that the static energy can be rewritten as

$$E_{\text{static}} = \left(\int d^d x \sum_{i=1}^{N} \Phi_i \left(\vec{\phi}, \partial \vec{\phi} \right) \right) + E_{\text{BPS}}, \quad (1)$$

where $\{\Phi_i\}$ is a set of positive-semidefinite functions and E_{BPS} is the boundary contributions defined by $E_{\text{BPS}} = -\int d^d x \mathscr{L}_{\text{BPS}}$. Neglecting the contribution from boundary terms in $\mathscr{L}_{\text{BPS}}$, as they do not affect the Euler-Lagrange equations, configurations that minimize the static energy are also solutions of the Euler-Lagrange equations and they are given by $\{\Phi_i = 0\}$ known as BPS equations. Rewriting the static energy to be in the form of (1) is not always an easy task. However it was argued in [16] that one does not need to know the explicit form of (1) in order to obtain the BPS equations. By realizing that, in the BPS limit, where the BPS equations are assumed to exist, remaining terms in the total static energy are in the form of boundary terms, $E_{\text{static}} = E_{\text{BPS}}$. Therefore we may conclude that BPS equations are solutions of $\mathscr{L} - \mathscr{L}_{BPS} = \sum_{i=1}^{N} \Phi_i(\vec{\phi}, \partial\vec{\phi}) = 0$.

Now let us see in detail what is inside $\mathscr{L}_{\text{BPS}}$. Suppose that in spherical coordinates the system effectively depends on only radial coordinate r. As shown by the On-Shell method on models of vortices [21], the total static energy in the BPS limit can be defined as

$$E_{\text{BPS}} = Q(r \longrightarrow \infty) - Q(r \longrightarrow 0) = \int_{r \longrightarrow 0}^{r \longrightarrow \infty} dQ, \quad (2)$$

where Q is called BPS energy function. The BPS energy function Q does not depend on the coordinate r explicitly; however in general it can also depend on r explicitly in accordance with the chosen ansatz. In most of the cases if we choose the ansatz that does not depend explicitly on coordinate r, then we would have $Q \neq Q(r)$. Hence, with a suitable ansatz, we could write $Q = Q(\tilde{\phi}_1, \dots, \tilde{\phi}_N)$ in which $\tilde{\phi}_i$ is the effective field of ϕ_i as a function of coordinate r only. Assume that Q can be treated with separation of variables

$$Q \equiv \prod_{i=1}^{N} Q_i \left(\tilde{\phi}_i \right), \quad (3)$$

this give us a pretty simple expression of E_{BPS}, i.e.,

$$E_{\text{BPS}} = \int \sum_{i=1}^{N} \frac{\partial Q}{\partial \tilde{\phi}_i} \frac{d\tilde{\phi}_i}{dr} dr, \quad (4)$$

and we could obtain $\mathscr{L}_{\text{BPS}}$ in terms of the effective fields and their first-derivative.

Now we proceed to find the Φ_is from $\mathscr{L} - \mathscr{L}_{BPS} = \sum_{i=1}^{N} \Phi_i(\vec{\phi}, \partial\vec{\phi})$. As we mentioned Φ_i must be positive-semidefinite function and we restrict that it has to be a function of $\vec{\tilde{\phi}}$ and $\partial_r\tilde{\phi}_i$ for each $i = 1,\ldots,N$. The BPS equation $\Phi_i = 0$ gives solutions to $\partial_r\tilde{\phi}_i$ as follows:

$$\partial_r\tilde{\phi}_i = \left\{F_i^{(1)}, F_i^{(2)}, \ldots, F_i^{(m)}\right\}, \tag{5}$$

with $F_i^{(k)} = F_i^{(k)}(\vec{\tilde{\phi}}; r)$ $(k = 1,\ldots,m)$. Positive-semidefinite condition fixes m to be an even number and further there must be even number of equal solutions in $\{F_i^{(k)}\}$. As an example if $m = 2$ for all i, then $\Phi_i = 0$ is a quadratic equation in $\partial_r\tilde{\phi}_i$ and so we will have $F_i^{(1)} = F_i^{(2)}$. The restriction on $\Phi_i \equiv \Phi_i(\partial_r\tilde{\phi}_i)$ forces us to rewrite the function $\mathscr{L} - \mathscr{L}_{BPS}$ into partitions $\sum_{i=1}^N \Phi_i$ explicitly. This is difficult to apply on more general forms of Lagrangian, since there exists a possibility that there are terms with $\partial\tilde{\phi}_i\partial\tilde{\phi}_j$ where $i \neq j$. Another problem is ambiguity in choosing which terms contain nonderivative of fields that should belong to which partitions Φ_i.

For more general situations, the BPS equations can be obtained by procedures explained in [16] which we describe below. On a closer look, we can consider $\mathscr{L} - \mathscr{L}_{\mathrm{BPS}} = 0$ as a polynomial equation of first-derivative fields. Seeing it as the polynomial equation of $\partial_r\tilde{\phi}_1$, whose maximal power is m_1, its roots are

$$\partial_r\tilde{\phi}_1 = \left\{G_1^{(1)}, G_1^{(2)}, \ldots, G_1^{(m_1)}\right\}, \tag{6}$$

with $G_1^{(k)} = G_1^{(k)}(\vec{\tilde{\phi}}, \partial_r\tilde{\phi}_2, \ldots, \partial_r\tilde{\phi}_N; r)$ and $k = 1,\ldots,m_1$. Then we have

$$\Phi_1 \propto \left(\partial_r\tilde{\phi}_1 - G_1^{(1)}\right)\left(\partial_r\tilde{\phi}_1 - G_1^{(2)}\right)\ldots\left(\partial_r\tilde{\phi}_1 - G_1^{(m_1)}\right). \tag{7}$$

As we mentioned before here m_1 must be an even number and to ensure positive-definiteness at least two or more even number of roots must be equal. This will result in some constraint equations that are polynomial equations of the remaining first-derivative fields $(\partial_r\tilde{\phi}_2, \ldots, \partial_r\tilde{\phi}_N)$. Repeate the previous procedures for $\partial_r\tilde{\phi}_2$ until $\partial_r\tilde{\phi}_N$ whose Φ_N is

$$\Phi_N \propto \left(\partial_r\tilde{\phi}_N - G_N^{(1)}\right)\left(\partial_r\tilde{\phi}_N - G_N^{(2)}\right)\cdots\left(\partial_r\tilde{\phi}_N - G_N^{(m_N)}\right), \tag{8}$$

with m_N being also an even number. Now all $G_N^{(k)}$ are only functions of $\vec{\tilde{\phi}}$ and equating some of the roots will become constraint equations that we can solve order by order for each power series of r. As an example let us take $N = 2$ and $m_1, m_2 = 2$. Then the constraint $G_1^{(1)} - G_1^{(2)} = 0$ can be seen as a quadratic equation of $\partial_r\tilde{\phi}_2$. This give us the last constraint $G_2^{(1)} - G_2^{(2)} = 0$. Since the model is valid for all r, we could write the constraint as $G_2^{(1)} - G_2^{(2)} = \sum_n a_n r^n$, where all a_ns are independent of $\partial_r\tilde{\phi}_1$ and $\partial_r\tilde{\phi}_2$. Then all a_ns need to be zero and from them we can find each $Q_i(\tilde{\phi}_i)$. Then the BPS equations for $\partial_r\tilde{\phi}_i$ can be found.

We can see that this more general method is straightforward for any Lagrangian. This will be used throughout this paper, since we will later use some DBI-type Lagrangian that contains terms inside square root which is not easy to write the partitions explicitly. In [16], with particular ansatz for the fields, writing $Q = 2\pi F(f)A(a)$ is shown to be adequate for some models of vortices. Here, we show that the method is also able to do the job, at least for some known models of magnetic monopoles and dyons, using the well-known 't Hooft-Polyakov ansatz.

3. The 't Hooft-Polyakov Monopole and Julia-Zee Dyon

The model is described in a flat $(1+3)$-dimensional space-time whose Minkowskian metric is $\eta_{\mu\nu} = \mathrm{diag}(1,-1,-1,-1)$. The standard Lagrangian for BPS monopole, or the $SU(2)$ Yang-Mills-Higgs model, has the following form [1, 2]:

$$\mathscr{L}_s = \frac{1}{2}\mathscr{D}_\mu\phi^a\mathscr{D}^\mu\phi^a - \frac{1}{4}F^{a\mu\nu}F^a_{\mu\nu} - V(|\phi|), \tag{9}$$

with $SU(2)$ gauge group symmetry and ϕ^a, $a = 1,2,3$, being a triplet real scalar field in adjoint representation of $SU(2)$. The potential V is a function of $|\phi| = \phi^a\phi^a$ which is invariant under $SU(2)$ gauge transformations. Here we use Einstein summation convention for repeated index. The definitions of covariant derivative and field strength tensor of the $SU(2)$ Yang-Mills gauge field are as follows:

$$\mathscr{D}_\mu\phi^a = \partial_\mu\phi^a + e\epsilon^{abc}A^b_\mu\phi^c, \tag{10a}$$

$$F^a_{\mu\nu} = \partial_\mu A^a_\nu - \partial_\nu A^a_\mu + e\epsilon^{abc}A^b_\mu A^c_\nu, \tag{10b}$$

with e being the gauge coupling and ϵ^{abc} being the Levi-Civita symbol. The Latin indices (a,b,c) denote the "vector components" in the vector space of $SU(2)$ algebra with generators $T_a = (1/2)\sigma_a$, where σ_a is Pauli's matrix. The generators satisfy commutation relation $[T_a, T_b] = i\epsilon_{abc}T_c$ and their trace is $\mathrm{tr}(T_aT_b) = (1/2)\delta_{ab}$. With these generators, the scalar field, gauge field, adjoint covariant derivative, and field strength tensor can then be rewritten in a compact form, respectively, as $\phi = \phi^aT_a$, $A_\mu = A^a_\mu T_a$,

$$\mathscr{D}_\mu\phi = \mathscr{D}_\mu\phi^aT_a = \partial_\mu\phi - ie\left[A_\mu, \phi\right], \tag{11a}$$

$$F_{\mu\nu} = F^a_{\mu\nu}T_a = \partial_\mu A_\nu - \partial_\nu A_\mu - ie\left[A_\mu, A_\nu\right]. \tag{11b}$$

These lead to the Lagrangian

$$\mathscr{L}_s = \mathrm{tr}\left(\mathscr{D}_\mu\phi\mathscr{D}^\mu\phi - \frac{1}{2}F^{\mu\nu}F_{\mu\nu}\right) - V(|\phi|). \tag{12}$$

Varying (9) with respect to the scalar field and the gauge field yields

$$\mathscr{D}_\mu\left(\mathscr{D}^\mu\phi^b\right) = -\frac{\partial V}{\partial\phi^b}, \tag{13a}$$

$$\mathscr{D}_\nu F^{b\mu\nu} = e\epsilon^{bca}\phi^c\mathscr{D}^\mu\phi^a, \tag{13b}$$

with additional Bianchi identity

$$\mathscr{D}_\mu\tilde{F}^{a\mu\nu} = 0, \tag{14}$$

where $\widetilde{F}^{a\mu\nu} = (1/2)\epsilon^{\mu\nu\kappa\lambda}F^a_{\kappa\lambda}$. Throughout this paper, we will consider only static configurations. The difference between monopole and dyon is whether A^a_0 is zero or nonzero, respectively. For monopole, the Bianchi identity becomes

$$\mathscr{D}_i B^a_i = 0. \tag{15}$$

Here $B^a_i = (1/2)\epsilon_{ijk}F_{jk}$ and $i, j, k = 1, 2, 3$ are the spatial indices. For dyon, $A^a_0 \neq 0$, there are additional equations of motion for "electric" part since the Gauss law is nontrivial,

$$\mathscr{D}_i E^b_i = -e\epsilon^{bca}\phi^c\mathscr{D}_0\phi^a, \tag{16}$$

where $E^a_i = F^a_{0i}$.

We could write the energy-momentum tensor $T_{\mu\nu}$ by varying the action with respect to the space-time metric. The energy density is then given by T_{00} component,

$$T_{00} = \frac{1}{2}\left(\mathscr{D}_0\phi^a\mathscr{D}_0\phi^a + \mathscr{D}_i\phi^a\mathscr{D}_i\phi^a + E^a_iE^a_i + B^a_iB^a_i\right) + V(|\phi|). \tag{17}$$

In [5], it is possible to obtain the exact solutions of the Euler-Lagrange equations in the BPS limit, i.e., $V = 0$, but still maintaining the asymptotic boundary conditions of ϕ, and we define a new parameter α such that

$$\begin{aligned} T_{00} &= \frac{1}{2}\left(\mathscr{D}_0\phi^a\mathscr{D}_0\phi^a + \mathscr{D}_i\phi^a\mathscr{D}_i\phi^a\sin^2\alpha + E^a_iE^a_i \right. \\ &\left. + \mathscr{D}_i\phi^a\mathscr{D}_i\phi^a\cos^2\alpha + B^a_iB^a_i\right) = \frac{1}{2}\left(\left(\mathscr{D}_0\phi^a\right)^2 \right. \\ &\left. + \left(\mathscr{D}_i\phi^a\sin\alpha \mp E^a_i\right)^2 + \left(\mathscr{D}_i\phi^a\cos\alpha \mp B^a_i\right)^2\right) \\ &\pm E^a_i\mathscr{D}_i\phi^a\sin\alpha \pm B^a_i\mathscr{D}_i\phi^a\cos\alpha. \end{aligned} \tag{18}$$

The last two terms can be converted to total derivative

$$E^a_i\mathscr{D}_i\phi^a = \partial_i\left(E^a_i\phi^a\right) - \left(\mathscr{D}_iE^a_i\right)\phi^a = \partial_i\left(E^a_i\phi^a\right), \tag{19a}$$

$$B^a_i\mathscr{D}_i\phi^a = \partial_i\left(B^a_i\phi^a\right) - \left(\mathscr{D}_iB^a_i\right)\phi^a = \partial_i\left(B^a_i\phi^a\right), \tag{19b}$$

after employing the Gauss law (16) and Bianchi identity (15). They are related to the "Abelian" electric and magnetic fields identified in [1], respectively. Since the total energy is $E = \int d^3xT_{00}$, the total derivative terms can be identified as the electric and magnetic charges accordingly

$$\mathscr{Q}_E = \int dS^iE^a_i\phi^a, \tag{20a}$$

$$\mathscr{Q}_B = \int dS^iB^a_i\phi^a, \tag{20b}$$

with dS^i denoting integration over the surface of a 2-sphere at $r \longrightarrow \infty$. Therefore the total energy is $E \geq \pm(\mathscr{Q}_E\sin\alpha + \mathscr{Q}_B\cos\alpha)$ since the other terms are positive semidefinite. The total energy is saturated if the BPS equations are satisfied as follows [24]:

$$\mathscr{D}_0\phi^a = 0, \tag{21a}$$

$$\mathscr{D}_i\phi^a\sin\alpha = E^a_i, \tag{21b}$$

$$\mathscr{D}_i\phi^a\cos\alpha = B^a_i. \tag{21c}$$

Solutions to these equations are called BPS dyons; they are particularly called BPS monopoles for $\alpha = 0$. The energy of this BPS configuration is simply given by

$$E_{BPS} = \pm\left(\mathscr{Q}_E\sin\alpha + \mathscr{Q}_B\cos\alpha\right). \tag{21d}$$

Adding the constant α contained in $\sin\alpha$ and $\cos\alpha$ is somehow a bit tricky. We will show later using BPS Lagrangian method that this constant comes naturally as a consequence of identifying two of the effective fields.

Employing the 't Hooft-Polyakov, together with Julia-Zee, ansatz [1, 2, 4]

$$\phi^a = f(r)\frac{x^a}{r}, \tag{22a}$$

$$A^a_0 = \frac{j(r)}{e}\frac{x^a}{r}, \tag{22b}$$

$$A^a_i = \frac{1-a(r)}{e}\epsilon^{aij}\frac{x^j}{r^2}, \tag{22c}$$

where $x^a \equiv (x, y, z)$ and $x^i \equiv (x, y, z)$ as well, denotes the Cartesian coordinate. Notice that the Levi-Civita symbol ϵ^{aij} in ((22a), (22b), and (22c)) mixes the space-index and the group-index. Substituting the ansatz ((22a), (22b), and (22c)) into Lagrangian (9) we can arrive at the following effective Lagrangian:

$$\mathscr{L}_s = -\frac{f'^2}{2} - \left(\frac{af}{r}\right)^2 + \frac{j'^2}{2e^2} + \left(\frac{aj}{er}\right)^2 - \left(\frac{a'}{er}\right)^2 - \frac{1}{2}\left(\frac{a^2-1}{er^2}\right)^2 - V(f), \tag{23}$$

where $' \equiv \partial/\partial r$; otherwise it means taking derivative over the argument. As shown in the effective Lagrangian above there is no dependency over angles coordinates ϕ and θ despite the fact that the ansatz ((22a), (22b), and (22c)) depends on ϕ and θ. Thus we may derive the Euler-Lagrange equations from the effective Lagrangian (23) which are given by

$$-\frac{1}{r^2}\left(r^2f'\right)' + \frac{2a^2f}{r^2} = -V'(f), \tag{24a}$$

$$-\frac{\left(r^2j'\right)'}{er^2} + \frac{2a^2j}{er^2} = 0, \tag{24b}$$

$$\frac{a\left(a^2-1\right)}{r^2} + a\left(e^2f^2 - j^2\right) - a'' = 0. \tag{24c}$$

Later we will also consider the case for generalize Lagrangian of (9) by adding scalar-dependent couplings to the kinetic terms as follows [25]:

$$\mathscr{L}_G = -\frac{1}{4}w(|\phi|)F^a_{\mu\nu}F^{a\mu\nu} + \frac{1}{2}G(|\phi|)\mathscr{D}_\mu\phi^a\mathscr{D}^\mu\phi^a - V(|\phi|). \tag{25}$$

The equations of motions are now given by

$$\mathscr{D}_\mu\left(G\,\mathscr{D}^\mu\phi^b\right) = -\frac{\partial V}{\partial\phi^b} + \frac{1}{2}\frac{\partial G}{\partial\phi^b}\mathscr{D}_\mu\phi^a\mathscr{D}^\mu\phi^a - \frac{1}{4}\frac{\partial w}{\partial\phi^b}F^a_{\mu\nu}F^{a\mu\nu}, \tag{26a}$$

$$\mathscr{D}_\nu\left(w\,F^{b\mu\nu}\right) = e\epsilon^{bca}\phi^c G\mathscr{D}^\mu\phi^a. \tag{26b}$$

In [25, 26], they found BPS monopole equations and a constraint equation $G = w^{-1}$. Using our method in the following sections, we obtain the similar BPS monopole equations and constraint equation. Furthermore, we generalize it to BPS dyon equations with a more general constraint equation.

There are other forms of Lagrangian for BPS monopole and dyon which were presented in the Born-Infeld type of action by Nakamula and Shiraishi in [14]. The Lagrangian for BPS monopole is different from the BPS dyon. The Lagrangians are defined such that the BPS equations ((21a), (21b), (21c), and (21d)) satisfy the Euler-Lagrange equations in the usual BPS limit. The Lagrangian for monopole and dyon is given, respectively, by [14]

$$\mathscr{L}_{\text{NSm}} = -b^2\,\text{tr}\left(\sqrt{1-\frac{2}{b^2}\mathscr{D}_\mu\phi\mathscr{D}^\mu\phi}\sqrt{1+\frac{1}{b^2}F_{\mu\nu}F^{\mu\nu}} - 1\right) - V\left(|\phi|\right), \tag{27}$$

$$\mathscr{L}_{\text{NSd}} = -b^2\,\text{tr}\left(\left\{1-\frac{2}{b^2}\mathscr{D}_\mu\phi\mathscr{D}^\mu\phi + \frac{1}{b^2}F_{\mu\nu}F^{\mu\nu} - \frac{1}{4b^4}\left(F_{\mu\nu}\tilde{F}^{\mu\nu}\right)^2 + \frac{4}{b^4}\tilde{F}_\mu{}^\nu\tilde{F}^{\mu\lambda}\mathscr{D}_\nu\phi\mathscr{D}_\lambda\phi\right\}^{1/2} - 1\right) - V\left(|\phi|\right), \tag{28}$$

with b^2 being the Born-Infeld parameter and the potential V is taken to be the same as in (9). It is apparent that, even though $E^a_i = 0$, $\mathscr{L}_{\text{NSd}} \neq \mathscr{L}_{\text{NSm}}$. Using the ansatz ((22a), (22b), and (22c)), both Lagrangians can be effectively written as

$$\mathscr{L}_{\text{NSm}} = -2b^2\left(\sqrt{1+\frac{1}{2b^2}\left(f'^2+\frac{2a^2f^2}{r^2}\right)}\sqrt{1+\frac{1}{2b^2}\left(\frac{2a'^2}{e^2r^2}+\frac{\left(a^2-1\right)^2}{e^2r^4}\right)} - 1\right) - V(f), \tag{29}$$

$$\mathscr{L}_{\text{NSd}} = -2b^2\left(\left\{1+\frac{1}{2b^2}\left(f'^2+\frac{2a^2f^2}{r^2}+\frac{2a'^2}{e^2r^2}+\frac{\left(a^2-1\right)^2}{e^2r^4}-\frac{j'^2}{e^2}-\frac{2a^2j^2}{e^2r^2}\right) + \frac{1}{4b^4}\left(-\left(-\frac{\left(a^2-1\right)j'}{e^2r^2}-\frac{2aja'}{e^2r^2}\right)^2+\left(\frac{\left(a^2-1\right)f'}{er^2}+\frac{2afa'}{er^2}\right)^2\right)\right\}^{1/2} - 1\right) - V(f). \tag{30}$$

We can see immediately that $\mathscr{L}_{\text{NSd}}(j=0) \neq \mathscr{L}_{\text{NSm}}$. However, by assuming the BPS equations $B^a_i = \pm\mathscr{D}_i\phi^a$ is valid beforehand we would get $\mathscr{L}_{\text{NSd}} = \mathscr{L}_{\text{NSm}}$. Hence from both Lagrangians, we could obtain the same BPS equations when we turn off the "electric" part for monopole.

4. BPS Equations in $SU(2)$ Yang-Mills-Higgs and Nakamula-Shiraishi Models

Here we will show that the BPS Lagrangian method [16] can also be used to obtain the known BPS equations for monopole and dyon in the $SU(2)$ Yang-Mills model (9) and the Nakamula-Shiraishi models, (27) and (28). To simplify our calculations, from here on we will set the gauge coupling to unity, $e = 1$.

4.1. BPS Monopole and Dyon in SU(2) Yang-Mills-Higgs Model. Writing the ansatz ((22a), (22b), and (22c)) in spherical coordinates,

$$\phi^a \equiv f\left(\cos\varphi\sin\theta, \sin\varphi\sin\theta, \cos\theta\right), \tag{31a}$$

$$A^a_0 \equiv j\left(\cos\varphi\sin\theta, \sin\varphi\sin\theta, \cos\theta\right), \tag{31b}$$

$$A^a_r \equiv (0,0,0), \tag{31c}$$

$$A^a_\theta \equiv (1-a)\left(\sin\varphi, -\cos\varphi, 0\right), \tag{31d}$$

$$A^a_\varphi \equiv (1-a)\sin\theta\left(\cos\varphi\cos\theta, \sin\varphi\cos\theta, -\sin\theta\right), \tag{31e}$$

we find that there is no explicit r dependent in all fields above. Therefore we propose that the BPS energy function for the case of monopole, where $j = 0$, should take the following form:

$$Q(a,f) = 4\pi F(f)\,A(a). \tag{32}$$

Since $\int d^3x\mathscr{L}_{\text{BPS}} = -\int dQ$, we have the BPS Lagrangian

$$\mathscr{L}_{\text{BPS}} = -\frac{FA'(a)}{r^2}a' - \frac{F'(f)A}{r^2}f'. \tag{33}$$

Before showing our results, for convenience we define through all calculations in this article $x = f'$, $y = a'$, $Q_a = F\,A'(a)$, and $Q_f = F'(f)A$.

Employing $\mathscr{L}_s - \mathscr{L}_{\text{BPS}} = 0$, where $\mathscr{L}_s$ is (23) and $\mathscr{L}_{\text{BPS}}$ is (33), we can consider it as a quadratic equation of either a' or f'. Here we show the roots of f' (or x) first which are

$$f'_\pm = \frac{Q_f \pm \sqrt{Q_f^2 - a^4 - 2a^2(f^2r^2 - 1) - 2r^2y(y - Q_a) - 2r^4V - 1}}{r^2}. \quad (34)$$

The two roots will be equal, $f_+ = f_-$, if the terms inside the square root is zero, which later can be considered as a quadratic equation for a' (or y) with roots

$$a'_\pm = \frac{1}{2}Q_a \pm \frac{1}{2r} \cdot \sqrt{-2a^4 + a^2(4 - 4f^2r^2) + Q_a^2r^2 + 2Q_f^2 - 4r^4V - 2}. \quad (35)$$

Again, we need the terms inside the square root to be zero for two roots to be equal, $a_+ = a_-$. The last equation can be written in power series of r,

$$\left(2Q_f^2 - 2\left(a^2 - 1\right)^2\right) + \left(Q_a^2 - 4a^2f^2\right)r^2 - 4Vr^4 = 0, \quad (36)$$

Demanding it is valid for all values of r, we may take $V = 0$, which is just the same BPS limit in [5]. From the terms with quadratic and zero power of r, we obtain

$$FA'(a) = \pm 2af, \quad (37)$$

$$F'(f)A = \pm\left(a^2 - 1\right), \quad (38)$$

which implies

$$FA = \pm\left(a^2 - 1\right)f. \quad (39)$$

Inserting this into (34) and (35), we reproduce the known BPS equations for monopole,

$$f' = \pm\frac{a^2 - 1}{r^2}, \quad (40a)$$

$$a' = \pm af. \quad (40b)$$

Now let us take $j(r) \neq 0$ and consider the BPS limit, $V \longrightarrow 0$. In this BPS limit, we can easily see from the effective Lagrangian (23) that the Euler-Lagrange equations for both fields f and j are equal. Therefore it is tempted to identify $j \propto f$. Let us write it explicitly as

$$j(r) = \beta f(r), \quad (41)$$

where β is a real-valued constant. With this identification, we can again use (32) as the BPS energy function for dyon and hence give the same BPS Lagrangian (33). Now the only difference, from the previous monopole case, is the effective Lagrangian (23) which takes a simpler form

$$\mathscr{L}_s = -\left(1 - \beta^2\right)\left(\frac{f'^2}{2} + \left(\frac{af}{r}\right)^2\right) - \left(\frac{a'}{r}\right)^2 - \frac{1}{2}\left(\frac{a^2 - 1}{r^2}\right)^2 - V. \quad (42)$$

Here we still keep the potential V and we will show later that V must be equal to zero in order to get the BPS equations using the BPS Lagrangian method.

Applying (41) and solving $\mathscr{L}_s - \mathscr{L}_{\text{BPS}} = 0$ as quadratic equation for f' (or x) give us two roots

$$f'_\pm = \frac{Q_f \pm \sqrt{\text{DD}}}{\left(1 - \beta^2\right)r^2}, \quad (43)$$

with

$$\text{DD} = \left(\beta^2 - 1\right)\left(a^4 - 2a^2\left(\left(\beta^2 - 1\right)f^2r^2 + 1\right) + 2r^2y\left(y - Q_a\right) + 2r^4V + 1\right) + Q_f^2. \quad (44)$$

Next, requiring $f'_+ = f'_-$, we obtain

$$a'_\pm = \frac{1}{2}\left(Q_a \pm \sqrt{\frac{\text{DDD}}{\left(\beta^2 - 1\right)r^2}}\right), \quad (45)$$

where we arrange DDD in power series of r, i.e.,

$$\text{DDD} = 2\left(\left(1 - a^2\right)^2\left(1 - \beta^2\right) - Q_f^2\right) + \left(1 - \beta^2\right)\left(4a^2\left(1 - \beta^2\right)f^2 - Q_a^2\right)r^2 + 4V\left(1 - \beta^2\right)r^4. \quad (46)$$

Again, $a'_- = a'_+$; we get $DDD = 0$. Solving the last equation, which must be valid for all values of r, we conclude $V = 0$ from r^4-terms, for nontrivial solution, and from the remaining terms we have

$$FA'(a) = \pm 2af\sqrt{1 - \beta^2}, \quad (47a)$$

$$F'(f)A = \pm\left(a^2 - 1\right)\sqrt{1 - \beta^2}, \quad (47b)$$

which give us

$$FA = \pm\left(a^2 - 1\right)f\sqrt{1 - \beta^2}. \quad (48)$$

The BPS equations are then

$$f'\sqrt{1 - \beta^2} = \pm\frac{a^2 - 1}{r^2}, \quad (49a)$$

$$a' = \pm af\sqrt{1 - \beta^2}. \quad (49b)$$

Since f' and a' are real valued, β should take values $|\beta| < 1$. They become the BPS equations for monopole ((40a) and (40b)) when we set $\beta = 0$. We can see that this constant is analogous to the constant α, or precisely $\beta = -\sin\alpha$, in (18); see [27] for detail. Substituting $\beta = -\sin\alpha$ into ((49a) and (49b)), we get the same BPS equations as in [5, 24]. Here we can see the constant β is naturally bounded as required by the BPS equations ((49a) and (49b)).

4.2. BPS Monopole and Dyon in Nakamula-Shiraishi Model. In this subsection we will show that the Lagrangians (27) and (28) of Nakamula-Shiraishi model do indeed possess the BPS equations ((40a) and (40b) and (49a) and (49b)), respectively, after employing the BPS Lagrangian method. Substituting (29) and (32) into $\mathscr{L}_{\text{NSm}} - \mathscr{L}_{\text{BPS}} = 0$ and following the same procedures as the previous subsection give us the roots of a',

$$a'_{\pm} = \frac{Q_a r^4\left(2b^2 - V\right) + Q_a Q_f r^2 x \pm \sqrt{r^2\left(2a^2 f^2 + r^2\left(2b^2 + x^2\right)\right)\text{DD}}}{r^2\left(4a^2 f^2 + 2r^2\left(2b^2 + x^2\right) - Q_a^2\right)}, \tag{50}$$

where

$$\begin{aligned}\text{DD} = {} & -4a^6 f^2 + a^4\left(-2r^2\left(2b^2 + x^2\right) + 8f^2 + Q_a^2\right) \\ & - 2a^2\left(f^2\left(4b^2 r^4 + 2\right) - 2r^2\left(2b^2 + x^2\right) + Q_a^2\right) \\ & + 2b^2 r^4\left(Q_a^2 + 4Q_f x - 2r^2\left(2V + x^2\right)\right) - 4b^2 r^2 \\ & + Q_a^2 \\ & + 2r^2\left(-Q_f x + r^2 V + x\right)\left(r^2 V - \left(Q_f + 1\right)x\right).\end{aligned} \tag{51}$$

Solving DD = 0 gives us

$$f'_{\pm} = \frac{2Q_f r^4\left(2b^2 - V\right) \pm \sqrt{-2r^2\left(\left(a^2 - 1\right)^2 + 2b^2 r^4\right)\text{DDD}}}{2r^2\left(a^4 - 2a^2 + 2b^2 r^4 - Q_f^2 + 1\right)}, \tag{52}$$

where

$$\begin{aligned}\text{DDD} = {} & 2b^2 r^4\left(4a^2 f^2 - Q_a^2\right) \\ & + 4b^2 r^2\left(\left(a^2 - 1\right)^2 - Q_f^2\right) \\ & + \left(\left(a^2 - 1\right)^2 - Q_f^2\right)\left(4a^2 f^2 - Q_a^2\right) \\ & + 2r^6 V\left(4b^2 - V\right).\end{aligned} \tag{53}$$

Then the last equation DDD = 0 gives us $V = 0$, or $V = 4b^2$,

$$FA'(a) = \pm 2af, \tag{54a}$$

$$F'(f)A = \pm\left(a^2 - 1\right), \tag{54b}$$

which again gives us

$$FA = \pm\left(a^2 - 1\right)f, \tag{55}$$

and thus we have a' and f', with $V = 0$,

$$f' = \pm\frac{a^2 - 1}{r^2}, \tag{56a}$$

$$a' = \pm af. \tag{56b}$$

the same BPS equations ((40a) and (40b)) for monopole. The other choice of potential $V = 4b^2$ will result in the same BPS equations with opposite sign relative to the BPS equations of $V = 0$,

$$f' = \mp\frac{a^2 - 1}{r^2}, \tag{57a}$$

$$a' = \mp af. \tag{57b}$$

For dyon, using the same identification (41), we have the effective Lagrangian (30) shortened to

$$\begin{aligned}\mathscr{L}_{\text{NSd}} = {} & -2b^2\left(\left\{1 + \frac{1-\beta^2}{2b^2}\left(f'^2 + \frac{2a^2 f^2}{r^2}\right)\right.\right. \\ & + \frac{1}{2b^2}\left(\frac{2a'^2}{r^2} + \frac{\left(a^2 - 1\right)^2}{r^4}\right) \\ & \left.\left. + \frac{1-\beta^2}{4b^4}\left(\frac{\left(a^2 - 1\right)f'}{r^2} + \frac{2afa'}{r^2}\right)^2\right\}^{1/2} - 1\right) \\ & - V.\end{aligned} \tag{58}$$

Equating the above effective Lagrangian with $\mathscr{L}_{\text{BPS}}$, using the same BPS energy density (32), and solving this for a' give us

$$a'_{\pm} = \frac{-2a^3\beta^2 fx + 2a^3 fx + 2a\beta^2 fx - 2afx - 2b^2 Q_a r^2 - Q_a Q_f x + Q_a r^2 V \pm \sqrt{\text{DD}}}{4a^2\left(\beta^2 - 1\right)f^2 - 4b^2 r^2 + Q_a^2}, \tag{59}$$

where

$$\begin{aligned}\mathrm{DD} = {} & \left(2a\left(a^2-1\right)\left(\beta^2-1\right)fx + 2b^2Q_ar^2 \right.\\ & \left. + Q_a\left(Q_fx - r^2V\right)\right)^2 + \left(4a^2\left(\beta^2-1\right)f^2 - 4b^2r^2 \right.\\ & \left. + Q_a^2\right) \times \left[2b^2\left(a^4 - 2a^2\left(\left(\beta^2-1\right)f^2r^2+1\right)\right.\right.\\ & \left. - 2Q_fr^2x + r^4\left(2V - \beta^2x^2 + x^2\right) + 1\right)\\ & - x^2\left(a^4\left(\beta^2-1\right) - 2a^2\left(\beta^2-1\right) + \beta^2 + Q_f^2 - 1\right)\\ & \left. + 2Q_fr^2Vx - r^4V^2\right].\end{aligned} \tag{60}$$

Solving $\mathrm{DD} = 0$ for f' gives us

$$f'_{\pm} = \frac{\mathrm{K} \pm \sqrt{2}\sqrt{\mathrm{M\ DDD}}}{\mathrm{L}} \tag{61}$$

where

$$\begin{aligned}\mathrm{K} = {} & 2a^3\left(\beta^2-1\right)fQ_ar^2\left(2b^2-V\right) - 4a^2\left(\beta^2-1\right)\\ & \cdot f^2Q_fr^2\left(2b^2-V\right) - 2a\left(\beta^2-1\right)fQ_ar^2\left(2b^2\right.\\ & \left. - V\right) + 8b^4Q_fr^4 - 4b^2Q_fr^4V,\end{aligned} \tag{62}$$

$$\begin{aligned}\mathrm{L} = {} & -4a\left(a^2-1\right)\left(\beta^2-1\right)fQ_aQ_f\\ & - 4\left[a^4b^2\left(\beta^2-1\right)r^2\right.\\ & - a^2\left(\beta^2-1\right)\left(f^2\left(2b^2\left(\beta^2-1\right)r^4 + Q_f^2\right) + 2b^2r^2\right)\\ & \left. + b^2r^2\left(\left(\beta^2-1\right)\left(2b^2r^4+1\right) + Q_f^2\right)\right] + \left(\beta^2-1\right)\\ & \cdot Q_a^2\left(a^4 - 2a^2 + 2b^2r^4 + 1\right),\end{aligned} \tag{63}$$

$$\begin{aligned}\mathrm{M} = {} & -b^2\left(a^4 - 2a^2\left(\left(\beta^2-1\right)f^2r^2+1\right) + 2b^2r^4 + 1\right)\\ & \cdot \left(4a^2\left(\beta^2-1\right)f^2 - 4b^2r^2 + Q_a^2\right),\end{aligned} \tag{64}$$

$$\begin{aligned}\mathrm{DDD} = {} & -2r^4\left(b^2\left(\beta^2-1\right)\left(4a^2\left(\beta^2-1\right)f^2 + Q_a^2\right)\right)\\ & - \left(\beta^2-1\right)\left(\left(a^2-1\right)Q_a - 2afQ_f\right)^2\\ & + 4b^2r^2\left(a^4\left(\beta^2-1\right) - 2a^2\left(\beta^2-1\right) + \beta^2 + Q_f^2\right.\\ & \left. - 1\right) + 2\left(\beta^2-1\right)r^6V\left(4b^2 - V\right).\end{aligned} \tag{65}$$

We may set $M = 0$, but this will imply $b^2 = 0$ which is not what we want. Requiring $\mathrm{DDD} = 0$ valid for all values of r, the terms with r^6 give us $V = 0$ or $V = 4b^2$. The terms with r^0 imply

$$FA'(a) = \frac{2afF'(f)A}{a^2-1}. \tag{66}$$

This is indeed solved by the remaining terms which imply

$$FA'(a) = \pm 2af\sqrt{1-\beta^2}, \tag{67}$$

$$F'(f)A = \pm\left(a^2-1\right)\sqrt{1-\beta^2}. \tag{68}$$

This again gives us

$$FA = \pm\left(a^2-1\right)f\sqrt{1-\beta^2}, \tag{69}$$

and hence, for $V = 0$,

$$f' = \pm\frac{a^2-1}{\sqrt{1-\beta^2r^2}}, \tag{70a}$$

$$a' = \pm af\sqrt{1-\beta^2}, \tag{70b}$$

the same BPS equations ((49a) and (49b)) for dyon. Similar to the monopole case choosing $V = 4b^2$ will switch the sign in the BPS equations. It is apparent that, in the limit of $\beta \longrightarrow 0$, the BPS equations for dyon become the ones for monopole. This indicates that, in the BPS limit and $\beta \longrightarrow 0$, $\mathscr{L}_{\mathrm{NSd}} \longrightarrow \mathscr{L}_{\mathrm{NSm}}$, since in general, even though in the limit of $\beta \longrightarrow 0$, $\mathscr{L}_{\mathrm{NSd}} \not\longrightarrow \mathscr{L}_{\mathrm{NSm}}$.

Now we know that the method works. In the next sections, we use it in some generalized Lagrangian whose BPS equations, for monopole or dyon, may or may not be known.

5. BPS Equations in Generalized $SU(2)$ Yang-Mills-Higgs Model

In this section, we use the Lagrangian (25) whose effective Lagrangian is given by

$$\begin{aligned}\mathscr{L}_{\mathrm{G}} = {} & -G\left(\frac{f'^2}{2} + \frac{a^2f^2}{r^2}\right) + w\left(\frac{j'^2}{2} + \frac{a^2j^2}{r^2}\right)\\ & - w\left(\frac{a'^2}{r^2} + \frac{\left(a^2-1\right)^2}{2r^4}\right) - V.\end{aligned} \tag{71}$$

We will see later that it turns out that G and w are related to each other by some constraint equations.

5.1. BPS Monopole Case. In this case, the BPS equations are already known [25, 26]. Setting $j = 0$ and employing $\mathscr{L}_{\mathrm{G}} - \mathscr{L}_{\mathrm{BPS}} = 0$ we get

$$f'_{\pm} = \frac{Q_fr^2 \pm \sqrt{-r^4\left(G\left(a^4w + 2a^2\left(f^2Gr^2 - w\right) + 2r^2y\left(wy - Q_a\right) + 2r^4V + w\right) - Q_f^2\right)}}{Gr^4}, \tag{72}$$

and from $f'_+ = f'_-$ we have the roots of a' (or y)

$$a'_\pm = \frac{GQ_a r^2 - \sqrt{Gr^2\left\{G\left(Q_a^2 r^2 - 2w\left(2r^2\left(a^2 f^2 G + r^2 V\right) + \left(a^2-1\right)^2 w\right)\right) + 2Q_f^2 w\right\}}}{2Gr^2 w}. \quad (73)$$

The terms inside the curly bracket in the square root must be zero in which after rearranging in power series of r

$$2w\left(Q_f^2 - \left(a^2-1\right)^2 Gw\right) + Gr^2\left(Q_a^2 - 4a^2 f^2 Gw\right) - 4r^4\left(GVw\right) = 0, \quad (74)$$

we obtain $V = 0$,

$$FA'(a) = \pm 2af\sqrt{Gw}, \quad (75)$$

$$F'(f)A = \pm\left(a^2-1\right)\sqrt{Gw}. \quad (76)$$

These imply

$$\frac{\partial}{\partial f}\left(f\sqrt{Gw}\right) = \sqrt{Gw}, \quad (77)$$

and hence

$$w = \frac{c}{G}, \quad (78)$$

where c is a positive constant. The BPS equations are given by

$$f' = \pm\frac{\left(a^2-1\right)}{r^2}\sqrt{\frac{w}{G}}, \quad (79a)$$

$$a' = \pm af\sqrt{\frac{G}{w}}, \quad (79b)$$

with a constraint equation $wG = c$, where c is a positive constant. This constant can be fixed to one, $c = 1$, by recalling that in the corresponding nongeneralized version, in which $G = w = 1$, we should get back the same BPS equations of (40a) and (40b).

5.2. BPS Dyon Case. As previously setting $j = \beta f$ and employing $\mathscr{L}_G - \mathscr{L}_{BPS} = 0$ we get

$$f'_\pm = \frac{Q_f r^2 \pm \sqrt{r^4 DD}}{r^4\left(G - \beta^2 w\right)}, \quad (80)$$

with

$$\begin{aligned} DD = Q_f^2 - \left(G-\beta^2 w\right)\left(a^4 w\right. &+ 2a^2\left(f^2 r^2\left(G-\beta^2 w\right) - w\right) + 2r^2 y\left(wy - Q_a\right) \\ &\left. + 2r^4 V + w\right), \end{aligned} \quad (81)$$

and from $DD = 0$ we have the roots of a'

$$a'_\pm = \frac{Q_a \pm \sqrt{DDD/r^2\left(G-\beta^2 w\right)}}{2w}, \quad (82)$$

where

$$\begin{aligned} DDD = r^2\left(G-\beta^2 w\right)\left(4a^2 f^2 w\left(\beta^2 w - G\right) + Q_a^2\right) &+ 2w\left(\left(a^2-1\right)^2 w\left(\beta^2 w - G\right) + Q_f^2\right) \\ &+ 4r^4 Vw\left(\beta^2 w - G\right). \end{aligned} \quad (83)$$

Requiring $DDD = 0$ we obtain $V = 0$,

$$FA'(a) = \pm 2af\sqrt{w\left(G-\beta^2 w\right)}, \quad (84)$$

$$F'(f)A = \pm\left(a^2-1\right)\sqrt{w\left(G-\beta^2 w\right)}. \quad (85)$$

Similar to the monopole case these imply

$$w\left(G-\beta^2 w\right) = c, \quad (86)$$

where c is a positive constant and it can also be fixed to $c = 1-\beta^2$ demanding that at $G = w = 1$ we should get the same BPS equations ((49a) and (49b)). At $\beta \longrightarrow 0$, we get back the constraint equation (78) for monopole case. These give us the BPS equations

$$f' = \pm\frac{\left(a^2-1\right)}{r^2}\sqrt{\frac{w}{G-\beta^2 w}}, \quad (87a)$$

$$a' = \pm af\sqrt{\frac{G-\beta^2 w}{w}}, \quad (87b)$$

in which at $\beta \longrightarrow 0$ we again get back the BPS equations for monopole case ((79a) and (79b)).

6. BPS Equations in Generalized Nakamula-Shiraishi Model

Here we present the generalized version of the Nakamula-Shiraishi models (27) and (28) for both monopole and dyon, respectively.

6.1. BPS Monopole Case. A generalized version of (27) is defined by

$$\mathscr{L}_{NSmG} = -b^2 \cdot \mathrm{tr}\left(\sqrt{1 - \frac{2}{b^2}G\left(|\phi|\right)\mathscr{D}_\mu\phi\mathscr{D}^\mu\phi}\sqrt{1 + \frac{1}{b^2}w\left(|\phi|\right)\mathscr{F}_{\mu\nu}\mathscr{F}^{\mu\nu}} - 1\right) - V\left(|\phi|\right), \quad (88)$$

where after inserting the ansatz, we write its effective Lagrangian as

$$\mathscr{L}_{\mathrm{NSmG}} = -2b^2\left(\sqrt{1+\frac{G}{2b^2}\left(f'^2+\frac{2a^2f^2}{r^2}\right)}\sqrt{1+\frac{w}{2b^2}\left(\frac{2a'^2}{r^2}+\frac{\left(a^2-1\right)^2}{r^4}\right)}-1\right)-V. \quad (89)$$

Using the similar BPS Lagrangian (33), we solve $\mathscr{L}_{\mathrm{NSmG}} - \mathscr{L}_{\mathrm{BPS}} = 0$ as a quadratic equation of a' (or y) first as such the roots are given by

$$a' = \frac{Q_a r^4\left(2b^2-V\right)+Q_aQ_f r^2 x \pm \sqrt{r^2\left(2a^2f^2G+r^2\left(2b^2+Gx^2\right)\right)\mathrm{DD}}}{r^2\left(2w\left(2a^2f^2G+r^2\left(2b^2+Gx^2\right)\right)-Q_a^2\right)}, \quad (90)$$

with

$$\begin{aligned}\mathrm{DD} = {} & w\left(\left(a^2-1\right)^2\left(Q_a^2-2Gw\left(2a^2f^2+r^2x^2\right)\right)\right.\\ & +2\left(r^3V-Q_f rx\right)^2\Big)+2b^2r^2\left(Q_a^2r^2-2w\left(a^4w\right.\right.\\ & +2a^2\left(f^2Gr^2-w\right)+r^4\left(Gx^2+2V\right)-2Q_f r^2x\\ & \left.\left.+w\right)\right).\end{aligned} \quad (91)$$

Taking DD = 0, we obtain the roots for f',

$$f' = \frac{2Q_f r^4 w\left(2b^2-V\right) \pm \sqrt{2r^2w\left(\left(a^2-1\right)^2w+2b^2r^4\right)\mathrm{DDD}}}{2r^2w\left(\left(a^2-1\right)^2Gw+2b^2Gr^4-Q_f^2\right)}, \quad (92)$$

with

$$\begin{aligned}\mathrm{DDD} = {} & 2b^2Gr^4\left(Q_a^2-4a^2f^2Gw\right)\\ & +4b^2r^2w\left(Q_f^2-\left(a^2-1\right)^2Gw\right)\\ & +\left(Q_f^2-\left(a^2-1\right)^2Gw\right)\left(4a^2f^2Gw-Q_a^2\right)\\ & +2Gr^6Vw\left(V-4b^2\right).\end{aligned} \quad (93)$$

Requiring DDD = 0, we obtain from the terms with r^6 that $V = 0$ or $V = 4b^2$. The remaining terms are also zero if $Q_a = \pm 2af\sqrt{Gw}$ and $Q_f = \pm(a^2-1)\sqrt{Gw}$. These again imply

$$G = \frac{1}{w}, \quad (94)$$

which is equal to the constraint equation (78) for monopole in Generalized $SU(2)$ Yang-Mills-Higgs model. Then the BPS equations, with $V = 0$, are

$$f' = \pm\frac{\left(a^2-1\right)}{r^2}\sqrt{\frac{w}{G}}, \quad (95a)$$

$$a' = \pm af\sqrt{\frac{G}{w}}, \quad (95b)$$

which are equal to BPS equations ((79a) and (79b)) for monopole in the Generalized $SU(2)$ Yang-Mills-Higgs model.

6.2. BPS Dyon Case. The generalization of Lagrangian (28) is defined as

$$\begin{aligned}\mathscr{L}_{\mathrm{NSdG}} = {} & -b^2\,\mathrm{tr}\left(\left\{1-\frac{2}{b^2}G\left(|\phi|\right)\mathscr{D}_\mu\phi\mathscr{D}^\mu\phi\right.\right.\\ & +\frac{1}{b^2}w\left(|\phi|\right)\mathscr{F}_{\mu\nu}\mathscr{F}^{\mu\nu}-\frac{1}{4b^4}G_1\left(|\phi|\right)\left(\mathscr{F}_{\mu\nu}\tilde{\mathscr{F}}^{\mu\nu}\right)^2\\ & \left.\left.+\frac{4}{b^4}G_2\left(|\phi|\right)\tilde{\mathscr{F}}_\mu{}^\nu\tilde{\mathscr{F}}^{\mu\lambda}\mathscr{D}_\nu\phi\mathscr{D}_\lambda\phi\right\}^{1/2}-1\right)\\ & -V\left(|\phi|\right).\end{aligned} \quad (96)$$

Employing the relation $j(f) = \beta f$, its effective Lagrangian is

$$\begin{aligned}\mathscr{L}_{\mathrm{NSdG}} = {} & -2b^2\left(\left\{1+\frac{G-w\beta^2}{2b^2}\left(f'^2+\frac{2a^2f^2}{r^2}\right)\right.\right.\\ & +\frac{w}{2b^2}\left(\frac{2a'^2}{r^2}+\frac{\left(a^2-1\right)^2}{r^4}\right)\\ & \left.\left.+\frac{G_2-G_1\beta^2}{4b^4}\left(\frac{\left(a^2-1\right)f'}{r^2}+\frac{2afa'}{r^2}\right)^2\right\}^{1/2}-1\right)\\ & -V.\end{aligned} \quad (97)$$

Employing $\mathscr{L}_{\text{NSdG}} - \mathscr{L}_{\text{BPS}} = 0$, with the same BPS Lagrangian (33), and solving it as a quadratic equation of a', first we get

$$a' = \frac{2a^3\beta^2 fG_1x - 2a^3 fG_2x - 2a\beta^2 fG_1x + 2afG_2x + 2b^2Q_ar^2 + Q_aQ_fx - Q_ar^2V \pm (1/2)\sqrt{\text{DD}}}{4a^2 f^2 (G_2 - \beta^2G_1) + 4b^2wr^2 - Q_a^2}, \tag{98}$$

where

$$\begin{aligned} \text{DD} = {} & \left(-4a\left(a^2-1\right)fx\left(G_2-\beta^2G_1\right)+4b^2Q_ar^2 \right. \\ & \left. + Q_a\left(2Q_fx-2r^2V\right)\right)^2 \\ & - 4\left(-4a^2f^2\left(G_2-\beta^2G_1\right)-4b^2wr^2+Q_a^2\right)\text{H}, \end{aligned} \tag{99}$$

$$\begin{aligned} \text{H} = {} & x^2\left(\left(a^2-1\right)^2\beta^2G_1-\left(a^2-1\right)^2G_2+Q_f^2\right) \\ & - 2b^2\text{J} - 2Q_fr^2Vx + r^4V^2, \end{aligned} \tag{100}$$

$$\begin{aligned} \text{J} = {} & r^2\left(2a^2f^2G-2Q_fx+r^2\left(2V+Gx^2\right)\right)+w\left(a^4\right. \\ & \left. - 2a^2\left(\beta^2f^2r^2+1\right)-\beta^2r^4x^2+1\right). \end{aligned} \tag{101}$$

Then from $\text{DD} = 0$ we have a quadratic equation of f' whose roots are

$$f' = \frac{\text{K} \pm (1/2)\sqrt{\text{DDD}}}{\text{L}}, \tag{102}$$

where

$$\begin{aligned} \text{K} = {} & -4a^3b^2\beta^2fG_1Q_ar^2 + 4a^3b^2fG_2Q_ar^2 \\ & + 2a^3\beta^2fG_1Q_ar^2V - 2a^3fG_2Q_ar^2V \\ & + 8a^2b^2\beta^2f^2G_1Q_fr^2 - 8a^2b^2f^2G_2Q_fr^2 \\ & - 4a^2\beta^2f^2G_1Q_fr^2V + 4a^2f^2G_2Q_fr^2V \\ & + 4ab^2\beta^2fG_1Q_ar^2 - 4ab^2fG_2Q_ar^2 \\ & - 2a\beta^2fG_1Q_ar^2V + 2afG_2Q_ar^2V \\ & - 8b^4wQ_fr^4 + 4b^2wQ_fr^4V, \end{aligned} \tag{103}$$

$$\begin{aligned} \text{L} = {} & -\beta^2G_1\left(a^2\left(-Q_a\right)+2afQ_f+Q_a\right)^2+2b^2r^2\text{M} \\ & + G_2\text{N} + 8b^4wr^6\left(\beta^2w-G\right), \end{aligned} \tag{104}$$

where in L we define M and N as

$$\begin{aligned} \text{M} = {} & r^2G\left(4a^2\beta^2f^2G_1+Q_a^2\right)+w\left(2Q_f^2\right. \\ & \left. - \beta^2\left(Q_a^2r^2-2G_1\left(a^4-2a^2\left(\beta^2f^2r^2+1\right)+1\right)\right)\right), \end{aligned} \tag{105}$$

$$\begin{aligned} \text{N} = {} & -4a^4b^2wr^2 + 4a^2\left(f^2\left(Q_f^2-2b^2r^4\left(G-\beta^2w\right)\right)\right. \\ & \left. + 2b^2wr^2\right) - 4\left(a^2-1\right)afQ_aQ_f + \left(a^2-1\right)^2Q_a^2 \\ & - 4b^2wr^2, \end{aligned} \tag{106}$$

and

$$\begin{aligned} \text{DDD} = {} & T_0 - 8T_2r^2 + 8T_4r^4 + 16b^2T_6r^6 + T_8r^8 \\ & - 32T_{10}r^{10} + T_{12}r^{12}, \end{aligned} \tag{107a}$$

where

$$\begin{aligned} T_0 = {} & 8\left(a^2-1\right)^2b^2w\left(G_2-\beta^2G_1\right)\left(\left(a^2-1\right)Q_a\right. \\ & \left. - 2afQ_f\right)^2\left(4a^2f^2\left(G_2-\beta^2G_1\right)-Q_a^2\right), \end{aligned} \tag{107b}$$

$$\begin{aligned} T_2 = {} & -4\left(a^2-1\right)^2b^4w^2\left(\left(a^2-1\right)^2\beta^2G_1-\left(a^2\right.\right. \\ & \left.\left. -1\right)^2G_2+Q_f^2\right)\left(4a^2f^2\left(G_2-\beta^2G_1\right)-Q_a^2\right) \\ & - b^2\left(G_2-\beta^2G_1\right)\left(\left(a^2-1\right)Q_a-2afQ_f\right)^2 \\ & \times\left(4\left(a^2-1\right)^2b^2w^2-2a^2f^2\left(G-\beta^2w\right)\left(Q_a^2\right.\right. \\ & \left.\left. - 4a^2f^2\left(G_2-\beta^2G_1\right)\right)\right), \end{aligned} \tag{107c}$$

$$\begin{aligned} T_4 = {} & 2\left(a^2-1\right)^2b^4w\left(\beta^2w-G\right)\left(Q_a^2-4a^2f^2\left(G_2\right.\right. \\ & \left.\left. -\beta^2G_1\right)\right)^2 + 2a^2f^2\left(V-2b^2\right)^2\left(G_2-\beta^2G_1\right)^2 \\ & \cdot\left(\left(a^2-1\right)Q_a-2afQ_f\right)^2 + 4b^4w\left(\left(a^2-1\right)^2\right. \\ & \left. \cdot\beta^2G_1-\left(a^2-1\right)^2G_2+Q_f^2\right)\times\left(4\left(a^2-1\right)^2\right. \\ & \cdot b^2w^2-2a^2f^2\left(G-\beta^2w\right)\left(Q_a^2\right. \\ & \left.\left. - 4a^2f^2\left(G_2-\beta^2G_1\right)\right)\right) - 2\left(G_2-\beta^2G_1\right)\left(\left(a^2\right.\right. \\ & \left.\left. -1\right)Q_a-2afQ_f\right)^2\times\left(b^4\left(4a^2f^2w\left(\beta^2w-G\right)\right.\right. \\ & \left. + Q_a^2\right) - 4a^2b^2f^2V\left(G_2-\beta^2G_1\right)+a^2f^2V^2\left(G_2\right. \\ & \left.\left. -\beta^2G_1\right)\right), \end{aligned} \tag{107d}$$

$$\begin{aligned} T_6 = {} & -4\left(a^2-1\right)^2b^4w^2\left(G-\beta^2w\right)\left(4a^2f^2\left(G_2\right.\right. \\ & \left.\left. -\beta^2G_1\right)-Q_a^2\right)+b^2\left(G-\beta^2w\right)\left(Q_a^2\right. \\ & \left. - 4a^2f^2\left(G_2-\beta^2G_1\right)\right)\times\left(4\left(a^2-1\right)^2b^2w^2\right. \\ & \left. - 2a^2f^2\left(G-\beta^2w\right)\left(Q_a^2-4a^2f^2\left(G_2-\beta^2G_1\right)\right)\right) \end{aligned}$$

$$- 4afwQ_f \left(V - 2b^2\right)^2 \left(G_2 - \beta^2 G_1\right) \left(\left(a^2 - 1\right) Q_a - 2afQ_f\right) + wV \left(4b^2 - V\right) \left(G_2 - \beta^2 G_1\right) \left(\left(a^2 - 1\right) Q_a - 2afQ_f\right)^2 + 4w \left(\left(a^2 - 1\right)^2 \beta^2 (-G_1) + \left(a^2 - 1\right)^2 G_2 - Q_f^2\right) \times \left(b^4 \left(4a^2 f^2 w \left(\beta^2 w - G\right) + Q_a^2\right) - 4a^2 b^2 f^2 V \left(G_2 - \beta^2 G_1\right) + a^2 f^2 V^2 \left(G_2 - \beta^2 G_1\right)\right), \tag{107e}$$

$$T_8 = 64b^4 w^2 Q_f^2 \left(V - 2b^2\right)^2 - 8 \left(8b^6 w \left(G - \beta^2 w\right) \cdot \left(4 \left(a^2 - 1\right)^2 b^2 w^2 - 2a^2 f^2 \left(G - \beta^2 w\right) \cdot \left(Q_a^2 - 4a^2 f^2 \left(G_2 - \beta^2 G_1\right)\right)\right) + 4b^2 \left(G - \beta^2 w\right) \cdot \left(Q_a^2 - 4a^2 f^2 \left(G_2 - \beta^2 G_1\right)\right) \cdot \left(b^4 \left(4a^2 f^2 w \left(\beta^2 w - G\right) + Q_a^2\right) - 4a^2 b^2 f^2 V \left(G_2 - \beta^2 G_1\right) + a^2 f^2 V^2 \left(G_2 - \beta^2 G_1\right)\right) + 8b^4 w^2 V \left(4b^2 - V\right) \cdot \left(\left(a^2 - 1\right)^2 \beta^2 (-G_1) + \left(a^2 - 1\right)^2 G_2 - Q_f^2\right)\right), \tag{107f}$$

$$T_{10} = b^4 w \left(\beta^2 w - G\right) \left(4b^4 \left(4a^2 f^2 w \left(\beta^2 w - G\right) + Q_a^2\right) + 4b^2 V \left(Q_a^2 - 8a^2 f^2 \left(G_2 - \beta^2 G_1\right)\right) + V^2 \left(8a^2 f^2 \left(G_2 - \beta^2 G_1\right) - Q_a^2\right)\right), \tag{107g}$$

$$T_{12} = 128b^6 w^2 V \left(4b^2 - V\right) \left(\beta^2 w - G\right). \tag{107h}$$

In order to find V, Q_f, Q_a, G_1, and G_2, we have to solve equation DDD $= 0$. Since the model is valid for all r, then each T_0 until T_{12} must be equal to zero. From $T_{12} = 0$ we need either $V = 0$ or $V = 4b^2$. This verifies that the BPS limit is indeed needed to obtain the BPS equations. Putting $V = 0$ into DDD we simplify a little (107a), (107b), (107c), (107d), (107e), (107f), (107g), and (107h). From $T_0 = 0$ we have

$$Q_f = \frac{\left(a^2 - 1\right) Q_a}{2af}, \tag{108}$$

which we input into DDD again. From $T_2 = 0$ we obtain $Q_a = \pm 2af\sqrt{G_2 - \beta^2 G_1}$. Now we will input each into two separate cases.

(1) Setting $Q_a = -2af\sqrt{G_2 - \beta^2 G_1}$, only T_8 and T_{10} are not vanished. Both can vanish if $\beta^2(w^2 - G_1) + (G_2 - wG) = 0$; hence we have $G_2 - \beta^2 G_1 = wG - w^2\beta^2$.

(2) Setting $Q_a = 2af\sqrt{G_2 - \beta^2 G_1}$, we also arrive at the same destination.

From these steps, we obtain that

$$FA'(a) = \pm 2af\sqrt{w\left(G - \beta^2 w\right)}, \tag{109}$$

$$F'(f)\,A = \pm\left(a^2 - 1\right)\sqrt{w\left(G - \beta^2 w\right)}, \tag{110}$$

which again imply that

$$w\left(G - \beta^2 w\right) = 1 - \beta^2, \tag{111}$$

which is equal to the constraint equation (86) for dyon in the Generalized $SU(2)$ Yang-Mills-Higgs model. Substituting everything, we obtain the BPS equations, with $V = 0$,

$$f' = \pm\frac{\left(a^2 - 1\right)}{r^2}\sqrt{\frac{w}{G - \beta^2 w}}, \tag{112}$$

$$a' = \pm af\sqrt{\frac{G - \beta^2 w}{w}}, \tag{113}$$

which is again equal to the BPS equations ((87a) and (87b)) for dyon in the Generalized $SU(2)$ Yang-Mills-Higgs model.

7. Discussion

We have shown that the BPS Lagrangian method, which was used before in [16] for BPS vortex, can also be applied to the case of BPS monopole and dyon in $SU(2)$ Yang-Mills-Higgs model (9). One main reason is because the effective Lagrangian (23) only depends on radial coordinate similar to the case of BPS vortex. We also took similar ansatz for the BPS Lagrangian (33) in which the BPS energy function Q (32) does not depend on the radial coordinate explicitly and it is a separable function of f and a. This due to no explicit dependent over radial coordinate on the ansatz for the fields written in spherical coordinates as in (31a), (31b), (31c), (31d), and (31e).

The BPS dyon could be obtained by identifying the effective field of the time-component gauge fields j to be proportional with the effective field of the scalars f by a constant β, $j = \beta f$. This identification seems natural by realizing that both effective fields give the same Euler-Lagrange equation in the BPS limit. Fortunately we found that the BPS Lagrangian method forced us to take this limit when solving the last equation with explicit power of radial coordinate order by order, which are also the case for all other models considered in this article. In this article we used this simple identification which gives us the known result of BPS dyon [5]. It turns out that the constant β takes values $|\beta| < 1$ and it will be equal to BPS dyon in [5, 24] if we set $\beta = -\sin\alpha$, with α being a constant. There is also a possibility where both effective fields are independent, or having no simple relation, but this will be discussed elsewhere.

Applying the BPS Lagrangian method to Born-Infeld extensions of the $SU(2)$ Yang-Mills-Higgs model, which is called Nakumula-Shiraishi models, we obtained the same BPS equations as shown in [14]. Those BPS equations switch the sign if we shift the potential to a nonzero constant $4b^2$, $V \longrightarrow V + 4b^2$ in which the BPS limit now becomes $V \longrightarrow 4b^2$, as shown in (53). Therefore adding infinitesimally the energy density up to a constant $4b^2$ seems to be related to a transition from monopole (dyon) to antimonopole (antidyon) and vice versa. Since this transition is between BPS monopoles, or dyons, it would be interesting to study continuous transitions by adding the energy density slowly from 0 to $4b^2$, which we would guess to be transition from BPS monopole(dyon) to non-BPS monopole and then to the corresponding BPS antimonopole (antidyon) with higher energy. This transition also appears in all Born-Infeld type of action discussed in this article and we wonder if this transition is generic in all other types of Born-Infeld actions at least with the ones possessing BPS monopole (dyon) in the BPS limit. However, this kind of transition does not appear in $SU(2)$ Yang-Mills-Higgs model and its generalized version since it would correspond to taking $b \longrightarrow \infty$ in the Nakamula-Shiraishi models, which means adding an infinite potential energy to the Lagrangians.

In particular case of monopole Lagrangian (27), we might try to use the identification $j = \beta f$, as previously, into the Lagrangian (27) and look for the BPS equations for dyon from it. However, there is no justification for this identification because the Euler-Lagrange equations for f and j are not identical even after substituting $j = \beta f$ into both Euler-Lagrange equations in the BPS limit. We might also try to consider f and j independently by adding a term that is proportional to j' in the BPS Lagrangian (33), but it will turn out that this term must be equal to zero and thus forces us to set $j = 0$. Surprisingly, for the case of dyon Lagrangian (28), the effective action (30) gives the identical Euler-Lagrange equations for f and j upon substituting $j = \beta f$ in the BPS limit. Therefore it is valid to use this identification for particular Born-Infeld type action of (28) for dyon.

We also applied the BPS Lagrangian method to the generalized version of $SU(2)$ Yang-Mills-Higgs model (25) in which the effective action is given by (71). For monopole case, we found there is a constraint between the scalar-dependent couplings of gauge kinetic term w and of scalar kinetic term G, which is $G = 1/w$, similar to the one obtained in [25]. The BPS equations are also modified and depend explicitly on these scalar-dependent couplings. For dyon case, the constraint is generalized to $w(G - \beta^2 w) = 1 - \beta^2$, with $\beta < |1|$, and the BPS equations are modified as well. This is relatively new result compared to [25, 26] in which they did not discuss dyon. As previously assumed $w, G > 0$, the constraint leads to $w_\pm = (1/2\beta^2)(G \pm \sqrt{G^2 - 4\beta^2(1-\beta^2)})$. Reality condition on $w_\pm$ gives lower bound to G as such $G \geq |2\beta\sqrt{1-\beta^2}|$ in all values of radius r. The generalized version of Nakamula-Shiraishi model for monopole, with Lagrangian (88) and effective Lagrangian (89), has also been computed. The results are similar to the generalized version of $SU(2)$ Yang-Mills-Higgs model for monopole in the BPS limit. In the case for generalized version of Nakamula-Shiraishi model for dyon, with Lagrangian (96) and effective Lagrangian (97), the results are similar to the generalized version of $SU(2)$ Yang-Mills-Higgs model for dyon, even though there are two additional scalar-dependent couplings G_1 and G_2. These additional couplings are related to the kinetic terms' couplings by $G_2 - \beta^2 G_1 = w(G - \beta^2 w)$. In the appendix, based on our results, we write down explicitly the complete square-forms of all effective Lagrangians (29), (30), (89), and (97).

Appendix

A. Complete Square-Forms for Monopoles in Nakamula-Shiraishi Model

For $V = 0$, the effective Lagrangian (29) can be rewritten in complete square-forms as the following:

$$\begin{aligned}\mathscr{L}_{\mathrm{NSm}} = &-\frac{b^2}{\sqrt{\left(1+(1/2b^2)\left(2a'^2/r^2+\left(a^2-1\right)^2/r^4\right)\right)\left(1+(1/2b^2)\left(f'^2+2a^2f^2/r^2\right)\right)}}\\ &\times\left(\frac{\left(2b^2+f'^2\right)}{2b^4r^2}\left(a'-af\frac{\left(a^2-1\right)f'\pm 2b^2r^2}{r^2\left(2b^2+f'^2\right)}\right)^2+\frac{\left(2a^2f^2+r^2\left(2b^2+f'^2\right)\right)}{2b^2r^2\left(2b^2+f'^2\right)}\left(f'\mp\frac{a^2-1}{r^2}\right)^2\right.\\ &+\left.\left(\sqrt{\left(1+\frac{1}{2b^2}\left(\frac{2a'^2}{r^2}+\frac{\left(a^2-1\right)^2}{r^4}\right)\right)\left(1+\frac{1}{2b^2}\left(f'^2+\frac{2a^2f^2}{r^2}\right)\right)}-1\mp\left(\frac{a^2-1}{2b^2r^2}f'+\frac{af}{b^2r^2}a'\right)\right)^2\right)\\ &\mp\left(\frac{2af}{r^2}a'+\frac{a^2-1}{r^2}f'\right).\end{aligned} \tag{A.1}$$

The above expression is different from the one presented in [14].

For $V = 4b^2$, it becomes

$$
\begin{aligned}
\mathscr{L}_{\mathrm{NSm}} = &-\frac{b^2}{\sqrt{\left(1+(1/2b^2)\left(2a'^2/r^2+(a^2-1)^2/r^4\right)\right)\left(1+(1/2b^2)\left(f'^2+2a^2f^2/r^2\right)\right)}}\\
&\times\left(\frac{\left(2b^2+f'^2\right)}{2b^4r^2}\left(a'-af\frac{\left(a^2-1\right)f'\mp 2b^2r^2}{r^2\left(2b^2+f'^2\right)}\right)^2+\frac{\left(2a^2f^2+r^2\left(2b^2+f'^2\right)\right)}{2b^2r^2\left(2b^2+f'^2\right)}\left(f'\pm\frac{a^2-1}{r^2}\right)^2\right.\\
&\left.+\left(\sqrt{\left(1+\frac{1}{2b^2}\left(\frac{2a'^2}{r^2}+\frac{\left(a^2-1\right)^2}{r^4}\right)\right)\left(1+\frac{1}{2b^2}\left(f'^2+\frac{2a^2f^2}{r^2}\right)\right)}-1\pm\left(\frac{a^2-1}{2b^2r^2}f'+\frac{af}{b^2r^2}a'\right)\right)^2\right)\\
&\pm\left(\frac{2af}{r^2}a'+\frac{a^2-1}{r^2}f'\right)-4b^2.
\end{aligned}
\tag{A.2}
$$

Its general expression can be written as

$$
\begin{aligned}
\mathscr{L}_{\mathrm{NSm}} = &-\frac{2b^2\left(\left(V/2b^2-1\right)^2+1\right)^{-1}}{\sqrt{\left(1+(1/2b^2)\left(2a'^2/r^2+(a^2-1)^2/r^4\right)\right)\left(1+(1/2b^2)\left(f'^2+2a^2f^2/r^2\right)\right)}}\times\left(\frac{\left(2b^2+f'^2\right)}{2b^4r^2}\left(a'\right.\right.\\
&\left.-af\frac{\left(a^2-1\right)f'\pm\left(2b^2-V\right)r^2}{r^2\left(2b^2+f'^2\right)}\right)^2+\frac{\left(2a^2f^2+r^2\left(2b^2+f'^2\right)\right)}{2b^2r^2\left(2b^2+f'^2\right)}\left(f'\mp\frac{a^2-1}{2b^2r^2}\left(2b^2-V\right)\right)^2\\
&+\left(\left(\frac{V}{2b^2}-1\right)\left(\sqrt{\left(1+\frac{1}{2b^2}\left(\frac{2a'^2}{r^2}+\frac{\left(a^2-1\right)^2}{r^4}\right)\right)\left(1+\frac{1}{2b^2}\left(f'^2+\frac{2a^2f^2}{r^2}\right)\right)}-1\right)\right.\\
&\left.\left.\pm\left(\frac{a^2-1}{2b^2r^2}f'+\frac{af}{b^2r^2}a'\right)\right)^2+\frac{V}{4b^4}\left(4b^2-V\right)\left(\frac{\left(a^2-1\right)^2}{2b^2r^4}+1\right)\left(\frac{2a^2f^2}{r^2\left(2b^2+f'^2\right)}+1\right)\right)\pm\frac{2\left(V/2b^2-1\right)}{\left(V/2b^2-1\right)^2+1}\left(\frac{2af}{r^2}\right.\\
&\left.\cdot a'+\frac{a^2-1}{r^2}f'\right)-\frac{V^2\left(V-2b^2\right)}{8b^4-4b^2V+V^2},
\end{aligned}
\tag{A.3}
$$

which is valid only if $V = 0$ or $V = 4b^2$.

B. Complete Square-Forms for Dyons in Nakamula-Shiraishi Model

General expression for the complete square-forms of effective Lagrangian (30) is given by

$$\mathscr{L}_{\mathrm{NSd}}$$

$$= -\frac{(2b^2)\left(\left(1 - V/2b^2\right)^2 + 1\right)^{-1}}{\sqrt{1 + (1-\beta^2)\left(f'^2 + 2a^2 f^2/r^2\right)/2b^2 + \left(\left(a^2-1\right)^2/r^4 + 2a'^2/r^2\right)/2b^2 + (1-\beta^2)\left(\left(a^2-1\right)f'/r^2 + 2aa' f/r^2\right)^2/4b^4}}$$
$$\times\left(\frac{(1-\beta^2)}{2b^2}\left(f' \pm \frac{(a^2-1)r^2(V-2b^2)}{2b^2\sqrt{1-\beta^2 r^4}}\right)^2 + \frac{1}{b^2r^2}\left(a' \pm af\left(\frac{V}{2b^2}-1\right)\sqrt{1-\beta^2}\right)^2\right.$$
$$+\frac{V(4b^2-V)}{8b^6r^4}\left(2r^2\left(a^2(1-\beta^2)f^2\right) + \left(a^2-1\right)^2 + 2b^2r^4\right)$$
$$+\left(\left(\frac{V}{2b^2}-1\right)\sqrt{1 + \frac{\left(f'^2 + 2a^2f^2/r^2\right)}{2b^2(1-\beta^2)^{-1}} + \frac{\left(\left(a^2-1\right)^2/r^4 + 2a'^2/r^2\right)}{2b^2} + \frac{\left(\left(a^2-1\right)f'/r^2 + 2aa'f/r^2\right)^2}{4b^4(1-\beta^2)^{-1}}} + 1 - \frac{V}{2b^2}\right.$$
$$\left.\left.\pm\sqrt{1-\beta^2}\left(\frac{af}{b^2r^2}a' + \frac{(a^2-1)}{2b^2r^2}f'\right)\right)^2\right) \mp \frac{2\left(1-V/2b^2\right)\sqrt{1-\beta^2}}{\left(1-V/2b^2\right)^2+1}\left(\frac{2af}{r^2}a' + \frac{(a^2-1)}{r^2}f'\right) + \frac{V^2(2b^2-V)}{8b^4-4b^2V+V^2}, \tag{B.1}$$

which is valid only if $V = 0$ or $V = 4b^2$.

C. Complete Square-Forms for Monopoles in Generalized Nakamula-Shiraishi Model

General expression for the complete square-forms of effective Lagrangian (89) is given by

$$\mathscr{L}_{\text{NSmG}} = -\frac{2b^2\left(\left(1-V/2b^2\right)^2+1\right)^{-1}}{\sqrt{\left(1 + (w/2b^2)\left(\left(a^2-1\right)^2/r^4 + 2a'^2/r^2\right)\right)\left(1 + (G/2b^2)\left(2a^2f^2/r^2 + f'^2\right)\right)}} \times \left(\frac{G\left(a'^2w + b^2r^2\right)}{2b^4r^2}\left(f'\right.\right.$$
$$\left. - \frac{2a\left(a^2-1\right)a'fGw \pm \left(a^2-1\right)\sqrt{Gw}r^2\left(2b^2-V\right)}{2Gr^2\left(a'^2w+b^2r^2\right)}\right)^2 + \frac{w\left(w\left(\left(a^2-1\right)^2 + 2a'^2r^2\right) + 2b^2r^4\right)}{2b^2r^4\left(a'^2w+b^2r^2\right)}\left(a'\right.$$
$$\left.\mp\frac{2af\sqrt{Gw}\left(2b^2-V\right)}{4b^2w}\right)^2 + V\left(4b^2-V\right)\frac{\left(w\left(\left(a^2-1\right)^2 + 2a'^2r^2\right) + 2b^2r^4\right)\left(a^2f^2G + b^2r^2\right)}{8b^6r^4\left(a'^2w+b^2r^2\right)}$$
$$+\left(\left(\frac{V}{2b^2}-1\right)\left(\sqrt{\left(1 + \frac{w\left(\left(a^2-1\right)^2/r^4 + 2a'^2/r^2\right)}{2b^2}\right)\left(1 + \frac{G\left(2a^2f^2/r^2 + f'^2\right)}{2b^2}\right)} - 1\right)\right. \tag{C.1}$$
$$\left.\left.\pm\left(\frac{af\sqrt{Gw}}{b^2r^2}a' + \frac{\left(a^2-1\right)\sqrt{Gw}}{2b^2r^2}f'\right)\right)^2\right) \mp \frac{2\left(1-V/2b^2\right)}{\left(1-V/2b^2\right)^2+1}\left(\frac{2af\sqrt{Gw}}{r^2}a' + \frac{\left(a^2-1\right)\sqrt{Gw}}{r^2}f'\right)$$
$$+\frac{V^2\left(2b^2-V\right)}{8b^4-4b^2V+V^2},$$

which is valid only if $V = 0$ or $V = 4b^2$.

D. Complete Square-Forms for Dyons in Generalized Nakamula-Shiraishi Model

General expression for the complete square-forms of effective Lagrangian (97) is given by

$$\begin{aligned}
&\mathscr{L}_{\mathrm{NSdG}} \\
&= -\frac{2b^2\left(\left(1 - V/2b^2\right)^2 + 1\right)^{-1}}{\sqrt{1 + (G - \beta^2 w)\left(f'^2 + 2a^2 f^2/r^2\right)/2b^2 + w\left(2a'^2/r^2 + \left(a^2 - 1\right)^2/r^4\right)/2b^2 + (G_2 - \beta^2 G_1)\left(\left(a^2 - 1\right)f'/r^2 + 2aa' f/r^2\right)^2/4b^4}} \\
&\times \Bigg(\frac{(G - \beta^2 w)}{2b^2}\left(f' \mp \frac{\left(a^2 - 1\right)\sqrt{G_2 - \beta^2 G_1}\left(2b^2 - V\right)}{2b^2 r^2 (G - \beta^2 w)}\right)^2 + \frac{w}{b^2 r^2}\left(a' \mp \frac{\left(2b^2 - V\right)(4af)\sqrt{G_2 - \beta^2 G_1}}{8b^2 w}\right)^2 + \Big(\beta^2\left(G_1 - w^2\right) \\
&- (G_2 - Gw)\Big)\frac{\left(2a^2 f^2 r^2 (G - \beta^2 w) + \left(a^2 - 1\right)^2 w\right)}{8b^6 r^4 w (G - \beta^2 w)\left(V - 2b^2\right)^{-2}} + V\left(4b^2 - V\right)\frac{\left(2a^2 f^2 r^2 (G - \beta^2 w) + \left(a^2 - 1\right)^2 w\right)}{8b^6 r^4} \qquad \text{(D.1)} \\
&+ \Bigg(\left(\frac{V}{2b^2} - 1\right)\sqrt{1 + \frac{\left(f'^2 + 2a^2 f^2/r^2\right)}{2b^2 (G - \beta^2 w)^{-1}} + \frac{\left(2a'^2/r^2 + \left(a^2 - 1\right)^2/r^4\right)}{2b^2 w^{-1}} + \frac{\left(\left(a^2 - 1\right)f'/r^2 + 2aa' f/r^2\right)^2}{4b^4 (G_2 - \beta^2 G_1)^{-1}}} + 1 - \frac{V}{2b^2} \\
&\pm \left(\frac{af\sqrt{G_2 - \beta^2 G_1}}{b^2 r^2}a' + \frac{\left(a^2 - 1\right)\sqrt{G_2 - \beta^2 G_1}}{2b^2 r^2}f'\right)\Bigg)\Bigg)^2\Bigg) \mp \frac{2\left(1 - V/2b^2\right)\sqrt{G_2 - \beta^2 G_1}}{\left(1 - V/2b^2\right)^2 + 1}\left(\frac{2af}{r^2}a' + \frac{\left(a^2 - 1\right)}{r^2}f'\right) \\
&+ \frac{V^2\left(2b^2 - V\right)}{8b^4 - 4b^2 V + V^2},
\end{aligned}$$

which is valid only if $V = 0$ or $V = 4b^2$, and $\beta^2(G_1 - w^2) = G_2 - Gw$.

Conflicts of Interest

The authors declare that they have no conflicts of interest.

Acknowledgments

We would like to thank Handhika Satrio Ramadhan during the initial work of this article. Ardian Nata Atmaja would like to thank CERN for hospitality during a visit, where the initial writing of this article was done. The visit was supported under RISET-PRO Non-Degree 2017 Programme by Indonesian Ministry of Research, Technology and Higher Education.

References

[1] G. 't Hooft, "Magnetic monopoles in unified gauge theories," *Nuclear Physics B*, vol. 79, pp. 276–284, 1974.

[2] A. M. Polyakov, "Particle Spectrum in the Quantum Field Theory," *Journal of Experimental and Theoretical Physics*, vol. 20, p. 194, 1974.

[3] P. A. M. Dirac, "Quantized Singularities in the Electromagnetic Field," *Proceedings of the Royal Society of London. Series A, Containing Papers of a Mathematical and Physical Character*, vol. 133, p. 60, 1931.

[4] B. Julia and A. Zee, "Poles with both magnetic and electric charges in non-Abelian gauge theory," *Physical Review D: Particles, Fields, Gravitation and Cosmology*, vol. 11, no. 8, pp. 2227–2232, 1975.

[5] M. Prasad and C. Sommerfield, "An Exact Classical Solution for the 't Hooft Monopole and the Julia-Zee Dyon," *Physical Review Letters*, vol. 35, no. 12, p. 760, 1975.

[6] E. B. Bogomolny, "Stability of classical solutions," *Soviet Journal of Nuclear Physics*, vol. 24, pp. 449–870, 1976.

[7] E. Witten and D. Olive, "Supersymmetry algebras that include topological charges," *Physics Letters B*, vol. 78, no. 1, pp. 97–101, 1978.

[8] A. Abouelsaood, C. G. Callan, C. R. Nappi, and S. A. Yost, "Open strings in background gauge fields," *Nuclear Physics B*, vol. 280, pp. 599–624, 1987.

[9] A. A. Tseytlin, "On non-abelian generalisation of the Born-Infeld action in string theory," *Nuclear Physics B*, vol. 501, no. 1, pp. 41–52, 1997.

[10] A. Hashimoto and I. Taylor, "Fluctuation spectra of tilted and intersecting D-branes from the Born-Infeld action," *Nuclear Physics B*, vol. 503, no. 1-2, pp. 193–219, 1997.

[11] D. J. Gross, A. Hashimoto, and I. R. Klebanov, "The Spectrum of a large N gauge theory near transition from confinement to screening," *Physical Review D: Particles, Fields, Gravitation and Cosmology*, vol. 57, no. 10, pp. 6420–6428, 1998.

[12] D. Brecher, "BPS states of the non-abelian Born-Infeld action," *Physics Letters B: Particle Physics, Nuclear Physics and Cosmology*, vol. 442, no. 1-4, pp. 117–124, 1998.

[13] S. Gonorazky, F. A. Schaposnik, and G. Silva, "Supersymmetric non-abelian Born-Infeld theory," *Physics Letters B: Particle Physics, Nuclear Physics and Cosmology*, vol. 449, no. 3-4, pp. 187–193, 1999.

[14] A. Nakamula and K. Shiraishi, "Born-Infeld Monopoles and Instantons," *Hadronic Journal*, vol. 14, no. 5, pp. 369–375, 1991.

[15] N. Grandi, R. L. Pakman, F. A. Schaposnik, and G. Silva, "Monopoles, dyons, and the theta term in Dirac-Born-Infeld theory," *Physical Review D: Particles, Fields, Gravitation and Cosmology*, vol. 60, no. 12, Article ID 125014, 1999.

[16] A. Nata Atmaja, "A method for BPS equations of vortices," *Physics Letters B*, vol. 768, pp. 351–358, 2017.

[17] C. Adam, L. A. Ferreira, E. da Hora, A. Wereszczynskiand, and W. J. Zakrzewski, "Some aspects of self-duality and generalised BPS theories," *Journal of High Energy Physics*, vol. 2013, article 62, 2013.

[18] D. Bazeia, C. B. Gomes, L. Losano, and R. Menezes, "First-order formalism and dark energy," *Physics Letters B: Particle Physics, Nuclear Physics and Cosmology*, vol. 633, no. 4-5, pp. 415–419, 2006.

[19] D. Bazeia, L. Losano, R. Menezes, and J. C. R. E. Oliveira, "Generalized global defect solutions," *The European Physical Journal C*, vol. 51, no. 4, pp. 953–962, 2007.

[20] D. Bazeia, L. Losano, J. Rodrigues, and R. Rosenfeld, "First-order formalism for dark energy and dust," *The European Physical Journal C*, vol. 55, no. 1, pp. 113–117, 2008.

[21] A. N. Atmaja and H. S. Ramadhan, "Bogomol'nyi equations of classical solutions," *Physical Review D: Particles, Fields, Gravitation and Cosmology*, vol. 90, no. 10, 2014.

[22] K. Sokalski, T. Wietecha, and Z. Lisowski, "A concept of strong necessary condition in nonlinear field theory," *Acta Physica Polonica B*, vol. 32, pp. 17–28, 2001.

[23] C. Adam and F. Santamaria, "The first-order Euler-Lagrange equations and some of their uses," *Journal of High Energy Physics*, vol. 47, no. 12, 2016.

[24] S. Coleman, S. Parke, A. Neveu, and C. M. Sommerfield, "Can one dent a dyon?" *Physical Review D: Particles, Fields, Gravitation and Cosmology*, vol. 15, no. 2, pp. 544-545, 1977.

[25] R. Casana, M. M. Ferreira Jr., and E. da Hora, "Generalized BPS magnetic monopoles," *Physical Review D: Particles, Fields, Gravitation and Cosmology*, vol. 86, no. 8, Article ID 085034, 2012.

[26] R. Casana, M. M. Ferreira, E. da Hora, and C. dos Santos, "Analytical self-dual solutions in a nonstandard Yang-Mills-Higgs scenario," *Physics Letters B*, vol. 722, no. 1–3, pp. 193–197, 2013.

[27] E. J. Weinberg, *Classical Solutions in Quantum Field Theory: Solitons and Instantons in High Energy Physics*, Cambridge University Press, 2012.

10

Simulation Study for the Energy Resolution Performances of Homogenous Calorimeters with Scintillator-Photodetector Combinations

G. Aydın

Department of Physics, Mustafa Kemal University, 31034 Hatay, Turkey

Correspondence should be addressed to G. Aydın; guralaydin@gmail.com

Academic Editor: Lorenzo Bianchini

The scintillating properties of active materials used in high energy and particle physics experiments play an important role regarding the performances of both calorimeters and experiments. Two scintillator materials, a scintillating glass and an inorganic crystals, were examined to be used for collider experiments showing good optical and scintillating properties. This paper discusses the simulated performances of two materials of interest assembled in a scintillator-photodetector combination. The computational study was carried out with Geant4 simulation program to determine energy resolutions of such calorimeter with different beam energies and calorimeter sizes.

1. Introduction

Scintillator materials are used in high energy physics experiments as active materials of calorimeters to measure energy and position of particles passing through calorimeters generating photons proportional to incoming beam energies. Two types of calorimeters could be constructed: sampling or homogenous [1, 2]. A sampling calorimeter consists of an absorber and an active material in alternating layers resulted in absorption of only some part of incident beam energy in active materials. A homogenous calorimeter is entirely made of an active material with no absorber, thus leading to absorption of the most of incident energy depending on thickness and radiation length of the material. Sampling calorimeters serve for both electromagnetic and hadronic interactions but homogenous calorimeters are used for only electromagnetic interactions due to their long interaction length. In such detector systems, several properties of active materials affect the performance of calorimeters and experiments. First of all, light yield of a scintillator should be high enough to obtain required energy resolution. Next, the rate of data taking is important when considering short time intervals between collisions. Therefore, the response time of detectors should be as fast as possible to detect even rare events. Decay times of scintillator materials affect the time interval of signal formation and thus they are key factors for data taking rate in calorimeters. High density in scintillating materials increases stopping power and it is important in two ways. One is that it increases energy and spatial resolutions and it enables construction of more compact systems. Moreover, a scintillator with a good optical transmission has significant impact on the formation of proper electrical signal in photodetectors to which photon pulses produced in the scintillator are directed. Beyond these facts, scintillator materials could show very good properties in some features but could also have some drawbacks for the remaining aspects. In summary, scintillating materials with high densities, required light yield, producing fast, and short light pulses have crucial role to build detector systems which are compact, enabling fast data taking and achieving required energy and spatial resolutions. For example, lead tungsten crystals (PWO) with high density and fast decay times are used in electromagnetic part of the Compact Muon Selenoid (CMS) to measure incident electron or photon beam energies which is used to search Higgs boson [1, 2]. New experiments are also searching for scintillating materials with good optical and scintillating properties as much as possible as an active material in calorimeter designs. This report presents a computational study concentrated on energy resolution performances of different scintillating materials which could be used as an active material of a homogenous calorimeter in particle physics experiments. The interested materials are Ce doped HfG (Hafnium Fluoride Glass) [3, 4] and Ce doped $Gd_2Y_1Ga_{2.7}Al_{2.3}O_{12}$ [5] due to their good optical and scintillating properties. Here, HfG is a scintillating glass and the other is inorganic crystal. To the best to our knowledge, both scintillating materials were not used in a high energy physics experiment as an active material of a homogenous calorimeter or their simulation studies were not presented for energy resolution calculation belonging to a certain size or sets of calorimeter setups. These materials have mass production capabilities. Generally, scintillation glasses are potentially more homogenous compared to crystal scintillators and light yield of scintillation glasses could be increased by changing their elemental compositions. The selected glass material has been preferred among some heavy metal fluoride glass due to its optical and scintillation properties. On the other hand, these improvements in crystals could be achieved by increasing purity of the crystals and with better understanding of luminescence mechanisms. Scintillation glasses have less light yield compared to crystals but generally fast decay times [6].

This study determines energy resolution of homogenous calorimeter setups with certain sizes to see the performance of the selected materials. The detailed explanation of the materials belonging to their physical and scintillation properties and the simulation procedure are given in detail in Section 2.

2. Materials and Methods

The energy resolution of a scintillator could be characterized with four parameters: lateral part, photostatistics contribution, constant, and noise term. The energy resolution is the quadratic summation of all four terms as indicated in

$$\frac{\sigma(E)}{E} = \frac{a_{lateral}}{E^{1/4}} \oplus \frac{a_{pe}}{\sqrt{E}} \oplus b \oplus \frac{c}{E} \quad (1)$$

Here, $a_{lateral}$, a_{pe}, b, and c refer to lateral part, photoelectron statistics contribution, constant, and noise terms, respectively. The lateral part represents fluctuations of shower development inside the scintillating material ($a_{lateral}$) belonging to lateral shower containment and contribution from the statistics of the photoelectrons produced in a photodetector, which converts photons reaching its active area into electrons in terms of its wavelength dependent quantum efficiency and its internal gain, is represented with a_{pe}. The total energy resolution is calculated as the quadratic summation of all terms excluding noise term in this study. The constant term refers to other inhomogeneities of the material such as variations of reflectance of different surfaces of the scintillator or impurities in the material. This is also some portion of the energy deposition fluctuation. The contribution from photoelectron statistics is given by [2]

$$a_{pe} = \sqrt{\frac{\overline{F}}{N_{pe}}} \quad (2)$$

Here, N_{pe} is the number of photoelectrons per GeV and F is the emission weighted excess noise factor due to avalanche gain process. N_{pe} depends on the number of photons reaching the rear edge of the scintillator, active area of photodetectors, and wavelength dependent quantum efficiencies of the photodetectors. In the presented study, noise term is not included in energy resolution calculation. Four different photodetectors were used in the study: two PIN diodes and two avalanche photo diodes (APD). Recently, several experiments used the pin diode: Babar [7], BELLE [8], and BesIII [9] used Hamamatsu S2744-08 PIN diode to collect photons from crystals. APD S8664-55, on the other hand, are used in CMS [10] with PWO calorimeter. The pin diode S2744-08 and APD S8664-55 have spectral ranges of 340 to 1100 nm and 320 nm and 1100 nm, respectively [11]. These were used with $Gd_2Y_1Ga_{2.7}Al_{2.3}O_{12}$ scintillator. Another Si APD and PIN diode were used for HfG by considering their emission spectra which peak at relatively lower wavelength. It is Si APD S5345 whose spectral range is between 200 nm and 1000 nm and Si PIN diode S1227-1010BQ with 190 to 1000 nm spectral range [11]. The photodetectors used in the presented study have the active areas of 1 cm x 2 cm, 5 mm x 5 mm, and 5 mm in diameter, and 10 mm x 10 mm for S2744-08 pin diode, APD S8664-55, and ADP S5345, and S1227-1010BQ pin, respectively. The quantum efficiency of the pin diode S2744-08 is around 10% at the wavelength of 300 nm, 50% at 400 nm, and reaches 83% at 580 nm wavelength of the photon emission. On the other hand, those efficiency values are seen for APD S8664-55 as 23%, 70%, and 85% indicating obviously that APD is more efficient at relatively higher wavelengths. Si photodiodes have fast response, high sensitivity, and low noise. PIN diodes have no internal gain, so they do not contribute to photoelectron statistics due to excess noise factor. Since APD has avalanche gain process, it contributes to the photoelectron statistics term as excess noise factor due to fluctuations in gain process. This factor is wavelength dependent and the excess noise factor is determined as 2 for the emission wavelength below 500 nm [12]. This value was used for HfG scintillator-photodetector systems. On the other hand, this factor was calculated as 2.346 for the $Gd_2Y_1Ga_{2.7}Al_{2.3}O_{12}$ by taking into account its emission spectrum together with excess noise factor distribution of an APD as a function of wavelength [12] according to

$$\overline{F} = \frac{\int F(\lambda)\, Em(\lambda)\, d\lambda}{\int Em(\lambda)\, d\lambda} \quad (3)$$

where $F(\lambda)$ is the excess noise factor as a function of wavelength and $Em(\lambda)$ is the emission weights of the spectrum for a given scintillator material.

The scintillating process in Geant4 [13–15] is as follows: energy lost for each step determines the number of optical photons which has Gaussian distribution shape and statistical fluctuations occur around average light yield entered as scintillation yield. Photons are generated along beam direction emitted uniformly into 4π. They are emitted according to random linear polarization and scintillation time components. This process produces optical photons which are directly used in a desired application. Instead of working directly with optical photons, the following method is reasonable by taking into account required information belonging the material and electronics coupled to it.

The number of optical photons produced in a scintillation process is proportional to energy deposition in the material during the process. If there is no self-absorption in the material, these produced photons will be directed to the rear back of the material in theoretical limits. Therefore, the following procedure to calculate energy resolution is an appropriate method by using Geant4 program: First of all, the distribution of energy deposition event to event is obtained for the related beam energies. Fitting this distribution with suitable function will give energy resolution value for the interested calorimeter setup. This value refers to the contribution to energy resolution due to event to event fluctuations of the number of produced optical photons. The remaining part of the fluctuation is due to photoelectron statistics. In this study, since pin diodes have no internal gain, and the contribution from variance of the gain process in the photodetector is neglected for pin diodes. Here, the calculation of the average number of photoelectrons produced in the photodetectors determines the photoelectron statistics contribution. Clearly, it depends on the number of photons produced in the material and reaching to the active areas of the photodetectors. The light yield is the main scintillation property of a scintillating material. In this study, the yield values per MeV were used to determine the average number of photons produced in the optical process. Then, the number of photons reaching the rear edge of the material was calculated by taking account of transmission spectra of the interested materials. The next step was to determine the number of photons hitting the photodetector active area by taking account of the ratio of the photodetector area to the total back face area of the scintillator. The final step was to determine the number of photoelectrons produced in the photodetectors according to their emission weighted quantum efficiencies.

In the present study, two beam facing areas (20 mm x 20 mm and 25 mm x 25 mm) were selected for each scintillator forming 5 x 5 matrix geometry. In this way, the total beam facing area was either 100 mm x 100 mm or 125 mm x 125 mm. The areas of beam facing and back face of each scintillator were set equally in the simulation. Five different thicknesses (17 cm, 20 cm, 23 cm, 25 cm, and 27 cm) were tested for calorimeter performances. Totally, ten geometric configurations were examined to see the changes of calorimeter performances with detector sizes and to compare obtained results with previous experimental or simulation results in certain sizes. Gamma was used as an incident beam with different energies ranging from 100 MeV to 2 GeV by directing the beam to the center of the matrix. The ratios of the active area of the APDs S8664-55 and S5345 to the total area of the calorimeter back face were calculated as 0,125 and 0,0982 for 20 mm x 20 mm and 0,08 and 0,0628 for 25 mm x 25 mm back face areas of the detectors, respectively, taking into account the fact that each scintillators locates a pair of diodes at the back faces. These values were calculated for pin S2744 and pin S1227 as 1,0 and 0,5 with 20 mm x 20 mm and 0,64 and 0,32 with 25 mm x 25 mm back face areas of the detectors, respectively.

The previous studies with CsI(Tl), PWO, and LYSO crystals showed that the mentioned simulation procedure gives compatible and reasonable results compared to experimental results [16, 17]. It was shown that number of average photoelectrons determined at the end of whole process is reasonable if considering it with experimental ones and energy resolution values obtained are in agreement. In this study, the standard electromagnetic process was used to obtain energy deposition distribution per event. In Geant4, the processes belonging to the interactions of beams with matter are determined in seven categories: electromagnetic, hadronic, decay, photolepton-hadron, optics, parameterization, and transportation. The electromagnetic processes could be summarized as follows. Photon processes include gamma conversion or pair production, photoelectric effect, Compton scattering, Rayleigh scattering, and muon pair production. Electron/positron processes cover ionization and delta-ray production, Bremsstrahlung radiation, electron-positron pair production, annihilation to two gammas of a positron, multiple scattering, the annihilation to two muons of a positron, and annihilation to two hadrons of a positron. On the other hand muon processes include the following processes: ionization and delta-ray production, Bremsstrahlung radiation, electron/positron pair production, and multiple scattering. Hadron and ions also include ionization for hadron and ions in addition to the standard electromagnetic processes. Coulomb scattering in the model is considered different for ions and charged particles. In addition, the production of optical photons is determined with Cherenkov and scintillation processes. The standard electromagnetic package uses physics tables which are reconstructed between 100 eV and 100 TeV energy range.

HfG (Hafnium Fluoride Glass) is based on HfF_4-BaF_2-NaF-AlF_3-YF_3 system with molar mass fractions of 0.56, 0.28, 0.12, 0.02, and 0.02, respectively. 2.5% Ce doped HfG was used in the present study since it shows good transparency within emission spectra indicating no self-absorption. Ce doped fluorohafnate glass showed very fast decay time with short and long time constants of 8 ns and 25 ns, respectively. Its emission spectra range between 290 nm and 400 nm peaking at the wavelength of 310 nm. It has a density of 5.95 g/cm^3 with the refractive index of 1.495. Its radiation length is 1.6 cm and light yield of 150 photon/MeV. When it is compared to newly produced scintillation glass of Ce doped DSB [18], the following expressions could be stated: DSB glass lower stopping power. Its density and radiation length are 3.8 g/cm^3 and, 3.3 cm, respectively. DSB has fast decay time of 30 ns and additionally slower decay time of 180 ns. On the other hand, its light output is about five times larger than that of PWO. HfG's emission weighted transmission

TABLE 1: Number of photoelectrons per MeV (N_{pe}/MeV) produced for different scintillator back face areas and photodetector combinations.

Material	Number of photoelectrons per MeV (N_{pe}/MeV)			
	APD		PIN	
	Area (20 mm x 20 mm)	Area (25 mm x 25 mm)	Area (20 mm x 20 mm)	Area (25 mm x 25 mm)
GdY	5402	3457	40881	26164
HfG	5,1	3,3	33	21

TABLE 2: The photostatistics parts of the parameterized energy resolution function (a_{pe}) with different scintillator back face areas and photodetector combinations.

Material	Photodetector signal fluctuations (a_{pe} as % in the unit of $GeV^{1/2}$)			
	APD		PIN	
	Area (20 mm x 20 mm)	Area (25 mm x 25 mm)	Area (20 mm x 20 mm)	Area (25 mm x 25 mm)
GdY	0.066	0.082	0.016	0.020
HfG	1.973	2.467	0.548	0.685

rate was determined as 80%. Emission weighted quantum efficiencies with APD and PIN were calculated as 43.7% and 55.5%, respectively. Transmission spectra and emission weighted quantum efficiencies together with APD and PIN diode indicate that 34.9% and 44.4% of the produced photons create an electron in the photodetector without considering photodetector active areas.

Ce1%:$Gd_2Y_1Ga_{2.7}Al_{2.3}O_{12}$ is a new single crystal grown by Czochralski method. Its emission spectra range between 490 nm and 590 nm peaking at 530 nm. It reached the 65000 photons/MeV with two decay time constants of 93.5 ns and 615 ns, the relative intensities of which is 40.2% and 50.8%, respectively. Its good optical and scintillation properties together with its relatively high density of 6.3 g/cm^3 make it a good alternative for gamma-ray detection and nuclear nonproliferation applications. It was seen from the report that its transmission spectra are well within its emission spectra indicating no significant self-absorption. Additionally, its emission spectra are well matched with pin diode efficiencies and APD quantum efficiency spectra. The emission weighted transmission value was determined as 79% for Ce1%:$Gd_2Y_1Ga_{2.7}Al_{2.3}O_{12}$ scintillator. Both APD and PIN diode were used as a photodetector with Ce1%:$Gd_2Y_1Ga_{2.7}Al_{2.3}O_{12}$ scintillator. Emission weighted quantum efficiencies were calculated as 84.1% and 79.5% with APD and PIN diode, respectively. When the spectra of the quantum efficiencies are considered with transmission spectra it is found that 66.5% and 62.9% of the produced photons contribute the production of photoelectron in APD and PIN diode, respectively. This will decrease when the active areas of the photodetectors are taken into account.

The simulation study was performed with Geant4 high energy physics simulation package to determine the energy resolution of the interested scintillating materials as homogenous calorimeters. The intrinsic energy resolution caused by event to event energy deposition fluctuation was defined as the ratio of the sigma to the mean value of the logarithmic Gaussian fit function on the distributions of energy deposition in scintillator material per event. The fit function is given with (4) [19]. Later, photodetector signal fluctuations were calculated with the appropriate process mentioned above.

$$F(x) \equiv N \exp\left(-\frac{1}{2\sigma_0^2}\ln^2\left(1-\frac{x-x_p}{\sigma_E}\eta\right)-\frac{\sigma_0^2}{2}\right) \quad (4)$$

where $\sigma_0 = 2/\xi \sinh^{-1}(\eta\xi/2)$ and $\xi = 2\sqrt{\ln 4}$. In the formula, x_p is the peak value, η is the asymmetry parameter, N is the normalization factor, and σ_E is the full width at half maximum (FWHM) divided by ξ. The energy resolution was defined as the ratio of σ_E to the peak value x_p.

3. Results and Discussion

After this point, the name of the Ce1%:$Gd_2Y_1Ga_{2.7}Al_{2.3}O_{12}$ scintillator will be abbreviated as GdY in histograms and in the text. First of all, photoelectron production rates and the ratios of the active areas of the photodetectors to scintillator back face area were evaluated together and the average number of photoelectrons (N_{pe}) per MeV produced at the photodetector in an event was obtained for different scintillator backface detector geometries and photodetector combinations. This is given in Table 1.

As expected HfG will give the lowest photoelectrons and this will cause significant contribution to energy resolution. The contributions from photodetector signal fluctuations (a_{pe}), which was calculated with (2), are given in Table 2 for different detector combinations.

As it is seen, they are very low and negligible for GdY material with both photodetectors. Those are significant for HfG scintillator. Indeed, Pin diode S1227 makes this contribution less harmful with its size and high UV sensitivity. Here, it can be said that new pin diode technology could make a scintillator more efficient compared to older photodetectors with unmatched scintillator emission spectrum at relatively lower wavelengths. A typical fit to the energy deposition distribution to obtain intrinsic energy resolution for a certain beam energy and detector geometry is shown in Figure 1. It is for the 1 GeV beam energy on the HfG calorimeter with the size of 25 mm x 25 mm back face and 27 cm in thickness.

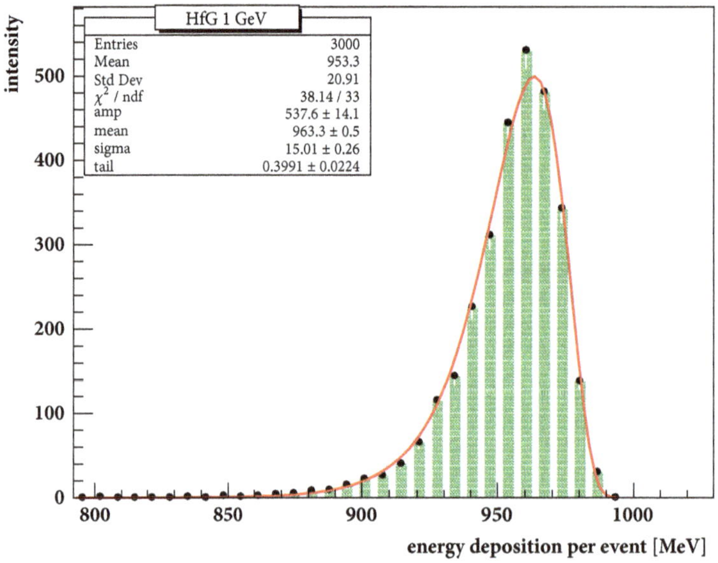

FIGURE 1: Typical energy resolution fitting belonging to HfG scintillator for 1 GeV beam energy.

Figures 2–5 show intrinsic energy resolution results for all detector geometries as a function of beam energies. First of all, with evaluating all histograms the thicknesses of 17 cm and 20 cm will not be considered as material thicknesses since they do not follow a good shape with beam energies even the resolutions decrease with beam energies. For GdY and 20 mm × 20 mm cross area of each scintillator, the energy resolutions were determined as 2.14% and 1.76% for the material thicknesses of 25 cm and 27 cm, respectively, at 2 GeV/c beam energy.

For GdY and 25 mm × 25 mm back face area, the resolution values has been obtained as 1.55% for 27 cm material thickness. In the case of 25 cm thick GdY, energy resolution has no proper shape fluctuating around a line with beam energy. For HfG and 20 mm × 20 mm back face area, energy resolution reached 2.23%, 1.83%, and 1.58% for 23 cm, 25 cm, and 27 cm calorimeter thicknesses, respectively. For HfG and 25 mm × 25 mm cross section area of each scintillator, these values were determined as 2.11%, 1.69%, and 1.41%. It is seen that the resolutions increase with back face areas and calorimeter thicknesses. It can be stated that the resolutions belonging to the geometries of 25 mm × 25 mm beam facing area and calorimeter thicknesses of 25 cm or 27 cm give more compatible results with previous studies [16, 17]. Therefore, energy resolution functions were parameterized with these detector geometries. Considering the selected material thicknesses and detector back face geometries, the parameters of the total energy resolution function were obtained with the function given in (1) excluding noise term (c). Here, photoelectron statistics contribution was determined according to (2). These parameters are given in Tables 3 and 4 for 25 mm × 25 mm and 20 mm × 20 mm back face detector geometries, respectively. These values could not be determined for GdY material with 25 mm x 25 mm cross area and 25 mm thickness since the energy resolution values do not follow good shape with beam energies fluctuating around a line. Figures 6–8 show parameterized energy resolution functions and resolution values calculated with the related parameterized function at certain beam energies for the scintillators with back face area of 25 mm × 25 mm and the thicknesses of 25 cm and 27 cm with APD and PIN. It should be noted that the resolutions are quite different especially at lower beam energies below 1 GeV/c whether PIN or APD is used with HfG. Finally, the followings are the best parameters obtained over the examined detector combinations with the scintillator sizes of 25 mm × 25 mm back face area and 27 cm thickness:

$$\begin{aligned}
\frac{\sigma}{E} &= \frac{0.84\%}{E^{1/4}} \oplus \frac{0.08\%}{\sqrt{E}} \oplus 1.39\% \quad \textit{for GdY + APD} \\
\frac{\sigma}{E} &= \frac{0.84\%}{E^{1/4}} \oplus \frac{0.02\%}{\sqrt{E}} \oplus 1.39\% \quad \textit{for GdY + PIN} \\
\frac{\sigma}{E} &= \frac{1.17\%}{E^{1/4}} \oplus \frac{2.47\%}{\sqrt{E}} \oplus 1.05\% \quad \textit{for HfG + APD} \\
\frac{\sigma}{E} &= \frac{1.17\%}{E^{1/4}} \oplus \frac{0.69\%}{\sqrt{E}} \oplus 1.05\% \quad \textit{for HfG + PIN}
\end{aligned} \tag{5}$$

After this point, the same parameterized results were obtained as 3x3 matrix with the optimized detector geometries of 27 cm in thickness and 25 mm × 25 mm back face area for scintillator-PIN diode combinations. The results were shown together with 5x5 matrices for GdY and HfG in Figures 9 and 10, respectively. It is obvious that energy resolutions increase with transverse sizes. It is also seen that transverse size is more effective at lower beam energies below 1 GeV/c.

4. Conclusions

A computational study was carried out to determine energy resolutions of two different scintillator materials to be used as

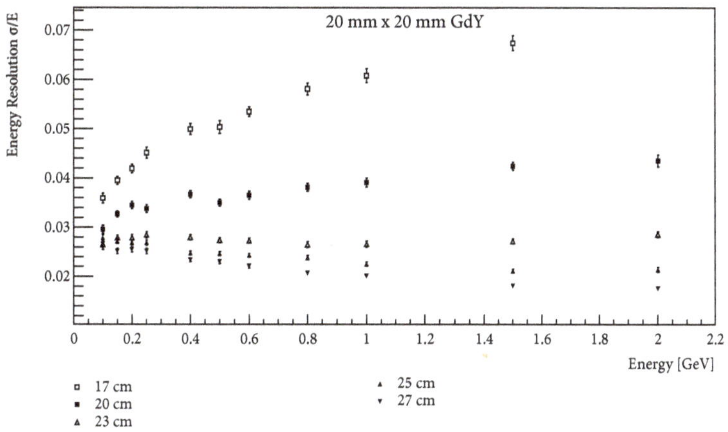

FIGURE 2: Energy resolutions as a function of beam energy for GdY with 20 mm x 20 mm back face area and five different thicknesses.

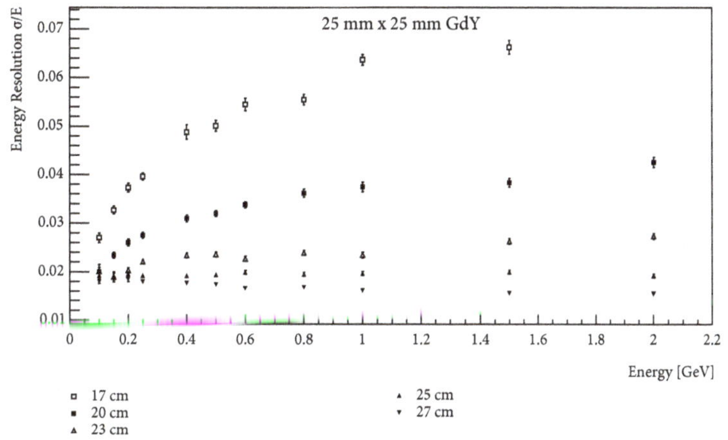

FIGURE 3: Energy resolutions as a function of beam energy for GdY with 25 mm x 25 mm back face area and five different thicknesses.

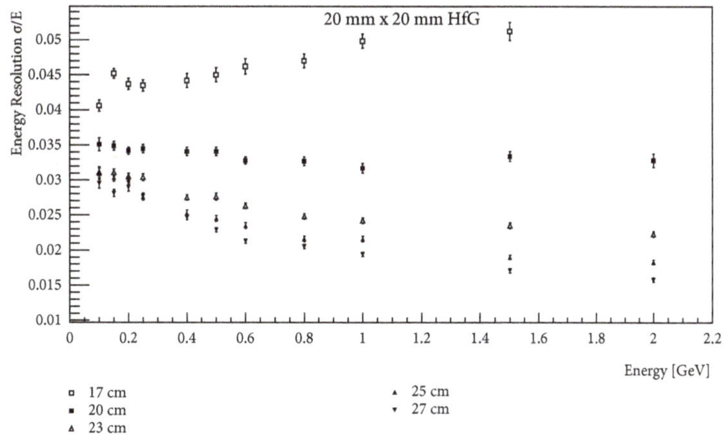

FIGURE 4: Energy resolutions as a function of beam energy for HfG with 20 mm x 20 mm back face area and five different thicknesses.

a homogenous calorimeter in particle physics experiments. Since GdY has very high light yield and its emission spectrum matches well with two photodetectors, photostatistics contribution to the total energy resolution is negligible. For HfG, in both case there will be significant contribution but with APD this will be enormous resulting in huge decrease on energy resolution especially at lower beam energies below 1 GeV/c. Above 1 GeV/c, this contribution could be thought as reasonable. On the other hand, it could be stated that the resolution will increase significantly at lower beam energies below 1 GeV/c if the number of photodiodes are increased. In addition, the calculation procedure of the average number

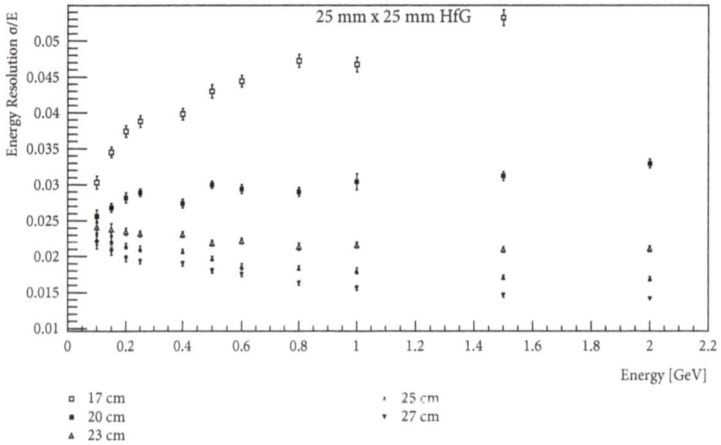

FIGURE 5: Energy resolutions as a function of beam energy for HfG with 25 mm x 25 mm back face area and five different thicknesses.

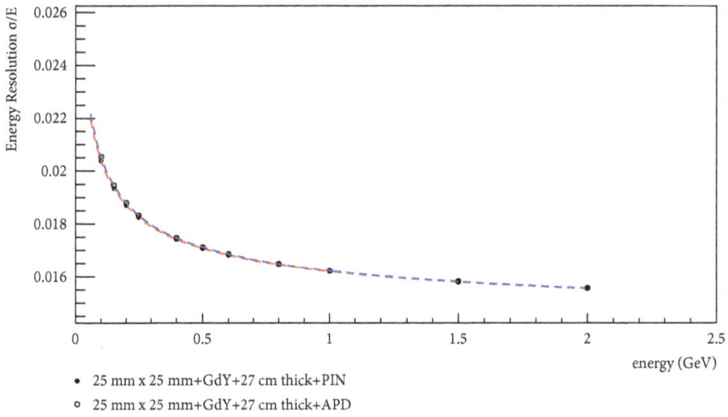

FIGURE 6: Parameterized energy resolution function (dashed lines) and energy resolution values at certain energies (points) for scintillator-photodetector combinations belonging to GdY with 25 mm x 25 mm back face area of each scintillator and 27 cm thickness.

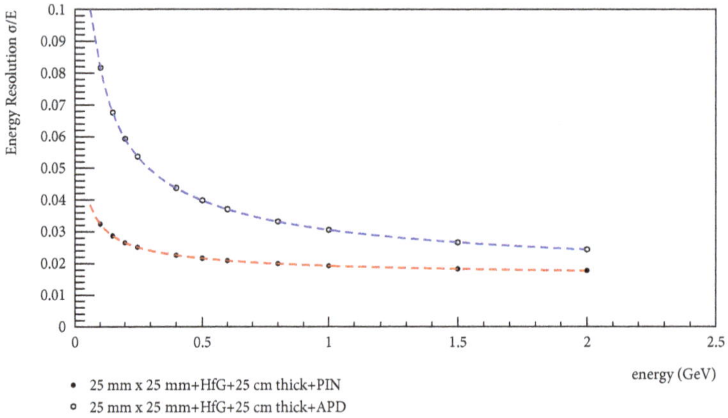

FIGURE 7: Parameterized energy resolution function (dashed lines) and energy resolution values at certain energies (points) for scintillator-photodetector combinations belonging to HfG with 25 mm x 25 mm back face area and 25 cm thickness.

of photoelectrons will give the estimation of the minimum number of photons detected. In a real experiment, the possibility of detection of the number of the photons will increase due to randomly polarized photons and scattering via surface reflectors. If PIN diode is used with HfG, this contribution will be very reasonable especially at relatively high energies. It could be stated that both scintillators will give very compatible results for material thicknesses of 25 or 27 cm and with appropriate photodetectors when compared to previous studies. This allows being stated that these

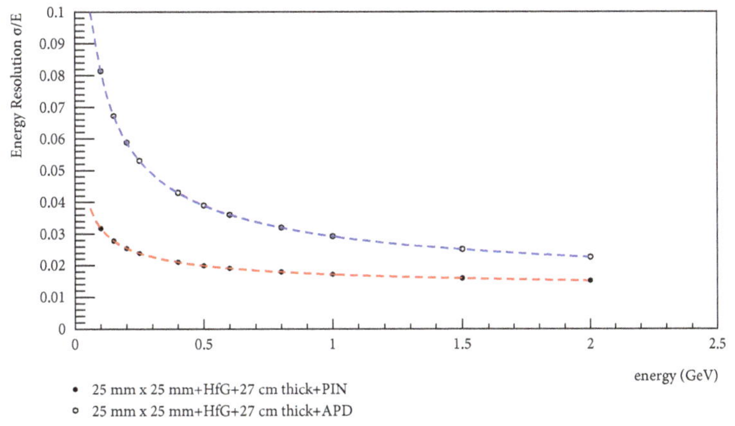

FIGURE 8: Parameterized energy resolution function (dashed lines) and energy resolution values at certain energies (points) for scintillator-photodetector combinations belonging to HfG with 25 mm x 25 mm back face area and 27 cm thickness.

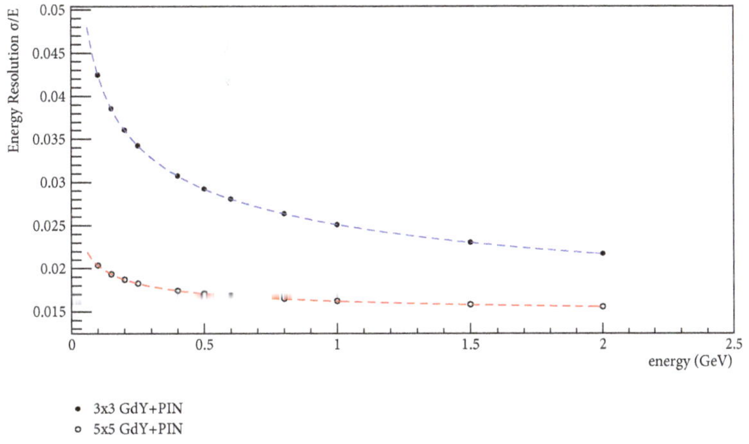

FIGURE 9: Parameterized energy resolution function for GdY as 3x3 and 5x5 matrices. Each scintillator has 27 cm thickness and back face area of 25 mm × 25 mm.

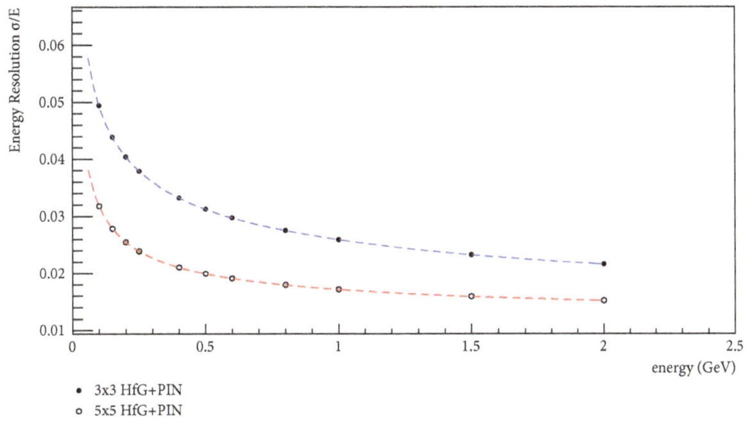

FIGURE 10: Parameterized energy resolution function for HfG as 3x3 and 5x5 matrices. Each scintillator has 27 cm thickness and back face area of 25 mm × 25 mm.

materials could be seen as alternatives in particle physics experiments by taking into account their advantages. GdY's very high light yield could make it preferable especially for relatively low beam energies. The main advantage of HfG is its very fast decay times of 8 ns and 25 ns. Finally, their relatively high densities allow reaching compatible resolutions results with smaller detector sizes compared to scintillators with lower densities.

TABLE 3: Parameters of the energy resolution functions of the scintillators with different scintillator geometries and photodetector combinations.

Material	Parameters of the energy resolution function with different scintillator thicknesses and photodetector combinations. Each scintillator has 25 mm × 25 mm back face area.											
	APD						PIN					
	25 cm			27 cm			25 cm			27 cm		
	$a_{lateral}$	a_{pe}	b	$a_{lateral}$	a_{pe}	b	$a_{lateral}$	a_{pe}	b	$a_{lateral}$	a_{pe}	b
GdY	-	-	-	0.84	0.08	1.39	-	-	-	0.84	0.02	1.39
HfG	1.09	2.47	1.44	1.17	2.47	1.05	1.09	0.69	1.44	1.17	0.69	1.05

TABLE 4: Parameters of the energy resolution functions of the scintillators with different scintillator geometries and photodetector combinations.

Material	Parameters of the energy resolution function with different scintillator thicknesses and photodetector combinations. Each scintillator has 20 mm × 20 mm back face area.											
	APD						PIN					
	25 cm			27 cm			25 cm			27 cm		
	$a_{lateral}$	a_{pe}	b	$a_{lateral}$	a_{pe}	b	$a_{lateral}$	a_{pe}	b	$a_{lateral}$	a_{pe}	b
GdY	1.25	0.07	1.91	1.41	0.07	1.40	1.25	0.02	1.91	1.41	0.02	1.40
HfG	1.80	1.97	1.07	1.83	1.98	0.57	1.80	0.55	1.07	1.83	0.55	0.57

Conflicts of Interest

The author declares that there are no conflicts of interest regarding the publication of this paper.

Acknowledgments

The author would like to thank to Dr. Ugur Akgun for his valuable comments on this work.

References

[1] R. Paramatti, "Design options for the upgrade of the CMS electromagnetic calorimeter," *Nuclear and Particle Physics Proceedings*, vol. 273-275, pp. 995–1001, 2016.

[2] CMS Collaboration, "The CMS experiment at the CERN LHC," *JINST*, vol. 3, Article ID S08004, 2008.

[3] E. Auffray, D. Bouttet, I. Dafinei et al., "Cerium doped heavy metal fluoride glasses, a possible alternative for electromagnetic calorimetry," *Nuclear Instruments and Methods in Physics Research Section A: Accelerators, Spectrometers, Detectors and Associated Equipment*, vol. 380, no. 3, pp. 524–536, 1996.

[4] I. Dafinei, E. Auffray, P. Lecoq, and M. Schneegans, *MRS Proceedings Series (Scintillators and Phosphors)*, vol. 348, 1994.

[5] C. Wang, Y. Wu, D. Ding et al., "Optical and scintillation properties of Ce-doped $(Gd_2Y_1)Ga_{2.7}Al_{2.3}O_{12}$ single crystal grown by Czochralski method," *Nuclear Instruments and Methods in Physics Research Section A: Accelerators, Spectrometers, Detectors and Associated Equipment*, vol. 820, pp. 8–13, 2016.

[6] R. Y. Zhu, "Crystal calorimeters in the next decade," *Journal of Physics: Conference Series*, vol. 160, Article ID 012017, 2009.

[7] B. Lewandowski, "The BaBar electromagnetic calorimeter," *Nuclear Instruments and Methods in Physics Research Section A: Accelerators, Spectrometers, Detectors and Associated Equipment*, vol. 494, p. 303, 2002.

[8] K. Miyabayashi, "Belle electromagnetic calorimeter," *Nuclear Instruments and Methods in Physics Research Section A: Accelerators, Spectrometers, Detectors and Associated Equipment*, vol. 494, p. 298, 2002.

[9] M. Ablikim, Z. H. An, J. Z. Bai et al., "Design and construction of the BesIII detector," *Nuclear Instruments and Methods in Physics Research Section A: Accelerators, Spectrometers, Detectors and Associated Equipment*, vol. 614, p. 345, 2010.

[10] D. Renker, "Properties of avalanche photodiodes for applications in high energy physics, astrophysics and medical imaging," *Nuclear Instruments and Methods in Physics Research Section A: Accelerators, Spectrometers, Detectors and Associated Equipment*, vol. 486, no. 1-2, pp. 164–169, 2002.

[11] http://www.hamamatsu.com.

[12] E. Pilicer, F. Kocak, and I. Tapan, "Excess noise factor of neutron-irradiated silicon avalanche photodiodes," *Nuclear Instruments and Methods in Physics Research Section A: Accelerators, Spectrometers, Detectors and Associated Equipment*, vol. 552, no. 1-2, pp. 146–151, 2005.

[13] S. Agostinelli, J. Allison, K. Amako et al., "Geant4—a simulation toolkit," *Nuclear Instruments and Methods in Physics Research Section A: Accelerators, Spectrometers, Detectors and Associated Equipment*, vol. 506, pp. 250–303, 2003.

[14] J. Allison, K. Amako, and J. Apostolakis, "Geant4 developments and applications," *IEEE Transactions on Nuclear Science*, vol. 53, no. 1, pp. 270–278, 2006.

[15] J. Allison, K. Amakoca, and J. Apostolakis, "Recent developments in Geant4," *Nuclear Instruments and Methods in Physics Research Section A: Accelerators, Spectrometers, Detectors and Associated Equipment*, vol. 835, pp. 186–225, 2016.

[16] F. Kocak, "Simulation studies of crystal-photodetector assemblies for the Turkish accelerator center particle factory electromagnetic calorimeter," *Nuclear Instruments and Methods in Physics Research Section A: Accelerators, Spectrometers, Detectors and Associated Equipment*, vol. 787, pp. 144–147, 2015.

[17] F. Kocak and I. Tapan, "Simulation of LYSO crystal for the TAC-PF electromagnetic calorimeter," *Acta Physica Polonica A*, vol. 131, no. 3, pp. 527–529, 2017.

[18] E. Auffray, N. Akchurin, A. Benaglia et al., "DSB:Ce3+ scintillation glass for future," *Journal of Physics: Conference Series*, vol. 587, Article ID 012062, 2015.

[19] H. Ikeda, A. Satpathy, and B. S. Ahn, "A detailed test of the CsI(Tl) calorimeter for BELLE with photon beams of energy between 20 MeV and 5.4 GeV," *Nuclear Instruments and Methods in Physics Research Section A: Accelerators, Spectrometers, Detectors and Associated Equipment*, vol. 441, pp. 401–426, 2000.

11

A Probability Distribution for Quantum Tunneling Times

José T. Lunardi[1] and Luiz A. Manzoni[2]

[1]*Department of Mathematics & Statistics, State University of Ponta Grossa, Avenida Carlos Cavalcanti 4748, 84030-900 Ponta Grossa, PR, Brazil*
[2]*Department of Physics, Concordia College, 901 8th St. S., Moorhead, MN 56562, USA*

Correspondence should be addressed to José T. Lunardi; jttlunardi@gmail.com

Academic Editor: Neelima G. Kelkar

We propose a general expression for the probability distribution of real-valued tunneling times of a localized particle, as measured by the Salecker-Wigner-Peres quantum clock. This general expression is used to obtain the distribution of times for the scattering of a particle through a static rectangular barrier and for the tunneling decay of an initially bound state after the sudden deformation of the potential, the latter case being relevant to understand tunneling times in recent attosecond experiments involving strong field ionization.

1. Introduction

The search for a proper definition of quantum tunneling times for massive particles, having well-behaved properties for a wide range of parameters, has remained an important and open *theoretical* problem since, essentially, the inception of quantum mechanics (see, e.g., [1, 2] and references therein). However, such tunneling times were beyond the experimental reach until recent advances in ultrafast physics have made possible measurements of time in the attosecond scale, opening up the experimental possibility of measuring electronic tunneling times through a classically forbidden region [3–6] and reigniting the discussion of tunneling times. Still, the intrinsic experimental difficulties associated with both the measurements and the interpretation of the results have, so far, prevented an elucidation of the problem and, in fact, contradictory results persist, with some experiments obtaining a finite nonzero result [3, 6] and others compatible with instantaneous tunneling [4]. It should be noticed that the similarity between Schrödinger and Helmholtz equations allows for analogies between quantum tunneling of massive particles and photons [7], and a noninstantaneous tunneling time is supported by this analogy and experiments measuring photonic tunneling times [8], as well as by many theoretical calculations based on both the Schrödinger (for reviews see, e.g., [1, 2]) and the Dirac equations (e.g., [9–16]).

The conceptual difficulty in obtaining an unambiguous and well-defined tunneling time is associated with the impossibility of obtaining a self-adjoint time operator in quantum mechanics [17], therefore leading to the need for operational definitions of time. Several such definitions exist, such as phase time [18], dwell time [19], the Larmor times [20–23], and the Salecker-Wigner-Peres (SWP) time [17, 24], and in some situations these lead to different, or even contradictory, results. This is not surprising, since by their own nature operational definitions can only describe limited aspects of the phenomena of tunneling, and it is unlikely that any one definition will be able to provide a unified description of the quantum tunneling times in a broad range of situations. Nevertheless, it remains an important task to obtain a well-defined and *real* time scale that accurately describes the recent experiments [3–6, 25–27].

It is important to notice that the time-independent approach to tunneling times (i.e., for incident particles with sharply defined energy), which comprises the vast majority of the literature, is ill-suited to accomplish the above-mentioned goal, since it ignores the essential role of localizability in defining a time scale [23, 28]; see, however, [29], which applies the time defined in [30] to investigate the half-life of α-decaying nuclei. A few works (e.g., [23, 28, 31, 32]) address the issue of localizability and, consequently, arrive at a probabilistic definition of tunneling times (that is, an average time). In particular, in [28] the SWP clock was used to obtain an average tunneling time of transmission (reflection) for an incident wave packet, and such time was employed to investigate the Hartman effect [33] for a particle scattered off a square barrier and it was shown that it does not saturate in the opaque regime [28, 34].

The tunneling time scales considered in [23, 28, 31] involve taking an average over the spectral components of the transmitted wave packet and, thus, obscure the interpretation of the resulting average time. In this paper, we take as a starting point the real-valued average tunneling time obtained in [28], using the SWP quantum clock, and obtain a *probability distribution of transmission times*, by using a standard transformation between random variables. In addition to providing a more accurate time characterization of the tunneling process, this should provide a clearer connection with the experiments (which measure a distribution of tunneling times; see, e.g., Figure 4 in [3]). It is worth noting that some approaches using Feynman's path integrals address the problem of obtaining a probabilistic distribution of the tunneling times (see, e.g., [35]). However, these methods in general result in a complex time (or, equivalently, multiple time scales), and some *arbitrary* procedure is needed to select the physically meaningful real time *a posteriori*.

After obtaining a general formula for the distribution of tunneling times, which is the main result of this work, we apply it to two specific cases. First, to illustrate the formalism in a simple scenario, we consider the situation of a particle tunneling through a rectangular barrier. Then, we consider a slight modification of the model proposed in [36] for the tunneling decay of an initially bound state, after the sudden deformation of the binding potential by the application of a strong external field; the modification considered here allows us to investigate the whole range of possibilities for the tunneling times, without having an "upper cutoff", as is the case in the original model. Finally, some additional comments on the results are reserved for the last section.

2. The SWP Clock's Average Tunneling Time

We start by briefly reviewing the time-dependent application of the SWP clock to the scattering of a massive particle off a localized static potential barrier in one dimension (for details see [28]) which is appropriate, since it follows from the three-dimensional Schrödinger equation for this problem that the dynamics is essentially one-dimensional [3].

The SWP clock is a quantum rotor weakly coupled to the tunneling particle and that runs only when the particle is within the region in which $V(x) \neq 0$, where $V(x)$ is the potential energy. The Hamiltonian of the particle-clock system is given by (we use $\hbar = 2\mu = 1$, where μ is the particle's mass) [17]

$$H = -\frac{\partial^2}{\partial x^2} + V(x) + \mathscr{P}(x) H_c, \tag{1}$$

where $\mathscr{P}(x) = 1$ if $V(x) \neq 0$ and zero otherwise. The clock's Hamiltonian is $H_c = -i\omega(\partial/\partial\theta)$, where the angle $\theta \in [0, 2\pi)$ is the clock's coordinate and $\omega = 2\pi/(2j+1)\vartheta$ is the clock's angular frequency, with j being a nonnegative integer or half-integer giving the clock's total angular momentum, and ϑ is the clock's resolution. The weak coupling condition amounts to assume that ϑ is large, in such a way that the clock's energy eigenvalues, $\eta_m \equiv m\omega\,(-j < m < j)$, are very small compared to the barrier height and the particle's energy. It is assumed that, at $t = 0$, well before it reaches the barrier, the particle is well-localized far to the left of the barrier and the wave function of the system is a product state of the form

$$\Phi(\theta, x, t = 0) = \psi(x)\, v_0(\theta), \tag{2}$$

where $\psi(x)$ is the particle's initial state, represented by a wave packet centered around an energy E_0, and the clock initial state is assumed to be "in the zero-th hour" [17]

$$v_0(\theta) = \frac{1}{\sqrt{2j+1}} \sum_{m=-j}^{j} u_m(\theta), \tag{3}$$

where $u_m(\theta) = e^{im\theta}/\sqrt{2\pi}$ are the clock's eigenfunctions corresponding to the energy eigenvalues η_m.

The state $v_0(\theta)$ is strongly peaked at $\theta = 0$, thus allowing the interpretation of the angle θ as the clock's hand, since for a freely running clock the peak evolves to ωt_c, where t_c is the time measured by the clock [17]. Since here clock and particle are coupled according to (1), when the particle passes through the region $V(x) \neq 0$ it becomes entangled with the clock, with the wave function for the entire system given by

$$\Phi(\theta, x, t) = \frac{1}{\sqrt{2j+1}} \sum_{m=-j}^{j} \Psi^{(m)}(x, t)\, u_m(\theta),$$
$$\Psi^{(m)}(x, t) = \int_0^\infty dk A(k)\, \psi_k^{(m)}(x)\, e^{-iEt}, \tag{4}$$

where E is the incident particle's energy, $k = \sqrt{E}$, and $A(k)$ is the Fourier spectral decomposition of the initial wave packet $\psi(x)$ in terms of the free particle eigenfunctions (we are assuming delta-normalized eigenfunctions). The functions $\psi_k^{(m)}(x)$ satisfy a time-independent Schrödinger equation with a constant potential η_m in the barrier region. Outside the potential barrier region and for a particle incident from the left, the (unnormalized) solution $\psi_k^{(m)}(x)$ of the time-independent Schrödinger equation is given by [28]

$$\psi_k^{(m)}(x) = \begin{cases} e^{ikx} + R^{(m)}(k)\, e^{-ikx}, & x \le -L \\ T^{(m)}(k)\, e^{ikx}, & x \ge L, \end{cases} \tag{5}$$

where $T^{(m)}(k)$ $[R^{(m)}(k)]$ stands for the transmission (reflection) coefficient, and it is assumed, without loss of generality, that the potential is located in the region $-L < x < L$. Considering only the transmitted solution in (5) and substituting it into the time-dependent solution (4), it can be shown that for weak coupling

$$\Phi_{tr}(\theta, x, t) = \int_0^\infty dk A(k)\, T(k)\, e^{i(kx-Et)} v_0\left(\theta - \omega t_c^T(k)\right), \tag{6}$$

where

$$t_c^T(k) = -\left(\frac{\partial \varphi_T^{(m)}}{\partial \eta_m}\right)_{\eta_m=0} \quad (7)$$

is the stationary transmission clock time corresponding to the wave number component k [17, 37]. The transmission coefficient $T(k)$ corresponds to the stationary problem in the absence of the clock.

For *tunneling* times one is interested only in the clock's reading for the *postselected* asymptotically transmitted wave packet. Thus, tracing out the particle's degrees of freedom, the expectation value of the clock's measurement can be defined, resulting in the average tunneling time [28]

$$\langle t_c^T \rangle = \int dk\, \rho(k)\, t_c(k), \quad \rho(k) = N\,|A(k)\,T(k)|^2, \quad (8)$$

where $N = 1/\int dk|A(k)T(k)|^2$ is a normalization constant and $\rho(k)$ is the probability density of finding the component k in the transmitted wave packet. Similar expressions can be obtained for the reflection time.

3. The Tunneling Times Distribution

An important aspect of the average tunneling time considered in the previous section is that it emphasizes the *probabilistic nature of the tunneling process*. However, since the average in (8) is over the time taken by the *spectral components* of the wave packet, it does not lend itself to an easy interpretation, given the spectral components of the wave packet tunnel with different times. Thus, instead of (8), one would rather obtain an average over (*real*) times of the form

$$\langle t_c \rangle = \int_0^\infty d\tau\, \tau\, \rho_t(\tau), \quad (9)$$

where $\rho_t(\tau)$ stands for the probability density for observing a particular tunneling time τ for the asymptotically transmitted wave packet. This can be easily be achieved by noticing that in probability theory (8) and (9), which must be equal, are related by a standard transformation between the two random variables k and τ through a function $t_c^T(k)$. It follows that the probability distribution of times is given by

$$\rho_t(\tau) = \int \rho(k)\, \delta\left(\tau - t_c^T(k)\right) dk \quad (10)$$

which, in essence, is the statement that all the k-components in the transmitted packet for which $t_c^T(k) = \tau$ must contribute to the value of $\rho_t(\tau)$ with a weight $\rho(k)$. Finally, using the properties of the Dirac delta function (specifically, we use the fact that $\delta(g(x)) = \sum_j(\delta(x - x_j)/|g'(x_j)|)$, where $\{x_j\}$ is the set of zeros of the function $g(x)$ and the prime indicates a derivative with respect to the independent variable), we obtain

$$\rho_t(\tau) = \sum_j \frac{\rho(k_j(\tau))}{\left|t_c^{T\prime}(k_j(\tau))\right|}, \quad (11)$$

where $\{k_j(\tau)\}$ is the set of zeros of the function $g(k) \equiv t_c^T(k) - \tau$ and $t_c^{T\prime}$ is the derivative of $t_c^T(k)$ with respect to k.

A similar definition of the distribution of tunneling times given in (10)-(11) can be obtained for any time scale which is probabilistic in nature, that is, of the form (8). Although several other probabilistic tunneling times exist in the literature (e.g., [23, 31, 32, 35]), the SWP clock has proven to yield well-behaved *real* times both in the time-independent [17, 37, 38] and time-dependent approaches [28, 34, 39] and it provides a simple procedure to *derive* the probabilistic expression (8). In addition, the role exerted by circularly polarized light in attoclock experiments [3, 25] seems to provide a natural possibility for interpretation in terms of the SWP clock.

As will be illustrated below, for the simple application of this formalism to the problem of a wave packet scattered off a rectangular potential barrier, the distribution of times (10)-(11) cannot, in general, be obtained analytically even for the simplest cases, except in trivial cases such as for a single Dirac delta potential barrier [40–42], in which case $t_c^T(k) = 0$ and $\rho_t(\tau) = \delta(\tau)\int dk\rho(k)$.

It should also be noticed that, despite the fact that the derivation of the previous section leading to (8) and, thus (10)-(11), assumed a scattering situation, these expressions can be shown to be valid for any situation involving preselection of an initial state localized to the left of a potential "barrier" followed by postselection of an asymptotic transmitted wave packet. This allows us to obtain the distribution of times for a model that simulates the tunneling decay of an initially bound particle by ionization induced by the sudden application of a strong external field; the model considered below is a variant of that introduced in [36].

4. The Distribution of Tunneling Times for a Rectangular Barrier

As a first illustration of the formalism developed above, let us consider a rectangular barrier of height V_0 located in the region $x \in (-L, L)$. The particle's initial state $\phi_0(x) \equiv \psi(x, t = 0)$ is assumed to be a Gaussian wave packet

$$\phi_0(x) = \frac{1}{(2\pi)^{1/4}\sqrt{\sigma}} \exp\left[ik_0 x - \frac{(x - x_0)^2}{4\sigma^2}\right], \quad (12)$$

where the parameters x_0, σ, and k_0 are chosen such that the wave packet is sharply peaked in a tunneling wave number $k_0 = \sqrt{E_0} < \sqrt{V_0}$ and is initially well-localized around $x = x_0$, far to the left of the barrier; in the calculations that follow we take $x_0 = -8\sigma$, such that at $t = 0$ the probability of finding the particle within or to the right of the barrier is negligible. The transmission coefficient $T(k)$ and the spectral function $A(k)$ are well-known and given by

$$T(k) = \frac{2ikq e^{-2ikL}}{(k^2 - q^2)\sinh(2Lq) + 2ikq\cosh(2Lq)} \quad (13)$$

$$A(k) = \left(\frac{2}{\pi}\right)^{1/4} \sqrt{\sigma} \exp\left[4k\sigma\left(k_0\sigma + 4i\right) - \sigma\left(k + k_0\right)\left(k\sigma + k_0\sigma + 8i\right)\right], \tag{14}$$

where $q = \sqrt{V_0 - k^2}$. The stationary transmission clock time (7) is [22, 28]

$$t_c^T(k) = \frac{k}{q} \cdot \frac{\left(q^2 + k^2\right)\tanh(2qL) + 2qL\left(q^2 - k^2\right)\operatorname{sech}^2(2qL)}{4q^2k^2 + \left(q^2 - k^2\right)^2 \tanh^2(2qL)}, \tag{15}$$

with tunneling times corresponding to real values of q (i.e., $V_0 > k^2$). Figure 1 shows a plot for the stationary transmission times $t_c^T(k)$, the distribution of wave numbers $\rho(k)$ in the transmitted wave packet, and the distribution $|A(k)|^2$ of wave numbers (momenta) in the incident packet, for two values of the barrier width. For the chosen parameters and barrier widths both the incident and the transmitted wave packets have an energy distribution very strongly peaked in a tunneling component (in the bottom plot of Figure 1 the barrier is much more opaque than that in the top plot and we can observe that—despite being with a negligible probability for the parameters chosen for this plot—in this situation some above-the-barrier components start to appear in the distribution of the transmitted wave packet. So, in order to consider mainly transmission by tunneling we must restrict the barrier widths to not too large ones). We also observe the very well-known fact that the transmitted wave packet "speeds up" when compared to the incident particle [28]. As a general rule, the larger is the barrier width (i.e., the more opaque is the barrier), the greater is the translation of the central component towards higher momenta. In what concerns the off-resonance stationary transmission time, it initially grows with the barrier width, and saturates for very opaque barriers (the Hartman effect); on the other hand, it presents peaks at resonant wave numbers that grow and narrow with the barrier width; for a detailed discussion see [28]).

Figure 2 shows plots of the probability distribution $\rho_t(\tau)$ of the tunneling times according to (10)-(11), corresponding to both the barrier widths shown in Figure 1 [to obtain these plots we used a Monte Carlo procedure to generate a large number of k outcomes from the distribution $\rho(k)$, which afterwards were transformed into τ values by using the function $\tau = t_c^T(k)$]. The vertical grey lines in these plots correspond to the time the light takes to cross the barrier distance. It is observed that for the two distributions shown in Figure 2 the probability to observe superluminal tunneling times is negligible. It is also observed that these distributions have a shape that resembles that of the k distribution, albeit with a more pronounced skewness. This shape could be inferred from Figure 1 and from (11), since $t_c^{T\prime}(k)$ grows very smoothly in the region were $\rho(k)$ is nonvanishing. Furthermore, a comparison between the two plots in Figure 2 shows that the tunneling times do not grow linearly with the barrier width and, therefore, the distribution

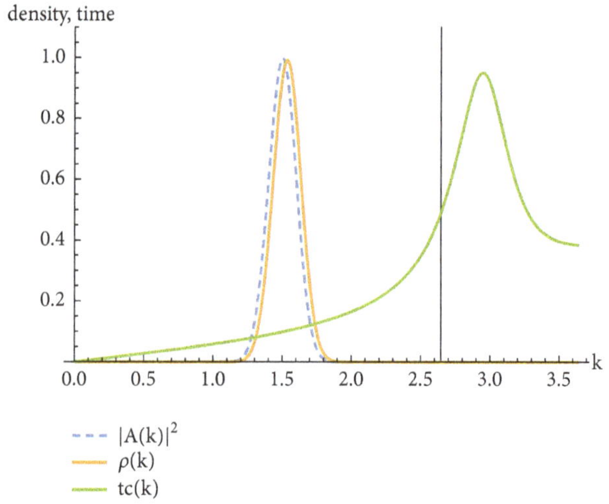

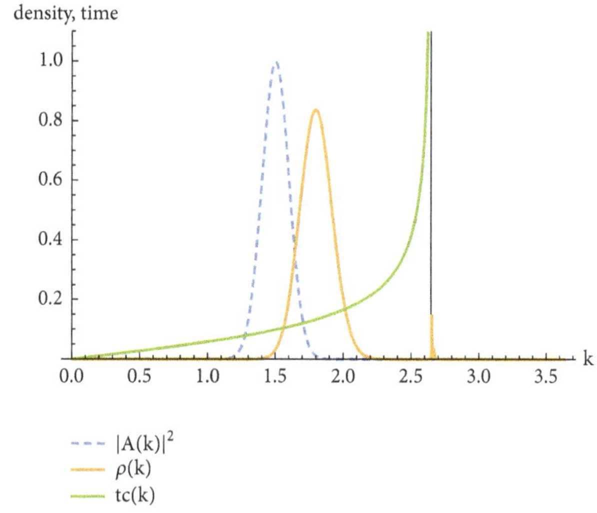

Figure 1: Stationary transmission clock time $t_c^T(k)$ (green) and the distributions $\rho(k)$ (orange and arbitrary scale) and $|A(k)|^2$ (blue, dashed, and arbitrary scale) for the transmitted and incident wave packets, respectively. Rydberg atomic units $\hbar = 2\mu = 1$ are used in all plots; the tunneling energies correspond to $0 < k < \sqrt{7}$, corresponding to a barrier height $V_0 = 7$ (the maximum tunneling wave number $\sqrt{7}$ is shown by a vertical grey line in the plots). In both plots the incident wave packet parameters are $k_0 = 1.5, \sigma = 5, x_0 = -8\sigma$. *Top:* barrier width $2L = 2$. *Bottom:* barrier width $2L = 16$.

in the bottom plot of Figure 2 is "closer" to the light time than the distribution shown in the top plot; [28] already observed that for intermediate values of barrier widths the average transmission time—corresponding to the mean of the distribution ρ_t—reaches a plateau.

5. Distribution of Ionization Tunneling Times

In this section we obtain a distribution for tunneling times for a particle that is initially in a bound state of a given binding potential. The potential is then suddenly deformed in such a way that the particle can escape from the initially confining region by tunneling. The model considered here is a slight modification of that proposed by Ban *et al.* [36] to simulate,

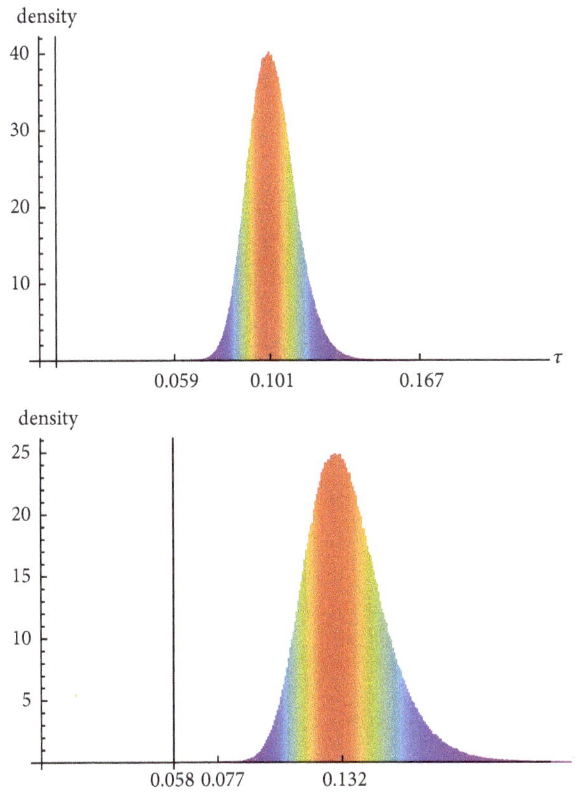

FIGURE 2: Probability distribution $\rho_t(\tau)$ for the tunneling times τ, obtained by Monte Carlo samplings of k-values from the distribution $\rho(k)$ and then transforming these to time values through $\tau = t_c^T(k)$. The parameters are the same as in the corresponding plots in Figure 1 and are all expressed in Rydberg atomic units. *Top*: $2L = 2$. *Bottom*: $2L = 16$. These plots are in the same range and scale and can be compared. The vertical grey line in both plots corresponds to the time the light takes to traverse the barrier distance. The ticks in the horizontal axes correspond to the light time, the minimum, the median and the maximum values of τ in the histogram (in the bottom plot the maximum τ is out of the plot's range).

in a simple scenario, key features of the decay of a localized state by tunneling ionization induced by the application of a strong external field with a finite duration.

In [36], for $t < 0$, the particle is in an eigenstate of a semi-infinite square-well potential $V_1(x)$,

$$V_1(x) = \begin{cases} +\infty & x < 0 \\ 0 & 0 \le x \le a \\ V_0 & x > a, \end{cases} \tag{16}$$

and, therefore, it cannot decay by tunneling. At $t = 0$ the potential is suddenly deformed to $V_2(x)$,

$$V_2(x) = \begin{cases} +\infty & x < 0 \\ 0 & 0 \le x \le a \\ V_0 & a < x < b \\ 0 & x \ge b, \end{cases} \tag{17}$$

such that the particle can now tunnel through the potential barrier; it is assumed that the wave function does not change during the sudden change of the potential. Finally, after a finite time t_0 the potential returns to its original configuration, $V_1(x)$, and tunneling terminates. The cutoff time t_0 mimics the natural upper bound for tunneling times measured in recent attoclock experiments (see, e.g., [3, 6] and references therein), since the opening and closing of the tunneling channel in these experiments occur in intervals of half the laser field's period.

Here, we deviate from [36] by setting $t_0 \longrightarrow \infty$; i.e., once deformed the potential does not return to its original form and, after a long enough time, the particle will be transmitted with unit probability; thus, by eliminating the cutoff (which is just an experimental limitation) we are able to explore the whole range of possibilities for the ionization tunneling time. In addition, for $t \ge 0$, the particle is assumed to be coupled to a SWP quantum clock running only in the region (a, b), so that the clock's readings for the asymptotic transmitted wave packet give the time the particle spent within the barrier after $t = 0$. Following [36], we assume that for $t < 0$ the particle is in the ground state of the potential $V_1(x)$, whose stationary wave function is given by

$$\phi_0(x) = N \begin{cases} \sin(k_0 x), & 0 < x \le a \\ \sin k_0 e^{q_0(a-x)}, & x > a, \end{cases} \tag{18}$$

where N is a normalization constant, $k_0 = \sqrt{E_0}$, E_0 is the ground state energy, and $q_0 = \sqrt{V_0 - k_0^2}$. It is also assumed, as in [36], that immediately after the sudden deformation of the potential from $V_1(x)$ to $V_2(x)$, at $t = 0$, the wave function does not change. However, for $t \ge 0$ the particle state, which is no longer an energy eigenstate, is given by a superposition of the energy eigenstates $\psi_k(x)$ ($k = \sqrt{E}$) of the potential $V_2(x)$, i.e., [36]

$$\psi(x, t = 0) = \phi_0(x) = \int_0^\infty S(k)\,\psi_k(x)\,dk, \tag{19}$$

where

$$S(k) = \int_0^\infty \phi_0(x)\,\psi_k^*(x)\,dx, \tag{20}$$

with

$$\psi_k(x) = \begin{cases} A(k)\sin(kx), & 0 < x \le a \\ C(k)e^{qx} + D(k)e^{-qx}, & a < x \le b \\ \sqrt{\dfrac{2}{\pi}}\cos[k(x-b) + \Omega(k)], & x > b, \end{cases} \tag{21}$$

where $q = \sqrt{V_0 - k^2}$ and the coefficients $A(k), C(k), D(k)$ and the phase $\Omega(k)$ are determined by the usual boundary conditions at $x = a$ and $x = b$ and are such that the normalization $\langle \psi_k(x), \psi_{k'}(x) \rangle = \delta(k - k')$ holds [36]. From the above expressions it follows that, without any loss of generality, we can take $S(k)$ and all the eigenfunctions (21) to be real.

In order to consider the coupling with the SWP clock for times $t \geq 0$ we proceed as follows. At $t = 0$ the system particle+clock is described by the product state $\psi(x,0)\nu_0(\theta)$, where $\psi(x,0)$ is the state (19) and $\nu_0(\theta)$ is the initial clock state given by (3). After $t = 0$ the particle and the clock states become entangled. For the procedure of postselection of the asymptotically transmitted wave function we notice that the role of the transmission coefficient for the wave function (21) is played by $\sqrt{2/\pi}\, e^{i(-kb+\Omega^{(m)}(k))}$, where the superscript m indicates the weak coupling with the clock. The right moving *asymptotic* wave packet representing the coupled system formed by the transmitted particle and the clock is

$$\Phi_{tr}(\theta, x, t) = \int_0^\infty dk S(k)\, e^{i[k(x-b)+\Omega^{(m)}(k)-Et]} \times \nu_0\left[\theta - \omega t_c^T(k)\right], \qquad (22)$$

where, as before, $t_c^T(k) = -(\partial\Omega^{(m)}(k)/\partial\eta_m)_{\eta_m=0} = -(1/2q)(\partial\Omega/\partial q)$ [with quantities without the subscript "(m)" representing the limit $\eta_m \longrightarrow 0$]. By following the same steps described in [28], we trace out the clock's degree of freedom in the asymptotic transmitted wave packet in order to obtain the distribution $\rho(k)$ of the wave numbers for the asymptotically transmitted wave packet, which in this case is simply given by

$$\rho(k) = |S(k)|^2; \qquad (23)$$

i.e., the probability to find a wave number k in the asymptotic transmitted wave packet is the same as in the initial state, which is as expected, since after a long enough time the initial wave packet will be transmitted with probability unit, as mentioned earlier.

The general behavior of $t_c^T(k)$ and $\rho(k)$ is illustrated in Figures 3 and 4, corresponding to two barriers with different opacities ($b - a = 2$ and 4, respectively). These plots show, as expected, that the distribution $\rho(k)$ is strongly peaked at the wave number k_0, corresponding to the energy of the initially bound state and is negligible for nontunneling components. For tunneling wave numbers ($k < \sqrt{V_0}$) the function $t_c^T(k)$ is also strongly peaked at the same wave number k_0, which corresponds to a local maximum (for nontunneling wave numbers there are several other resonance peaks). From (11) we would expect that the peaks in the tunneling times distribution $\rho_t(\tau)$ would occur for times $\tau = t_c^T(k)$ corresponding to values of k for which $t_c^{T\prime}(k) \approx 0$—which occur at points of local maxima and minima of the function $t_c^T(k)$—*and* corresponding to nonnegligible $\rho(k)$. Therefore, from the plots in Figures 3 and 4 one could expect the first peak of the tunneling time distribution $\rho_t(\tau)$ at $\tau \approx 0.105$ *a.u.* (the local minimum of $t_c^T(k)$, which is similar for both barrier widths, since nonresonant times $t_c^T(k)$ change little with the barrier width for opaque barriers, as is the case in Figures 3 and 4); a second peak in $\rho_t(\tau)$ is expected to occur around the local maximum of $t_c^T(k)$, which corresponds to $\tau \approx t_c^T(k_0)$ (this local maximum—corresponding to resonant wave numbers—changes significantly with the barrier widths; see, e.g., [28]). On the other hand, peaks in

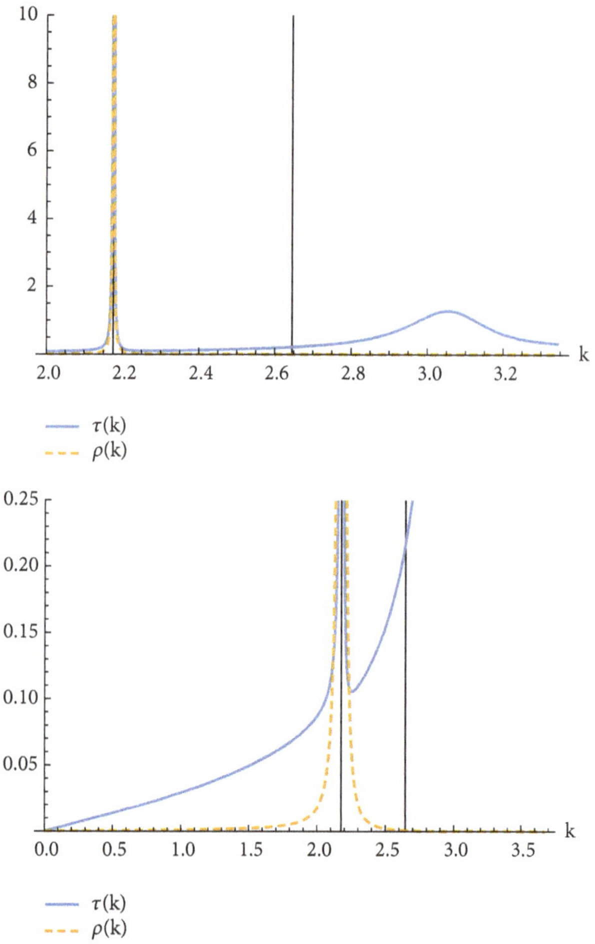

FIGURE 3: *Top*: the stationary transmission clock time $t_c^T(k)$ (blue) and the wave number distribution $\rho(k) = |S(k)|^2$ (orange, dashed, and arbitrary scale), for $V_0 = 7$, $a = 1$, $b = 3$, and $k_0 \approx 2.175932$), with the initial state given by (18). *Bottom:* close view of the above plot for small times. The vertical grey lines in the plots correspond to $k = k_0$ and $k = \sqrt{V_0}$. The regions in which $t_c^{T\prime}(k) \approx 0$ (around the local maximum and minimum of $t_c^T(k)$) correspond to times $\tau \approx t_c^T(k_0)$ and $\tau \approx 0.105$ *a.u.* Rydberg atomic units were used in all the plots.

$\rho_t(\tau)$ coming from local maxima (resonances) and minima associated with nontunneling values of k are suppressed, since $\rho(k) \approx 0$ in these cases. Figure 5 confirm these claims. For both barrier widths considered, the distribution of tunneling times is "U" shaped, having peaks at the times corresponding to the local maxima and minima of the stationary time $t_c^T(k)$ inside the tunneling region. It should be observed that the larger is the barrier width, the broader is the tunneling time distribution, due to the strong increase of the resonant tunneling time with the barrier width.

Figures 6 and 7 show close views of the tunneling time distributions $\rho_t(\tau)$ for small and large tunneling times (Figure 6 corresponds to the plot at the top of Figure 5, while Figure 7 corresponds to the plot at the bottom of Figure 5). In the top plots of these Figures we can clearly observe the first peak around the local minimum of $t_c(k)$ in the tunneling

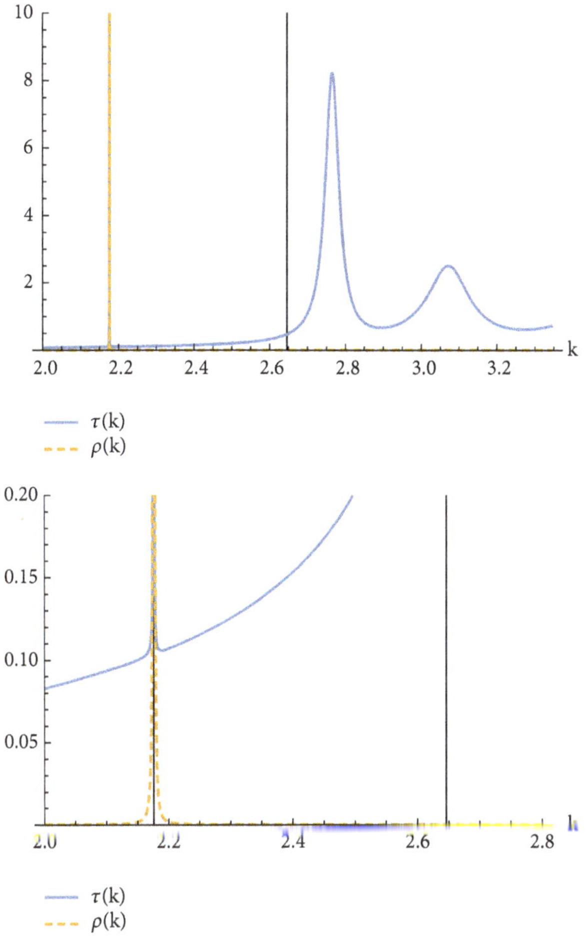

FIGURE 4: *Top*: the stationary transmission clock time $t_c^T(k)$ (blue) and the wave number distribution $\rho(k) = |S(k)|^2$ (orange, dashed, and arbitrary scale), for $V_0 = 7$, $a = 1$, $b = 5$, and $k_0 \approx 2.175932$), with the initial state given by (18). *Bottom:* close view of the above plot for small times. The vertical grey lines in the plots correspond to $k = k_0$ and $k = \sqrt{V_0}$. The region of relatively slow growth of the derivative $t^{T'}(k)$ correspond to times around 0.105 *a.u.* Rydberg atomic units were used in all the plots.

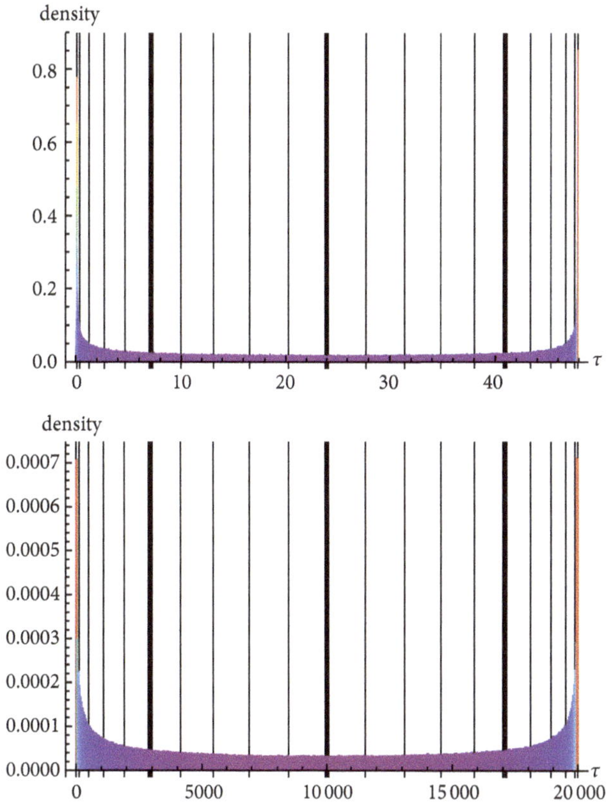

FIGURE 5: Distributions of tunneling decay times $\rho_t(\tau)$ through the barrier of the potential $V_2(x)$ for the initial bound state $\phi_0(x)$ given by (18). The histograms were built by using the Monte Carlo procedure described in Figure 2 and in the main text. The vertical grey lines indicate percentiles of the distribution (the first and the last correspond to 1% and 99%, the remaining ones range from 5% to 95%, in steps of 5%); the three thick vertical lines indicate the first quartile (percentile 25%), the median, and the third quartile (percentile 75%). Rydberg atomic units were used in all the plots. *Top:* barrier width $b - a = 2$ and bin length $\approx 0.0031 a.u.$ ($\approx$ 0.15 attoseconds). *Bottom:* barrier width $b-a = 4$ and bin length $\approx 40 a.u.$ ($\approx$ 1,935 attoseconds).

region, which in both plots corresponds to almost the same value $\tau \approx 0.105$ *a.u* $\approx$ 5.1 attoseconds. The top plot of Figure 6 shows that for the less opaque barrier there exists a (very small) probability to observe a superluminal tunneling time. Even if this possibility cannot be precluded in principle (see, e.g., [16]), in the present case the possibility of emergence of such small times was expected, since at $t = 0$ there was a significant portion of the wave packet (roughly 27%) penetrating the whole distance of the barrier, and this has an important contribution to the emergence of small times in the clock's readings associated with the transmitted particle. On the other hand, the top plot of Figure 7 shows that for the thicker barrier the probability for superluminal times is negligible; the portion of the wave packet already inside the barrier at $t = 0$ is the same (~27%), but the wave packet penetrates proportionally a smaller distance inside the barrier and, thus, it does not contribute in a significant way to the emergence of very small times in the clock readings. We note that the introduction of the cutoff t_0, as in [36], would result in a time distribution similar to the truncated distributions shown in the top plots of Figures 6 and 7.

It is also worth observing that, for small times, the distributions obtained here resemble qualitatively those in Figure 4 of [3], except for the presence of several peaks at discrete values of the time in the latter. The considerations above, relating the peaks of the distribution of clock times $\rho_t(\tau)$ to the local maxima and minima of the stationary time $t_c^T(k)$ and the magnitude of distribution $\rho(k)$ in the neighborhood of these points, suggest a scenario in which such multiple peaks at discrete values of time can appear in the distribution $\rho_t(\tau)$ of transmission times. Indeed, if above-the-barrier wave numbers had a significant contribution to the initial wave packet, then the several local maxima and minima present in the vicinities of the resonant *nontunneling* components will also contribute in a significant way to build

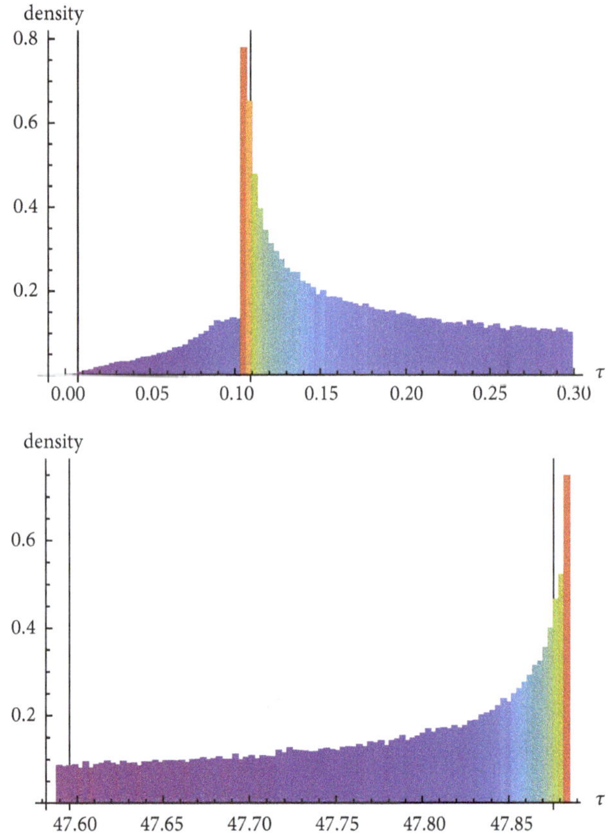

FIGURE 6: Close views of the plot at the top of Figure 5, corresponding to the barrier width $b - a = 2$, with the bin length $\approx 0.0031 a.u.$ (≈ 0.15 attoseconds). *Top*: small tunneling times. The vertical grey line in the left of this plot corresponds to the time the light takes to travel the barrier distance. The second grey vertical line corresponds to the percentile 1%. *Bottom:* large tunneling times. The vertical grey lines correspond to the percentiles 95% and 99%, respectively.

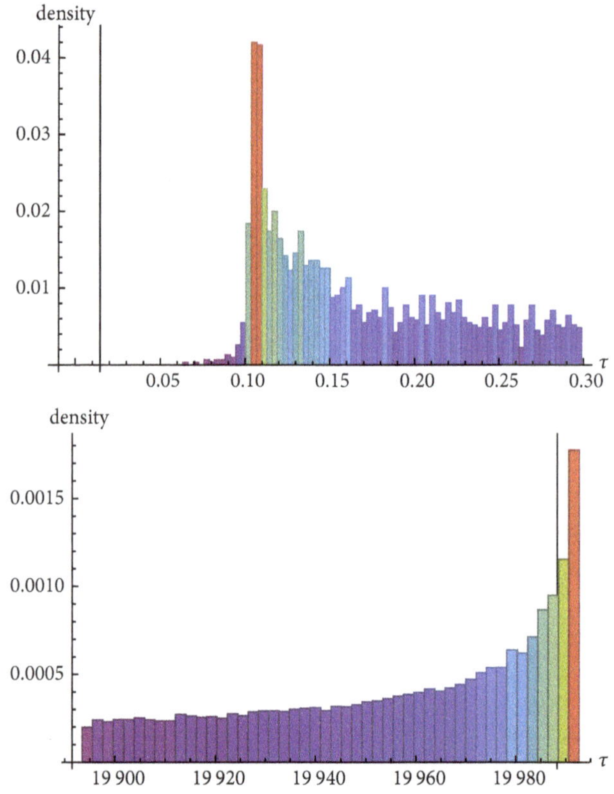

FIGURE 7: Close views of the plot at the bottom of Figure 5, corresponding to the barrier width $b - a = 4$. *Top*: small tunneling times and bin length $\approx 0.0031 a.u.$ (≈ 0.15 attoseconds). The vertical grey line in the left of this plot corresponds to the time the light takes to travel the barrier distance. The percentile 1% (corresponding to ≈ 5.1 *a.u.* ≈ 247 attoseconds) is out of the range of this plot. *Bottom:* large tunneling times, with bin length ≈ 2 *a.u* ≈ 100 attoseconds. The vertical grey line corresponds to the percentile 99%.

multiple peaks in the distribution of transmission times; these peaks, however, could not be associated with the tunneling process. We can consider such a scenario by choosing as the initial state a tightly localized state given by $\psi(x, 0) = \sqrt{2}\sin k_0 x$, with $k_0 = \pi$ and the barrier parameters $a = 1$, $b = 2$, and $V_0 = 11$, in Rydberg atomic units. In this situation the initial wave function is perfectly confined to the left of the barrier ($0 < x < 1$), and above-the-barrier components contribute in a significant way to build the wave packet, as can be seen from $\rho(k)$ in the top plot of Figure 8 (in this case the probability of finding a nontunneling k component in the wave packet is approximately 75%). In this plot we can also observe that all the local maxima and minima of $t_c^T(k)$ shown occur in neighborhoods of wave numbers k for which $\rho(k)$ is nonnegligible; therefore, all these local maxima and minima contribute significantly to build multiple peaks in the distribution of transmission times $\rho_t(\tau)$. The middle and the bottom plots of Figure 8 confirm this statement: all the peaks of the distribution of transmission times correspond very closely to the local maxima and minima of $t_c(k)$, as can be seen by comparing the plots in the top and the bottom of this figure (except for the first, all the other significant peaks in the bottom plot are associated with nontunneling components).

6. Conclusions

Taking as a starting point the *probabilistic* (average) tunneling time obtained in [28] with the use of a SWP clock [17, 24, 37], we obtained a *probability distribution of times* (10)-(11). An important advantage of using the SWP clock, in addition to those already mentioned, is that by running only when the particle is inside the barrier it allows us to address the concept of *tunneling* time in a proper way, since the time spent by the particle standing in the well *before* penetrating the barrier is *not* computed. A clear advantage of having a probability distribution of transmission (tunneling) times is that, in addition to the usual expectation value, we can obtain *all* the statistical properties of this time, such as its most probable values (peaks of the distribution), the dispersion around the mean value, and the probability to observe extreme outcomes (superluminal times, for instance).

As an initial test, the distribution of times (10)-(11) was applied to the simple problem of a particle tunneling through

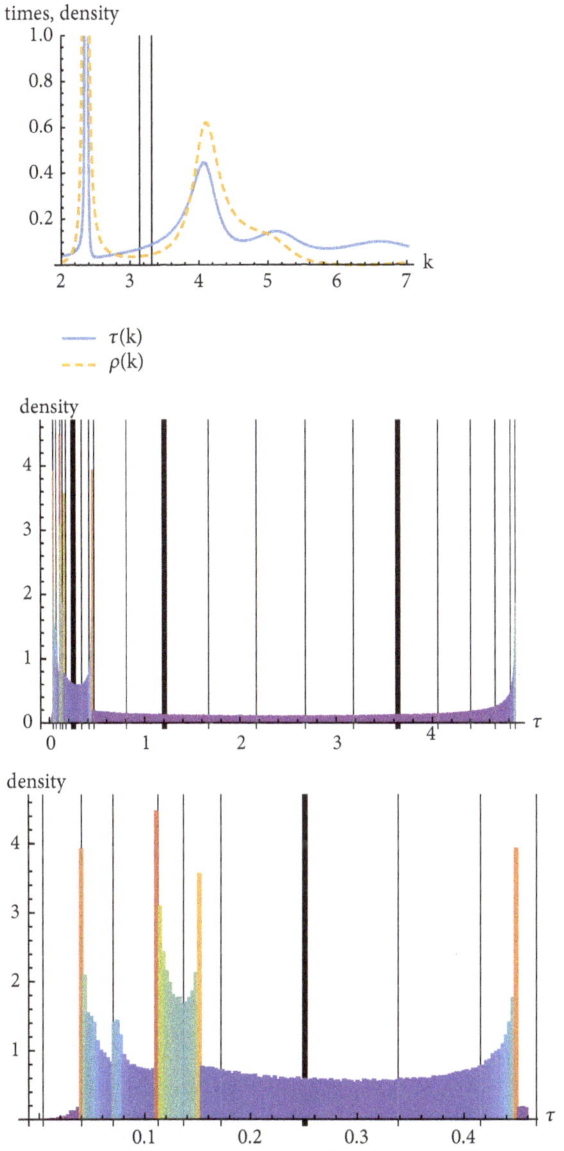

FIGURE 8: *Top*: the stationary time $t_c^T(k)$ and the wave number distribution $\rho(k)$, for $V_0 = 11$, barrier width $b - a = 36$, $k_0 = \pi$, and an initial state $\psi(x, 0) = \sqrt{2} \sin k_0 x$, with $k_0 = \pi$. The vertical lines correspond to $k = k_0$ and $k = \sqrt{V_0}$. *Middle:* distribution $\rho_t(\tau)$ for the transmission times, with the histogram built by the Monte Carlo procedure described in the text. Vertical lines indicate the percentiles, as in the Figure 5. *Bottom:* close view of the above histogram for the range of small times. The first vertical line at the left indicates the time the light takes to cross the barrier distance. In both the histograms we used a bin length ≈ 0.0031 *a.u.* ≈ 0.15 attoseconds. Rydberg atomic units were used in all these plots.

a rectangular barrier. Unsurprisingly, it revealed behavior similar to that already known from previous works using a distribution of wave numbers (momentum)—see, e.g., [28]—although, using $\rho_t(\tau)$ these conclusions are much more transparent. For example, one could answer the question about the possibility of superluminal tunneling by direct calculation from the probability distribution $\rho_t(\tau)$. In the nonstationary case—which is the correct to address this question—this problem is usually answered by considering just the average tunneling time. But, given its probabilistic nature, an answer based only on the average time may not be satisfactory, especially if the dispersion of the distribution of tunneling times is large, which is often the case when one deals with well-localized particles, as suggested by the two situations addressed in the present work.

As a main application of (10)-(11), we considered a slight modification of the problem considered in [36] to model strong field ionization by tunneling. The modification considered here was the elimination of the cutoff time that was introduced in [36] to simulate the upper bound that arises in attoclock experiments [3, 6] due to the opening and closing of the tunneling channel, naturally associated with the oscillations in the laser field intensity. This cutoff is not a fundamental requirement, but rather it is associated with the experimental methods employed—in any case, its implementation is rather trivial, since it just truncates the distribution of times. The consideration of the full range of the distribution of times allowed us to show that an important contribution to $\rho_t(\tau)$ comes from very large times associated with the resonance peaks in the tunneling region; these very long tunneling times occur with a probability comparable to very short ones, thus having an important impact on the average tunneling times and, therefore, cause difficulties when comparing theoretical predictions based on an average time with the outcomes of experiments presenting a natural cutoff in the possible time measurements. In particular, in the attoclock experiments the relevant measure is often associated with the peak of the tunneling time, which may be promptly identified once one knows the probability distribution for all possible times. A remark is in place; the distribution of times proposed here, built on the SWP clock readings, refers to the time the particle *dwells within* the barrier, while the tunneling times often measured in recent attoclock experiments actually refer to *exit* times [43].

In sum, the approach introduced above and resulting in (10)-(11) builds upon the already (conceptually) well tested SWP clock to provide a real-valued distribution of times that, in the simple models considered here, was demonstrated to have physically sound properties and, in fact, (rough) similarities with the time distribution obtained in recent experiments [3], therefore warrantying further investigation with more realistic potentials.

Conflicts of Interest

The authors declare that they have no conflicts of interest.

References

[1] H. G. Winful, "Tunneling time, the Hartman effect, and superluminality: a proposed resolution of an old paradox," *Physics Reports*, vol. 436, no. 1-2, pp. 1–69, 2006.

[2] H. G. Winful, "The meaning of group delay in barrier tunnelling: a re-examination of superluminal group velocities," *New Journal of Physics*, vol. 8, Article ID 101, 2006.

[3] A. S. Landsman, M. Weger, J. Maurer et al., "Ultrafast resolution of tunneling delay time," *Optica*, vol. 1, no. 5, pp. 343–349, 2014.

[4] L. Torlina, F. Morales, J. Kaushal et al., "Interpreting attoclock measurements of tunnelling times," *Nature Physics*, vol. 11, no. 6, pp. 503–508, 2015.

[5] O. Pedatzur, G. Orenstein, V. Serbinenko et al., "Attosecond tunnelling interferometry," *Nature Physics*, vol. 11, no. 10, pp. 815–819, 2015.

[6] T. Zimmermann, S. Mishra, B. R. Doran, D. F. Gordon, and A. S. Landsman, "Tunneling time and weak measurement in strong field ionization," *Physical Review Letters*, vol. 116, no. 23, Article ID 233603, 2016.

[7] R. Chiao, P. Kwiat, and A. Steinberg, "Analogies between electron and photon tunneling: a proposed experiment to measure photon tunneling times," *Physica B: Condensed Matter*, vol. 175, no. 1–3, pp. 257–262, 1991.

[8] A. M. Steinberg, P. G. Kwiat, and R. Y. Chiao, "Measurement of the single-photon tunneling time," *Physical Review Letters*, vol. 71, no. 5, pp. 708–711, 1993.

[9] C. R. Leavens and R. Sala Mayato, "Are predicted superluminal tunneling times an artifact of using nonrelativistic Schrödinger equation?" *Annalen der Physik*, vol. 7, no. 7-8, pp. 662–670, 1999.

[10] P. Krekora, Q. Su, and R. Grobe, "Effects of relativity on the time-resolved tunneling of electron wave packets," *Physical Review A: Atomic, Molecular and Optical Physics*, vol. 63, no. 3, Article ID 032107, 2001.

[11] C.-F. Li and X. Chen, "Traversal time for Dirac particles through a potential barrier," *Annalen der Physik*, vol. 11, no. 12, pp. 916–925, 2002.

[12] V. Petrillo and D. Janner, "Relativistic analysis of a wave packet interacting with a quantum-mechanical barrier," *Physical Review A: Atomic, Molecular and Optical Physics*, vol. 67, no. 1, Article ID 012110, 2003.

[13] X. Chen and C. Li, "Negative group delay for Dirac particles traveling through a potential well," *Physical Review A: Atomic, Molecular and Optical Physics*, vol. 68, no. 6, Article ID 052105, 2003.

[14] S. De Leo and P. Rotelli, "Dirac equation studies in the tunneling energy zone," *The European Physical Journal C*, vol. 51, no. 1, pp. 241–247, 2007.

[15] J. T. Lunardi and L. A. Manzoni, "Relativistic tunneling through two successive barriers," *Physical Review A: Atomic, Molecular and Optical Physics*, vol. 76, no. 4, Article ID 042111, 2007.

[16] J. T. Lunardi, L. A. Manzoni, A. T. Nystrom, and B. M. Perreault, "Average transmission times for the tunneling of wave packets," *Journal of Russian Laser Research*, vol. 32, no. 5, pp. 431–438, 2011.

[17] A. Peres, "Measurement of time by quantum clocks," *American Journal of Physics*, vol. 48, no. 7, pp. 552–557, 1980.

[18] E. P. Wigner, "Lower limit for the energy derivative of the scattering phase shift," *Physical Review*, vol. 98, no. 1, p. 145, 1955.

[19] F. T. Smith, "Lifetime matrix in collision theory," *Physical Review*, vol. 118, no. 1, p. 349, 1960.

[20] A. I. Baz', "Lifetime of intermediate states," *Soviet Journal of Nuclear Physics*, vol. 4, p. 182, 1967.

[21] V. F. Rybachenko, "Time penetration of a particle through a potential barrier," *Soviet Journal of Nuclear Physics*, vol. 5, p. 635, 1967.

[22] M. Büttiker, "Larmor precession and the traversal time for tunneling," *Physical Review B*, vol. 27, no. 10, p. 6178, 1983.

[23] J. P. Falck and E. H. Hauge, "Larmor clock reexamined," *Physical Review B: Condensed Matter and Materials Physics*, vol. 38, no. 5, pp. 3287–3297, 1988.

[24] H. Salecker and E. P. Wigner, "Quantum limitations of the measurement of space-time distances," *Physical Review*, vol. 109, pp. 571–577, 1958.

[25] P. Eckle, A. N. Pfeiffer, C. Cirelli et al., "Attosecond ionization and tunneling delay time measurements in helium," *Science*, vol. 322, no. 5907, pp. 1525–1529, 2008.

[26] G. Orlando, C. R. McDonald, N. H. Protik, G. Vampa, and T. Brabec, "Tunnelling time, what does it mean?" *Journal of Physics B: Atomic, Molecular and Optical Physics*, vol. 47, no. 20, Article ID 204002, 2014.

[27] A. S. Landsman and U. Keller, "Attosecond science and the tunnelling time problem," *Physics Reports*, vol. 547, pp. 1–24, 2015.

[28] J. T. Lunardi, L. A. Manzoni, and A. T. Nystrom, "Salecker–Wigner–Peres clock and average tunneling times," *Physics Letters A*, vol. 375, no. 3, pp. 415–421, 2011.

[29] N. G. Kelkar, H. M. Castañeda, and M. Nowakowski, "Quantum time scales in alpha tunneling," *Europhysics Letters*, vol. 85, no. 2, Article ID 20006, 2009.

[30] M. Goto, H. Iwamoto, V. De Aquino, V. C. Aguilera-Navarro, and D. H. Kobe, "Relationship between dwell, transmission and reflection tunnelling times," *Journal of Physics A: Mathematical and General*, vol. 37, no. 11, pp. 3599–3606, 2004.

[31] S. Brouard, R. Sala, and J. G. Muga, "Systematic approach to define and classify quantum transmission and reflection times," *Physical Review A: Atomic, Molecular and Optical Physics*, vol. 49, no. 6, pp. 4312–4325, 1994.

[32] V. Petrillo and V. S. Olkhovsky, "Time asymptotic expansion of the tunneled wave function for a double-barrier potential," *Europhysics Letters*, vol. 74, no. 2, p. 327, 2006.

[33] T. E. Hartman, "Tunneling of a wave packet," *Journal of Applied Physics*, vol. 33, pp. 3427–3433, 1962.

[34] B. A. Frentz, J. T. Lunardi, and L. A. Manzoni, "Average clock times for scattering through asymmetric barriers," *European Physical Journal Plus*, vol. 129, no. 1, Article ID 5, 2014.

[35] N. Turok, "On quantum tunneling in real time," *New Journal of Physics*, vol. 16, Article ID 063006, 2014.

[36] Y. Ban, E. Y. Sherman, J. G. Muga, and M. Büttiker, "Time scales of tunneling decay of a localized state," *Physical Review A: Atomic, Molecular and Optical Physics*, vol. 82, no. 6, Article ID 062121, 2010.

[37] M. Calçada, J. T. Lunardi, and L. A. Manzoni, "Salecker-Wigner-Peres clock and double-barrier tunneling," *Physical Review A: Atomic, Molecular and Optical Physics*, vol. 79, no. 1, Article ID 012110, 2009.

[38] C.-S. Park, "Transmission time of a particle in the reflectionless Sech-squared potential: quantum clock approach," *Physics Letters A*, vol. 375, no. 38, pp. 3348–3354, 2011.

[39] C.-S. Park, "Barrier interaction time and the Salecker-Wigner quantum clock: wave-packet approach," *Physical Review A*, vol. 80, no. 1, Article ID 012111, 2009.

[40] Y. Aharonov, N. Erez, and B. Reznik, "Superoscillations and tunneling times," *Physical Review A: Atomic, Molecular and Optical Physics*, vol. 65, no. 5, Article ID 052124, 2002.

[41] Y. Aharonov, N. Erez, and B. Reznik, "Superluminal tunnelling times as weak values," *Journal of Modern Optics*, vol. 50, no. 7, pp. 1139–1149, 2003.

[42] M. A. Lee, J. T. Lunardi, L. A. Manzoni, and E. A. Nyquist, "Double General Point Interactions: Symmetry and Tunneling Times," *Frontiers in Physics*, vol. 4, Article ID 10, 2016.

[43] N. Teeny, E. Yakaboylu, H. Bauke, and C. H. Keitel, "Ionization time and exit momentum in strong-field tunnel ionization," *Physical Review Letters*, vol. 116, no. 6, Article ID 063003, 2016.

12

Analysis of *CP* Violation in $D^0 \longrightarrow K^+K^-\pi^0$

Hang Zhou,[1] Bo Zheng,[1,2] and Zhen-Hua Zhang[1]

[1]*School of Nuclear Science and Technology, University of South China, Hengyang, Hunan 421001, China*
[2]*Helmholtz Institute Mainz, Johann-Joachim-Becher-Weg 45, D-55099 Mainz, Germany*

Correspondence should be addressed to Bo Zheng; zhengbo_usc@163.com and Zhen-Hua Zhang; zhangzh@usc.edu.cn

Guest Editor: Tao Luo

We study the *CP* violation induced by the interference between two intermediate resonances $K^*(892)^+$ and $K^*(892)^-$ in the phase space of singly-Cabibbo-suppressed decay $D^0 \longrightarrow K^+K^-\pi^0$. We adopt the factorization-assisted topological approach in dealing with the decay amplitudes of $D^0 \longrightarrow K^\pm K^*(892)^\mp$. The *CP* asymmetries of two-body decays are predicted to be very tiny, which are $(-1.27 \pm 0.25) \times 10^{-5}$ and $(3.86 \pm 0.26) \times 10^{-5}$, respectively, for $D^0 \longrightarrow K^+K^*(892)^-$ and $D^0 \longrightarrow K^-K^*(892)^+$, while the differential *CP* asymmetry of $D^0 \longrightarrow K^+K^-\pi^0$ is enhanced because of the interference between the two intermediate resonances, which can reach as large as 3×10^{-4}. For some NPs which have considerable impacts on the chromomagnetic dipole operator O_{8g}, the global *CP* asymmetries of $D^0 \longrightarrow K^+K^*(892)^-$ and $D^0 \longrightarrow K^-K^*(892)^+$ can be then increased to $(0.56\pm0.08)\times10^{-3}$ and $(-0.50\pm0.04)\times10^{-3}$, respectively. The regional *CP* asymmetry in the overlapped region of the phase space can be as large as $(1.3 \pm 0.3) \times 10^{-3}$.

1. Introduction

Charge-Parity (*CP*) violation, which was first discovered in *K* meson system in 1964 [1], is one of the most important phenomena in particle physics. In the Standard Model (SM), *CP* violation originates from the weak phase in the Cabibbo-Kobayashi-Maskawa (CKM) matrix [2, 3] and the unitary phases which usually arise from strong interactions. One reason for the smallness of *CP* violation is that the unitary phase is usually small. Nevertheless, *CP* violation can be enhanced in three-body decays of heavy hadrons, when the corresponding decay amplitudes are dominated by overlapped intermediate resonances in certain regions of phase space. Owing to the overlapping, a regional *CP* asymmetry can be generated by a relative strong phase between amplitudes corresponding to different resonances. This relative strong phase has nonperturbative origin. As a result, the regional *CP* asymmetry can be larger than the global one. In fact, such kind of enhanced *CP* violation has been observed in several three-body decay channels of *B* meson [4–7], which was followed by a number of theoretical works [8–19].

The study of *CP* violation in singly-Cabibbo-suppressed (SCS) *D* meson decays provides an ideal test of the SM and exploration of New Physics (NP) [20–23]. In the SM, *CP* violation is predicted to be very small in charm system. Experimental researches have shown that there is no significant *CP* violation so far in charmed hadron decays [24–33]. *CP* asymmetry in SCS *D* meson decay can be as small as

$$A_{CP} \sim \frac{\left|V_{cb}^*V_{ub}\right|}{\left|V_{cs}^*V_{us}\right|}\frac{\alpha_s}{\pi} \sim 10^{-4}, \qquad (1)$$

or even less, due to the suppression of the penguin diagrams by the CKM matrix as well as the smallness of Wilson coefficients in penguin amplitudes. The SCS decays are sensitive to new contributions to the $\Delta C = 1$ QCD penguin and chromomagnetic dipole operators, while such contributions can affect neither the Cabibbo-favored (CF) ($c \longrightarrow s\bar{d}u$) nor the doubly-Cabibbo-suppressed (DCS) ($c \longrightarrow d\bar{s}u$) decays [34]. Besides, the decays of charmed mesons offer a unique opportunity to probe *CP* violation in the up-type quark sector.

Several factorization approaches have been wildly used in nonleptonic *B* decays. In the naive factorization approach [35, 36], the hadronic matrix elements were expressed as a product of a heavy to light transition form factor and a decay constant. Based on Heavy Quark Effect Theory, it is shown in the QCD factorization approach that the corrections to the hadronic matrix elements can be expressed in terms of short-distance coefficients and meson light-cone distribution amplitudes [37, 38]. Alternative factorization approach based on QCD factorization is often applied in study of quasi two-body hadronic B decays [19, 39, 40], where they introduced unitary meson-meson form factors, from the perspective of unitarity, for the final state interactions. Other QCD-inspired approaches, such as the perturbative approach (pQCD) [41] and the soft-collinear effective theory (SCET) [42], are also wildly used in B meson decays.

However, for D meson decays, such QCD-inspired factorization approaches may not be reliable since the charm quark mass, which is just above 1 GeV, is not heavy enough for the heavy quark expansion [43, 44]. For this reason, several model-independent approaches for the charm meson decay amplitudes have been proposed, such as the flavor topological diagram approach based on the flavor $SU(3)$ symmetry [44–47] and the factorization-assisted topological-amplitude (FAT) approach with the inclusion of flavor $SU(3)$ breaking effect [48, 49]. One motivation of these aforementioned approaches is to identify as complete as possible the dominant sources of nonperturbative dynamics in the hadronic matrix elements.

In this paper, we study the CP violation of SCS D meson decay $D^0 \longrightarrow K^+K^-\pi^0$ in the FAT approach. Our attention will be mainly focused on the region of the phase space where two intermediate resonances, $K^*(892)^+$ and $K^*(892)^-$, are overlapped. Before proceeding, it will be helpful to point out that direct CP asymmetry is hard to be isolated for decay process with CP-eigen-final-state. When the final state of the decay process is CP eigenstate, the time integrated CP violation for $D^0 \longrightarrow f$, which is defined as

$$a_f \equiv \frac{\int_0^\infty \Gamma\left(D^0 \longrightarrow f\right) dt - \int_0^\infty \Gamma\left(\overline{D}^0 \longrightarrow f\right) dt}{\int_0^\infty \Gamma\left(D^0 \longrightarrow f\right) dt + \int_0^\infty \Gamma\left(\overline{D}^0 \longrightarrow f\right) dt}, \quad (2)$$

can be expressed as [34]

$$a_f = a_f^d + a_f^m + a_f^i, \quad (3)$$

where a_f^d, a_f^m, and a_f^i are the CP asymmetries in decay, in mixing, and in the interference of decay and mixing, respectively. As is shown in [34, 50, 51], the indirect CP violation $a^{\text{ind}} \equiv a^m + a^i$ is universal and channel-independent for two-body CP-eigenstate. This conclusion is easy to be generalized to decay processes with three-body CP-eigenstate in the final state, such as $D^0 \longrightarrow K^+K^-\pi^0$. In view of the universality of the indirect CP asymmetry, we will only consider the direct CP violations of the decay $D^0 \longrightarrow K^+K^-\pi^0$ throughout this paper.

The remainder of this paper is organized as follows. In Section 2, we present the decay amplitudes for various decay channels, where the decay amplitudes of $D^0 \longrightarrow K^\pm K^*(892)^\mp$ are formulated via the FAT approaches. In Section 3, we study the CP asymmetries of $D^0 \longrightarrow K^\pm K^*(892)^\mp$ and the CP asymmetry of $D^0 \longrightarrow K^+K^-\pi^0$ induced by the interference between different resonances in the phase space. Discussions and conclusions are given in Section 4. We list some useful formulas and input parameters in the Appendix.

2. Decay Amplitude for $D^0 \longrightarrow K^+K^-\pi^0$

In the overlapped region of the intermediate resonances $K^*(892)^+$ and $K^*(892)^-$ in the phase space, the decay process $D^0 \longrightarrow K^+K^-\pi^0$ is dominated by two cascade decays, $D^0 \longrightarrow K^+K^*(892)^- \longrightarrow K^+K^-\pi^0$ and $D^0 \longrightarrow K^-K^*(892)^+ \longrightarrow K^-K^+\pi^0$, respectively. Consequently, the decay amplitude of $D^0 \longrightarrow K^+K^-\pi^0$ can be expressed as

$$\mathcal{M}_{D^0 \longrightarrow K^+K^-\pi^0} = \mathcal{M}_{K^{*+}} + e^{i\delta}\mathcal{M}_{K^{*-}} \quad (4)$$

in the overlapped region, where $\mathcal{M}_{K^{*+}}$ and $\mathcal{M}_{K^{*-}}$ are the amplitudes for the two cascade decays and δ is the relative strong phase. Note that nonresonance contributions have been neglected in (4).

The decay amplitude for the cascade decay $D^0 \longrightarrow K^+K^*(892)^- \longrightarrow K^+K^-\pi^0$ can be expressed as

$$\mathcal{M}_{K^{*-}} = \frac{\sum_\lambda \mathcal{M}^\lambda_{K^{*-} \longrightarrow K^-\pi^0} \cdot \mathcal{M}^\lambda_{D^0 \longrightarrow K^{*-}K^+}}{s_{\pi^0 K^-} - m^2_{K^{*-}} + i m_{K^{*-}}\Gamma_{K^{*-}}}, \quad (5)$$

where $\mathcal{M}^\lambda_{K^{*-} \longrightarrow K^-\pi^0}$ and $\mathcal{M}^\lambda_{D^0 \longrightarrow K^+K^{*-}}$ represent the amplitudes corresponding to the strong decay $K^{*-} \longrightarrow K^-\pi^0$ and weak decay $D^0 \longrightarrow K^+K^{*-}$, respectively, λ is the helicity index of K^{*-}, $s_{\pi^0 K^-}$ is the invariant mass square of $\pi^0 K^-$ system, and $m_{K^{*-}}$ and $\Gamma_{K^{*-}}$ are the mass and width of $K^*(892)^-$, respectively. The decay amplitude for the cascade decay, $D^0 \longrightarrow K^-K^*(892)^+ \longrightarrow K^-K^+\pi^0$, is the same as (5) except replacing the subscripts K^{*-} and $K^\pm$ with K^{*+} and $K^\mp$, respectively.

For the strong decays $K^*(892)^\pm \longrightarrow \pi^0 K^\pm$, one can express the decay amplitudes as

$$\mathcal{M}_{K^{*\pm} \longrightarrow \pi^0 K^\pm} = g_{K^{*\pm}K^\pm\pi^0}\left(p_{\pi^0} - p_{K^\pm}\right) \cdot \varepsilon_{K^{*\pm}}(p, \lambda), \quad (6)$$

where p_{π^0} and $p_{K^\pm}$ represent the momentum for π^0 and $K^\pm$ mesons, respectively, and $g_{K^{*\pm}K^\pm\pi^0}$ is the effective coupling constant for the strong interaction, which can be extracted from the experimental data via

$$g^2_{K^{*\pm}K^\pm\pi^0} = \frac{6\pi m^2_{K^{*\pm}} \Gamma_{K^{*\pm} \longrightarrow K^\pm\pi^0}}{\lambda^3_{K^{*\pm}}}, \quad (7)$$

with

$$\lambda_{K^{*\pm}} = \frac{1}{2m_{K^{*\pm}}} \cdot \sqrt{\left[m^2_{K^{*\pm}} - \left(m_{\pi^0} + m_{K^\pm}\right)^2\right] \cdot \left[m^2_{K^{*\pm}} - \left(m_{\pi^0} - m_{K^\pm}\right)^2\right]}, \quad (8)$$

and $\Gamma_{K^{*\pm} \longrightarrow K^\pm\pi^0} = \text{Br}(K^{*\pm} \longrightarrow K^\pm\pi^0) \cdot \Gamma_{K^{*\pm}}$. The isospin symmetry of the strong interaction implies that $\Gamma_{K^{*\pm} \longrightarrow K^\pm\pi^0} \simeq (1/3)\Gamma_{K^{*\pm}}$.

The decay amplitudes for the weak decays, $D^0 \longrightarrow K^+K^*(892)^-$ and $D^0 \longrightarrow K^-K^*(892)^+$, will be handled with

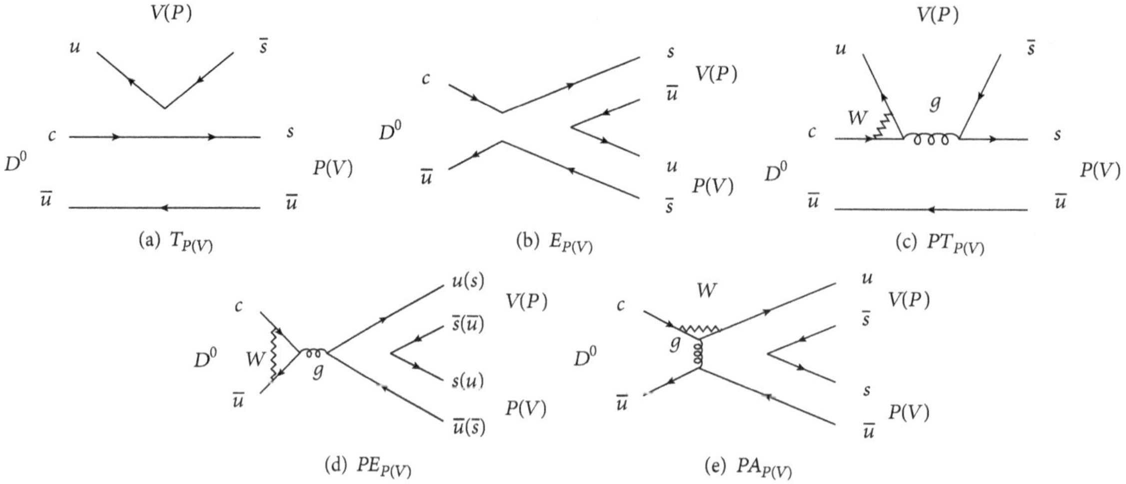

FIGURE 1: The relevant topological diagrams for $D \longrightarrow PV$ with (a) the color-favored tree amplitude $T_{P(V)}$, (b) the W-exchange amplitude $E_{P(V)}$, (c) the color-favored penguin amplitude $PT_{P(V)}$, (d) the gluon-annihilation penguin amplitude $PE_{P(V)}$, and (e) the gluon-exchange penguin amplitude $PA_{P(V)}$.

the aforementioned FAT approach [48, 49]. The relevant topological tree and penguin diagrams for $D \longrightarrow PV$ are displayed in Figure 1, where P and V denote a light pseudoscalar and vector meson (representing $K^{\pm}$ and $K^{*\pm}$ in this paper), respectively.

The two tree diagrams in first line of Figure 1 represent the color-favored tree diagram for $D \longrightarrow P(V)$ transition and the W-exchange diagram with the pseudoscalar (vector) meson containing the antiquark from the weak vertex, respectively. The amplitudes of these two diagrams will be, respectively, denoted as $T_{P(V)}$ and $E_{P(V)}$.

According to these topological structures, the amplitudes of the color-favored tree diagrams $T_{P(V)}$, which are dominated by the factorizable contributions, can be parameterized as

$$T_P = \frac{G_F}{\sqrt{2}}\lambda_s a_2(\mu) f_V m_V F_1^{D\longrightarrow P}\left(m_V^2\right) 2\left(\varepsilon^* \cdot p_D\right), \tag{9}$$

and

$$T_V = \frac{G_F}{\sqrt{2}}\lambda_s a_2(\mu) f_P m_V A_0^{D\longrightarrow V}\left(m_P^2\right) 2\left(\varepsilon^* \cdot p_D\right), \tag{10}$$

respectively, where G_F is the Fermi constant, $\lambda_s = V_{us}V_{cs}^*$, with V_{us} and V_{cs} being the CKM matrix elements, $a_2(\mu) = c_2(\mu) + c_1(\mu)/N_c$, with $c_1(\mu)$ and $c_2(\mu)$ being the scale-dependent Wilson coefficients, and the number of color $N_c = 3$, $f_{V(P)}$ and $m_{V(P)}$ are the decay constant and mass of the vector (pseudoscalar) meson, respectively, $F_1^{D\longrightarrow P}$ and $A_0^{D\longrightarrow V}$ are the form factors for the transitions $D \longrightarrow P$ and $D \longrightarrow V$, respectively, ε is the polarization vector of the vector meson, and p_D is the momentum of D meson. The scale μ of Wilson coefficients is set to energy release in individual decay channels [52, 53], which depends on masses of initial and final states and is defined as [48, 49]

$$\mu = \sqrt{\Lambda m_D\left(1 - r_P^2\right)\left(1 - r_V^2\right)}, \tag{11}$$

with the mass ratios $r_{V(P)} = m_{V(P)}/m_D$, where Λ represents the soft degrees of freedom in the D meson, which is a free parameter.

For the W-exchange amplitudes, since the factorizable contributions to these amplitudes are helicity-suppressed, only the nonfactorizable contributions need to be considered. Therefore, the W-exchange amplitudes are parameterized as

$$E^q_{P,V} = \frac{G_F}{\sqrt{2}}\lambda_s c_2(\mu)\chi^E_q e^{i\phi^E_q} f_D m_D \frac{f_P f_V}{f_\pi f_\rho}\left(\varepsilon^* \cdot p_D\right), \tag{12}$$

where m_D is the mass of D meson, f_D, f_π, and f_ρ are the decay constants of the D, π, and ρ mesons, respectively, and χ^E_q and ϕ^E_q characterize the strengths and the strong phases of the corresponding amplitudes, with $q = u, d, s$ representing the strongly produced q quark pair. The ratio of $f_P f_V$ over $f_\pi f_\rho$ indicates that the flavor $SU(3)$ breaking effects have been taken into account from the decay constants.

The penguin diagrams shown in the second line of Figure 1 represent the color-favored, the gluon-annihilation, and the gluon-exchange penguin diagrams, respectively, whose amplitudes will be denoted as $PT_{P(V)}$, $PE_{P(V)}$, and $PA_{P(V)}$, respectively.

Since a vector meson cannot be generated from the scalar or pseudoscalar operator, the amplitude PT_P does not include contributions from the penguin operator O_5 or O_6. Consequently, the color-favored penguin amplitudes PT_P and PT_V can be expressed as

$$PT_P = -\frac{G_F}{\sqrt{2}}\lambda_b a_4(\mu) f_V m_V F_1^{D\longrightarrow P}\left(m_V^2\right) 2\left(\varepsilon^* \cdot p_D\right), \tag{13}$$

and

$$PT_V = -\frac{G_F}{\sqrt{2}}\lambda_b\left[a_4(\mu) - r_\chi a_6(\mu)\right] f_P m_V A_0^{D\longrightarrow V}\left(m_P^2\right) \cdot 2\left(\varepsilon^* \cdot p_D\right), \tag{14}$$

respectively, where $\lambda_b = V_{ub}V_{cb}^*$ with V_{ub} and V_{cb}^* being the CKM matrix elements, $a_{4,6}(\mu) = c_{4,6}(\mu) + c_{3,5}(\mu)/N_c$, with $c_{3,4,5,6}$ being the Wilson coefficients, and r_χ is a chiral factor, which takes the form

$$r_\chi = \frac{2m_P^2}{\left(m_u + m_q\right)\left(m_q + m_c\right)}, \tag{15}$$

with $m_{u(c,q)}$ being the masses of $u(c,q)$ quark. Note that the quark-loop corrections and the chromomagnetic-penguin contribution are also absorbed into $c_{3,4,5,6}$ as shown in [49].

Similar to the amplitudes $E_{P,V}$, the amplitudes PE only include the nonfactorizable contributions as well. Therefore, the amplitudes $PE_{P,V}$, which are dominated by O_4 and O_6 [48], can be parameterized as

$$PE_{P,V}^q = -\frac{G_F}{\sqrt{2}}\lambda_b\left[c_4(\mu) - c_6(\mu)\right]\chi_q^E e^{i\phi_q^E} f_D m_D \cdot \frac{f_P f_V}{f_\pi f_\rho}\left(\varepsilon^* \cdot p_D\right). \tag{16}$$

For the amplitudes PA_P and PA_V, the helicity suppression does not apply to the matrix elements of $O_{5,6}$, so the factorizable contributions exist. In the pole resonance model [54], after applying the Fierz transformation and the factorization hypothesis, the amplitudes PA_P and PA_V can be expressed as

$$PA_P^q = -\frac{G_F}{\sqrt{2}}\lambda_b\left[(-2)a_6(\mu)(2g_S) \cdot \frac{1}{m_D^2 - m_{P^*}^2}\left(f_{P^*}m_{P^*}^0\right)\left(f_D\frac{m_D^2}{m_c}\right) + c_3(\mu) \cdot \chi_q^A e^{i\phi_q^A} f_D m_D \frac{f_P f_V}{f_\pi f_\rho}\right]\left(\varepsilon^* \cdot p_D\right), \tag{17}$$

and

$$PA_V^q = -\frac{G_F}{\sqrt{2}}\lambda_b\left[(-2)a_6(\mu)(-2g_S) \cdot \frac{1}{m_D^2 - m_{P^*}^2}\left(f_{P^*}m_{P^*}^0\right)\left(f_D\frac{m_D^2}{m_c}\right) + c_3(\mu) \cdot \chi_q^A e^{i\phi_q^A} f_D m_D \frac{f_P f_V}{f_\pi f_\rho}\right]\left(\varepsilon^* \cdot p_D\right), \tag{18}$$

respectively, where g_S is an effective strong coupling constant obtained from strong decays, e.g., $\rho \longrightarrow \pi\pi$, $K^* \longrightarrow K\pi$, and $\phi \longrightarrow KK$, and is set as $g_S = 4.5$ [54] in this work, m_{P^*} and f_{P^*} are the mass and decay constant of the pole resonant pseudoscalar meson P^*, respectively, and χ_q^A and ϕ_q^A are the strengths and the strong phases of the corresponding amplitudes.

From Figure 1, the decay amplitudes of $D^0 \longrightarrow K^+K^*(892)^-$ and $D^0 \longrightarrow K^-K^*(892)^+$ in the FAT approach can be easily written down

$$\mathcal{M}_{D^0\longrightarrow K^+K^{*-}}^\lambda = T_{K^{*-}} + E_{K^+}^u + PT_{K^{*-}} + PE_{K^{*-}}^s + PE_{K^+}^u + PA_{K^{*-}}^s, \tag{19}$$

and

$$\mathcal{M}_{D^0\longrightarrow K^-K^{*+}}^\lambda = T_{K^-} + E_{K^{*+}}^u + PT_{K^-} + PE_{K^-}^s + PE_{K^{*+}}^u + PA_{K^-}^s, \tag{20}$$

respectively, where λ is the helicity of the polarization vector $\varepsilon(p,\lambda)$. In the FAT approach, the fitted nonperturbative parameters, $\chi_{q,s}^E$, $\phi_{q,s}^E$, $\chi_{q,s}^A$, $\phi_{q,s}^A$, are assumed to be universal and can be determined by the data [49].

In Table 1, we list the magnitude of each topological amplitude for $D^0 \longrightarrow K^+K^*(892)^-$ and $D^0 \longrightarrow K^-K^*(892)^+$ by using the global fitted parameters for $D \longrightarrow PV$ in [49]. One can see from Table 1 that the penguin contributions are greatly suppressed. PT is dominant in the penguin contributions of $D^0 \longrightarrow K^-K^*(892)^+$, while PT is small in $D^0 \longrightarrow K^+K^*(892)^-$, which is even smaller than the amplitude PA. This difference is because of the chirally enhanced factor contained in (14) while not in (13). The very small PE do not receive the contributions from the quark-loop and chromomagnetic penguins, since these two contributions to c_4 and c_6 are canceled with each other in (16). Besides, the relations $PE_V^s = PE_P^s$, $PE_V^u = PE_P^u$, and $PE_V^s \neq PE_V^u$ can be read from Table 1, this is because that the isospin symmetry and the flavor $SU(3)$ breaking effect have been considered.

Since the form factors are inevitably model-dependent, we list in Table 2 the branching ratios of $D^0 \longrightarrow K^+K^*(892)^-$ and $D^0 \longrightarrow K^-K^*(892)^+$ predicted by the FAT approach, by various form factor models. The pole, dipole, and covariant light-front (CLF) models are adopted. The uncertainties in Table 2 mainly come from decay constants. The CLF model agrees well with the data for both decay channels, and other models are also consistent with the data. However, the model-dependence of form factor leads to large uncertainty of the branching fraction, as large as 20%. Because of the smallness of the Wilson coefficients and the CKM-suppression of the penguin amplitudes, the branching ratios are dominated by the tree amplitudes. Therefore, there is no much difference for the branching ratios whether we consider the penguin amplitudes or not.

3. *CP* Asymmetries for $D^0 \longrightarrow K^\pm K^*(892)^\mp$ and $D^0 \longrightarrow K^+K^-\pi^0$

The direct *CP* asymmetry for the two-body decay $D \longrightarrow PV$ is defined as

$$A_{CP}^{D\longrightarrow PV} = \frac{\left|\mathcal{M}_{D\longrightarrow PV}\right|^2 - \left|\mathcal{M}_{\overline{D}\longrightarrow\overline{PV}}\right|^2}{\left|\mathcal{M}_{D\longrightarrow PV}\right|^2 + \left|\mathcal{M}_{\overline{D}\longrightarrow\overline{PV}}\right|^2}, \tag{21}$$

where $\mathcal{M}_{\overline{D}\longrightarrow\overline{PV}}$ represents the decay amplitude of the *CP* conjugate process $\overline{D} \longrightarrow \overline{PV}$, such as $\overline{D}^0 \longrightarrow K^+K^*(892)^-$ or $\overline{D}^0 \longrightarrow K^-K^*(892)^+$. In the framework of FAT approach,

TABLE 1: The magnitude of tree and penguin contributions (in unit of 10^{-3}) corresponding to the topological amplitudes in (19) and (20). The factors "$(G_F/\sqrt{2})\lambda_s(\varepsilon^* \cdot p_D)$" and "$-(G_F/\sqrt{2})\lambda_b(\varepsilon^* \cdot p_D)$" are omitted in this table.

Decay modes	$T_{K^{*-}}$	$E^u_{K^+}$	$PT_{K^{*-}}$	$PE^s_{K^{*-}}$	$PE^u_{K^+}$	$PA^s_{K^{*-}}$
$D^0 \longrightarrow K^+K^*(892)^-$	0.23	$-0.02 + 0.15i$	$3.83 + 4.32i$	$0.96 - 0.03i$	$0.13 - 0.81i$	$6.73 + 8.22i$
	T_{K^-}	$E^u_{K^{*+}}$	PT_{K^-}	$PE^s_{K^-}$	$PE^u_{K^{*+}}$	$PA^s_{K^-}$
$D^0 \longrightarrow K^-K^*(892)^+$	0.44	$-0.02 + 0.15i$	$-23.3 - 19.3i$	$0.96 - 0.03i$	$0.13 - 0.81i$	$-8.53 - 5.53i$

TABLE 2: Branching ratios (in unit of 10^{-3}) of singly-Cabibbo-suppressed decays $D^0 \longrightarrow K^+K^*(892)^-$ and $D^0 \longrightarrow K^-K^*(892)^+$. Both experimental data [55–57] and theoretical predictions of FAT approach of the branching ratios are listed.

Form factors	$\mathrm{Br}(D^0 \longrightarrow K^+K^*(892)^-)$	$\mathrm{Br}(D^0 \longrightarrow K^-K^*(892)^+)$
Pole	1.57 ± 0.04	3.73 ± 0.17
Dipole	1.69 ± 0.04	4.02 ± 0.19
CLF	1.45 ± 0.04	4.44 ± 0.20
Exp.	1.56 ± 0.12	4.38 ± 0.21

we predict very small direct *CP* asymmetries of $D^0 \longrightarrow K^+K^*(892)^-$ and $D^0 \longrightarrow K^-K^*(892)^+$ presented in Table 3. The uncertainties induced by the model-dependence of form factor to the *CP* asymmetries of $D^0 \longrightarrow K^+K^*(892)^-$ and $D^0 \longrightarrow K^-K^*(892)^+$ are about 30% and 10%, respectively.

The differential *CP* asymmetry of the three-body decay $D^0 \longrightarrow K^+K^-\pi^0$, which is a function of the invariant mass of $s_{\pi^0K^+}$ and $s_{\pi^0K^-}$, is defined as

$$A_{CP}^{D^0\longrightarrow K^+K^-\pi^0}\left(s_{\pi^0K^+}, s_{\pi^0K^-}\right) = \frac{\left|\mathcal{M}_{D^0\longrightarrow K^+K^-\pi^0}\right|^2 - \left|\mathcal{M}_{\overline{D}^0\longrightarrow K^-K^+\pi^0}\right|^2}{\left|\mathcal{M}_{D^0\longrightarrow K^+K^-\pi^0}\right|^2 + \left|\mathcal{M}_{\overline{D}^0\longrightarrow K^-K^+\pi^0}\right|^2}, \quad (22)$$

where the invariant mass $s_{\pi^0K^\pm} = (p_{\pi^0} + p_{K^\pm})^2$. As can be seen from (4), the differential *CP* asymmetry $A_{CP}^{D^0\longrightarrow K^+K^-\pi^0}$ depends on the relative strong phase δ, which is impossible to be calculated theoretically because of its nonperturbative origin. Despite this, we can still acquire some information of this relative strong phase δ from data. By using a Dalitz plot technique [55, 58, 59], the phase difference $\delta^{\exp}$ between D^0 decays to $K^+K^*(892)^-$ and $K^-K^*(892)^+$ can be extracted from data. One should notice that $\delta^{\exp}$ is not the same as the strong phase δ defined in (4). The strong phase δ is the relative phase between the decay amplitudes of $D^0 \longrightarrow K^+K^*(892)^-$ and $D^0 \longrightarrow K^-K^*(892)^+$. On the other hand, the phase $\delta^{\exp}$ is defined through

$$\mathcal{M}_{D^0\longrightarrow K^+K^-\pi^0} = \left(\left|\mathcal{M}_{K^{*+}}\right| + e^{i\delta^{\exp}}\left|\mathcal{M}_{K^{*-}}\right|\right)e^{i\delta_{K^{*+}}} \quad (23)$$

in the overlapped region of the phase space, where $\delta_{K^{*\pm}}$ is the phase of the amplitude $\mathcal{M}_{K^{*\pm}}$:

$$\mathcal{M}_{K^{*\pm}} = \left|\mathcal{M}_{K^{*\pm}}\right|e^{i\delta_{K^{*\pm}}}. \quad (24)$$

Therefore, neglecting the CKM suppressed penguin amplitudes, $\delta^{\exp}$ and δ can be related by

$$\delta^{\exp} - \delta \approx \delta^{K^{*-}K^+} - \delta^{K^{*+}K^-}, \quad (25)$$

where $\delta^{K^{*\mp}K^\pm} = \arg(T_{K^{*\mp}} + E^u_{K^\pm})$ are the phases in tree-level amplitudes of $D^0 \longrightarrow K^\pm K^*(892)^\mp$ and are equivalent to $\delta_{K^{*\mp}}$ if the penguin amplitudes are neglected. With the relation of (25), and $\delta^{\exp} = -35.5^\circ \pm 4.1^\circ$ measured by the BABAR Collaboration [56], we have $\delta \approx -51.85^\circ \pm 4.1^\circ$.

In Figure 2, we present the differential *CP* asymmetry of $D^0 \longrightarrow K^+K^-\pi^0$ in the overlapped region of $K^*(892)^-$ and $K^*(892)^+$ in the phase space, with $\delta = -51.85^\circ$. Namely, we will focus on the region $m_{K^*} - 2\Gamma_{K^*} < \sqrt{s_{\pi^0K^-}}, \sqrt{s_{\pi^0K^+}} < m_{K^*} + 2\Gamma_{K^*}$ of the phase space. One can see from Figure 2 that the differential *CP* asymmetry of $D^0 \longrightarrow K^+K^-\pi^0$ can reach 3.0×10^{-4} in the overlapped region, which is about 10 times larger than the *CP* asymmetries of the corresponding two-body decay channels shown in Table 3.

The behavior of the differential *CP* asymmetry of $D^0 \longrightarrow K^+K^-\pi^0$ in Figure 2 motivates us to separate this region into four areas, area A ($m_{K^*} < \sqrt{s_{\pi^0K^-}} < m_{K^*} + 2\Gamma_{K^*}, m_{K^*} - 2\Gamma_{K^*} < \sqrt{s_{\pi^0K^+}} < m_{K^*}$), area B ($m_{K^*} < \sqrt{s_{\pi^0K^-}} < m_{K^*} + 2\Gamma_{K^*}, m_{K^*} < \sqrt{s_{\pi^0K^+}} < m_{K^*} + 2\Gamma_{K^*}$), area C ($m_{K^*} - 2\Gamma_{K^*} < \sqrt{s_{\pi^0K^-}} < m_{K^*}, m_{K^*} - 2\Gamma_{K^*} < \sqrt{s_{\pi^0K^+}} < m_{K^*}$), and area D ($m_{K^*} - 2\Gamma_{K^*} < \sqrt{s_{\pi^0K^-}} < m_{K^*}, m_{K^*} < \sqrt{s_{\pi^0K^+}} < m_{K^*} + 2\Gamma_{K^*}$). We further consider the observable of regional *CP* asymmetry in areas A, B, C, and D displayed in Table 4, which is defined by

$$A_{CP}^{\Omega} = \frac{\int_\Omega \left(\left|\mathcal{M}_{\text{tot}}\right|^2 - \left|\overline{\mathcal{M}}_{\text{tot}}\right|^2\right) ds_{\pi^0K^-}s_{\pi^0K^+}}{\int_\Omega \left(\left|\mathcal{M}_{\text{tot}}\right|^2 + \left|\overline{\mathcal{M}}_{\text{tot}}\right|^2\right) ds_{\pi^0K^-}s_{\pi^0K^+}}, \quad (26)$$

where Ω represents a certain region of the phase space.

Comparing with the *CP* asymmetries of two-body decays, the regional *CP* asymmetries, from Table 4, are less sensitive to the models we have used. We would like to use only the CLF model for the following discussion. The uncertainties in Table 4 come from decay constants as well as the relative phase $\delta^{\exp}$. In addition, if we focus on the right part of area A, that is, $m_{K^*} < \sqrt{s_{\pi^0K^-}} < m_{K^*} + 2\Gamma_{K^*}, m_{K^*} - \Gamma_{K^*} < \sqrt{s_{\pi^0K^+}} < m_{K^*}$, the regional *CP* violation will be $(1.09 \pm 0.16) \times 10^{-4}$.

The energy dependence of the propagator of the intermediate resonances can lead to a small correction to *CP* asymmetry. For example, if we replace the Breit-Wigner

TABLE 3: *CP* asymmetries (in unit of 10^{-5}) of $D^0 \longrightarrow K^+K^*(892)^-$ and $D^0 \longrightarrow K^-K^*(892)^+$ predicted by the FAT approach with pole, dipole, and CLF models adopted. The uncertainties in this table are mainly from decay constants.

Form factors	$A_{CP}(D^0 \longrightarrow K^+K^*(892)^-)$	$A_{CP}(D^0 \longrightarrow K^-K^*(892)^+)$
Pole	-1.45 ± 0.25	3.60 ± 0.23
Dipole	-1.63 ± 0.26	3.70 ± 0.24
CLF	-1.27 ± 0.25	3.86 ± 0.26

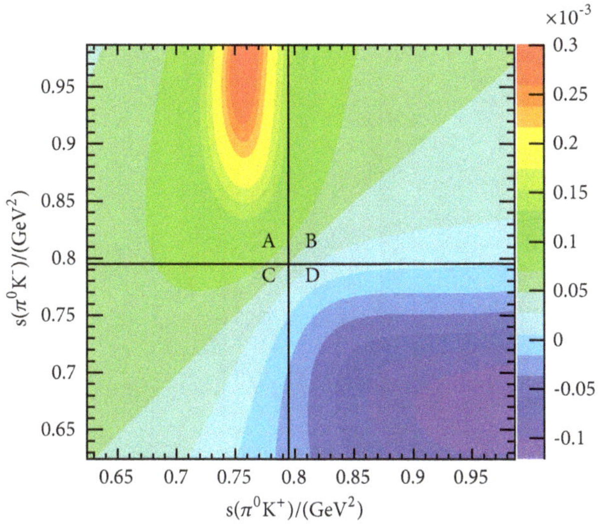

FIGURE 2: The differential *CP* asymmetry distribution of $D^0 \longrightarrow K^+K^-\pi^0$ in the overlapped region of $K^*(892)^-$ and $K^*(892)^+$ in the phase space.

propagator by the Flatté Parametrization [60], the correction to the regional *CP* asymmetry will be about 1%.

Since the *CP* asymmetry of $D^0 \longrightarrow K^+K^-\pi^0$ is extremely suppressed, it should be more sensitive to the NP. For example, some NPs have considerable impacts on the chromomagnetic dipole operator O_{8g} [34, 61–66]. Consequently, the *CP* violation in SCS decays may be further enhanced. In practice, the NP contributions can be absorbed into the corresponding effective Wilson coefficient c_{8g}^{eff} [67, 68]. For comparison, we first consider a relative small value of c_{8g}^{eff} (as in [48, 64]) lying within the range (0, 1) and the global *CP* asymmetry of $D^0 \longrightarrow K^*(892)^{\pm}K^{\mp}$ are no larger than 5×10^{-5}. Moreover, if we follow [49] taking $c_{8g}^{\text{eff}} \approx 10$ (while $c_{8g}^{\text{eff}} = 10$, which is extracted from ΔA_{CP} measured by LHCb [69], is a quite large quantity even for the coefficients corresponding tree-level operators, however, such large contribution can be realized if some NPs effects are pulled in. For example, the up squark-gluino loops in supersymmetry (SUSY) can arise significant contributions to c_{8g}. More details about the squark-gluino loops and other models in SUSY can be found in [34, 62, 70–72]), the global *CP* asymmetries of $D^0 \longrightarrow K^+K^*(892)^-$ and $D^0 \longrightarrow K^-K^*(892)^+$ are then $(0.56\pm0.08)\times 10^{-3}$ and $(-0.50 \pm 0.04) \times 10^{-3}$, respectively.

We further display the *CP* asymmetry of $D^0 \longrightarrow K^+K^-\pi^0$ in the overlapped region of $K^*(892)^-$ and $K^*(892)^+$ in Figures 3(a) and 3(b) for $c_{8g}^{\text{eff}} = 1$ and $c_{8g}^{\text{eff}} = 10$, respectively. After taking the interference effect into account, the differential *CP* asymmetry of $D^0 \longrightarrow K^+K^-\pi^0$ can be increased as large as 5.5×10^{-4} and 2.8×10^{-3} for $c_{8g}^{\text{eff}} = 1$ and $c_{8g}^{\text{eff}} = 10$, respectively. The regional ones (in phase space of $\sqrt{0.74}$ GeV $< \sqrt{s_{\pi^0K^-}} < \sqrt{0.81}$ GeV, $\sqrt{0.84} < \sqrt{s_{\pi^0K^+}} < m_{K^*} + 2\Gamma_{K^*}$) can reach $(2.7\pm0.5)\times10^{-4}$ and $(1.3\pm0.3)\times10^{-3}$ for $c_{8g}^{\text{eff}} = 1$ and $c_{8g}^{\text{eff}} = 10$, respectively.

4. Discussion and Conclusion

In this work, we studied *CP* violations in $D^0 \longrightarrow K^*(892)^{\pm}K^{\mp} \longrightarrow K^+K^-\pi^0$ via the FAT approach. The *CP* violations in two-body decay processes $D^0 \longrightarrow K^+K^*(892)^-$ and $D^0 \longrightarrow K^-K^*(892)^+$ are very small, which are $(-1.27 \pm 0.25) \times 10^{-5}$ and $(3.86 \pm 0.26) \times 10^{-5}$, respectively. Our discussion shows that the *CP* violation can be enhanced by the interference effect in three-body decay $D^0 \longrightarrow K^+K^-\pi^0$. The differential *CP* asymmetry can reach 3.0×10^{-4} when the interference effect is taken into account, while the regional one can be as large as $(1.09 \pm 0.16) \times 10^{-4}$.

Besides, since the chromomagnetic dipole operator O_{8g} is sensitive to some NPs, the inclusion of this kind of NPs will lead to a much larger global *CP* asymmetries of $D^0 \longrightarrow K^+K^*(892)^-$ and $D^0 \longrightarrow K^-K^*(892)^+$, which are $(0.56 \pm 0.08) \times 10^{-3}$ and $(-0.50 \pm 0.04) \times 10^{-3}$, respectively, while the regional *CP* asymmetry of $D^0 \longrightarrow K^+K^-\pi^0$ can be also increased to $(1.3 \pm 0.3) \times 10^{-3}$ when considering the interference effect in the phase space. Since the $\mathcal{O}(10^{-3})$ of

TABLE 4: Three from factor models: the pole, dipole, and CLF models are used for the regional CP asymmetries (in unit of 10^{-4}) in the four areas, A, B, C, and D, of the phase space.

Form factors	A_{CP}^{A}	A_{CP}^{B}	A_{CP}^{C}	A_{CP}^{D}	A_{CP}^{All}
Pole	0.87 ± 0.11	0.42 ± 0.08	0.39 ± 0.07	-0.30 ± 0.08	0.33 ± 0.05
Dipole	0.87 ± 0.11	0.41 ± 0.08	0.38 ± 0.07	-0.30 ± 0.08	0.32 ± 0.05
CLF	0.84 ± 0.10	0.45 ± 0.08	0.42 ± 0.07	-0.25 ± 0.08	0.36 ± 0.06

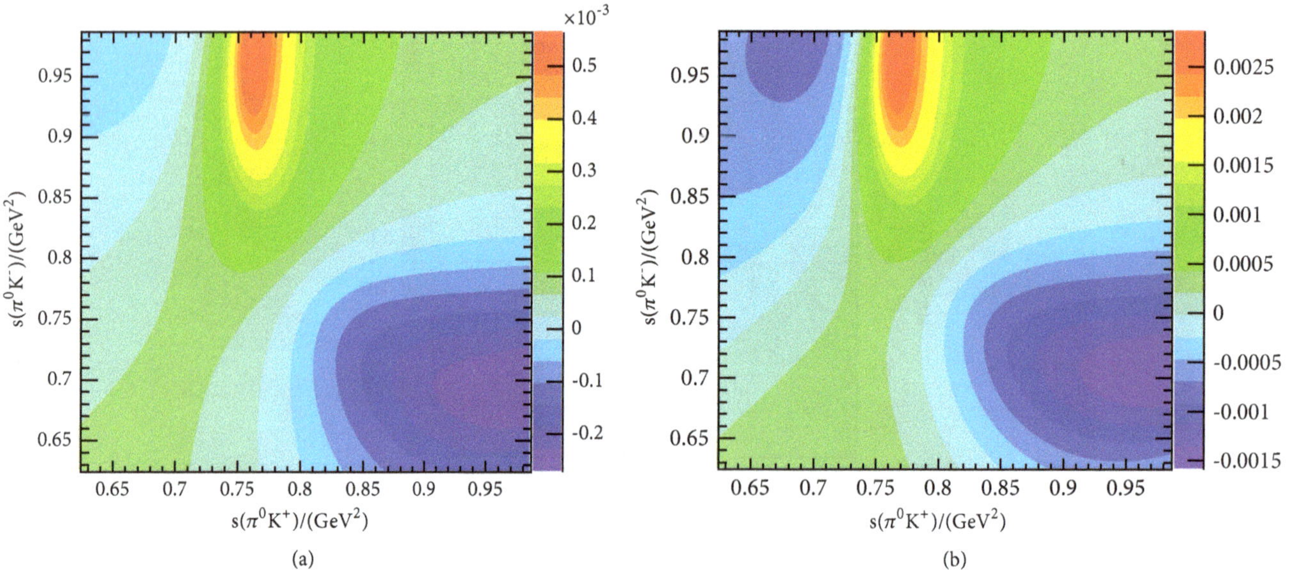

FIGURE 3: The differential CP asymmetry distribution of $D^0 \longrightarrow K^+K^-\pi^0$ for (a) $c_{8g}^{\text{eff}} = 1$ and (b) $c_{8g}^{\text{eff}} = 10$, in the overlapped region of $K^*(892)^-$ and $K^*(892)^+$ in the phase space.

CP asymmetry is attributed to the large c_{8g}^{eff}, which is almost impossible for the SM to generate such large contribution, it will indicate NP if such CP violation is observed. Here, we roughly estimate the number of $D^0\overline{D}^0$ needed for testing such kind of asymmetries, which is about $(1/Br)(1/A_{CP}^2) \sim 10^9$. This could be observed in the future experiments at Belle II [73, 74], while the current largest $D^0\overline{D}^0$ yields are about 10^8 at BABAR and Belle [75, 76] and 10^7 at BESIII [77].

Appendix

Some Useful Formulas and Input Parameters

(1) Effective Hamiltonian and Wilson Coefficients. The weak effective Hamiltonian for SCS D meson decays, based on the Operator Product Expansion (OPE) and Heavy Quark Effective Theory (HQET), can be expressed as [78]

$$\mathcal{H}_{\text{eff}} = \frac{G_F}{\sqrt{2}}\left[\sum_{q=d,s}\lambda_q\left(c_1O_1^q + c_2O_2^q\right) - \lambda_b\left(\sum_{i=3}^{6}c_iO_i + c_{8g}O_{8g}\right)\right] + h.c., \quad \text{(A.1)}$$

where G_F is the Fermi constant, $\lambda_q = V_{uq}V_{cq}^*$, c_i $(i = 1,\ldots,6)$ is the Wilson coefficient, and O_1^q, O_2^q, O_i $(i = 1,\ldots,6)$, and O_{8g} are four-fermion operators which are constructed from different combinations of quark fields. The four-fermion operators take the following form:

$$\begin{aligned}
O_1^q &= \bar{u}_\alpha\gamma_\mu\left(1-\gamma_5\right)q_\beta\bar{q}_\beta\gamma^\mu\left(1-\gamma_5\right)c_\alpha,\\
O_2^q &= \bar{u}\gamma_\mu\left(1-\gamma_5\right)q\bar{q}\gamma^\mu\left(1-\gamma_5\right)c,\\
O_3 &= \bar{u}\gamma_\mu\left(1-\gamma_5\right)c\sum_{q'}\bar{q}'\gamma^\mu\left(1-\gamma_5\right)q',\\
O_4 &= \bar{u}_\alpha\gamma_\mu\left(1-\gamma_5\right)c_\beta\sum_{q'}\bar{q}'_\beta\gamma^\mu\left(1-\gamma_5\right)q'_\alpha,\\
O_5 &= \bar{u}\gamma_\mu\left(1-\gamma_5\right)c\sum_{q'}\bar{q}'\gamma^\mu\left(1+\gamma_5\right)q',\\
O_6 &= \bar{u}_\alpha\gamma_\mu\left(1-\gamma_5\right)c_\beta\sum_{q'}\bar{q}'_\beta\gamma^\mu\left(1+\gamma_5\right)q'_\alpha,\\
O_{8g} &= -\frac{g_s}{8\pi^2}m_c\bar{u}\sigma_{\mu\nu}\left(1+\gamma_5\right)G^{\mu\nu}c,
\end{aligned} \quad \text{(A.2)}$$

where α and β are color indices and $q' = u, d, s$. Among all these operators, O_1^q and O_2^q are tree operators, $O_3 - O_6$ are QCD penguin operators, and O_{8g} is chromomagnetic dipole operator. The electroweak penguin operators are neglected in practice. One should notice that SCS decays receive contributions from all aforementioned operators while only tree operators can contribute to CF decays and DCS decays.

The Wilson coefficients used in this paper are evaluated at $\mu = 1\text{GeV}$, which can be found in [48].

(2) CKM Matrix. We use the Wolfenstein parameterization for the CKM matrix elements, which up to order $\mathcal{O}(\lambda^8)$ read [79, 80]

$$\begin{aligned}
V_{us} &= \lambda - \frac{1}{2}A^2\lambda^7\left(\rho^2+\eta^2\right),\\
V_{cs} &= 1 - \frac{1}{2}\lambda^2 - \frac{1}{8}\lambda^4\left(1+4A^2\right)\\
&\quad - \frac{1}{16}\lambda^6\left(1-4A^2+16A^2\left(\rho+i\eta\right)\right)\\
&\quad - \frac{1}{128}\lambda^8\left(5-8A^2+16A^4\right),\\
V_{ub} &= A\lambda^3\left(\rho-i\eta\right),\\
V_{cb} &= A\lambda^2 - \frac{1}{2}A^3\lambda^8\left(\rho^2+\eta^2\right),
\end{aligned} \tag{A.3}$$

where A, ρ, η, and λ are the Wolfenstein parameters, which satisfy following relation:

$$\rho + i\eta = \frac{\sqrt{1-A^2\lambda^4}\left(\overline{\rho}+i\overline{\eta}\right)}{\sqrt{1-\lambda^2}\left[1-A^2\lambda^4\left(\overline{\rho}+i\overline{\eta}\right)\right]}. \tag{A.4}$$

Numerical values of Wolfenstein parameters which have been used in this work are as follows:

$$\begin{aligned}
\lambda &= 0.22548^{+0.00068}_{-0.00034},\\
A &= 0.810^{+0.018}_{-0.024},\\
\overline{\rho} &= 0.145^{+0.013}_{-0.007},\\
\overline{\eta} &= 0.343^{+0.011}_{-0.012}.
\end{aligned} \tag{A.5}$$

(3) Decay Constants and Form Factors. In (17) and (18), the pole resonance model was employed for the matrix element $\langle PV|\overline{q}_1 q_2|0\rangle$ in the annihilation diagrams. By considering angular momentum conservation at weak vertex and all conservation laws are preserved at strong vertex, the matrix element $\langle PV|\overline{q}_1 q_2|0\rangle$ is therefore dominated by a pseudoscalar resonance [54],

$$\begin{aligned}
\langle PV|\overline{q}_1 q_2|0\rangle &= \langle PV \mid P^*\rangle\langle P^*|\overline{q}_1 q_2|0\rangle\\
&= g_{P^*PV}\frac{m_{P^*}}{m_D^2 - m_{P^*}^2}f_{P^*},
\end{aligned} \tag{A.6}$$

where g_{P^*PV} is a strong coupling constant and m_{P^*} and f_{P^*} are the mass and decay constant of the pseudoscalar resonance P^*. Therefore, η and η' are the dominant resonances for the final states of $K^{*\pm}K^{\mp}$, which can be expressed as flavor mixing of η_q and η_s,

$$\begin{pmatrix}\eta\\ \eta'\end{pmatrix} = \begin{pmatrix}\cos\phi & -\sin\phi\\ \sin\phi & \cos\phi\end{pmatrix}\begin{pmatrix}\eta_q\\ \eta_s\end{pmatrix} \tag{A.7}$$

where ϕ is the mixing angle and η_q and η_s are defined by

$$\begin{aligned}
\eta_q &= \frac{1}{\sqrt{2}}\left(u\overline{u}+d\overline{d}\right),\\
\eta_s &= s\overline{s}.
\end{aligned} \tag{A.8}$$

The decay constants of η and η' are defined by

$$\begin{aligned}
\langle 0|\overline{u}\gamma_\mu\gamma_5 u|\eta(p)\rangle &= if_\eta^u p_\mu,\\
\langle 0|\overline{u}\gamma_\mu\gamma_5 u|\eta'(p)\rangle &= if_{\eta'}^u p_\mu,\\
\langle 0|\overline{d}\gamma_\mu\gamma_5 d|\eta(p)\rangle &= if_\eta^d p_\mu,\\
\langle 0|\overline{d}\gamma_\mu\gamma_5 d|\eta'(p)\rangle &= if_{\eta'}^d p_\mu,\\
\langle 0|\overline{s}\gamma_\mu\gamma_5 s|\eta(p)\rangle &= if_\eta^s p_\mu,\\
\langle 0|\overline{s}\gamma_\mu\gamma_5 s|\eta'(p)\rangle &= if_{\eta'}^s p_\mu,
\end{aligned} \tag{A.9}$$

where

$$\begin{aligned}
f_\eta^u &= f_\eta^d = \frac{1}{\sqrt{2}}f_\eta^q,\\
f_{\eta'}^u &= f_{\eta'}^d = \frac{1}{\sqrt{2}}f_{\eta'}^q.
\end{aligned} \tag{A.10}$$

According to [81, 82], the decay constants of η and η' can be expressed as

$$\begin{aligned}
f_\eta^q &= f_q\cos\phi,\\
f_{\eta'}^q &= f_q\sin\phi,\\
f_\eta^s &= -f_s\sin\phi,\\
f_{\eta'}^s &= f_s\cos\phi,
\end{aligned} \tag{A.11}$$

where $f_q = (1.07 \pm 0.02)f_\pi$ and $f_s = (1.34 \pm 0.02)f_\pi$ [81], and the mixing angle $\phi = (40.4 \pm 0.6)^\circ$ [83]. Other decay constants used in this paper are listed in Table 5.

The transition form factors $A_0^{D^0\to K^{*-}}$ and $F_1^{D^0\to K^-}$, based on the relativistic covariant light-front quark model [85], are expressed as a momentum-dependent, 3-parameter form (the parameters can be found in Table 6):

$$F\left(q^2\right) = \frac{F(0)}{1 - a\left(q^2/m_D^2\right) + b\left(q^2/m_D^2\right)^2}. \tag{A.12}$$

(4) Decay Rate. The decay width takes the form

$$\Gamma_{D\to KK^*} = \frac{|\mathbf{p}_1|^3}{8\pi m_{K^*}^2}\left|\frac{\mathcal{M}_{D\to KK^*}}{\varepsilon^*\cdot p_D}\right|^2, \tag{A.13}$$

TABLE 5: The meson decay constants used in this paper (MeV) [57, 84].

f_{K^*}	f_ρ	f_K	f_π	f_D
220(5)	216(3)	156(0.4)	130(1.7)	208(10)

TABLE 6: The parameters of $D \longrightarrow K^*, K$ transitions form factors in (A.12).

Form factor	$A_0^{D\longrightarrow K^*}$	$F_1^{D\longrightarrow K}$
$F(0)$	0.69	0.78
a	1.04	1.05
b	0.44	0.23

where $\mathbf{p}_1$ represents the center of mass (c.m.) 3-momentum of each meson in the final state and is given by

$$|\mathbf{p}_1| = \frac{\sqrt{\left[\left(m_D^2 - (m_{K^*} + m_K)^2\right)\left(m_D^2 - (m_{K^*} - m_K)^2\right)\right]}}{2m_D}. \quad \text{(A.14)}$$

$\mathcal{M}$ is the corresponding decay amplitude.

Conflicts of Interest

The authors declare that there are no conflicts of interest regarding the publication of this paper.

Acknowledgments

This work was partially supported by National Natural Science Foundation of China (Project Nos. 11447021, 11575077, and 11705081), National Natural Science Foundation of Hunan Province (Project No. 2016JJ3104), the Innovation Group of Nuclear and Particle Physics in USC, and the China Scholarship Council.

References

[1] J. H. Christenson, J. W. Cronin, V. L. Fitch, and R. Turlay, "Evidence for the 2π Decay of the $K_2{}^0$ Meson," *Physical Review Letters*, vol. 13, no. 4, pp. 138–140, 1964.

[2] M. Kobayashi and T. Maskawa, "*CP*-violation in the renormalizable theory of weak interaction," *Progress of Theoretical and Experimental Physics*, vol. 49, pp. 652–657, 1973.

[3] N. Cabibbo, "Unitary symmetry and leptonic decays," *Physical Review Letters*, vol. 10, no. 12, pp. 531–533, 1963.

[4] R. Aaij, LHCb Collaboration et al., "Measurement of *CP* Violation in the Phase Space of $B^\pm \longrightarrow K^\pm\pi^+\pi^-$ and $B^\pm \longrightarrow K^\pm K^+K^-$ Decays," *Physical Review Letters*, vol. 111, no. 2, Article ID 101801, 2017.

[5] R. Aaij, LHCb Collaboration et al., "Measurement of *CP* Violation in the Phase Space of $B^\pm \longrightarrow K^+K^-\pi^\pm$ and $B^\pm \longrightarrow \pi^+\pi^-\pi^\pm$ Decays," *Physical Review Letters*, vol. 112, no. 2, Article ID 011801, 2014.

[6] R. Aaij, LHCb Collaboration et al., "Measurements of *CP* violation in the three-body phase space of charmless $B^\pm$ decays," *Physical Review D*, vol. 90, Article ID 112004, 2014.

[7] J. H. A. Nogueira, S. Amato, A. Austregesilo et al., "Summary of the 2015 LHCb workshop on multi-body decays of D and B mesons," https://arxiv.org/abs/1605.03889.

[8] Z.-H. Zhang, X.-H. Guo, and Y.-D. Yang, "*CP* violation in $B^\pm \longrightarrow \pi^\pm\pi^+\pi^-$ in the region with low invariant mass of one $\pi^+\pi^-$ pair," *Physical Review D*, vol. 87, Article ID 076007, 2013.

[9] I. Bediaga, T. Frederico, and O. Lourenço, "*CP* violation and *CPT* invariance in $B^\pm$ decays with final state interactions," *Physical Review D*, vol. 89, Article ID 094013, 2014.

[10] H.-Y. Cheng and C.-K. Chua, "Branching fractions and direct *CP* violation in charmless three-body decays of *B* mesons," *Physical Review D*, vol. 88, Article ID 114014, 2013.

[11] Z.-H. Zhang, X.-H. Guo, and Y.-D. Yang, "CP violation induced by the interference of scalar and vector resonances in three-body decays of bottom mesons," https://arxiv.org/abs/1308.5242.

[12] B. Bhattacharya, M. Gronau, and J. L. Rosner, "CP asymmetries in three-body $B^\pm$ decays to charged pions and kaons," *Physics Letters B*, vol. 726, no. 1-3, pp. 337–343, 2013.

[13] D. Xu, G.-N. Li, and X.-G. He, "Large SU(3) Breaking Effects and CP Violation in B^+ Decays Into Three Charged Octet Pseudoscalar Mesons," *International Journal of Modern Physics A*, vol. 29, Article ID 1450011, 2014.

[14] W.-F. Wang, H.-C. Hu, H.-N. Li, and C.-D. Lü, "Direct *CP* asymmetries of three-body *B* decays in perturbative QCD," *Physical Review D*, vol. 89, Article ID 074031, 2014.

[15] Z.-H. Zhang, C. Wang, and X.-H. Guo, "Possible large *CP* violation in three-body decays of heavy baryon," *Physics Letters B*, vol. 751, pp. 430–433, 2015.

[16] C. Wang, Z.-H. Zhang, Z.-Y. Wang, and X.-H. Guo, "Localized direct CP violation in $B^\pm \longrightarrow \rho^0(\omega)\pi^\pm \longrightarrow \pi^+\pi^-\pi^\pm$," *The European Physical Journal C*, vol. 75, p. 536, 2015.

[17] J. H. A. Nogueira, I. Bediaga, A. B. R. Cavalcante, T. Frederico, and O. Lourenço, "*CP* violation: Dalitz interference, *CPT*, and final state interactions," *Physical Review D*, vol. 92, Article ID 054010, 2015.

[18] J. Dedonder, A. Furman, R. Kamiński, L. Leśniak, and B. Loiseau, "S-, P- and D-wave $\pi\pi$ final state interactions and CP violation in $B^\pm \longrightarrow \pi^\pm\pi^\mp\pi^\pm$ decays," *Acta Physica Polonica B*, vol. 42, no. 9, Article ID 2013, 2011.

[19] B. El-Bennich, A. Furman, R. Kamiński, L. Leśniak, and B. Loiseau, "Interference between $f_0(980)$ and $\rho(770)^0$ resonances in $B \longrightarrow \pi^+\pi^-K$ decays," *Physical Review D*, vol. 74, Article ID 114009, 2006.

[20] I. Bigi and A. Sanda, "On $D^0\overline{D}^0$ mixing and *CP* violation," *Physics Letters B*, vol. 171, no. 2-3, pp. 320–324, 1986.

[21] G. Blaylock, A. Seiden, and Y. Nir, "The role of CP violation in $D^0\overline{D}^0$ mixing," *Physics Letters B*, vol. 335, no. 3, pp. 555–560, 1995.

[22] S. Bergmann, Y. Grossman, Z. Ligeti, Y. Nir, and A. A. Petrov, "Lessons from CLEO and FOCUS measurements of D^0-$\overline{D}^0$ mixing parameters," *Physics Letters B*, vol. 486, no. 3-4, pp. 418–425, 2000.

[23] U. Nierste and S. Schacht, "Neutral D⟶KK* decays as discovery channels for charm *CP* violation," *Physical Review Letters*, vol. 119, Article ID 251801, 2017.

[24] G. Bonvicini, CLEO Collaboration et al., "Search for CP violation in $D^0 \longrightarrow K_S^o\pi^0$, $D^0 \longrightarrow \pi^0\pi^0$ and $D^0 \longrightarrow K_S^oK_S^o$ decays," *Physical Review D*, vol. 63, Article ID 071101, 2001.

[25] J. M. Link, "Search for *CP* Violation in the decays $D^+ \longrightarrow K_S\pi^+$ and $D^+ \longrightarrow K_SK^+$," *Physical Review Letters*, vol. 88, Article ID 041602, 2002.

[26] T. Aaltonen, CDF Collaboration et al., "Measurement of *CP*-violating asymmetries in $D^0 \longrightarrow \pi^+\pi^-$ and $D^0 \longrightarrow K^+K^-$ decays at CDF," *Physical Review D*, vol. 85, Article ID 012009, 2012.

[27] R. Cenci, "Mixing and CP Violation in Charm Decays at BABAR," in *Proceedings of the 7th International Workshop on the CKM Unitarity Triangle (CKM 2012)*, Cincinnati, Ohio, USA, 2012.

[28] J. P. Lees, BABAR Collaboration et al., "Search for *CP* violation in the decays $D^\pm \longrightarrow K_S^oK^\pm$, $D_s^\pm \longrightarrow K_S^oK^\pm$, and $D_s^\pm \longrightarrow K_S^o\pi^\pm$," *Physical Review D*, vol. 87, Article ID 052012, 2013.

[29] M. Starič, A. Abdesselam, I. Adachi et al., "Measurement of D^0-$\overline{D}^0$ mixing and search for *CP* violation in $D^0 \longrightarrow K^+K^-, \pi^+\pi^-$ decays with the full Belle data set," *Physics Letter B*, vol. 753, pp. 412–418, 2016.

[30] R. Aaij, R. Aaij, B. Adeva et al., "Measurement of *CP* asymmetries in $D^\pm \longrightarrow \eta'\pi^\pm$ and $D_s^\pm \longrightarrow \eta'\pi^\pm$ decays," *Physics Letters B*, vol. 771, pp. 21–30, 2017.

[31] R. Aaij, LHCb Collaboration et al., "Measurements of charm mixing and *CP* violation using $D^0 \longrightarrow K^\pm\pi^\mp$ decays," *Physical Review D*, vol. 95, Article ID 052004, 2017.

[32] LHCb Collaboration, "Search for *CP* violation in the phase space of $D^0 \longrightarrow \pi^+\pi^-\pi^+\pi^-$ decays," *Physics Letters B*, vol. 769, pp. 345–356, 2017.

[33] V. Bhardwaj, "Latest Charm Mixing and CP results from B-factories," in *Proceedings of the 9th International Workshop on the CKM Unitarity Triangle (CKM2016)*, vol. 139, Mumbai, India, 2017.

[34] Y. Grossman, A. L. Kagan, Y. Nir et al., "New physics and CP violation in singly Cabibbo suppressed Ddecays," *Physical Review D*, vol. 75, Article ID 036008, 2007.

[35] J. D. Bjorken, "Topics in B-physics," *Nuclear Physics B (Proceedings Supplements)*, vol. 11, no. C, pp. 325–341, 1989.

[36] M. J. Dugan and B. Grinstein, "QCD basis for factorization in decays of heavy mesons," *Physics Letters B*, vol. 255, no. 4, pp. 583–588, 1991.

[37] M. Beneke, G. Buchalla, M. Neubert, and C. T. Sachrajda, "QCD factorization for $B \longrightarrow \pi\pi$ decays: strong phases and CP violation in the heavy quark limit," *Physical Review Letters*, vol. 83, no. 10, pp. 1914–1917, 1999.

[38] M. Beneke and M. Neubert, "QCD factorization for B⟶PP and B⟶PV decays," *Nuclear Physics B*, vol. 675, no. 1-2, pp. 333–415, 2003.

[39] D. R. Boito, J. Dedonder, B. El-Bennich, O. Leitner, and B. Loiseau, "Scalar resonances in a unitary," *Physical Review D*, vol. 96, Article ID 113003, 2017.

[40] A. Furmana, R. Kamińskia, L. Leśniaka, and B. Loiseaub, "Long-distance effects and final state interactions in $B \longrightarrow \pi\pi K$ and $B \longrightarrow K\overline{K}K$ decays," *Physics Letters B*, vol. 622, pp. 207–217, 2005.

[41] Y.-Y. Keum, H.-N. Li, A. I. Sanda et al., "Penguin enhancement and $\overrightarrow{B}K\pi$ decays in perturbative QCD," *Physical Review D*, vol. 63, Article ID 054008, 2001.

[42] C. W. Bauer, D. Pirjol, I. Z. Rothstein, and I. W. Stewart, "$B \longrightarrow M_1M_2$: Factorization, charming penguins, strong phases, and polarization," *Physical Review D*, vol. 70, Article ID 054015, 2004.

[43] B. Loiseau, "Theory overview on amplitude analyses with charm decays," in *Proceedings of the 8th International Workshop on Charm Physics (Charm 2016)*, Bologna, Italy, 2017.

[44] H.-Y. Cheng and C.-W. Chiang, "Direct CP violation in two-body hadronic charmed meson decays," *Physical Review D*, vol. 85, no. 3, Article ID 034036, 2012, Erratum: [Physical Review D, vol. 85, Article ID 079903, 2012].

[45] L.-L. Chau, "Quark mixing in weak interactions," *Physics Reports*, vol. 95, no. 1, pp. 1–94, 1983.

[46] B. Bhattacharya, M. Gronau, and J. L. Rosner, "Publisher's Note:," *Physical Review D*, vol. 85, Article ID 054014, 2012.

[47] H.-Y. Cheng, C.-W. Chiang, and A.-L. Kuo, "Global analysis of two-body D⟶VP decays within the framework of flavor symmetry," *Physical Review D*, vol. 93, Article ID 114010, 2016.

[48] H.-N. Li, C.-D. Lu, F.-S. Yu et al., "Branching ratios and direct CP asymmetries in D⟶PP decays," *Physical Review D*, vol. 86, Article ID 036012, 2012.

[49] Q. Qin, H.-N. Li, C.-D. Lü, and F.-S. Yu, "Branching ratios and direct CP asymmetries in D⟶PV decays," *Physical Review D*, vol. 89, Article ID 054006, 2014.

[50] B. Aubert (BABAR Collaboration) et al., "Limits on $D^0 - \overline{D}^0$ Mixing and CP Violation from the Ratio of Lifetimes for Decay to $K^-\pi^+$, K^-K^+, and $\pi^-\pi^+$," *Physical Review Letters*, vol. 91, Article ID 121801, 2003.

[51] K. Abe (Belle Collaboration) et al., "Measurement of the $D^0\overline{D}^0$ lifetime difference using $D^0 \longrightarrow K\pi/KK$ decays," in *Proceedings of the 21st International Symposium on Lepton and Photon Interactions at High Energies*, pp. 11–16, Batavia, ILL, USA, 2003, https://arxiv.org/abs/hep-ex/0308034.

[52] Y.-Y. Keum, H.-N. Li, and A. I. Sandac, "Fat penguins and imaginary penguins in perturbative QCD," *Physics Letters B*, vol. 504, pp. 6–14, 2001.

[53] C.-D. Lü, K. Ukai, and M.-Z. Yang, "Branching ratio and CP violation of $B \longrightarrow \pi\pi$ decays in the perturbative QCD approach," *Physical Review D*, vol. 63, Article ID 074009, 2001.

[54] F.-S. Yu, X.-X. Wang, and C.-D. Lü, "Nonleptonic two-body decays of charmed mesons," *Physical Review D*, vol. 84, Article ID 074019, 2011.

[55] C. Cawlfield (CLEO Collaboration) et al., "Measurement of interfering $K^{*+}K^-$ and $K^{*-}K^+$ amplitudes in the decay $D^0 \longrightarrow K^+K^-\pi^0$," *Physical Review D*, vol. 74, Article ID 031108, 2006.

[56] B. Aubert, L. L. Zhang, S. Chen et al., "Amplitude analysis of the decay $D^0 \longrightarrow K^-K^+\pi^0$," *Physical Review D*, vol. 76, Article ID 011102, 2007.

[57] C. Patrignani, K. Agashe, G. Aielli et al., "Review of Particle Physics," *Chinese Physics C*, vol. 40, no. 10, Article ID 100001, 2016.

[58] J. L. Rosner and D. A. Suprun, "Measuring the relative strong phase in $D^0 \longrightarrow K^{*+}K^-$ and $D^0 \longrightarrow K^{*-}K^+$ decays," *Physical Review D*, vol. 68, Article ID 054010, 2003.

[59] I. Bediaga, I. I. Bigi, A. Gomes, G. Guerrer, J. Miranda, and A. C. dos Reis, "On a CP anisotropy measurement in the Dalitz plot," *Physical Review D*, vol. 80, no. 9, Article ID 096006, 2009.

[60] S. M. Flatte, "Coupled-channel analysis of the $\pi\eta$ and $K\overline{K}$ systems near $K\overline{K}$ threshold," *Physics Letters B*, vol. 63, no. 2, pp. 224–227, 1976.

[61] M. Golden and B. Grinstein, "Enhanced CP violations in hadronic charm decays," *Physics Letters B*, vol. 222, no. 3-4, pp. 501–506, 1989.

[62] G. F. Giudice, G. Isidori, and P. Paradisi, "Direct CP violation in charm and flavor mixing beyond the SM," *Journal of High Energy Physics*, vol. 1204, Article ID 060, 2012.

[63] M. Gronau, "New physics in singly Cabibbo-suppressed D decays," *Physics Letters B*, vol. 738, pp. 136–139, 2014.

[64] J. Brod, A. L. Kagan, and J. Zupan, "Size of direct CP violation in singly Cabibbo-suppressed D decays," *Physical Review D*, vol. 86, Article ID 014023, 2012.

[65] Y. Grossman, A. L. Kagan, and J. Zupan, "Testing for new physics in singly Cabibbo suppressed D decays," *Physical Review D*, vol. 85, Article ID 114036, 2012.

[66] G. Isidori, J. F. Kamenik, Z. Ligetie, and G. Perez, "Implications of the LHCb evidence for charm CP violation," *Physics Letters B*, vol. 711, no. 1, pp. 46–51, 2012.

[67] M. Beneke, G. Buchalla, M. Neubert, and C. T. Sachrajda, "QCD factorization in $B \longrightarrow \pi K$, $\pi\pi$ decays and extraction of Wolfenstein parameters," *Nuclear Physics B*, vol. 606, no. 1-2, pp. 245–321, 2001.

[68] H.-n. Li, S. Mishima, and A. I. Sanda, "Resolution to the $B \longrightarrow \pi K$ puzzle," *Physical Review D*, vol. 72, Article ID 114005, 2005.

[69] R. Aaij (LHCb Collaboration) et al., "Evidence for CP violation in time-integrated $D^0 \longrightarrow h^- h^+$ decay rates," *Physical Review Letters*, vol. 108, Article ID 111602, 2012.

[70] F. Gabbiani, E. Gabrielli, A. Masiero, and L. Silvestrini, "A complete analysis of FCNC and CP constraints in general SUSY extensions of the standard model," *Nuclear Physics B*, vol. 477, no. 2, pp. 321–352, 1996.

[71] E. Gabrielli, A. Masiero, and L. Silvestrini, "Flavour changing neutral currents and CP violating processes in generalized supersymmetric theories," *Physics Letters B*, vol. 374, no. 1-3, pp. 80–86, 1996.

[72] J. S. Hagelin, S. Kelley, and T. Tanaka, "Supersymmetric flavor-changing neutral currents: exact amplitudes and phenomenological analysis," *Nuclear Physics B*, vol. 415, no. 2, pp. 293–331, 1994.

[73] G. De Pietro, "Charm physics prospects at Belle II," in *Proceedings of the European Physical Society Conference on High Energy Physics (EPS-HEP2017)*, vol. 314, Venice, Italy, 2017.

[74] T. Abe, I. Adachi, K. Adamczyk et al., "Belle II Technical Design Report," https://arxiv.org/abs/1011.0352.

[75] J. P. Lees, V. Poireau, V. Tisserand et al., "Measurement of the $D^0 \longrightarrow \pi^- e^+ \nu_e$ differential decay branching fraction as a function of q^2 and study of form factor parametrizations," *Physical Review D*, vol. 91, Article ID 052022, 2015.

[76] N. K. Nisar, G. B. Mohanty, K. Trabelsi et al., "Search for the rare decay $D^0 \longrightarrow \gamma\gamma$ at Belle," *Physical Review D*, vol. 93, Article ID 051102, 2016.

[77] M. Ablikim (BESIII Collaboration) et al., "Search for $D^0 \longrightarrow \gamma\gamma$ and improved measurement of the branching fraction for $D^0 \longrightarrow \pi^0\pi^0$," *Physicla Review D*, vol. 91, no. 11, Article ID 112015, 2015.

[78] G. Buchalla, A. J. Buras, and M. E. Lautenbacher, "Weak decays beyond leading logarithms," *Reviews of Modern Physics*, vol. 68, no. 4, pp. 1125–1244, 1996.

[79] A. J. Buras, M. E. Lautenbacher, and G. Ostermaier, "Waiting for the top quark mass, $K^+ \longrightarrow \pi^+ \nu\overline{\nu}$, $B_s^0 - \overline{B}_s^0$ mixing, and *CP* asymmetries in *B* decays," *Physical Review D*, vol. 50, Article ID 3433, 1994.

[80] J. Charles, A. Höcker, and H. Lacker, "*CP* violation and the CKM matrix: assessing the impact of the asymmetric *B* factories," *The European Physical Journal C*, vol. 41, no. 1, pp. 1–131, 2005.

[81] T. Feldmann, P. Kroll, and B. Stech, "Mixing and decay constants of pseudoscalar mesons," *Physical Review D*, vol. 58, Article ID 114006, 1998.

[82] T. Feldmann, P. Kroll, and B. Stech, "Mixing and decay constants of pseudoscalar mesons: the sequel," *Physics Letters B*, vol. 449, no. 3-4, pp. 339–346, 1999.

[83] F. Ambrosino, A. Antonelli, and M. Antonelli, "A global fit to determine the pseudoscalar mixing angle and the gluonium content of the η' meson," *Journal of High Energy Physics*, vol. 907, Article ID 105, 2009.

[84] P. Ball, G. W. Jones, and R. Zwicky, "$B \longrightarrow V\gamma$ beyond QCD factorization," *Physical Review D*, vol. 75, Article ID 054004, 2007.

[85] H.-Y. Cheng, C.-K. Chua, and C.-W. Hwang, "Covariant light-front approach for s-wave and p-wave mesons: Its application to decay constants and form factors," *Physical Review D*, vol. 69, Article ID 074025, 2004.

13

A New Distribution for Multiplicities in Leptonic and Hadronic Collisions at High Energies

Ridhi Chawla and M. Kaur

Physics Department, Panjab University, Chandigarh, 160014, India

Correspondence should be addressed to M. Kaur; manjit@pu.ac.in

Academic Editor: Smarajit Triambak

Charged particles' production in the e^+e^-, $\bar{p}p$, and pp collisions in full phase space as well as in the restricted phase space slices, at high energies, is described with predictions from shifted Gompertz distribution, a model of adoption of innovations. The distribution has been extensively used in diffusion theory, social networks, and forecasting. A two-component model in which PDF obtained from the superposition of two shifted Gompertz distributions is introduced to improve the fitting of the experimental distributions by several orders. The two components correspond to the two subgroups of a data set, one representing the soft interactions and the other semihard interactions. Mixing is done by appropriately assigning weights to each subgroup. Our first attempt to analyse the data with shifted Gompertz distribution has produced extremely good results. It is suggested that the distribution may be included in the host of distributions more often used for the multiplicity analyses.

1. Introduction

The shifted Gompertz distribution was introduced by Bemmaor [1] in 1994 as a model of adoption of innovations. It is the distribution of the largest of two independent random variables one of which has an exponential distribution with parameter b and the other has a Gumbel distribution, also known as log-Weibull distribtion, with parameters η and b. Several of its statistical properties have been studied by Jiménez and Jodrá [2] and Jiménez Torres [3]. In machine learning, the Gumbel distribution is also used to generate samples from the generalised Bernoulli distribution, which is a discrete probability distribution that describes the possible results of a random variable that can take on one of the K-possible elementary events, with the probability of each elementary event separately specified. The shifted Gompertz distribution has mostly been used in the market research and diffusion theory, social networks, and forecasting. It has also been used to predict the growth and decline of social networks and online services and shown to be superior to the Bass model and Weibull distribution [4]. It is interesting to study the statistical phenomena in high energy physics in terms of this distribution. Recently, Weibull distribution has been used to understand the multiplicity distributions in various particle-particle collisions at high energies and more recently [5] to explain the LHC data. Weibull models studied in the literature were appropriate for modelling a continuous random variable which assumes that the variable takes on real values over the interval $[0, \infty]$. In situations where the observed data values are very large, a continuous distribution is considered an adequate model for the discrete random variable; for example, in case of a particle collider, the luminosity during a fill decreases roughly exponentially. Therefore, the mean collision rate will likewise decrease. That decrease will be reflected in the number of observed particles per unit time. In the same way a photon detector counts photons in a continuous train of time bins. If the photons are antibunched in time, that is, they tend to be separated from each other, one will get a different distribution of photon counts than if the photons are bunched, that is, bunched together in time. By analyzing the photon counting statistics one can infer information about the continuous underlying distribution of the temporal spacing of photons. The shifted Gompertz distribution with non-negative fit parameters identified with the scale and shape parameters, can in this way be used for studying the distributions of particles produced in collisions at accelerators. One of the studies in statistics is when the variables take on discrete values. The idea was first introduced by Nakagawa and Osaki [6], as they introduced discrete Weibull distribution with two shape parameters q and β, where $0 < q < 1$ and $\beta > 0$. Models which assume only nonnegative integer values for modelling discrete random variables are useful for modelling the kind of problems mentioned above.

The charged-particle multiplicity is one of the simplest observables in collisions of high energy particles, yet it imposes important constraints on the dynamics of particle production. The particle production has been studied in terms of several theoretical, phenomenological, and statistical models. Each of these models has been reasonably successful in explaining the results from different experiments and useful for extrapolations to make predictions. Although Weibull distribution has been studied recently, no attempt has been made so far to analyze the high energy collision data in terms of shifted Gompertz distribution. Our first attempt to analyse the data produced good results and encouraged us to do a comprehensive analysis.

The aim of the present work is to introduce a statistical distribution, the shifted Gompertz distribution to investigate the multiplicity distributions of charged particles produced in e^+e^-, pp, and $\overline{p}p$ collisions at different center of mass energies in full phase space as well as in restricted phase space windows. Energy-momentum conservation strongly influences the multiplicity distribution for the full phase space. The distribution in restricted rapidity window, however, is less prone to such constraints and thus can be expected to be a more sensitive probe to the underlying dynamics of QCD, as inferred in references [7, 8].

In Section 2, details of Probability Distribution Function (PDF) of the shifted Gompertz distribution is discussed. For e^+e^- collisions a two-component model has been used and modification of distributions done in terms of these two components, one from soft events and another from semihard events. Superposition of distributions from these two components, by using appropriate weights, is done to build the full multiplicity distribution. When multiplicity distribution is fitted with the weighted superposition of two shifted Gompertz distributions, we find that the agreement between the data and the model improves considerably. The fraction of soft events, α for various energies have been taken from [9, 10] which use the K_T clustering algorithm, the most extensively used algorithm for LEP e^+e^- data analyses. The corresponding fractions for pp and $\overline{p}p$ are not available in different rapidity bins. For $\overline{p}p$ data at all energies under study, the α values for full phase space are taken from [11]. We also tried to fit the multiplicity distribution to find the best fit alpha value. It is found that the fit values agree very closely with values obtained from [11]. We thus fitted distributions in restricted rapidity windows for $\overline{p}p$ and pp data in terms of soft and semihard components to get the best fit α values.

In a recent publication, Wilk and Włodarczyk [12] have developed a method of retrieving additional information from the multiplicity distributions. They propose, in case of a conventional Negative Binomial Distribution fit [11], making the parameters dependent on the multiplicity in place of having a 2-component model. They demonstrated that the additional valuable information from the MDs, namely, the oscillatory behaviour of the counting statistics can be derived. In a future extension of the present work, we shall analyse the shifted Gompertz distribution, using the approach proposed and described by the authors [12].

Section 3 presents the analysis of experimental data and the results obtained by the two approaches. Discussion and conclusion are presented in Section 4.

2. Shifted Gompertz Distribution

The dynamics of hadron production can be probed using the charged particle multiplicity distribution. Measurements of multiplicity distributions provide relevant constraints for particle-production models. Charged particle multiplicity is defined as the average number of charged particles, n produced in a collision $\langle n\rangle = \sum_{n=0}^{n_{max}} nP_n$. Hadron production depends upon the center of mass energy available for particle production nearly independent of the types of particles undergoing collisions. Subsequently, it is the fragmentation of quarks and gluons which produce hadrons nonperturbatively. Thus the same PDFs can be used to describe behaviour of multiplicity distributions. In numerous works in the past, the most popular Negative Binomial Distribution has been successfully used for a wide variety of collisions [13]. Universality of multiparticle production in e^+e^-, pp, and $\overline{p}p$ has been discussed in several papers; a detailed paper amongst these is [14].

We briefly outline the probability density function (PDF) of the shifted Gompertz distribution used for studying the multiplicity distributions. Equations (1)-(3) define the PDF and the mean value of the distribution;

$$P(n \mid b,\eta) = be^{-bn}e^{-\eta e^{-bn}}\left[1+\eta\left(1-e^{-bn}\right)\right] \quad for\ n>0 \quad (1)$$

Mean of the distribution is given by

$$\left(-\frac{1}{b}\right)\left(E\left[\ln(X)\right]-\ln(\eta)\right) \quad where\ X=\eta e^{-bn} \quad (2)$$

and

$$E\left[\ln(X)\right] = \left[1+\frac{1}{\eta}\right]\int_0^{\infty} e^{-X}\left[\ln(X)\right]dX - \frac{1}{\eta}\int_0^{\infty} Xe^{-X}\left[\ln(X)\right]dX \quad (3)$$

where $b \geq 0$ is a scale parameter and $\eta \geq 0$ is a shape parameter. Similar to the Weibull distribution, shifted Gompertz distribution is also a two parameter distribution, in terms of its shape and scale.

2.1. Two-Component Approach. It is well established that at high energies, charged particle multiplicity distribution in full phase space becomes broader than a Poisson distribution.

This behaviour has been successfully described by a two parameters negative binomial (NB) distribution defined by

$$P(n \mid \langle n \rangle, k) = \frac{\Gamma(n+k)}{\Gamma(n+1)\Gamma(k)} \frac{(\langle n \rangle / k)^n}{(1 + \langle n \rangle / k)^{n+k}} \quad (4)$$

where k is related to the dispersion D by

$$\frac{D^2}{\langle n \rangle^2} = \frac{1}{\langle n \rangle} + \frac{1}{k} \quad (5)$$

k parameter of the distribution is negative in the lower energy domain, where the distribution is binomial like. k is positive in the higher energy domain and the distribution is truly NB, the two particle correlations dominate and $1/k$ is closely related to the integral over full phase space of the two particle correlation function. NB distribution was very successful until the results from UA5 collaboration [15] showed a shoulder structure in the multiplicity distribution on $\overline{p}p$ collisions. To explain this NB regularity violations, C. Fuglesang [16] suggested the violations as the effect of the weighted superposition of soft events (events without minijets) and semihard events (events with mini-jets), the weight α being the fraction of soft events. The multiplicity distribution of each component being NB. This idea was successfully implemented in several analyses at high energies to fit the multiplicity distributions with superposed NB functions.

Adopting this suggestion for the multiplicity distributions in e^+e^-, pp, and $\overline{p}p$ collisions at high energies, we have used a superposition of two shifted Gompertz components. The two components are interpreted as soft and hard components, as explained above. The Multiplicity distribution is produced by adding weighted superposition of multiplicity in soft events and multiplicity distribution in semi-hard events. This approach combines two classes of events, not two different particle-production mechanisms in the same event. Therefore, no interference terms are needed to be introduced. The final distribution is the sum of the two independent distributions, henceforth called modified shifted Gompertz distribution.

$$P(n) = \alpha P_{soft}^{shGomp}(n) + (1-\alpha) P_{semi\text{-}hard}^{shGomp}(n) \quad (6)$$

In this approach, the multiplicity distribution depends on five parameters as given below:

$$P_n(\alpha : b_1, \eta_1; b_2, \eta_2) = \alpha P_n(soft) + (1-\alpha) P_n(semi\text{-}hard) \quad (7)$$

As described by A. Giovannini et al. [11], the superimposed physical substructures in the cases of e^+e^- annihilation and hadron-hadron interactions are different, the weighted superposition mechanism is the same.

3. The Data

The data from different experiments and three collision types are considered:

(i) e^+e^- annihilations at different collision energies, from 91 GeV up to the highest energy of 206.2 GeV at LEP2, from two experiments L3 [17] and OPAL [18–21], are analysed.

(ii) pp collisions at LHC energies from 900 GeV, 2360 GeV, and 7000 GeV [22] are analysed in five restricted rapidity windows, $|y| = 0.5, 1.0, 1.5, 2.0$, and 2.4.

(iii) $\overline{p}p$ collisions at energies from 200 GeV, 540 GeV, and 900 GeV [15, 23] are analysed in full phase space as well as in restricted rapidity windows, $|y| = 0.5, 1.5, 3.0$, and 5.0.

3.1. Results and Discussion. The PDF defined by (1), (6) are used to fit the experimental data. Figures 1 and 2 show the shifted Gompertz function and the modified (two-component) shifted Gompertz function fits to the data e^+e^- from L3 and OPAL experiments. Parameters of the fits, χ^2/ndf, and the p-values are documented in Table 1. Figure 3 shows the ratio of data over modified shifted Gompertz fit plots for e^+e^- collisions at two energies. The plots correspond to the worst and the best fits depending upon the maximum and minimum χ^2/ndf values and show that fluctuations between the data and the fits are acceptably small, as the ratio is nearly one.

Figure 4 shows the modified shifted Gompertz distribution, equation (6) fitted to the $\overline{p}p$ data at energies from 200 GeV to 900 GeV in four rapidity windows. To avoid cluttering of figures, the plots for shifted Gompertz are not shown. Figure 5 shows the shifted Gompertz and modified shifted Gompertz distributions, fitted to the $\overline{p}p$ collisions in full phase space for the same energies. The comparison can be seen from the parameters of the fits, χ^2/ndf, and the p-values documented in Table 2. Figure 6 shows the ratio plots of the data over modified shifted Gompertz fit for $\overline{p}p$ collisions at different energies in full phase space. The plots show acceptable fluctuations with the ratio values around unity.

Figure 7 shows the modified shifted Gompertz distribution, (6) fitted to the pp collisions at LHC energies from 900 GeV to 7000 GeV in four rapidity windows. Again for restricting the number of figures, only the modified distributions are shown. Comparison between the two types of distributions can be seen from the parameters of the fits, χ^2/ndf and the p-values documented in Table 3. Figure 8 shows the ratio plots of data over modified shifted Gompertz fit for the collisions at different energies in full phase space. The plots show acceptable fluctuations with the ratio values around unity.

Comparison of the fits and the parameters shows that overall shifted Gompertz distribution is able to reproduce the data at most of the collision energies in full phase space as well as in the restricted rapidity windows for e^+e^-, $\overline{p}p$, and pp collisions. It does fail and is excluded statistically for some energies where p-value is $< 0.1\%$, in particular for 540 GeV $\overline{p}p$ data for some rapidity intervals and for LHC data at the highest energy of 7000 GeV. However, comparison of the fits and the parameters from the (two-component) modified shifted Gompertz distribution shows that though the data are very well reproduced in full phase space as well as in all rapidity intervals for all collision energies in e^+e^-, $\overline{p}p$, and pp

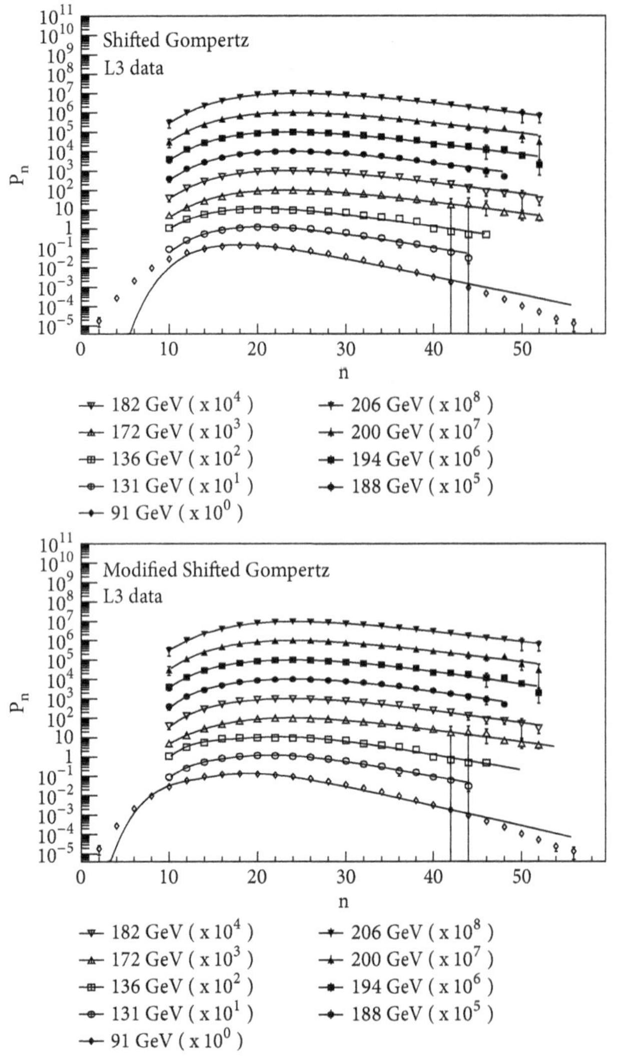

FIGURE 1: Charged multiplicity distributions from L3 experiment. Solid lines represent the Gompertz distributions.

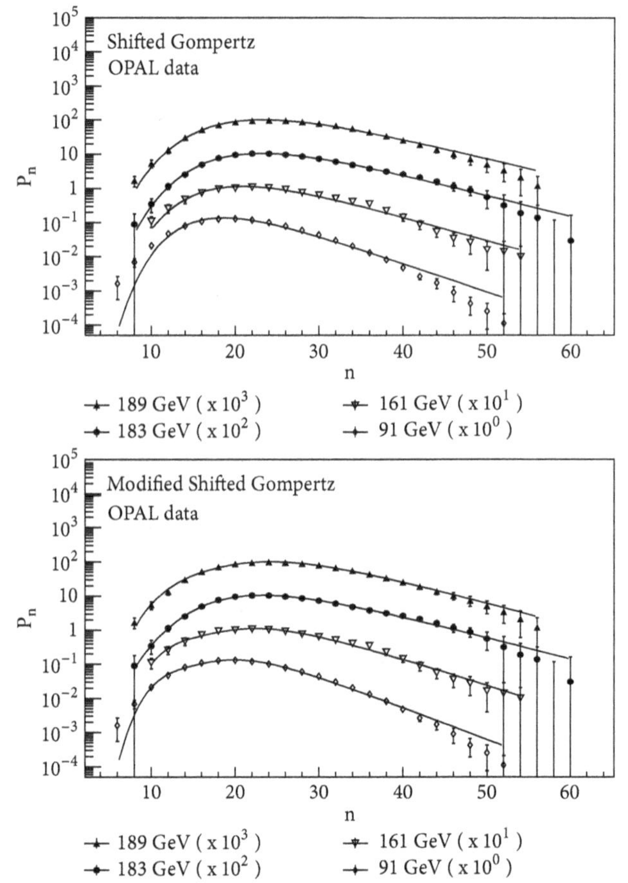

FIGURE 2: Charged multiplicity distributions from OPAL experiment. Solid lines represent the Gompertz distributions.

collisions, the distribution does fail for the e^+e^- collisions at 91 GeV. The χ^2/ndf value in each case reduces enormously, when modified shifted Gompertz fit is used. In each case the fit is accepted with p-value > 0.1%.

For shifted Gompertz distribution, the scale parameter b and the shape parameter η values are plotted in Figure 9 for e^+e^- interactions for LEP data from L3 and OPAL experiments. A power law is fitted to the data. It is observed that both b and η values decrease with increase in collision energy and are parametrised as

$$b = (1.514 \pm 0.084)\sqrt{s}^{(-0.459\pm 0.012)} \tag{8}$$

$$\eta = (357.693 \pm 96.837)\sqrt{s}^{(-0.524\pm 0.053)} \tag{9}$$

For minimisation of χ^2 for the fits, CERN library MINUIT2 has been used. In case of modified shifted Gompertz, the fit parameters are doubled while introducing the modification. This causes large error limits on the parameters resulting in the very large p values, particularly close to 1. In addition, the LEP data for e^+e^- collisions suffer from very small sample size at some energies, thereby adding to the errors on the fit parameters.

Using shifted Gompertz distribution, the multiplicity distribution for 500 GeV e^+e^- collisions at a future Collider is predicted, as shown in Figure 10. The value of mean multiplicity $\langle n \rangle$ is predicted to be the 37.14 ± 1.12. Figure 11 shows the dependence of mean multiplicity from experimental data on energy $\sqrt{s}$. The fitted curve in (10) represents Fermi-Landau model [24, 25] and fits the data reasonably well with a_1 = -10.609 ± 2.003 and b_1 = 10.156 ± 0.561

$$\langle n \rangle = a_1 + b_1\sqrt{s}^{1/4} \tag{10}$$

It may be observed that the value of $\langle n \rangle$ predicted from shifted Gompertz distribution at 500 GeV fits well on the curve, as shown in the Figure 11. A parameterization of the multiplicity data in e^+e^- collisions at the next-to-leading-order QCD was done by D.E. Groom et al [14, 26] and is given in (9) of the reference

$$\langle n(s) \rangle = a.\exp\left[\frac{4}{\beta_0}\sqrt{\frac{6\pi}{\alpha_s(s)}} + \left(\frac{1}{4} + \frac{10n_f}{27\beta_0}\right)\ln\alpha_s(s)\right] + c \tag{11}$$

TABLE 1: Parameters of shifted Gompertz and modified shifted Gompertz functions for e^+e^- collisions.

	shifted Gompertz	$\longrightarrow$			Modified shifted Gompertz	$\longrightarrow$					
Energy (GeV)	b	η	χ^2/ndf	p value	b_1	η_1	b_2	η_2	α	χ^2/ndf	p value
OPAL											
91	0.191 ± 0.001	33.920 ± 1.133	190.90/21	< 0.0001	0.213 ± 0.003	91.100 ± 7.541	0.265 ± 0.012	46.890 ± 8.189	0.657	51.88/19	< 0.0001
161	0.159 ± 0.004	26.590 ± 3.120	16.72/20	0.6711	0.178 ± 0.008	67.240 ± 16.290	0.244 ± 0.040	45.310 ± 24.960	0.716	7.13/18	0.9890
183	0.142 ± 0.003	25.370 ± 1.773	7.49/24	0.9995	0.145 ± 0.010	30.810 ± 6.324	0.143 ± 0.039	20.470 ± 9.522	0.675	7.06/22	0.9989
189	0.135 ± 0.002	21.930 ± 1.265	22.62/22	0.4234	0.149 ± 0.005	47.490 ± 8.789	0.177 ± 0.017	25.960 ± 6.152	0.662	8.68/20	0.9863
L3											
91	0.215 ± 0.001	45.170 ± 0.563	3989.00/25	< 0.0001	0.234 ± 0.001	103.100 ± 1.989	0.244 ± 0.003	28.330 ± 0.870	0.651	1094.00/23	< 0.0001
131	0.173 ± 0.004	31.430 ± 3.157	12.66/15	0.6285	0.194 ± 0.009	87.480 ± 27.270	0.242 ± 0.031	50.250 ± 21.510	0.654	5.32/13	0.9675
136	0.157 ± 0.004	23.140 ± 2.185	27.42/16	0.0370	0.183 ± 0.009	82.870 ± 25.990	0.258 ± 0.025	56.730 ± 19.510	0.657	19.73/14	0.1389
172	0.138 ± 0.003	22.760 ± 1.427	4.82/19	0.9996	0.146 ± 0.008	36.660 ± 12.880	0.169 ± 0.055	22.680 ± 16.590	0.767	2.61/17	1.0000
182	0.138 ± 0.003	23.120 ± 1.469	8.07/19	0.9860	0.148 ± 0.008	46.180 ± 12.560	0.192 ± 0.023	35.170 ± 11.900	0.668	5.25/17	0.9970
188	0.138 ± 0.002	23.860 ± 1.179	26.16/17	0.0716	0.156 ± 0.006	58.890 ± 10.230	0.199 ± 0.016	38.780 ± 9.268	0.670	10.77/15	0.7687
194	0.136 ± 0.003	22.900 ± 1.479	11.83/19	0.8928	0.149 ± 0.008	25.140 ± 3.422	0.164 ± 0.016	114.500 ± 67.31	0.772	6.59/17	0.9883
200	0.132 ± 0.003	22.220 ± 1.356	3.96/19	0.9999	0.134 ± 0.009	29.340 ± 12.030	0.169 ± 0.059	26.260 ± 17.710	0.779	3.73/17	0.9997
206	0.129 ± 0.003	22.320 ± 1.233	1.42/19	1.0000	0.135 ± 0.013	25.170 ± 5.682	0.114 ± 0.050	14.700 ± 12.630	0.790	1.32/17	1.0000

TABLE 2: Parameters of shifted Gompertz and modified shifted Gompertz functions for $\overline{p}p$ collisions.

		Shifted Gompertz	$\longrightarrow$			Modified shifted Gompertz	$\longrightarrow$					
Energy (GeV)	$\|y\|$	b	η	χ^2/ndf	p value	b_1	η_1	b_2	η_2	α	χ^2/ndf	p value
200	0.5	0.455 ± 0.013	0.652 ± 0.061	11.66/11	0.3897	0.518 ± 0.041	0.584 ± 0.105	0.572 ± 0.061	8.893 ± 4.092	0.850	3.79/9	0.9247
200	1.5	0.194 ± 0.005	1.570 ± 0.099	9.11/29	0.9998	0.171 ± 0.007	1.064 ± 0.127	0.338 ± 0.052	6.284 ± 2.489	0.812	5.39/27	1.0000
200	3.0	0.123 ± 0.003	2.606 ± 0.179	12.60/47	1.0000	0.159 ± 0.010	3.455 ± 0.332	0.108 ± 0.015	9.180 ± 5.148	0.800	2.98/45	1.0000
200	5.0	0.101 ± 0.002	3.381 ± 0.175	35.33/52	0.9628	0.133 ± 0.013	4.921 ± 0.774	0.097 ± 0.016	10.430 ± 7.851	0.760	3.48/46	1.0000
200	full	0.111 ± 0.003	5.033 ± 0.353	3.96/25	1.0000	0.107 ± 0.004	4.830 ± 0.641	0.197 ± 0.054	15.400 ± 12.430	0.900	4.59/26	1.0000
540	0.5	0.397 ± 0.005	0.682 ± 0.033	26.90/20	0.1381	0.490 ± 0.029	0.674 ± 0.074	0.407 ± 0.019	5.363 ± 1.598	0.810	21.29/18	0.2650
540	1.5	0.162 ± 0.002	1.387 ± 0.049	17.22/26	0.0370	0.224 ± 0.012	1.504 ± 0.109	0.155 ± 0.006	5.071 ± 1.088	0.700	11.04/24	0.9887
540	3.0	0.097 ± 0.001	2.308 ± 0.055	176.40/28	< 0.0001	0.089 ± 0.003	3.070 ± 0.389	0.158 ± 0.008	2.794 ± 0.205	0.640	21.29/23	0.5634
540	5.0	0.080 ± 0.001	3.489 ± 0.069	69.33/33	0.0002	0.081 ± 0.001	7.072 ± 0.566	0.126 ± 0.005	3.579 ± 0.165	0.580	48.50/31	0.0236
540	full	0.079 ± 0.001	4.088 ± 0.094	59.83/49	0.0226	0.079 ± 0.002	5.932 ± 0.649	0.116 ± 0.012	3.633 ± 0.345	0.730	58.97/47	0.1130
900	0.5	0.327 ± 0.008	0.616 ± 0.063	10.16/20	0.9652	0.431 ± 0.033	0.798 ± 0.116	0.284 ± 0.041	4.561 ± 2.874	0.845	5.13/18	0.9986
900	1.5	0.129 ± 0.003	1.083 ± 0.087	35.85/46	0.8593	0.191 ± 0.007	1.834 ± 0.155	0.127 ± 0.011	12.530 ± 3.698	0.841	5.77/44	1.0000
900	3.0	0.076 ± 0.002	1.834 ± 0.103	59.90/71	0.8234	0.129 ± 0.005	3.256 ± 0.274	0.075 ± 0.004	10.320 ± 2.117	0.725	8.01/69	1.0000
900	5.0	0.063 ± 0.001	3.226 ± 0.125	89.95/95	0.6272	0.104 ± 0.004	5.355 ± 0.492	0.057 ± 0.003	10.370 ± 2.132	0.660	17.98/93	1.0000
900	full	0.063 ± 0.001	4.043 ± 0.195	67.16/47	0.0283	0.101 ± 0.003	7.852 ± 0.618	0.061 ± 0.003	18.170 ± 3.879	0.710	13.46/45	1.0000

TABLE 3: Parameters of shifted Gompertz and modified shifted Gompertz functions for pp collisions.

		Shifted Gompertz	$\longrightarrow$			Modified Shifted Gompertz	$\longrightarrow$					
Energy (GeV)	$\|y\|$	b	η	χ^2/ndf	p value	b_1	η_1	b_2	η_2	α	χ^2/ndf	p value
900	0.5	0.320 ± 0.005	0.719 ± 0.113	3.57/19	1.0000	0.327 ± 0.007	1.250 ± 0.243	0.949 ± 0.370	1.259 ± 2.089	0.840	2.24/17	1.0000
900	1.0	0.193 ± 0.002	1.307 ± 0.109	84.97/36	< 0.0001	0.198 ± 0.002	2.260 ± 0.159	0.667 ± 0.079	3.366 ± 1.286	0.810	63.32/34	0.0017
900	1.5	0.131 ± 0.001	1.320 ± 0.092	66.98/48	0.0364	0.136 ± 0.002	2.553 ± 0.169	0.397 ± 0.036	3.114 ± 0.866	0.760	44.22/46	0.5471
900	2.0	0.101 ± 0.001	1.431 ± 0.084	55.41/58	0.5722	0.104 ± 0.001	2.660 ± 0.169	0.283 ± 0.024	3.123 ± 0.756	0.750	33.43/56	0.9928
900	2.4	0.087 ± 0.001	1.585 ± 0.081	72.26/64	0.2239	0.088 ± 0.001	2.662 ± 0.162	0.250 ± 0.021	3.816 ± 1.020	0.780	48.03/62	0.9036
2360	0.5	0.242 ± 0.005	0.516 ± 0.099	8.13/19	0.9853	0.424 ± 0.042	1.148 ± 0.433	0.253 ± 0.011	5.447 ± 1.320	0.720	5.70/17	0.9950
2360	1.0	0.133 ± 0.002	0.700 ± 0.091	24.30/34	0.8904	0.139 ± 0.003	2.472 ± 0.270	0.352 ± 0.033	1.823 ± 0.550	0.620	14.03/32	0.9975
2360	1.5	0.092 ± 0.002	0.819 ± 0.090	28.08/45	0.9773	0.099 ± 0.002	3.089 ± 0.303	0.249 ± 0.019	2.195 ± 0.499	0.590	10.03/43	1.0000
2360	2.0	0.071 ± 0.001	0.917 ± 0.088	39.83/55	0.9383	0.076 ± 0.002	3.342 ± 0.321	0.190 ± 0.013	2.526 ± 0.487	0.580	12.70/53	1.0000
2360	2.4	0.062 ± 0.001	1.122 ± 0.089	59.55/66	0.6993	0.137 ± 0.006	2.699 ± 0.379	0.070 ± 0.002	8.042 ± 0.832	0.610	14.49/64	1.0000
7000	0.5	0.184 ± 0.002	0.580 ± 0.073	117.50/37	< 0.0001	0.387 ± 0.013	1.376 ± 0.248	0.202 ± 0.003	7.402 ± 0.599	0.710	24.08/35	0.9178
7000	1.0	0.101 ± 0.001	0.846 ± 0.068	223.70/66	< 0.0001	0.229 ± 0.007	1.778 ± 0.250	0.110 ± 0.001	6.826 ± 0.409	0.650	48.85/64	0.9195
7000	1.5	0.068 ± 0.001	0.854 ± 0.062	247.90/88	< 0.0001	0.192 ± 0.006	2.419 ± 0.336	0.074 ± 0.001	4.459 ± 0.241	0.520	60.87/86	0.9817
7000	2.0	0.049 ± 0.001	0.667 ± 0.053	164.60/108	0.0004	0.136 ± 0.005	2.111 ± 0.276	0.055 ± 0.001	4.306 ± 0.229	0.530	27.79/106	1.0000
7000	2.4	0.042 ± 0.001	0.693 ± 0.051	179.70/123	0.0007	0.046 ± 0.001	3.871 ± 0.195	0.122 ± 0.004	2.396 ± 0.304	0.510	30.92/121	1.0000

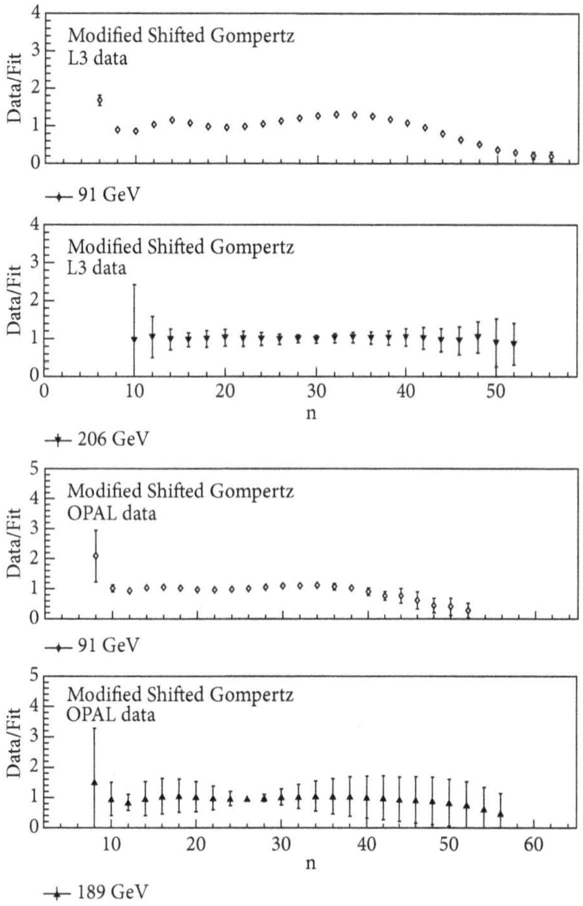

FIGURE 3: Ratio plots of data over modified shifted Gompertz fit values.

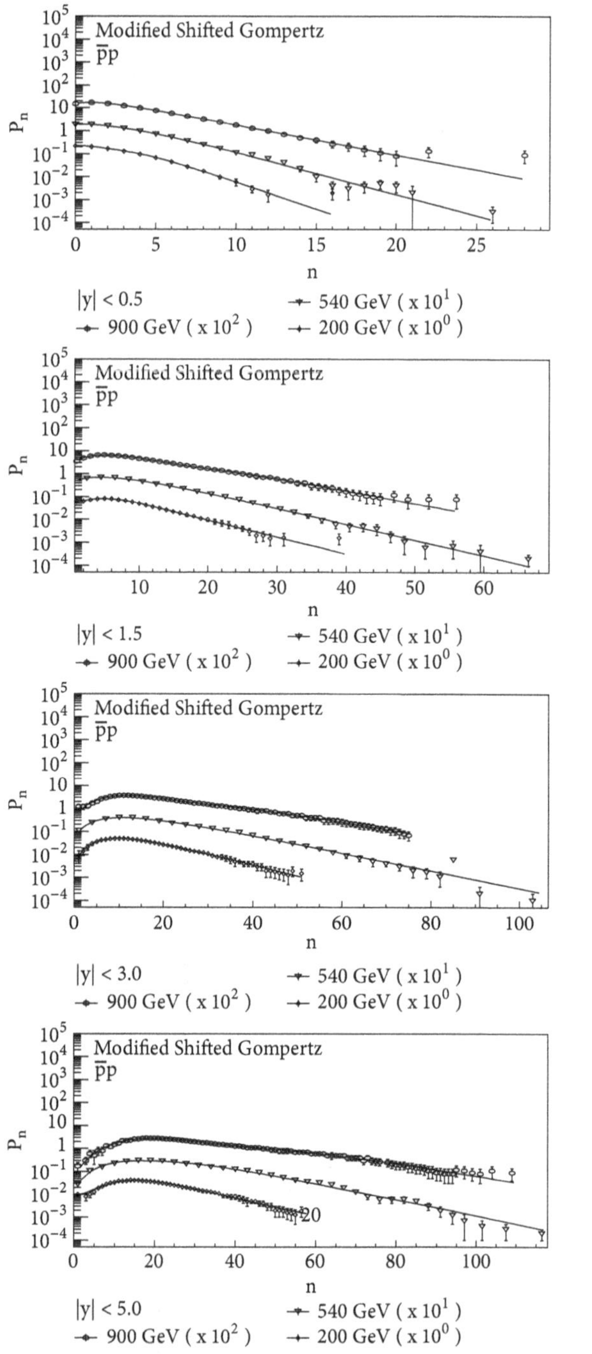

FIGURE 4: Charged multiplicity distributions in $\overline{p}p$ collisions in four rapidity windows. Solid lines represent modified shifted Gompertz function.

where a and c are constants and β_0 is defined in (9.4b). $\langle n \rangle$ versus $\sqrt{s}$ dependence was shown in [27]. Parameters a and c were fitted to the experimental data and a very good agreement was shown. It is observed that both formulae, (10) and (11), provide excellent extrapolations for $\sqrt{s} > 206$. The mean multiplicity $\langle n \rangle$ at 500 GeV is predicted to be 39.18 by NLO QCD equation. In the present work, the mean value predicted by the shifted Gompertz distribution as 37.14 ± 1.23 agrees very closely with the value derived from NLO QCD. This is a good test of the validity of the proposed distribution.

An interesting description of universality of multiplicity in e^+e^- and $p + p(\overline{p})$ has been discussed by Grosse-Oetringhaus et al. [14]. It is shown that although the multiplicity distributions differ between $p + p(\overline{p})$ and e^+e^- collisions, their average multiplicities as a function of $\sqrt{s}$ show similar trends that can be unified using the concepts of effective energy and inelasticity. It is also shown that the Fermi-Landau form $\langle n \rangle \sim s^{1/4}$ fails to describe the pp multiplicity data. But the data is well described by $\langle n \rangle = A + B \ln s + C \ln^2 s$. The universality appears to be valid at least up to Tevatron energies. The multiplicities in e^+e^- and $p + p(\overline{p})$ collisions become strikingly similar when the effective energy E_{eff} in $p + p(\overline{p})$ collisions, available for particle production is used.

$$E_{eff} = \sqrt{s} - \left(E_{lead}, 1 + E_{lead}, 2\right),$$
$$\left\langle E_{eff} \right\rangle = \sqrt{s} - 2\left\langle E_{leading} \right\rangle \tag{12}$$

where E_{lead} is the energy of the leading particle and the inelasticity K is defined as $K = E_{eff}/\sqrt{s}$. K is estimated in $p + p(\overline{p})$ collisions by comparing $p + p(\overline{p})$ with e^+e^- collisions. Given a parameterization $f_{ee}(\sqrt{s})$ of the $\sqrt{s}$ dependence of

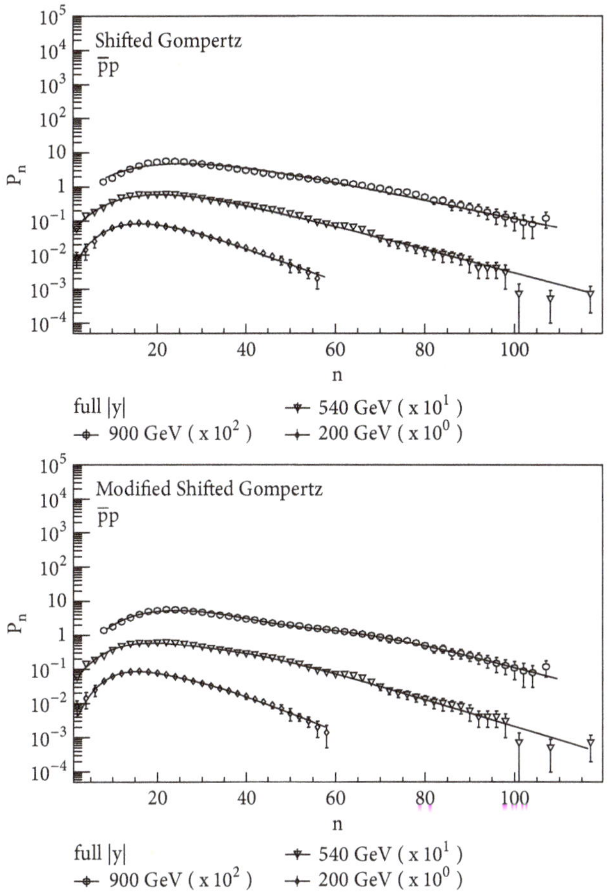

FIGURE 5: Charged multiplicity distributions in $\overline{p}p$ collisions in full phase space. Solid lines represent shifted Gompertz function (top) and modified shifted Gompertz distributions (bottom).

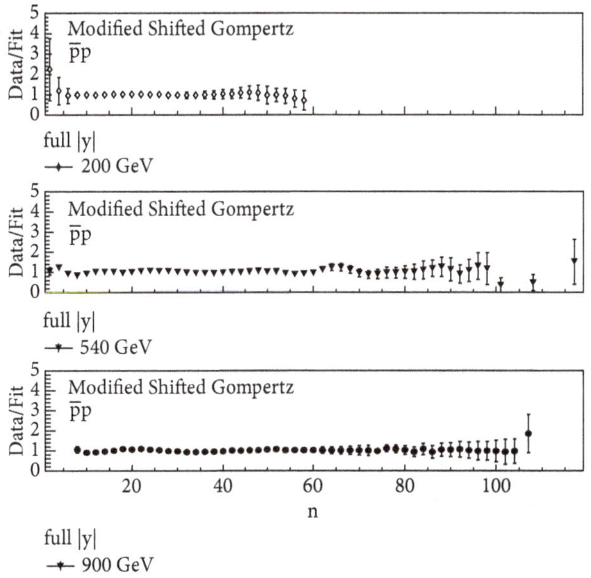

FIGURE 6: Ratio plots of data versus modified shifted Gompertz fit values for $\overline{p}p$ collisions in full phase space.

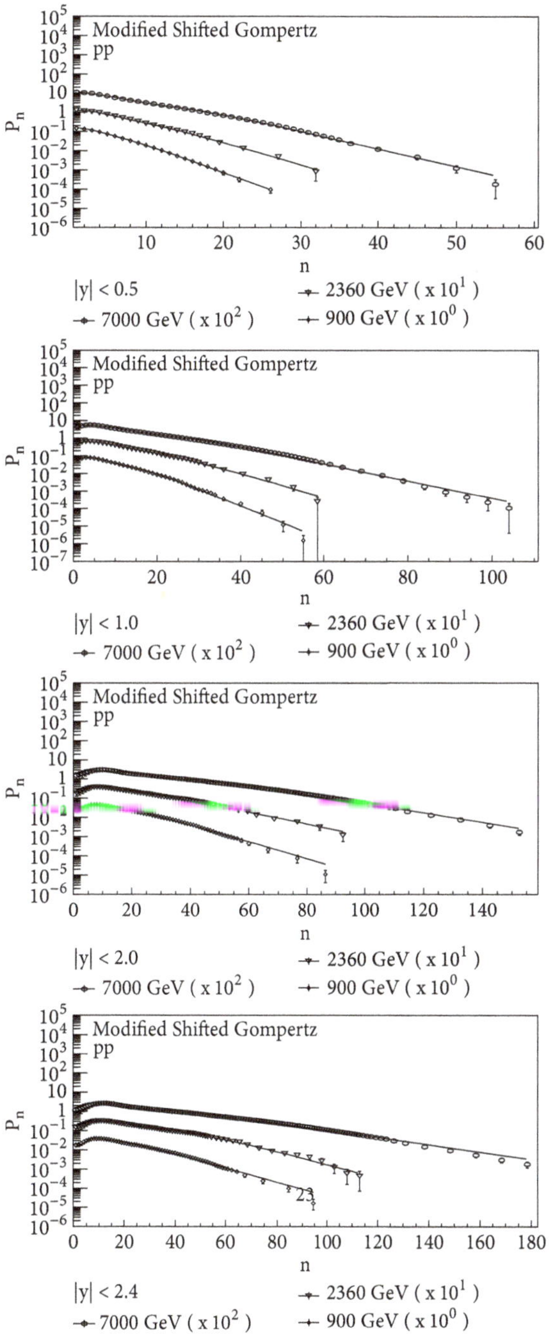

FIGURE 7: Charged multiplicity distributions in pp collisions in four rapidity windows for the modified shifted Gompertz distributions.

the charged multiplicity $\langle n \rangle$ in e^+e^- collisions, one can fit the $p + p(\overline{p})$ data with

$$f_{pp}\left(\sqrt{s}\right) = f_{ee}\left(K.\sqrt{s}\right) + n_0 \quad (13)$$

The parameter n_0 corresponds to the contribution from the two leading protons to the total multiplicity and is expected to be close to $n_0 = 2$. One can use this fit of $p + p(\overline{p})$ data to predict the multiplicities at the LHC. As described in [14], using a fit with (13), Jan Fiete Grosse-Oetringhaus et al. have estimated $K = 0.35 \pm 0.01$ and $n_0 = 2.2 \pm 0.19$. Under the

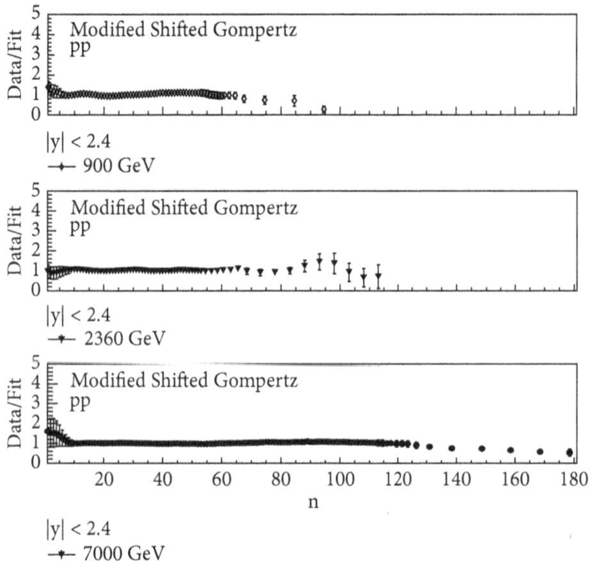

FIGURE 8: ratio plots of data versus modified shifted Gompertz fit values for the pp collisions in full phase space.

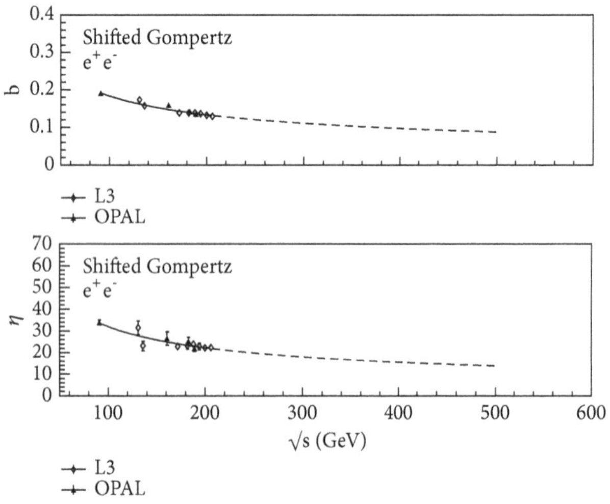

FIGURE 9: Scale parameter b and shape parameter η for e^+e^- data. Points represent the data; solid line is the power law fit and the dotted line the extension of power law fit.

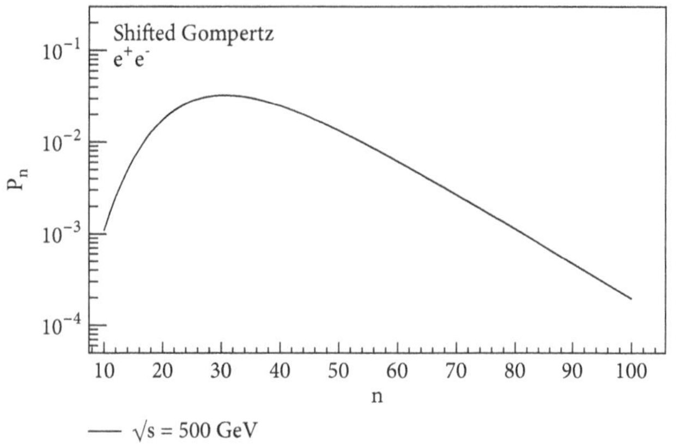

FIGURE 10: Probability distribution for e^+e^- collisions predicted from shifted Gompertz function.

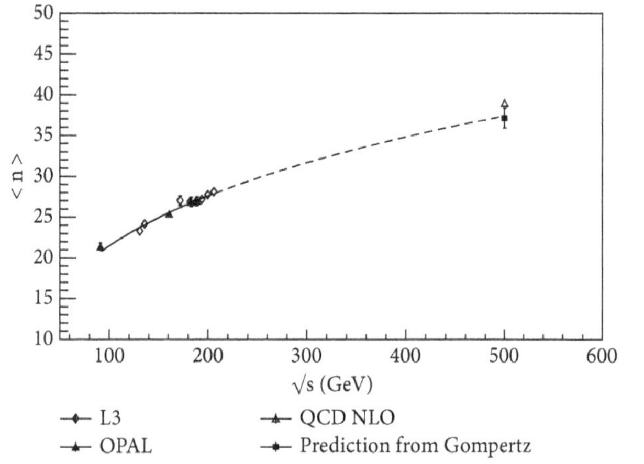

FIGURE 11: Average multiplicity in e^+e^- collisions as a function of $\sqrt{s}$. The fitted curve represents Fermi-Landau model.

assumptions that inelasticity remains constant at about 0.35 at LHC energies and that the extrapolation of the e^+e^- data with the QCD form is still reliable, authors fit the $p+p(\overline{p})$ data to predict the multiplicities at the LHC. They find $\langle n \rangle$=88.9. We use these values of inelasticity and average multiplicity to build the multiplicity distribution at $\sqrt{s}$=14 TeV using shifted Gompertz function.

Figure 12 shows the multiplicity distribution predicted from shifted Gompertz PDF for pp collisions at $\sqrt{s} = 14$ TeV at the LHC. The mean value of the multiplicity is predicted to be $\langle n \rangle \approx 89.2$. It is observed that, in general, at all energies for different types of collisions, the multiplicity distributions can be described by shifted Gompertz function. However the LHC data at 7000 GeV in the lower rapidity windows are an exception, whereby the fits are statistically excluded with $CL < 0.1\%$. At all energies, both the scale parameter b and the shape parameter η decrease with the collision energy in the center of momentum. In the rapidity windows, b decreases with the increase in the rapidity. The shape parameter η increases with rapidity as it determines the width of the distribution.

The fact that multiplicity distributions at higher energies show a shoulder structure is well established. In order to improve upon shifted Gompertz fits to the data, the multiplicity distribution is reproduced by a weighted superposition of two shifted Gompertz distributions corresponding to the soft component and the semi-hard component. It is observed that this modified shifted Gompertz distribution improves the fits excellently and the χ^2 values diminish by several orders. However, distributions fail for 91 GeV e^+e^- data.

The data for pp collisions at 7 TeV fail for shifted Gompertz distribution in three rapidity windows. But the modified shifted Gompertz distribution shows a very good agreement with the data for all rapidities, as shown in Table 3. For each of the rapidity bins, χ^2/dof values are reduced manifold with $CL > 0.1\%$.

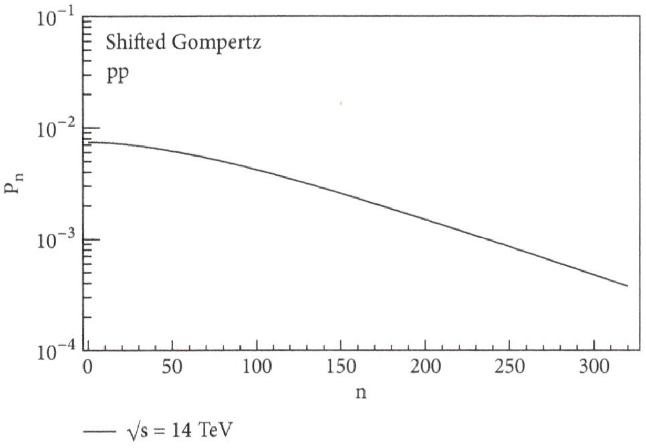

FIGURE 12: Probability distribution pp collisions predicted from shifted Gompertz function.

Using the shifted Gompertz function and the analysis, the multiplicity distributions at future collider and the mean multiplicity predicted for 500 GeV e^+e^- agree very well with the predictions from NLO QCD prediction and also with the Fermi-Landau model of particle production.

4. Conclusion

The aim of this paper is to propose the use of a new statistical distribution for studying the multiplicity distributions in high energy collisions; the shifted Gompertz distribution function often used in model of adoption of innovations describes the multiplicity data extremely well. A detailed analysis of data from e^+e^-, $\overline{p}p$, and pp collisions at high energies in terms of shifted Gompertz distribution shows that, in general, the distribution fits the data very well at most of the energies and in various rapidity intervals with the exception of a very few. Very similar to the Weibull distribution, which recently has been extensively used, it determines two nonnegative parameters measuring the scale and shape of the distribution. A power law dependence of the scale parameter and shape parameter on the collision energy is established for the e^+e^- data. The parametrisation as a power law is inspired by the observation that single particle energy distribution obeys a power law behaviour.

The occurrence of a shoulder structure in the multiplicity distribution (MD) of charged particles at high energy is well established. This affects the shape of the distribution fit. To improve upon the fits to the data, a weighted superposition of the distributions using shifted Gompertz function for the soft events (events with mini-jets) and the semihard events (events without mini-jets) is done. The concept of superposition originates from purely phenomenological and very simple considerations. The two fragments of the distribution suggest the presence of the substructure. The two-component shifted Gompertz distribution fits the data from different types of collisions at different energies, extremely well. Describing the MD in terms of soft and semihard components allows one to model, under simple assumptions the new energy domain. While predicting the multiplicity distribution using shifted Gompertz Distribution at 14 TeV, it remains interesting to determine the dependence of fraction of minijet events, α upon the rapidity windows compared to the events without minijets. To predict the more accurate multiplicity distributions in different rapidity windows at 14 TeV, modified shifted Gompertz PDF is required, for which α value in each rapidity window is needed. The analysis presented for 7 TeV data shows that the minijet fraction of events decreases with energy as well as with the increasing size of rapidity window. This trend has also been shown in [11] where the α fraction for full rapidity range of pp collisions at 14 TeV has been estimated as 0.30. When multiplicity distributions in full phase space at higher energies like 7 TeV become available, the extrapolations from the lower energy domain to the highest energies can be well established, as predicted in other works also, using different approaches [16, 28].

A good agreement between the mean multiplicity and the multiplicity dependence on energy, predicted by NLO QCD and the Fermi-Landau model of particle production, with the predictions made by shifted Gompertz distribution, serves as a good test of the validity of the proposed distribution.

The future extension of the present work shall focus on analysis of multiplicities from lower energy domains, in hadron-nucleus interactions and nucleus-nucleus interactions using shifted Gompertz distribution, and deriving the additional information from the oscillatory behaviour of the counting statistics, as suggested by Wilk and Włodarczyk [12].

Conflicts of Interest

The authors declare that they have no conflicts of interest.

Acknowledgments

The author M. Kaur is thankful to Harrison Prosper of Florida State University, US, for discussion on the distributions suitable to describe the collision data from the high energy accelerators. The author Ridhi Chawla is grateful to the Department of Science and Technology, Government of India, for the Inspire-Fellowship grant.

References

[1] A. C. Bemmaor, G. L. Laurent, B. Lilien, and B Pras, Eds., *Research Traditions in Marketing*, vol. 201, 1994.

[2] F. Jiménez and P. Jodrá, "A note on the moments and computer generation of the shifted Gompertz distribution," *Communications in Statistics—Theory and Methods*, vol. 38, no. 1-2, pp. 75–89, 2009.

[3] F. Jiménez Torres, "Estimation of parameters of the shifted Gompertz distribution using least squares, maximum likelihood and moments methods," *Journal of Computational and Applied Mathematics*, vol. 255, pp. 867–877, 2014.

[4] C. Bauckhage and K. Kristian, "Strong Regularities in Growth and Decline of Popularity of Social Media Services," https://arxiv.org/abs/1406.6529.

[5] S. Dash, "Multiplicity distributions for e+e− collisions using Weibull distribution," *Physical Review D*, vol. 94, no. 7, Article ID 074044, 2016.

[6] T. Nakagawa and S. Osaki, "The discrete weibull distribution," *IEEE Transactions on Reliability*, vol. 24, no. 5, pp. 300-301, 1975.

[7] K. Urmossy, G. Barnaföldi, and T. Biró, "Generalised Tsallis statistics in electron–positron collisions," *Physics Letters B*, vol. 107, no. 1, pp. 111–116, 2011.

[8] S. Hegyi, "Log-KNO scaling: a new empirical regularity at very high energies?" *Physics Letters B*, vol. 467, no. 1-2, pp. 126–131, 1999.

[9] S. Catani, L. Dokshitzer, and M. H. Seymour, "Longitudinally-invariant k⊥-clustering algorithms for hadron-hadron collisions," *Nuclear Physics B*, vol. 406, no. 1-2, pp. 187–224, 1993.

[10] G. Dissertori et al., "An Improved theoretical prediction for the two jet rate in e+ e- annihilation," *Physics Letters B*, vol. 361, p. 167, 1995.

[11] A. Giovannini and R. Ugoccioni, "Clan structure analysis and qcd parton showers in multiparticle dynamics: an intriguing dialog between theory and experiment," *International Journal of Modern Physics A*, vol. 20, no. 17, p. 3897, 2005.

[12] G. Wilk and Z. Włodarczyk, "How to retrieve additional information from the multiplicity distributions," *Journal of Physics G: Nuclear and Particle Physics*, vol. 44, no. 1, Article ID 015002, 2016.

[13] P. Carruthers et al., "The Phenomenological Analysis of Hadronic Multiplicity Distributions," *International Journal of Modern Physics A*, vol. 2, p. 179, 1987.

[14] J. Fiete Grosse-Oetringhaus and J. Klaus Reygers, "Charged-particle multiplicity in proton-proton collisions," *Journal of Physics G: Nuclear and Particle Physics*, vol. 37, Article ID 083001, 2010.

[15] R. E. Ansorge, B. Asman et al., "Charged particle multiplicity distribtions at 200 GeV and 900 GeV center-of-mass energy," *Zeitschrift für Physik*, vol. C43, 357, 1989, UA5 Collaboration.

[16] C. Fuglesang, *Multiparticle Dynamics*, vol. 193, World Scientific, Singapore, 1990.

[17] P. Achard et al., "Studies of hadronic event structure in e+e- annihilation from 30 GeV to 209 GeV with the L3 detector," *Physics Reports*, vol. 399, p. 71, 2004, L3 Collaboration.

[18] P. D. Acton et al., "A study of charged particle multiplicities in hadronic decays of the Z^0," *Zeitschrift für Physik*, vol. C53, 539, 1992, OPAL Collaboration.

[19] G. Alexander et al., "QCD studies with e+e- annihilation data at 130 and 136 GeV," *Zeitschrift für Physik*, vol. C72, 191, 1996, OPAL Collaboration.

[20] K. Ackerstaff et al., "QCD studies with e+e- annihilation data at 161 GeV," *Zeitschrift für Physik*, vol. C75, 193, 1997, OPAL Collaboration.

[21] G. Abbiendi et al., "QCD analysis and determination of α_s in e+e- annihilations at energies between 35 and 189 GeV," *The European Physical Journal*, vol. C17, 19, 2000.

[22] V. Khachatryan, A. M. Sirunyan et al., "Charged multiplicities in pp interactions at 0.9, 2.36 and 7 TeV," *Journal of High Energy Physics*, vol. 1, no. 79, 2011, CMS Collaboration.

[23] G. J. Alner et al., "An investigation of multiplicity distributions in different pseudorapidity intervals in pbarp reactions at a CMS energy of 540 GeV," *Physics Letters B*, vol. 160, 193, 1985.

[24] E. Fermi, "High Energy Nuclear Events," *Progress of Theoretical Physics*, vol. 5, 570, 1950.

[25] W. Cheuk-Yin, "Landau hydrodynamics reexamined," *Physical Review*, vol. C78, 054902, 2008.

[26] D. E. Groom, Particle Data Group et al., "Reviews, Tables, and Plots," *The European Physical Journal*, vol. C 15, 1, 2000.

[27] O. Biebel, P. Nason, and B. R. Webber, "Jet fragmentation in e+e- annihilation," https://arxiv.org/abs/hep-ph/0109282.

[28] A. Capella and E. G. Ferreiro, arXiv:1301.3339v1A.

14

Generalized Dirac Oscillator in Cosmic String Space-Time

Lin-Fang Deng,[1] Chao-Yun Long,[1,2] Zheng-Wen Long,[1] and Ting Xu[1]

[1]*Department of Physics, Guizhou University, Guiyang 550025, China*
[2]*Laboratory for Photoelectric Technology and Application, Guizhou University, Guiyang 550025, China*

Correspondence should be addressed to Chao-Yun Long; longchaoyun66@163.com

Academic Editor: Edward Sarkisyan-Grinbaum

In this work, the generalized Dirac oscillator in cosmic string space-time is studied by replacing the momentum p_μ with its alternative $p_\mu+m\omega\beta f_\mu(x_\mu)$. In particular, the quantum dynamics is considered for the function $f_\mu(x_\mu)$ to be taken as Cornell potential, exponential-type potential, and singular potential. For Cornell potential and exponential-type potential, the corresponding radial equations can be mapped into the confluent hypergeometric equation and hypergeometric equation separately. The corresponding eigenfunctions can be represented as confluent hypergeometric function and hypergeometric function. The equations satisfied by the exact energy spectrum have been found. For singular potential, the wave function and energy eigenvalue are given exactly by power series method.

1. Introduction

In quantum mechanics, there has been an increasing interest in finding the analytical solutions that play an important role for getting complete information about quantum mechanical systems [1–3]. The Dirac oscillator proposed in [4] is one of the important issues in this relativistic quantum mechanics recently. In this quantum model, the coupling proposed is introduced in such a way that the Dirac equation remains linear in both spatial coordinates and momenta and recovers the Schrödinger equation for a harmonic oscillator in the nonrelativistic limit of the Dirac equation [4–11]. As a solvable model of relativistic quantum mechanical system, the Dirac oscillator has many applications and has been studied extensively in different field such as high-energy physics [12–15], condensed matter physics [16–18], quantum Optics [19–25], and mathematical physics [26–33]. On the other hand, the analysis of gravitational interactions with a quantum mechanical system has recently attracted a great deal attention and has been an active field of research [5, 6, 34–43]. The study of quantum mechanical problems in curved space-time can be considered as a new kind of interaction between quantum matter and gravitation in the microparticle world. In recent years, the Dirac oscillator embedded in a cosmic string background has inspired a great deal of research such as the dynamics of Dirac oscillator in the space-time of cosmic string [44–47], Aharonov-Casher effect on the Dirac oscillator [5, 48], and noninertial effects on the Dirac oscillator in the cosmic string space-time [49–51]. It is worth mentioning that based on the coupling corresponding to the Dirac oscillator a new coupling into Dirac equation first has been proposed by Bakke *et al.* [52] and used in different fields [53–57]. This model is called the generalized Dirac oscillator which in special case is reduced to ordinary Dirac oscillator. Inspired by the above work, the main aim of this paper is to analyze the generalized Dirac oscillator model with the interaction functions $f_\mu(x_\mu)$ taken as Cornell potential, singular potential and exponential-type potential in cosmic string space-time and to find the corresponding energy spectrum and wave functions. This work is organized as follows. In Section 2, the new coupling is introduced in such a way that the Dirac equation remains linear in momenta, but not in spatial coordinates in a curved space-time. In Sections 3, 4, and 5, we concentrate our efforts in analytically solving the quantum systems with different function $f_\mu(x_\mu)$ and find the corresponding energy spectrum and spinors, respectively. In Section 6, we make a short conclusion.

2. Generalized Dirac Oscillator with a Topological Defect

In cosmic string space-time, the general form of the cosmic string metric in cylindrical coordinates read [41, 42, 44, 58, 59]

$$ds^2 = -dt^2 + d\rho^2 + \alpha^2\rho^2 d\varphi^2 + dz^2, \quad (1)$$

with $-\infty < (t, z) < +\infty$, $0 < \rho < +\infty$, and $0 < \varphi < 2\pi$. The parameter α is related to the linear mass density of string η by $\alpha = 1 - 4\eta$ and runs in the interval $(0, 1]$. In the limit as $\alpha \longrightarrow 1$ we get the line element of cylindrical coordinates. The Dirac equation in the curved space-time ($\hbar = c = 1$) reads

$$\left[i\gamma^{\mu}(x)\partial_{\mu} - i\gamma_{\mu}(x)\Gamma_{\mu}(x) - m\right]\psi(t,x) = 0, \quad (2)$$

where the γ^{μ} matrices are the generalized Dirac matrices defining the covariant Clifford algebra $\{\gamma^{\mu}, \gamma^{\nu}\} = 2g^{\mu\nu}$, m is mass of the particle, and Γ_{μ} is the spinor affine connection. We choose the basis tetrad e_a^{μ} as

$$e_a^{\mu} = \begin{pmatrix} 1 & 0 & 0 & 0 \\ 0 & \cos\varphi & \sin\varphi & 0 \\ 0 & \dfrac{\sin\varphi}{\alpha\rho} & \dfrac{\cos\varphi}{\alpha\rho} & 0 \\ 0 & 0 & 0 & 1 \end{pmatrix}, \quad (3)$$

then in this representation the matrices γ^{μ} [44] can be found to be

$$\begin{aligned} \gamma^0 &= \gamma^t, \\ \gamma^1 &= \gamma^{\rho} = \gamma^1\cos\varphi + \gamma^2\sin\varphi, \\ \gamma^2 &= \gamma^{\varphi} = -\gamma^1\sin\varphi + \gamma^2\cos\varphi, \\ \gamma^3 &= \gamma^z, \\ \gamma^{\mu}\Gamma_{\mu}(x) &= \frac{1-\alpha}{2\alpha\rho}\gamma^{\rho}. \end{aligned} \quad (4)$$

It is well known that, in both Minkowski space-time and curved space-time, usual Dirac oscillator can be obtained by the carrying out nonminimal substitution $p_{\mu} \longrightarrow p_{\mu} + m\omega\beta x_{\mu}$ in Dirac equation where m and ω are the mass and oscillator frequency. In the following, we will construct the generalized oscillator in curved space-time. To do this end, we can replace momenta p_{μ} in the Dirac equation of curved space-time by

$$p_{\mu} \longrightarrow p_{\mu} + m\omega\beta f_{\mu}(x_{\mu}), \quad (5)$$

where $f_{\mu}(x_{\mu})$ are undetermined functions of x_{μ}. It is to say that we introduce a new coupling in such a way that the Dirac equation remains linear in momenta but not in coordinates. In particular, in this work, we only consider the radial component the nonminimal substitution

$$f_{\mu}(x_{\mu}) = (0, f_{\rho}(\rho), 0, 0). \quad (6)$$

By introducing this new coupling (6) into (2) and with the help of (4), in cosmic string space-time the eigenvalue equation of generalized Dirac oscillator can be written as

$$\left\{-i\gamma^t\partial_t + i\gamma^{\rho}\left(\partial_{\rho} - \frac{1-\alpha}{2\alpha\rho} + m\omega\rho f(\rho)\right) + \frac{i\gamma^{\varphi}\partial_{\varphi}}{\alpha\rho} + i\gamma^z\partial_z - m\right\}\psi = 0. \quad (7)$$

We choose the following ansatz:

$$\psi = e^{-iEt+i(l+1/2-\Sigma^3/2)\varphi+ikz}\begin{pmatrix}\chi_1(\rho) \\ \chi_2(\rho)\end{pmatrix}, \quad (8)$$

then we have

$$\begin{aligned} &\left[\alpha_1\left(\frac{d}{d\rho} + \frac{1}{2\rho} - m\omega\rho f(\rho)\right) - \frac{\lambda}{\rho}\alpha_2 - k\alpha_3\right] \\ &\cdot\left[\alpha_1\left(\frac{d}{d\rho} + \frac{1}{2\rho} + m\omega f(\rho)\right) - \frac{\lambda}{\rho}\alpha_2 - k\alpha_3\right]\chi_1 \\ &= \left(E^2 - m^2\right)\chi_1, \end{aligned} \quad (9)$$

$$\begin{aligned} &\left[\alpha_1\left(\frac{d}{d\rho} + \frac{1}{2\rho} + m\omega f(\rho)\right) - \frac{\lambda}{\rho}\alpha_2 - k\alpha_3\right] \\ &\cdot\left[\alpha_1\left(\frac{d}{d\rho} + \frac{1}{2\rho} - m\omega f(\rho)\right) - \frac{\lambda}{\rho}\alpha_2 - k\alpha_3\right]\chi_2 \\ &= \left(E^2 - m^2\right)\chi_2, \end{aligned} \quad (10)$$

where

$$\begin{aligned} \alpha_1 &= i(\sigma_1\cos\varphi + \sigma_2\sin\varphi), \\ \alpha_2 &= -\sigma_1\sin\varphi + \sigma_2\cos\varphi, \\ \alpha_3 &= \sigma_3. \end{aligned} \quad (11)$$

It is straightforward to prove the following relations satisfied above matrices α_i:

$$\begin{aligned} \alpha_1^2 &= -\alpha_2^2 = \alpha_3^2 = -1, \\ \alpha_1\alpha_2 &= -\alpha_2\alpha_1 = i\sigma_1\sigma_2, \\ \alpha_1\alpha_3 &= -\alpha_3\alpha_1 = i(\sigma_1\sigma_3\cos\varphi + \sigma_2\sigma_3\sin\varphi), \\ \alpha_3 &= -\alpha_3\alpha_2 = -\sigma_1\sigma_3\sin\varphi + \sigma_2\sigma_3\cos\varphi. \end{aligned} \quad (12)$$

With help of (12) and simple algebraic calculus, (9) becomes

$$\begin{aligned} &\left\{\partial_{\rho}^2 + \frac{1}{\rho}\partial_{\rho} - \frac{1}{\rho^2}\left[\frac{1}{4} + i\lambda\sigma_1\sigma_2 + \lambda^2\right]\right\}\chi_1 + \left\{-2m\omega\right. \\ &\left.\cdot\frac{f(\rho)}{\rho}\left[ik\rho(\sigma_1\sigma_3\cos\varphi + \sigma_2\sigma_3\sin\varphi) + i\sigma_1\sigma_2\lambda\right]\right\} \\ &\cdot\chi_1 + \left\{m^2 + k^2 - E^2 + m\omega\frac{f(\rho)}{\rho} - m^2\omega^2 f^2(\rho)\right\} \\ &\cdot\chi_1 = 0. \end{aligned} \quad (13)$$

It is easy to prove the following relation [44]:

$$i\sigma_1\sigma_2\lambda + ik\rho\left(\sigma_1\sigma_3\cos\varphi + \sigma_2\sigma_3\sin\varphi\right) = -2\vec{s}.\vec{L}, \quad (14)$$

where $\vec{s} = \vec{\sigma}/2$. The eigenvalue of $\vec{s}.\vec{L}$ can be assumed as $(l + 1/2)/2\alpha$ and (13) reads

$$\frac{d^2\chi_1}{d\rho^2} + \frac{1}{\rho}\frac{d}{d\rho}\chi_1 - \left[\frac{\lambda^2}{\rho^2} + \mu\frac{f(\rho)}{\rho} - m\omega\frac{df(\rho)}{d\rho} + m^2\omega^2 f^2(\rho)\right]\chi_1 + \nu\chi_1 = 0, \quad (*a)$$

where

$$\lambda = \left(\frac{l + 1/2}{\alpha} - \frac{1}{2}\right),$$
$$\mu = \frac{-2(l + 1/2)m\omega}{\alpha}, \quad (15)$$
$$\nu = E^2 - m^2 - k^2.$$

For the component χ_2, from (10) an analog equation can be also obtained

$$\frac{d^2\chi_2}{d\rho^2} + \frac{1}{\rho}\frac{d}{d\rho}\chi_2 - \left[\frac{\lambda^2}{\rho^2} + \mu\frac{f(\rho)}{\rho} + m\omega\frac{df(\rho)}{d\rho} + m^2\omega^2 f^2(\rho)\right]\chi_2 + \nu\chi_2 = 0, \quad (*b)$$

where

$$\lambda = \left(\frac{l + 1/2}{\alpha} + \frac{1}{2}\right),$$
$$\mu = \frac{-2(l + 1/2)m\omega}{\alpha}, \quad (16)$$
$$\nu = E^2 - m^2 - k^2.$$

In particular, $(*a)$ and $(*b)$ will be reduced to the result obtained in [44] when the function $f(\rho)$ is taken as $f(\rho) = \rho$. As we can see, $(*a)$ and $(*b)$ have the same form. So without loss of generality in remaining parts of this work, our main tasks is only to solve the equation $(*a)$ with different functions $f(\rho)$ and find corresponding eigenvalue and eigenfunction. While with regard to $(*b)$, it is straightforward to obtain the corresponding solution.

3. The Solution with $f(\rho)$ to Be Cornell Potential

The Cornell potential that consists of Coulomb potential and linear potential has gotten a great deal of attention in particle physics and was used with considerable success in models describing systems of bound heavy quarks [60–62]. In Cornell potential, the short-distance Coulombic interaction arises from the one-gluon exchange between the quark and its antiquark, and the long-distance interaction is included to take into account confining phenomena.

Now we let the function $f(\rho)$ be Cornell potential

$$f(\rho) = a\rho - \frac{b}{\rho}, \quad (17)$$

where a and b are two constants. Substituting (17) into $(*a)$ and $(*b)$ leads to following equation:

$$\frac{d^2\chi}{d\rho^2} + \frac{1}{\rho}\frac{d}{d\rho}\chi + \left[\frac{-\tau_1^2}{\rho^2} - \tau_2\rho^2 + \tau_3\right]\chi = 0, \quad (18)$$

where

$$\tau_1^2 = \lambda^2 - \mu b + m^2\omega^2 b^2 - \omega m b,$$
$$\tau_2 = m^2\omega^2 a^2, \quad (19)$$
$$\tau_3 = \upsilon + 2abm^2\omega^2 - a\mu + m\omega a.$$

We make a change in variables $\xi = m\omega a\rho^2$ and then (18) can be rewritten as

$$\xi\frac{d^2\chi}{d\xi^2} + \frac{d}{d\xi}\chi + \left[\frac{-\tau_1^2}{4\xi} - \frac{1}{4}\xi + \frac{\tau_3}{4ma\omega}\right]\chi = 0. \quad (20)$$

Taking account of the boundary conditions satisfied by the wave function χ, i.e., $\chi \propto \xi^{\tau_1/2}$ for $\xi \longrightarrow 0$ and $\chi \propto e^{-\xi/2}$ for $\xi \longrightarrow \infty$, physical solutions χ can be expressed as [44, 60, 63, 64]

$$\chi = \xi^{|\tau_1|/2}e^{-\xi/2}g(\xi). \quad (21)$$

If we insert this wave function χ into (20), we have the second-order homogeneous linear differential equation in the following form:

$$\xi\frac{d^2 g}{d\xi^2} + \left(|\tau_1| + 1 - \xi\right)\frac{d}{d\xi}g + \left[\frac{\tau_3}{4ma\omega} - \frac{|\tau_1|}{2} - \frac{1}{2}\right]g = 0. \quad (22)$$

It is well known that (22) is the confluent hypergeometric equation and it is immediate to obtain the corresponding eigenvalues and eigenfunctions

$$g(\xi) = F\left[-\left(\frac{\tau_3}{4ma\omega} - \frac{|\tau_1| + 1}{2}\right), |\tau_1| + 1, \xi\right], \quad (23)$$

$$E_n^2 = \delta_1 - \frac{2(l + 1/2)m\omega a}{\alpha} + 4ma\omega\left[n + \frac{1}{2} + \frac{|\delta_2|}{2}\right], \quad (24)$$

with

$$\delta_1 = m^2 + k^2 - 2abm^2\omega^2 + m\omega a,$$
$$\delta_2 = \sqrt{\lambda^2 - \mu b + b^2 m^2\omega^2 - m\omega b}. \quad (25)$$

In particular, if we assume that $\alpha = 1$, from (24), the energy levels of generalized Dirac oscillator with $f(\rho)$ to be Cornell potential in the absence of a topological defect can be obtained. In addition if we let $a = 1$, $b = 0$ in (24) the energy levels given here will be reduced to that one obtained in [44].

4. The Solution with $f(\rho)$ to Be Singular Potential

The investigation of singular potentials in quantum mechanics is almost as old as quantum mechanics itself and covers a wide range of physical and mathematical interest because the real world interactions were likely to be highly singular [65]. The singular potentials of $v(r) \propto 1/r^n$ type, with $n \geq 2$, are of great current physical interest and are relevant to many problems such as the three-body problem in nuclear physics [66, 67], point-dipole interactions in molecular physics [68], the tensor force between nucleons [69], and the interaction between a charges and an induced dipole [70], respectively. Recently, in cosmic string background, singular inverse-square potential with a minimal length had been studied [71].

Next let us take $f(\rho)$ to be singular inverse-square-type potential [72]

$$f_\rho(\rho) = a + \frac{b}{\rho} + \frac{c}{\rho^2}. \tag{26}$$

Substituting (26) into (∗a), the corresponding radial equation reads

$$\left\{\frac{d^2}{d\rho^2} + \frac{1}{\rho}\frac{d}{d\rho} - \frac{\delta_1}{\rho} - \frac{\lambda^2 + \delta_2}{\rho^2} - \frac{\delta_3}{\rho^3} - \frac{\delta_4}{\rho^4} + \gamma\right\}\chi = 0, \tag{27}$$

where

$$\begin{aligned}
\gamma &= \upsilon - m^2a^2\omega^2\\
\delta_1 &= 2abm^2\omega^2 + a\mu,\\
\delta_2 &= \mu b + m\omega b + b^2m^2\omega^2 + 2acm^2\omega^2,\\
\delta_3 &= \mu c + 2m\omega c + 2bcm^2\omega^2,\\
\delta_4 &= c^2m^2\omega^2.
\end{aligned} \tag{28}$$

It is obvious that (27) has the same mathematical structure with the Schrödinger equation of fourth-order inverse-potential in [73]. So (27) can be solved by power series method.

We look for an exact solution of (27) via the following ansatz to the radial wave function [73–75]:

$$\begin{aligned}
\chi &= \Theta(\rho)\exp[g(\rho)],\\
g(\rho) &= -\frac{\delta_1}{2\rho} - \frac{\delta_2}{2}\rho - \frac{\delta_3}{2}\log\rho.
\end{aligned} \tag{29}$$

Thence, (27) can be transformed into the following form:

$$\begin{aligned}
&\left\{\frac{d^2}{d\rho^2} + \left[-\delta_1 + \frac{1-\delta_3}{\rho} + \frac{\delta_2}{\rho^2}\right]\frac{d}{d\rho} + \gamma + \frac{\delta_1^2}{4}\right.\\
&+ \frac{\delta_1(\delta_3 - 3)}{2\rho} + \frac{\delta_3^2 - 4\delta_2 - 2\delta_1\delta_2 - 4\lambda^2}{4\rho^2}\\
&\left.+ \frac{-\delta_2(1+\delta_3) - \delta_3}{2\rho^3} + \frac{\delta_2^2 - 4\delta_4}{4\rho^4}\right\}\Theta(\rho) = 0.
\end{aligned} \tag{30}$$

Now we take $\Theta(\rho)$ in following series form:

$$\Theta(\rho) = \sum_{n=0}^{\infty} a_n\rho^{n+\lambda+1/2}, \quad a_0 \neq 0,\ a_1 \neq 0. \tag{31}$$

Substituting (31) into (30) gives rise to following equation:

$$\begin{aligned}
&\sum_{n=0}^{\infty} a_n\left\{-\left[\delta_1\left(n+\lambda+2-\frac{\delta_3}{2}\right)\right]\rho^{n+\lambda-1/2} + \left[\gamma + \frac{\delta_1^2}{4}\right]\right.\\
&\cdot\rho^{n+\lambda+1/2} + \left[\frac{(2n+2\lambda+1)(2n+2\lambda+1-2\delta_3)}{4}\right.\\
&\left.- \frac{\delta_2(\delta_1+2)}{2} - \lambda^2 + \frac{\delta_3^2}{4}\right]\rho^{n+\lambda-3/2}\\
&- \left[\delta_1\left(n+\lambda+2-\frac{\delta_3}{2}\right)\right]\rho^{n+\lambda-1/2} + \left[\gamma + \frac{\delta_1^2}{4}\right]\\
&\left.\cdot\rho^{n+\lambda+1/2}\right\} = 0.
\end{aligned} \tag{32}$$

To make (32) be valid for all values of ρ, the coefficients of each term of the polynomial of ρ must be equal to zero separately. We, therefore, obtain

$$2(\delta_1+2)(n+\lambda+2) = 4\lambda^2 - 9 - 4, \tag{33a}$$

$$\delta_2 = -(n+\lambda+2), \tag{33b}$$

$$\delta_3 = 2(n+\lambda+2), \tag{33c}$$

$$2\delta_4 = (n+\lambda+2)^2, \tag{33d}$$

$$\gamma = -\frac{\delta_1^2}{4}. \tag{33e}$$

Using (∗a), (33a), and (33e) and after simple algebraic calculation, the corresponding energy can be written as

$$E_n^2 = m^2 + k^2 + m^2a^2\omega^2 - \frac{1}{16}\left(\frac{4\lambda^2 - 9}{n+\lambda+2}\right)^2. \tag{34}$$

The general radial wave functions corresponding to the energy spectra given in (34) are

$$\Theta(\rho) = \sum_{n=0}^{\infty} a_n\rho^{n+\lambda+1/2}\exp\left[-\frac{\delta_1}{2\rho} - \frac{\delta_2}{2}\rho - \frac{\delta_3}{2}\log\rho\right]. \tag{35}$$

With the help of (32) and (33a)–(33e), the expansion coefficients a_n in (35) satisfy the following recursion relation [73]:

$$a_{n+1} = \frac{4\lambda^2 - 9 - 4(n+\lambda+2)}{2(n+\lambda+2)^2} a_{n-1}. \tag{36}$$

From the recursion relation (36) we can determine the coefficients a_n $(n \neq 0, 1)$ of the power series in terms of a_0 and a_1. In addition the above recursion relation implies that (35) yields one solution as a power series in even powers of ρ and another in odd powers of ρ.

In addition, (27) can be also mapped to the double-confluent Heun equation by appropriate function transformation [76]. So when $f(\rho)$ is taken as singular inverse-square-type potential, the solutions of (27) can be also given by the solution of the double-confluent Heun equation [76, 77].

5. The Solution with $f(\rho)$ to Be Exponential-Type Potential

The exponential-type potentials are very important in the study of various physical systems, particularly for modeling diatomic molecules. The typical exponential-type potentials include Eckart potentials [78], the Morse potential [79, 80], the Wood–Saxon potential [81], and Hulthén potential [82, 83]. The research work on the Dirac equation with the above potential is mainly concentrated on Minkowski time and space. However, it has been noticed recently that it is also interesting to study this kind quantum systems in a cosmic string background [84]. In this section we will take the $f(\rho)$ as exponential-type function and solve the corresponding Dirac equation in cosmic string space-time.

As is known to all, the Dirac equation and Schrödinger equation have been studied by resorting different methods. A usual way is transforming the eigenvalue equation of quantum system considered into a solvable equation via suitable variable substitutions and function transformations [85–87]. In order to obtain solution for $f(\rho)$ being exponential-type potential, we firstly consider the following linear second-order differential equation

$$x^2(1-x)^2 \frac{d^2 y}{dx^2} + x(1-x)^2 \frac{dy}{dx} - \left(\mathscr{L}_1 + \mathscr{L}_2 x - \mathscr{L}_3 x^2\right) y = 0, \tag{37}$$

where $\mathscr{L}_i$, $(i = 1, 2, 3)$ are constants. It is known that singular points of a differential equation determine the form of solutions. In this equation, there are two singular points, i.e., $x = 0$ and $x = 1$. In order to remove these singularities and get physically acceptable solutions we use the following ansatz:

$$y = x^{\Omega}(1-x)^{\Lambda} R(x). \tag{38}$$

where Λ and Ω are two real parameters. Further we make this two parameters to satisfy following relationships:

$$\Omega = \pm\sqrt{\mathscr{L}_1}, \quad \Lambda = \frac{1}{2}\left[1 \pm \sqrt{1 - 4\left(\mathscr{L}_3 - \mathscr{L}_2 - \mathscr{L}_1\right)}\right], \tag{39}$$

and by substituting (38) into (37), the differential equation for $\chi(x)$ can be written as

$$(1-x)x\frac{\partial^2}{\partial x^2}R(x) + [2\Omega + 1 - (2\Omega + 2\Lambda + 1)]x\frac{\partial}{\partial x}R(x) - (\Omega + \Lambda + \Delta)(\Omega + \Lambda - \Delta)R(x) = 0, \tag{40}$$

with $\Delta = \pm\sqrt{-\mathscr{L}_3}$. In other words, (37) is reduced to the well-known hypergeometric equation, when condition (39) is imposed. Making use of the boundary conditions at $r = 0$ and $r = \infty$ [87, 88], we can find the equation obeyed by the energy eigenvalue:

$$\Omega + \Lambda + \Delta = -n, \tag{41}$$

and the corresponding eigenfunctions is given in terms of the Gauss hypergeometric functions below

$$R(x) = AF(\tau_1, \tau_2; 1 + 2\Omega; x), \quad \tau_1 = \Omega + \Lambda + \Delta, \quad \tau_2 = \Omega + \Lambda - \Delta, \tag{42}$$

where A is normalization constant. Next, we will use the results given here to obtain the solutions of Dirac equation exponential-type interaction in cosmic string space-time. As a direct application of the above method, let us take the function $f(\rho)$ to be as Yukawa potential, Hulthén-Type potential, and generalized Morse potential, respectively.

Case 1 ($f(\rho)$ being Yukawa potential). In Yukawa meson theory, the Yukawa potential firstly was introduced to describe the interactions between nucleons [89]. Afterwards, it has been applied to many different areas of physics such as high-energy physics [90, 91], molecular physics [92], and plasma physics [93]. In recent years, the considerable efforts have also been made to study the bound state solutions by using different methods.

Now let us choose $f(\rho)$ to be Yukawa potential

$$f(\rho) = \frac{a}{\rho}e^{-\beta\rho}, \tag{43}$$

then $(*a)$ takes the form

$$\frac{d^2\chi}{d\rho^2} + \frac{1}{\rho}\frac{d\chi}{d\rho} - \left[\frac{\lambda^2}{\rho^2} + \frac{m\omega a}{\rho^2}e^{-\beta\rho} + \frac{\mu a + am\omega\beta}{\rho}e^{-\beta\rho} + \frac{a^2 m^2\omega^2}{\rho^2}e^{-2\beta\rho}\right]\chi + \nu\chi = 0. \tag{44}$$

However, the radial equation (44) cannot accept exact solution due to the presence of the centrifugal term [86]. In order to find analytical solution, we have to use some approximation approaches for the centrifugal term potential. Following [87], the approximation for the centrifugal term reads

$$\frac{1}{\rho^2} \approx \frac{\beta^2}{\left(1-e^{-\beta\rho}\right)^2}, \qquad \frac{1}{\rho} \approx \frac{\beta}{1-e^{-\beta\rho}}. \tag{45}$$

It is worth mentioning that the above approximations are valid for $\beta\rho \ll 1$ [87]. So if we make the control parameter β small enough, then we can guarantee that the above approximations in (45) hold for larger values ρ. In other words, this approximation (45) is valid in our work.

Using the approximation in (45) and setting

$$\chi = \frac{1}{\sqrt{\rho}}\Theta(\rho), \qquad x = e^{-\beta\rho}, \tag{46}$$

in the new function χ and new variable x, (44) becomes

$$x^2(1-x)^2\frac{d^2\Theta}{dx^2} + x(1-x)^2\frac{d\Theta}{dx} - \left(\mathscr{L}_1 + \mathscr{L}_2 x - \mathscr{L}_3 x^2\right)\Theta = 0, \tag{47}$$

where

$$\mathscr{L}_1 = -\lambda^2\beta^2 + m^2 + k^2 - E^2,$$
$$\mathscr{L}_2 = -m\omega a\beta^2 - a\mu\beta + 2m^2 + 2k^2 - 2E^2, \tag{48}$$
$$\mathscr{L}_3 = m^2\omega^2 a^2 - a\mu\beta - am\omega\beta^2 - m^2 - k^2 + E^2.$$

Comparing (47) with (37) and using the results given in (41) and (42), it is not difficult to find the equation obeyed by eigenvalues and eigenfunctions and they can be given, respectively,

$$E_n^2 - q_1 - \sqrt{q_2 - E_n^2} + \left[n + \frac{1}{2}\left(1 + \sqrt{1 + q_3 - 16E_n^2}\right)\right] = 0, \tag{49}$$

$$\Theta(\rho) = Ae^{-\beta\Omega\rho}\left(1-e^{-\beta\rho}\right)^{\Lambda} F\left(\tau_1, \tau_2; 1+2\Omega; e^{-\beta\rho}\right),$$

where

$$q_1 = \lambda^2\beta^2 - m^2 - k^2,$$
$$q_2 = m^2 + k^2 + a\mu\beta + am\omega\beta^2 - m^2\omega^2 a^2,$$
$$q_3 = 4\left(4m^2 + 4k^2 - am^2\omega^2\right), \tag{50}$$
$$\tau_1 = \Omega + \Lambda + \Delta,$$
$$\tau_2 = \Omega + \Lambda - \Delta.$$

Case 2 ($f(\rho)$ being Hulthén-type potential). In this section, we are interested in considering the Hulthén potential that describes the interaction between two atoms and has been used in different areas of physics and attracted a great of interest for some decades [82, 83, 94]. Next we take the interaction function $f(\rho)$ being Hulthén-Type potential

$$f(\rho) = a + \frac{be^{-\beta\rho}}{1-e^{-\beta\rho}}, \tag{51}$$

where a, b, and β arc real constants. Inserting (45) and (51) into $(*a)$, then $(*a)$ can written as

$$\frac{d^2\chi}{d\rho^2} + \frac{1}{\rho}\frac{d\chi}{d\rho} - \left[\frac{\lambda^2}{\rho^2} + \frac{\mu a}{\rho} + a^2m^2\omega^2 + \left(\frac{\mu b}{\rho} + 2abm^2\omega^2\right)\frac{e^{-\beta\rho}}{1-e^{-\beta\rho}}\right]\chi + \left(b^2m^2\omega^2 e^{-\beta\rho} + m\omega b\beta\right)\frac{e^{-\beta\rho}}{\left(1-e^{-\beta\rho}\right)^2}\chi + \nu\chi = 0. \tag{52}$$

In the same way as in previous section, taking into consideration approximation (45) for the centrifugal term and using the variable transformation $x = e^{-\beta\rho}$ and function transformation $\chi = (1/\sqrt{\rho})\Theta(\rho)$, (51) changes

$$x^2(1-x)^2\frac{d^2\Theta}{dx^2} + x(1-x)^2\frac{d\Theta}{dx} - \left(\mathscr{L}_1 + \mathscr{L}_2 x - \mathscr{L}_3 x^2\right)\Theta = 0, \tag{53}$$

where

$$\mathscr{L}_1 = \lambda^2\beta^2 + m^2\omega^2a^2 + \beta\mu a + m^2 + k^2 - E^2,$$
$$\mathscr{L}_2 = m\omega b\beta + (b-a)(\mu\beta - 2m\omega) + 2m^2 + 2k^2 - 2E^2, \tag{54}$$
$$\mathscr{L}_3 = -m^2\omega^2(a-b)^2 - m^2 - k^2 + E^2.$$

With the help of (38), (41), and (42), the solutions for $f(\rho)$ being Hulthén-Type potential can be easily obtained and the corresponding eigenvalues and eigenfunctions are, respectively,

$$E_n^2 - q_4 - \sqrt{q_5 - E_n^2} + \left[n + \frac{1}{2}\left(1 + \sqrt{1 + q_6 - 16E_n^2}\right)\right] = 0, \tag{55}$$

$$\Theta(\rho) = Ae^{-\beta\Omega\rho}\left(1-e^{-\beta\rho}\right)^{\Lambda} F\left(\tau_1, \tau_2; 1+2\Omega; e^{-\beta\rho}\right), \tag{56}$$

where

$$q_4 = \lambda^2\beta^2 + m^2\omega^2a^2 + \beta\mu a + m^2 + k^2,$$
$$q_5 = m^2\omega^2(a-b)^2 + m^2 + k^2,$$
$$q_6 = 4\left(m^2+k^2\right) + 4m^2\omega^2\left(2a^2 - 2ab + b^2\right) + 4\beta^2\lambda^2 + 4\beta\mu a + 4m\omega b\beta + 4(a-b)(\mu\beta - 2m\omega), \quad (57)$$
$$\tau_1 = \Omega + \Lambda + \Delta,$$
$$\tau_2 = \Omega + \Lambda - \Delta.$$

Case 3 ($f(\rho)$ being generalized Morse potential). The Morse potential [79, 80] as an important molecular potential describes the interaction between two atoms. We choose the interaction function $f(\rho)$ being generalized Morse potential

$$f(\rho) = \frac{a}{\rho^2}\left(e^{-\beta\rho} - e^{-2\beta\rho}\right). \quad (58)$$

As before, substitution of form (59) into (∗a) and straightforward calculation lead to the following equation:

$$\frac{d^2\chi}{d\rho^2} + \frac{1}{\rho}\frac{d\chi}{d\rho} - \left[\frac{\lambda^2}{\rho^2} + \frac{\mu a + 2m\omega a\mu}{\rho^3}\left(e^{-\beta\rho} - e^{-2\beta\rho}\right) - \frac{a\beta m\omega}{\rho^2}\left(2e^{-2\beta\rho} - e^{-\beta\rho}\right)\right]\chi + \frac{a^2m^2\omega^2}{\rho^4}\left(e^{-\beta\rho} - e^{-2\beta\rho}\right)^2\chi + \nu\chi = 0. \quad (59)$$

Letting $x = e^{-\beta\rho}$ and $\chi = (1/\sqrt{\rho})\Theta(\rho)$, the above differential equation (59) changes into the form

$$x^2(1-x)^2\frac{d^2\Theta}{dx^2} + x(1-x)^2\frac{d\Theta}{dx} - \left(\mathscr{L}_1 + \mathscr{L}_2x - \mathscr{L}_3x^2\right)\Theta = 0, \quad (60)$$

where

$$\mathscr{L}_1 = \lambda^2\beta^2 + m^2 + k^2 - E^2,$$
$$\mathscr{L}_2 = \mu a(1+2m\omega) + 2am\omega\beta - 2m^2 - 2k^2 + 2E^2, \quad (61)$$
$$\mathscr{L}_3 = E^2 - m^2\omega^2a^2 + 2m\omega a\beta - m^2 - k^2.$$

It is easy to see that the differential equation (60) is also similar to (37). So again according to the quantization condition (40) the corresponding expression of eigenvalues can be written as

$$E_n^2 - q_7 - \sqrt{q_8 - E_n^2} + \left[n + \frac{1}{2}\left(1 + \sqrt{1 + q_9 - 16E_n^2}\right)\right] = 0,$$
$$q_7 = \lambda^2\beta^2 + m^2 + k^2, \quad (62)$$
$$q_8 = m^2\omega^2a^2 - 2m\omega a\beta + m^2 + k^2,$$
$$q_9 = 4\left[\mu a(1-2m\omega) - 2m\omega a\beta + a^2m^2\omega^2 + \lambda^2\beta^2\right].$$

The wave function in this case read

$$\Theta(\rho) = Ae^{-\beta\Omega\rho}\left(1 - e^{-\beta\rho}\right)^{\Lambda}F\left(\tau_1, \tau_2; 1 + 2\Omega; e^{-\beta\rho}\right), \quad (63)$$

where

$$\tau_1 = \Omega + \Lambda + \Delta,$$
$$\tau_2 = \Omega + \Lambda - \Delta. \quad (64)$$

The above results show that the radial equation of the generalized Dirac oscillator with interaction function $f_\mu(x_\mu)$ to be taken as the exponential-type potential can be mapped into the well-known hypergeometric equation and the analytical solutions can have been found.

6. Conclusion

In this work, the generalized Dirac oscillator has been studied in the presence of the gravitational fields of a cosmic string. The corresponding radial equation of generalized Dirac oscillator is obtained. In our generalized Dirac oscillator model, we take the interaction function $f_\mu(x_\mu)$ to be as Cornell potential, Yukawa potential, generalized Morse potential, Hulthén-Type potential, and singular potential, respectively. By solving the corresponding wave equations the corresponding energy eigenvalues and the wave functions have been obtained and we have showed how the cosmic string leads to modifications in the spectrum and wave function. Based on consideration that Dirac oscillator has been studied extensively in high-energy physics, condensed matter physics, quantum optics, mathematical physics, and even connection with Higgs symmetry it also makes sense to generalize the generalized Dirac oscillator to these fields.

Conflicts of Interest

The authors declare that they have no conflicts of interest.

Acknowledgments

This work is supported by the National Natural Science Foundation of China (Grant no. 11565009).

References

[1] C. Quigg and J. L. Rosner, "Quantum mechanics with applications to quarkonium," *Physics Reports*, vol. 56, no. 4, pp. 167–235, 1979.

[2] H. Hassanabadi, S. Zarrinkamar, and A. A. Rajabi, "A simple efficient methodology for Dirac equation in minimal length quantum mechanics," *Physics Letters B*, vol. 718, no. 3, pp. 1111–1113, 2013.

[3] E. Hackmann, B. Hartmann, C. Lämmerzahl, and P. Sirimachan, "Test particle motion in the space-time of a Kerr black

hole pierced by a cosmic string," *Physical Review D: Particles, Fields, Gravitation and Cosmology*, vol. 82, no. 4, 2010.

[4] M. Moshinsky and A. Szczepaniak, "The Dirac oscillator," *Journal of Physics A: Mathematical and General*, vol. 22, no. 17, pp. L817–L819, 1989.

[5] K. Bakke and C. Furtado, "On the interaction of the Dirac oscillator with the Aharonov–Casher system in topological defect backgrounds," *Annals of Physics*, vol. 336, pp. 489–504, 2013.

[6] K. Bakke, "Rotating effects on the Dirac oscillator in the cosmic string spacetime," *General Relativity and Gravitation*, vol. 45, no. 10, pp. 1847–1859, 2013.

[7] D. Itô, K. Mori, and E. Carriere, "An example of dynamical systems with linear trajectory," *Nuovo Cimento A*, vol. 51, p. 1119, 1967.

[8] H. Hassanabadi, Z. Molaee, and S. Zarrinkamar, "DKP oscillator in the presence of magnetic field in (1+2)-dimensions for spin-zero and spin-one particles in noncommutative phase space," *The European Physical Journal C*, vol. 72, article 2217, 2012.

[9] O. Aouadi, Y. Chargui, and M. S. Fayache, "The Dirac oscillator in the presence of a chain of delta-function potentials," *Journal of Mathematical Physics*, vol. 57, no. 2, 023522, 10 pages, 2016.

[10] A. Albrecht, A. Retzker, and M. B. Plenio, "Testing quantum gravity by nanodiamond interferometry with nitrogen-vacancy centers," *Physical Review A: Atomic, Molecular and Optical Physics*, vol. 90, no. 3, Article ID 033834, 2014.

[11] M. Presilla, O. Panella, and P. Roy, "Quantum phase transitions of the Dirac oscillator in the anti-Snyder model," *Physical Review D: Particles, Fields, Gravitation and Cosmology*, vol. 92, no. 4, 2015.

[12] F. M. Andrade, E. O. Silva, Ferreira. M. M., and E. C. Rodrigues, "On the κ-Dirac Oscillator revisited," *Physics Letters B*, vol. 731, no. 15, pp. 327–330, 2014.

[13] H. Hassanabadi, S. S. Hosseini, A. Boumali, and S. Zarrinkamar, "The statistical properties of Klein-Gordon oscillator in noncommutative space," *Journal of Mathematical Physics*, vol. 55, no. 3, Article ID 033502, 2014.

[14] J. Munarriz, F Dominguez-Adame, and R. P. A. Lima, "Spectroscopy of the Dirac oscillator perturbed by a surface delta potential," *Physics Letters A*, vol. 376, no. 46, pp. 3475–3478, 2012.

[15] J. Grineviciute and D. Halderson, "Relativistic," *Physical Review C: Nuclear Physics*, vol. 85, no. 5, Article ID 054617, 2012.

[16] A. Faessler, V. I. Kukulin, and M. A. Shikhalev, "Description of intermediate- and short-range NN nuclear force within a covariant effective field theory," *Annals of Physics*, vol. 320, no. 1, pp. 71–107, 2005.

[17] V. M. Villalba, "Exact solution of the two-dimensional Dirac oscillator," *Physical Review A: Atomic, Molecular and Optical Physics*, vol. 49, no. 1, 586 pages, 1994.

[18] C. Quesne and V. M. Tkachuk, "Dirac oscillator with nonzero minimal uncertainty in position," *Journal of Physics A: Mathematical and General*, vol. 38, no. 8, pp. 1747–1765, 2005.

[19] P. L. Knight, "Quantum Fluctuations and Squeezing in the Interaction of an Atom with a Single Field Mode," *Physica Scripta*, vol. 12, pp. 51–55, 1986.

[20] P. Rozmej and R. Arvieu, "The Dirac oscillator. A relativistic version of the Jaynes-Cummings model," *Journal of Physics A: Mathematical and General*, vol. 32, no. 28, pp. 5367–5382, 1999.

[21] F. M. Toyama, Y. Nogami, and F. A. Coutinho, "Behaviour of wavepackets of the "DIRac oscillator": DIRac representation versus Foldy-Wouthuysen representation," *Journal of Physics A: Mathematical and General*, vol. 30, no. 7, pp. 2585–2595, 1997 (Arabic).

[22] A. Bermudez, M. A. Martin-Delgado, and E. Solano, "Exact mapping of the 2 + 1 Dirac oscillator onto the Jaynes-Cummings model: Ion-trap experimental proposal," *Physical Review A: Atomic, Molecular and Optical Physics*, vol. 76, no. 4, Article ID 041801, 2007.

[23] J. M. Torres, E. Sadurni, and T. H. Seligman, "Two interacting atoms in a cavity: exact solutions, entanglement and decoherence," *Journal of Physics A: Mathematical and General*, vol. 43, no. 19, 192002, 8 pages, 2010.

[24] A. Bermudez, M. A. Martin-Delgado, and E. Solano, "Mesoscopic Superposition States in Relativistic Landau Levels," *Physical Review Letters*, vol. 99, Article ID 123602, 2007.

[25] L. Lamata, J. León, T. Schätz, and E. Solano, "Dirac Equation and Quantum Relativistic Effects in a Single Trapped Ion," *Physical Review Letters*, vol. 98, no. 25, 2007.

[26] M. Hamzavi, M. Eshghi, and S. M. Ikhdair, "Effect of tensor interaction in the Dirac-attractive radial problem under pseudospin symmetry limit," *Journal of Mathematical Physics*, vol. 53, no. 082101, 2012.

[27] C.-K. Lu and I. F. Herbut, "Supersymmetric Runge-Lenz-Pauli vector for Dirac vortex in topological insulators and graphene," *Journal of Physics A: Mathematical and Theoretical*, vol. 44, no. 29, Article ID 295003, 2011.

[28] S. Zarrinkamar, A. A. Rajabi, and H. Hassanabadi, "Dirac equation in the presence of coulomb and linear terms in (1+ 1) dimensions; the supersymmetric approach," *Annals of Physics*, vol. 325, pp. 1720–1726, 2010.

[29] Y. Chargui, A. Trabelsi, and L. Chetouani, "Bound-states of the (1+1)-dimensional DKP equation with a pseudoscalar linear plus Coulomb-like potential," *Physics Letters A*, vol. 374, no. 29, pp. 2907–2913, 2010.

[30] A. Bermudez, M. A. Martin-Delgado, and A. Luis, "Chirality quantum phase transition in the Dirac oscillator," *Physical Review A: Atomic, Molecular and Optical Physics*, vol. 77, no. 6, 2008.

[31] D. A. Kulikov, R. S. Tutik, and A. P. Yaroshenko, "Relativistic two-body equation based on the extension of the *SL(2,C)* group," *Modern Physics Letters B*, vol. 644, no. 5-6, pp. 311–314, 2007.

[32] A. S. de Castro, P. Alberto, R. Lisboa, and M. Malheiro, "Relating pseudospin and spin symmetries through charge conjugation and chiral transformations: The case of the relativistic harmonic oscillator," *Physical Review C*, vol. 73, Article ID 054309, 2006.

[33] J. A. Franco-Villafañe, E. Sadurní, S. Barkhofen, U. Kuhl, F. Mortessagne, and T. H. Seligman, "First Experimental Realization of the Dirac Oscillator," *Physical Review Letters*, vol. 111, Article ID 170405, 2013.

[34] L. C. N. Santos and C. C. Barros, "Scalar bosons under the influence of noninertial effects in the cosmic string spacetime," *Eur. Phys. J. C*, vol. 77, no. 186, 2017.

[35] K. Bakke and C Furtado, "Anandan quantum phase for a neutral particle with Fermi–Walker reference frame in the cosmic string background," *The European Physical Journal C*, vol. 69, pp. 531–539, 2010.

[36] M. Hosseinpour, F. M. Andrade, E. O. Silva, and H. Hassanabadi, "Erratum to: Scattering and bound states for the Hulthén

potential in a cosmic string background," *The European Physical Journal C*, vol. 77, no. 270, 2017.

[37] M. Falek, M. Merad, and T. Birkandan, "Duffin-Kemmer-Petiau oscillator with Snyder-de SITter algebra," *Journal of Mathematical Physics*, vol. 58, no. 2, 023501, 13 pages, 2017.

[38] K. Bakke and J. Plus, "Noninertial effects on the Dirac oscillator in a topological defect spacetime," *Eur. Phys. J. Plus*, vol. 127, no. 82, 2012.

[39] F. M. Andrade and E. O. Silva, "Effects of spin on the dynamics of the 2D Dirac oscillator in the magnetic cosmic string background," *European Physical Journal C*, vol. 74, no. 3187, 2014.

[40] E. R. Figueiredo Medeiros and E. R. Bezerra de Mello, "Relativistic quantum dynamics of a charged particle in cosmic string spacetime in the presence of magnetic field and scalar potential," *The European Physical Journal C*, vol. 72, article 2051, 2012.

[41] L. B. Castro, "Quantum dynamics of scalar bosons in a cosmic string background," *The European Physical Journal C*, vol. 75, no. 287, 2015.

[42] K. Jusufi, "Noninertial effects on the quantum dynamics of scalar bosons," *The European Physical Journal C*, vol. 76, no. 61, 2016.

[43] G. A. Marques, V. B. Bezerra, and S. G. Fernandes, "Exact solution of the dirac equation for a coulomb and scalar potentials in the gravitational field of a cosmic string," *Physics Letters A*, vol. 341, no. 1-4, pp. 39–47, 2005.

[44] J. Carvalho, C. Furtado, and F. Moraes, "Dirac oscillator interacting with a topological defect," *Physical Review A: Atomic, Molecular and Optical Physics*, vol. 84, no. 3, Article ID 032109, 2011.

[45] D. Chowdhury and B. Basu, "Effect of a cosmic string on spin dynamics," *Physical Review D*, vol. 90, Article ID 125014, 2014.

[46] H. Hassanabadi, A. Afshardoost, and S. Zarrinkamar, "On the motion of a quantum particle in the spinning cosmic string space-time," *Annals of Physics*, vol. 356, pp. 346–351, 2015.

[47] N. Ferkous and A. Bounames, "Energy spectrum of a 2D Dirac oscillator in the presence of the Aharonov—Bohm effect," *Physics Letters A*, vol. 325, no. 1, pp. 21–29, 2004.

[48] H. F. Mota and K. Bakke, "Noninertial effects on the ground state energy of a massive scalar field in the cosmic string spacetime," *Physical Review D*, vol. 89, Article ID 027702, 2014.

[49] K. Bakke, "Torsion and noninertial effects on a nonrelativistic Dirac particle," *Annals of Physics*, vol. 346, pp. 51–58, 2014.

[50] P. Strange and L. H. Ryder, "The Dirac oscillator in a rotating frame of reference," *Physics Letters A*, vol. 380, no. 42, pp. 3465–3468, 2016.

[51] J. Anandan, "Gravitational and rotational effects in quantum interference," *Physical Review D*, vol. 15, no. 1448, 1977.

[52] K. Bakke and C. Furtado, "On the confinement of a Dirac particle to a two-dimensional ring," *Physics Letters A*, vol. 376, no. 15, pp. 1269–1273, 2012.

[53] M. J. Bueno, J. Lemos de Melo, C. Furtado, and A. M. de M. Carvalho, "Quantum dot in a graphene layer with topological defects," *The European Physical Journal Plus*, vol. 129, no. 9, 2014.

[54] J. Amaro Neto, M. J. Bueno, and C. Furtado, "Two-dimensional quantum ring in a graphene layer in the presence of a Aharonov-Bohm flux," *Annals of Physics*, vol. 373, pp. 273–285, 2016.

[55] Neto. Jose Amaro, J. R. de S, Claudio. Furtadoa, and Sergei. Sergeenkov, "Quantum ring in gapped graphene layer with wedge disclination in the presence of a uniform magnetic field," *European Physical Journal Plus 133*, vol. 185, 2018.

[56] D. Dutta, O. Panella, and P. Roy, "Pseudo-Hermitian generalized Dirac oscillators," *Annals of Physics*, vol. 331, pp. 120–126, 2013.

[57] H. P. Laba and V. M. Tkachuk, 2018, https://arxiv.org/pdf/1804.06091.pdf.

[58] E. R. Bezerra de Mello and A. A. Saharian, "Fermionic current induced by magnetic flux in compactified cosmic string spacetime," *The European Physical Journal C*, vol. 73, no. 8, 2013.

[59] S. Bellucci, E. R. Bezerra de Mello, A. de Padua, and A. A. Saharian, "Fermionic vacuum polarization in compactified cosmic string spacetime," *The European Physical Journal C*, vol. 74, no. 1, 2014.

[60] J. Domenech-Garret and M. Sanchis-Lozano, "Spectroscopy, leptonic decays and the nature of heavy quarkonia," *Physics Letters B*, vol. 669, no. 1, pp. 52–57, 2008.

[61] E. Eichten, K. Gottfried, T. Kinoshita, K. D. Lane, and T. -. Yan, "Erratum: Charmonium: The model," *Physical Review D: Particles, Fields, Gravitation and Cosmology*, vol. 21, no. 1, pp. 313-313, 1980.

[62] E. Eichten, K. Gottfried, T. Kinoshita, K. D. Lane, and T. M. Yan, "Charmonium: comparison with experiment," *Physical Review D: Particles, Fields, Gravitation and Cosmology*, vol. 21, no. 1, pp. 203–233, 1980.

[63] J. Carvalho, A. M. M. Carvalho, E. Cavalcante, and C. Furtado, "Klein–Gordon oscillator in Kaluza–Klein theory," *The European Physical Journal C*, vol. 76, no. 7, p. 365, 2016.

[64] F. Ahmed, "The energy–momentum distributions and relativistic quantum effects on scalar and spin-half particles in a Gödel-type space–time," *The European Physical Journal C*, vol. 78, no. 7, p. 598, 2018.

[65] C. P. Burgess, P. Hayman, M. Rummel, M. Williams, and L. Zalavári, "Point-Particle Effective Field Theory III: Relativistic Fermions and the Dirac Equation," *Journal of High Energy Physics*, vol. 04, no. 106, 2017.

[66] M. Bawin and S. A. Coon, "Singular inverse square potential, limit cycles, and self-adjoint extensions," *Physical Review A: Atomic, Molecular and Optical Physics*, vol. 67, no. 4, 2003.

[67] S. R. Beane, P. F. Bedaque, L. Childress, A. Kryjevski, J. McGuire, and U. van Kolck, "Singular potentials and limit cycles," *Physical Review A: Atomic, Molecular and Optical Physics*, vol. 64, no. 4, 2001.

[68] J. Lévy-Leblond, "Electron Capture by Polar Molecules," *Physical Review Journals Archive*, vol. 153, no. 1, pp. 1–4, 1967.

[69] U. van Kolck and Prog. Part, "Prog. Part. Nucl. Phys. 43, 409, 1999".

[70] E. Vogt and G. H. Wannier, "Scattering of ions by polarization forces," *Physical Review A: Atomic, Molecular and Optical Physics*, vol. 95, no. 5, pp. 1190–1198, 1954.

[71] D. Bouaziz and M. Bawin, "Singular inverse square potential in arbitrary dimensions with a minimal length: Application to the motion of a dipole in a cosmic string background," *Physical Review A*, vol. 78, 2010.

[72] D. Bouaziz and M. Bawin, "Singular inverse-square potential: Renormalization and self-adjoint extensions for medium to weak coupling," *Physical Review A*, vol. 89, Article ID 022113, 2014.

[73] G. R. Khan, "Exact solution of N-dimensional radial Schrodinger equation for the fourth-order inverse-power potential," *The European Physical Journal D. Atomic, Molecular, Optical and Plasma Physics*, vol. 53, no. 2, pp. 123–125, 2009.

[74] S.-H. Dong, Z.-Q. Ma, and G. Esposito, "Exact solutions of the Schrödinger equation with inverse-power potential," *Foundations of Physics Letters*, vol. 12, no. 5, pp. 465–474, 1999.

[75] M. Hosseinpour, H. Hassanabadi, and M. Fabiano, "The DKP oscillator with a linear interaction in the cosmic string space-time," *Eur. Phys. J. C*, vol. 78, no. 93, 2018.

[76] B. D. B. Figueiredo, "Ince's limits for confluent and double-confluent Heun equations," *Journal of Mathematical Physics*, vol. 46, no. 11, Article ID 113503, 2005.

[77] L. J. El-Jaick and B. D. Figueiredo, "Solutions for confluent and double-confluent Heun equations," *Journal of Mathematical Physics*, vol. 49, no. 8, 083508, 28 pages, 2008.

[78] S.-H. Dong, W.-C. Qiang, G.-H. Sun, and V. B. Bezerra, "Analytical approximations to the l-wave solutions of the Schrodinger equation with the Eckart potential," *Journal of Physics A: Mathematical and General*, vol. 40, no. 34, pp. 10535–10540, 2007.

[79] P. Malik, "Application of the exact quantization rule to the relativistic solution of the rotational Morse potential with pseudospin symmetry," *Journal of Physics A*, vol. 40, pp. 1677–1685, 2007.

[80] O. Bayrak and I. Boztosun, "The pseudospin symmetric solution of the Morse potential for any κ state," *Journal of Physics A: Mathematical and Theoretical*, vol. 40, no. 36, pp. 11119–11127, 2007.

[81] O. C. Aydoğdu and R. Sever, "Pseudospin and spin symmetry in the Dirac equation with Woods-Saxon potential and tensor potential," *The European Physical Journal A*, vol. 43, article 73, 2010.

[82] A. Soylu, O. Bayrak, and I. Boztosun, "An approximate solution of Dirac-Hulthén problem with pseudospin and spin symmetry for any κ state," *Journal of Mathematical Physics*, vol. 48, Article ID 082302, 2007.

[83] G. Chen, "Spinless particle in the generalized hulthén potential," *Modern Physics Letters A*, vol. 19, no. 26, pp. 2009–2012, 2004.

[84] M. Hosseinpoura, F. M. Andradeb, E. O. Silvac, and H. Hassanabadid, "Scattering and bound states of Dirac Equation in presence of cosmic string for Hulthén potential," *Eur. Phys. J. C*, vol. 77, no. 270, 2017.

[85] J. J. Pena, J. Morales, and J. Garcia-Ravelo, "Bound state solutions of Dirac equation with radial exponential-type potentials," *Journal of Mathematical Physics*, vol. 58, Article ID 043501, 2017.

[86] F. J. Ferreira and V. B. Bezerra, "Some remarks concerning the centrifugal term approximation," *Journal of Mathematical Physics*, vol. 58, no. 10, 102104, 12 pages, 2017.

[87] A. Tas, O. Aydogdu, and M. Salti, "Dirac particles interacting with the improved Frost-Musulin potential within the effective mass formalism," *Annals of Physics*, vol. 379, pp. 67–82, 2017.

[88] B. C. Lutfuğlu, F. Akdeniz, and O. Bayrak, "Scattering, bound, and quasi-bound states of the generalized symmetric Woods-Saxon potential," *Journal of Mathematical Physics*, vol. 57, Article ID 032103, 2016.

[89] H. Yukawa, "On the Interaction of Elementary Particles. I," in *Proceedings of the Physico-Mathematical Society of Japan. 3rd Series*, vol. 17, pp. 48–57, 1935.

[90] A. Loeb and N. Weiner, "Cores in Dwarf Galaxies from Dark Matter with a Yukawa Potential," *Physical Review Letters*, vol. 106, Article ID 171302, 2011.

[91] F. Pakdel, A. A. Rajabi, and M. Hamzavi, "Scattering and Bound State Solutions of the Yukawa Potential within the Dirac Equation," *Advances in High Energy Physics*, vol. 2014, Article ID 867483, 7 pages, 2014.

[92] H. Bahlouli, M. S. Abdelmonem, and S. M. Al-Marzoug, "Analytical treatment of the oscillating Yukawa potential," *Chemical Physics*, vol. 393, no. 1, pp. 153–156, 2012.

[93] A. D. Alhaidari, H. Bahlouli, and M. S. Abdelmonem, "Taming the Yukawa potential singularity: improved evaluation of bound states and resonance energies," *Journal of Physics A: Mathematical and General*, vol. 41, no. 3, 032001, 9 pages, 2008.

[94] Altuğ Arda, "Solution of Effective-Mass Dirac Equation with Scalar-Vector and Pseudoscalar Terms for Generalized Hulthén Potential," *Advances in High Energy Physics*, vol. 2017, Article ID 6340409, 9 pages, 2017.

15

Chemical Potentials of Light Flavor Quarks from Yield Ratios of Negative to Positive Particles in Au+Au Collisions at RHIC

Ya-Qin Gao,[1] Hai-Ling Lao,[2] and Fu-Hu Liu[2]

[1]*Department of Physics, Taiyuan University of Science and Technology, Taiyuan, Shanxi 030024, China*
[2]*Institute of Theoretical Physics & State Key Laboratory of Quantum Optics and Quantum Optics Devices, Shanxi University, Taiyuan, Shanxi 030006, China*

Correspondence should be addressed to Ya-Qin Gao; gaoyaqin@tyust.edu.cn and Fu-Hu Liu; fuhuliu@163.com

Academic Editor: Xiaochun He

The transverse momentum spectra of π^-, π^+, K^-, K^+, $\bar{p}$, and p produced in Au+Au collisions at center-of-mass energy $\sqrt{s_{NN}}$ = 7.7, 11.5, 19.6, 27, 39, 62.4, 130, and 200 GeV are analyzed in the framework of a multisource thermal model. The experimental data measured at midrapidity by the STAR Collaboration are fitted by the (two-component) standard distribution. The effective temperature of emission source increases obviously with the increase of the particle mass and the collision energy. At different collision energies, the chemical potentials of up, down, and strange quarks are obtained from the antiparticle to particle yield ratios in given transverse momentum ranges available in experiments. With the increase of logarithmic collision energy, the chemical potentials of light flavor quarks decrease exponentially.

1. Introduction

The constructions of the Relativistic Heavy Ion Collider (RHIC) and the Large Hadron Collider (LHC) have been opening a new epoch for the studies of nuclear and quark matters. One of the major goals of the RHIC and LHC studies is to obtain information on the quantum chromodynamics (QCD) phase diagram [1]. The phase diagram includes at least a fundamental phase transition between the hadron gas and the quark-gluon plasma (QGP) or quark matter and is usually plotted as chemical freeze-out temperature (T_{ch}) versus baryon chemical potential (μ_{baryon}). Nowadays, the detailed characteristics of the phase diagram are not known yet. The experimental and theoretical nuclear physicists have been focusing their attentions on the searching for the critical end point and phase boundary. Lattice QCD calculations show that a system is produced at small μ_{baryon} or high energies through a crossover at the quark-hadron phase transition [2–4]. Based on the lattice QCD [5] and several QCD-based models calculations [6–9], as well as mathematical extensions of lattice techniques [10–13], researchers suggest that the transition at larger μ_{baryon} is the first order and the QCD critical end point is existent.

Pinpointing the phase boundary and the critical end point is the central issue to understand the properties of interacting matter under extreme conditions and to map the QCD phase diagram. The matter produced in high-energy heavy-ion collisions provides the opportunity to search for the phase boundary and the critical end point [6, 14]. To this end, the STAR Collaboration at the RHIC has undertaken the first phase of the beam energy scan (BES) program [15–17], starting the second phase from 2018 to 2019 [18]. The program is to vary the collision energy which enables a search for nonmonotonic excitation functions over a broad domain of the phase diagram. Before looking for an evidence for the existence of a critical end point and the phase boundary, it is important to know the (T_{ch}, μ_{baryon}) region of phase diagram one can access. The produced particles spectra and yield ratios allow us only to infer the values of T_{ch} and μ_{baryon} [19]. Furthermore, the bulk properties such as rapidity density dN/dy, mean transverse momentum $\langle p_T \rangle$, particle ratios, and freeze-out properties may provide an insight into the particle production mechanisms at BES energies. Therefore, it is very important to study these bulk properties systematically, which may reveal the evolution and the changes of the system created in high-energy heavy-ion collisions.

As one of the most important measured quantities, the transverse momentum (p_T) spectrum includes abundant information which is related to the excitation degree of the collision system. The spectra of identified particles can also provide useful information about temperature, particle ratio, and chemical potential by using thermal and statistical investigations [20]. For any system, one can determine the direction and limitation of mass transfer by comparing the chemical potentials of particles; that is to say, the chemical potential is a sign to mark the direction of spontaneous chemical reaction. The chemical potential can also be a criterion for determining whether thermodynamic equilibrium does exist in the interacting region in high-energy collisions [1]. Generally, a low absolute value of chemical potential corresponds to a high degree of thermodynamic equilibrium. Therefore, the chemical potential is also one of the major solutions for investigating the QGP. One can see that the chemical potentials of quarks are an important subject at high energy. Therefore, we are very interested in measuring the chemical potentials of quarks.

In this paper, we extract the chemical potentials of light flavor quarks from the yield ratios of negatively to positively charged particles. By using the (two-component) standard distribution, the p_T spectra of π^-, π^+, K^-, K^+, $\overline{p}$, and p produced in Au+Au collisions at center-of-mass energy (per nucleon pair) $\sqrt{s_{NN}}$ = 7.7, 11.5, 19.6, 27, 39, 62.4, 130, and 200 GeV measured by the STAR Collaboration in midrapidity interval ($|y| < 0.1$) [19, 21] are described. The considered energies stretch across a wide energy range which covers the main range of the RHIC at its BES.

2. The Model and Method

To extract the chemical potentials of quarks, we need to know the yield ratios of negatively to positively charged particles. Although we can have the values of yield ratios directly in experiments, they are not complete and comprehensive in some cases. Usually, the p_T spectra of charged particles are given in many experiments and we can get the yield ratios by fitting the available data. Then, the values of chemical potentials for the up, down, and strange quarks can be obtained from the yield ratios π^-/π^+, K^-/K^+, and $\overline{p}/p$ which are synthetically considered in special ways.

In this paper, the p_T spectra are analyzed in the framework of a multisource thermal model [22], which assumes that various sources are involved in high-energy collisions. These sources are divided into few groups by different interaction mechanisms, geometrical relations, or event samples. Each group of sources forms a relatively large emission source which stays in a local thermal equilibrium state at the chemical or kinetic freeze-out. Each emission source is considered to emit particles in its rest frame and treated as a thermodynamic system of relativistic and quantum ideal gas. This means that each emission source can be described by the thermal and statistical model or other similar models and distributions. The final-state distribution is attributed to all sources in the whole system, which results in a multi-characteristic emission process [22] if we use the standard distribution [23–26]. This also means that p_T spectrum can be described by a multicomponent standard distribution in which each component describes a given emission source.

We now structure the multicomponent standard distribution. It is assumed that there are l components to be considered. For the i-th component, the standard Boltzmann, Fermi-Dirac, and Bose-Einstein distributions [23–26] can be uniformly expressed as

$$\begin{aligned} f_i(p_T) &= \frac{1}{N}\frac{dN}{dp_T} \\ &= C_i p_T \sqrt{p_T^2 + m_0^2} \int_{y_{\min}}^{y_{\max}} \cosh y \\ &\quad \times \left[\exp\left(\frac{\sqrt{p_T^2 + m_0^2}\cosh y - \mu}{T_i} \right) + S \right]^{-1} dy, \end{aligned} \quad (1)$$

where C_i is the normalization constant which results in $\int_0^\infty f_i(p_T)dp_T = 1$; N, m_0, μ, and T_i denote the particle number, the rest mass of the considered particle, the chemical potential of the considered particle, and the effective temperature for the i-th component, respectively; $y_{\min}$ is the minimum rapidity and $y_{\max}$ is the maximum rapidity; the values of S are 0, +1, and −1, which denote the Boltzmann, Fermi-Dirac, and Bose-Einstein distributions, respectively. We neglect the existence of μ in (1) due to the fact that it has mainly effect on the normalization which can be redone, but not the trend of curve.

In the final state, p_T spectrum is resulting from l components; that is,

$$f(p_T) = \frac{1}{N}\frac{dN}{dp_T} = \sum_{i=1}^{l} w_i f_i(p_T), \quad (2)$$

where w_i ($i = 1, 2, \ldots, l$) is the relative weight resulting from the i-th component. Because of the probability distribution being acquiescently normalized to 1, the coefficient obeys the normalization condition of $\sum w_i = 1$. Considering the relative contribution of each component, we have the mean effective temperature to be $T_{eff} = \sum w_i T_i$, which reflects the mean excitation degree of different sources corresponding to different components and can be used to describe the effective temperature of whole interacting system. It should be noted that the effective temperature contains the contributions of transverse flow and thermal motion. It is not the "real" temperature of the interacting system.

According to [27, 28], the relation between antiproton to proton yield ratios can be written as

$$\frac{\overline{p}}{p} = \exp\left(-\frac{2\mu_p}{T_{ch}}\right) \approx \exp\left(-\frac{2\mu_{baryon}}{T_{ch}}\right), \quad (3)$$

where μ_p denotes the chemical potential of proton. In the framework of the statistical thermal model of noninteracting gas particles with the assumption of standard Maxwell-Boltzmann statistics, there is an empirical expression for T_{ch} [29–32]; one has

$$T_{ch} = T_{\lim} \frac{1}{1 + \exp\left[2.60 - \ln\left(\sqrt{s_{NN}}\right)/0.45\right]}, \quad (4)$$

where $\sqrt{s_{NN}}$ is in the units of GeV and the "limiting" temperature $T_{\lim} = 0.164$ GeV [29, 30].

In a similar way, the yield ratios of antiparticles to particles for other hadrons can be written as

$$\begin{aligned} k_\pi &\equiv \frac{\pi^-}{\pi^+} = \exp\left(-\frac{2\mu_\pi}{T_{ch}}\right), \\ k_K &\equiv \frac{K^-}{K^+} = \exp\left(-\frac{2\mu_K}{T_{ch}}\right), \\ k_p &\equiv \frac{\overline{p}}{p} = \exp\left(-\frac{2\mu_p}{T_{ch}}\right), \\ k_D &\equiv \frac{D^-}{D^+} = \exp\left(-\frac{2\mu_D}{T_{ch}}\right), \\ k_B &\equiv \frac{B^-}{B^+} = \exp\left(-\frac{2\mu_B}{T_{ch}}\right), \end{aligned} \quad (5)$$

where k_j ($j = \pi, K, p, D$, and B) denote the yield ratios of negatively to positively charged particles obtained from the normalization constants of p_T spectra. The symbols μ_π, μ_K, μ_D, and μ_B represent the chemical potentials of π, K, D, and B, respectively. In the above discussion, the symbol of a given particle is used for its yield for the purpose of simplicity. Furthermore, we have

$$\begin{aligned} \mu_\pi &= -\frac{1}{2} T_{ch} \cdot \ln\left(k_\pi\right), \\ \mu_K &= -\frac{1}{2} T_{ch} \cdot \ln\left(k_K\right), \\ \mu_p &= -\frac{1}{2} T_{ch} \cdot \ln\left(k_p\right), \\ \mu_D &= -\frac{1}{2} T_{ch} \cdot \ln\left(k_D\right), \\ \mu_B &= -\frac{1}{2} T_{ch} \cdot \ln\left(k_B\right). \end{aligned} \quad (6)$$

Let μ_q denote the chemical potential for quark flavor, where $q = u, d, s, c$, and b represent the up, down, strange, charm, and bottom quarks, respectively. In principle, we can use k_j to give relations among different μ_q. The values of μ_q are then expected from these relations. According to [33, 34], based on the same T_{ch}, k_j in terms of μ_q are

$$\begin{aligned} k_\pi &= \frac{\exp\left[-\left(\mu_u - \mu_d\right)/T_{ch}\right]}{\exp\left[\left(\mu_u - \mu_d\right)/T_{ch}\right]} = \exp\left[-\frac{2\left(\mu_u - \mu_d\right)}{T_{ch}}\right], \\ k_K &= \frac{\exp\left[-\left(\mu_u - \mu_s\right)/T_{ch}\right]}{\exp\left[\left(\mu_u - \mu_s\right)/T_{ch}\right]} = \exp\left[-\frac{2\left(\mu_u - \mu_s\right)}{T_{ch}}\right], \\ k_p &= \frac{\exp\left[-\left(2\mu_u + \mu_d\right)/T_{ch}\right]}{\exp\left[\left(2\mu_u + \mu_d\right)/T_{ch}\right]} \\ &= \exp\left[-\frac{2\left(2\mu_u + \mu_d\right)}{T_{ch}}\right], \\ k_D &= \frac{\exp\left[-\left(\mu_c - \mu_d\right)/T_{ch}\right]}{\exp\left[\left(\mu_c - \mu_d\right)/T_{ch}\right]} = \exp\left[-\frac{2\left(\mu_c - \mu_d\right)}{T_{ch}}\right], \\ k_B &= \frac{\exp\left[-\left(\mu_u - \mu_b\right)/T_{ch}\right]}{\exp\left[\left(\mu_u - \mu_b\right)/T_{ch}\right]} = \exp\left[-\frac{2\left(\mu_u - \mu_b\right)}{T_{ch}}\right]. \end{aligned} \quad (7)$$

Thus, we have

$$\begin{aligned} \mu_u &= -\frac{1}{6} T_{ch} \cdot \ln\left(k_\pi \cdot k_p\right), \\ \mu_d &= -\frac{1}{6} T_{ch} \cdot \ln\left(k_\pi^{-2} \cdot k_p\right), \\ \mu_s &= -\frac{1}{6} T_{ch} \cdot \ln\left(k_\pi \cdot k_K^{-3} \cdot k_p\right), \\ \mu_c &= -\frac{1}{6} T_{ch} \cdot \ln\left(k_\pi^{-2} \cdot k_p \cdot k_D^{3}\right), \\ \mu_b &= -\frac{1}{6} T_{ch} \cdot \ln\left(k_\pi \cdot k_p \cdot k_B^{-3}\right). \end{aligned} \quad (8)$$

As can be seen from (8), μ_q are obtained from k_j. In addition to the yield ratios π^-/π^+, K^-/K^+, and $\overline{p}/p$, other combinations can also give μ_q if the spectra in the numerator and denominator are under the same experimental conditions.

3. Results and Discussion

The energy dependent double-differential p_T spectra of π^-, π^+, K^-, K^+, $\overline{p}$, and p produced in central Au+Au collisions at $\sqrt{s_{NN}}$ = 7.7, 11.5, 19.6, 27, 39, 62.4, 130, and 200 GeV at the midrapidity $|y| < 0.1$ are presented in Figure 1, where the centrality interval at 130 GeV is 0–6% and at other energies is 0–5%. The different symbols represent the data measured by the STAR Collaboration [19, 21], and the curves are the results fitted here by the (two-component) standard distribution. Generally, the standard distribution is firstly used in the fit process. If it does not fit the data, the two-component standard distribution is used. It is because of the quality of the measurements that (two-component) standard distribution is used. In the case of using the two-component standard distribution, the first component results in narrow p_T region and the second component results in wide p_T regions. That is, in low p_T region both components contribute to the spectra, and in high p_T region only the second component contributes to the spectra. In the calculation, the values of the free parameters (T_1, w_1, and T_2), the normalization constant (N_0), and χ^2 obtained by fitting the data are listed in Table 1 including the degrees of freedom (dof). One can see that the data are well fitted by the (two-component) standard distribution. From the parameter values, one can see that

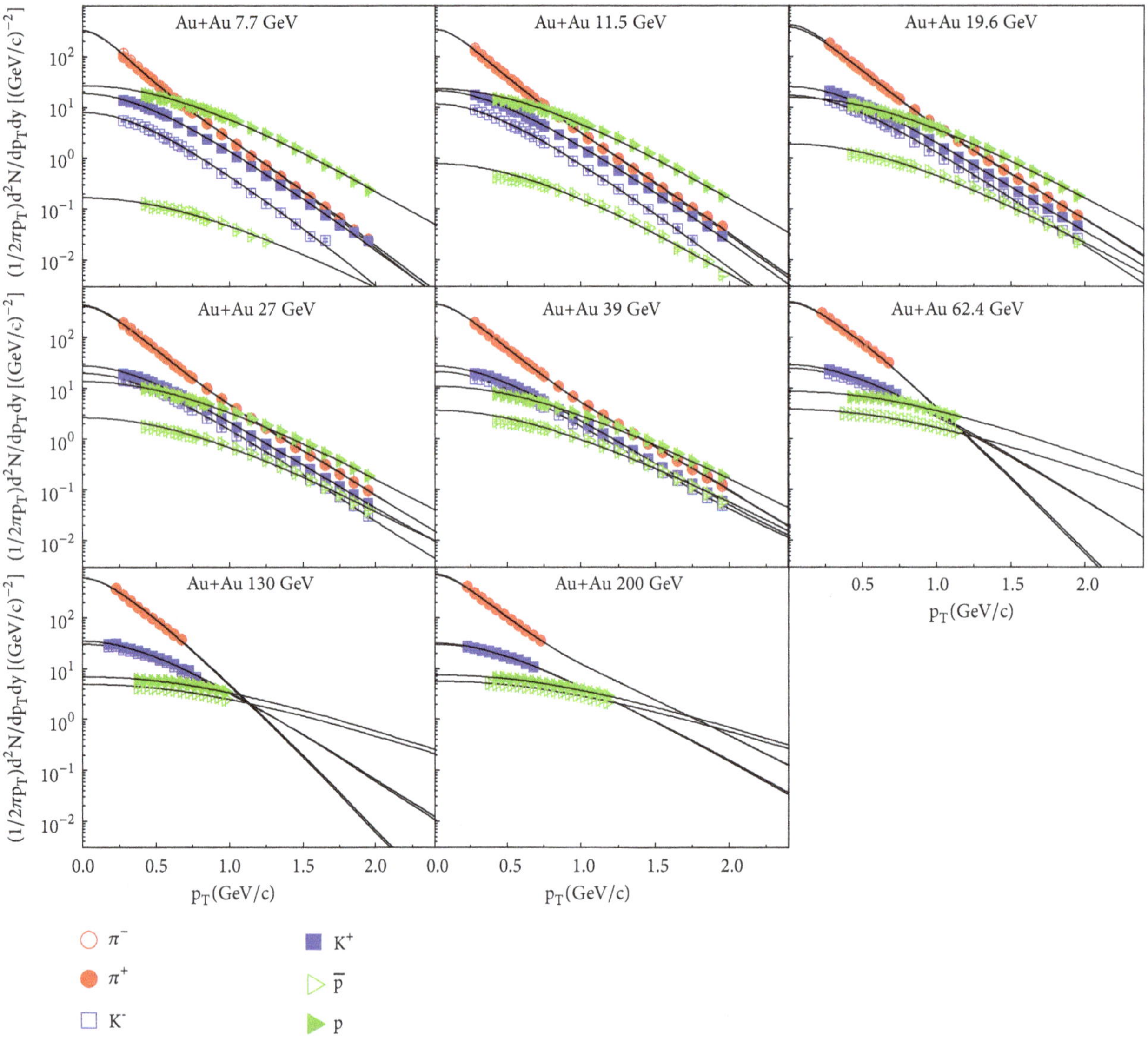

FIGURE 1: Midrapidity ($|y| < 0.1$) double-differential p_T spectra for π^-, π^+, K^-, K^+, $\bar{p}$, and p in central Au+Au collisions at $\sqrt{s_{NN}}$ = 7.7, 11.5, 19.6, 27, 39, 62.4, 130, and 200 GeV, where the centrality interval at 130 GeV is 0–6% and at other energies is 0–5%. The different symbols represent the measurements done by the STAR experiment [19, 21] and the curves represent the results fitted by the (two-component) standard distribution. The values of parameters can be found in Table 1.

the effective temperature increases with the increase of the particle mass and the collision energy for emissions of the six types of particles.

Based on the above successful fits of the p_T spectra of antiparticles and particles, we can use (8) and the p_T spectra in Figure 1 to study the dependence of μ_q on $\sqrt{s_{NN}}$. This is done by integrating the yield over the given p_T ranges available in experiments at different energies. Figure 2 shows the correlations between μ_q and $\sqrt{s_{NN}}$, where μ is used on the vertical axis to replace μ_q which are marked in the panel for different styles of symbols. With the increase of logarithmic $\sqrt{s_{NN}}$, an exponential decrease of μ_q is observed. Corresponding to the solid, dashed, and dotted curves which fit to the dependences of μ_u, μ_d, and μ_s on $\sqrt{s_{NN}}$, respectively, we have

$$\mu_u = (820.1 \pm 0.1)\left(\sqrt{s_{NN}}\,[\text{GeV}]\right)^{-(0.914\pm0.025)}\ \text{MeV},$$
$$\mu_d = (681.1 \pm 0.1)\left(\sqrt{s_{NN}}\,[\text{GeV}]\right)^{-(0.834\pm0.031)}\ \text{MeV}, \quad (9)$$
$$\mu_s = (420.7 \pm 0.2)\left(\sqrt{s_{NN}}\,[\text{GeV}]\right)^{-(1.004\pm0.063)}\ \text{MeV}$$

with the χ^2/dof = 1.67/6, 2.73/6, and 1.04/6, respectively.

The similarity in up and down quark masses renders the similarity in their chemical potentials. The difference between the chemical potentials of up (or down) and strange quarks is caused by the difference between their masses. At

Table 1: Values of T_1, w_1, T_2, N_0, χ^2, and dof corresponding to the curves in Figure 1. The values for positively and negatively charged particles are given using the slash (/), where the values for positive particles are shown before the slash and the values for negative particles after the slash. The values of N_0 are obtained due to the comparisons between the experimental $(1/2\pi p_T)d^2N/dp_T dy$ and the calculated $(1/2\pi p_T)N_0 f(p_T)/dy$, where $dy = 0.2$ and $f(p_T)$ is presented in (2).

$\sqrt{s_{NN}}$ (GeV)	Particles	T_1 (MeV)	w_1	T_2 (MeV)	N_0	χ^2	dof
7.7	$\pi^{\pm}$	$87 \pm 7/89 \pm 10$	$0.65 \pm 0.04/0.63 \pm 0.07$	$181 \pm 2/177 \pm 4$	$81.0 \pm 4.0/81.0 \pm 9.0$	0.38/0.45	22/22
	$K^{\pm}$	$185 \pm 6/170 \pm 8$	1.00/1.00	–	$9.5 \pm 0.8/3.8 \pm 0.5$	0.75/0.98	21/21
	$p/\bar{p}$	$224 \pm 7/257 \pm 20$	1.00/1.00	–	$19.0 \pm 0.2/0.13 \pm 0.01$	1.15/0.52	27/13
11.5	$\pi^{\pm}$	$102 \pm 10/105 \pm 10$	$0.70 \pm 0.04/0.69 \pm 0.03$	$193 \pm 4/190 \pm 4$	$92.0 \pm 8.0/93.0 \pm 8.0$	0.25/0.48	22/22
	$K^{\pm}$	$190 \pm 5/178 \pm 4$	1.00/1.00	–	$11.0 \pm 0.8/5.7 \pm 0.5$	0.49/0.97	23/21
	$p/\bar{p}$	$217 \pm 10/216 \pm 20$	1.00/1.00	–	$16.0 \pm 3.0/0.54 \pm 0.10$	1.70/3.07	26/21
19.6	$\pi^{\pm}$	$118 \pm 10/116 \pm 10$	$0.79 \pm 0.02/0.83 \pm 0.02$	$215 \pm 8/219 \pm 7$	$109.0 \pm 10.0/119.0 \pm 10.0$	0.19/0.40	22/22
	$K^{\pm}$	$181 \pm 7/181 \pm 5$	$0.88 \pm 0.04/0.89 \pm 0.05$	$260 \pm 20/239 \pm 20$	$13.0 \pm 1.4/8.7 \pm 0.8$	0.55/1.61	22/22
	$p/\bar{p}$	$234 \pm 20/237 \pm 10$	1.00/1.00	–	$11.5 \pm 2.0/1.4 \pm 0.1$	0.69/19.81	27/20
27	$\pi^{\pm}$	$117 \pm 8/118 \pm 7$	$0.79 \pm 0.03/0.81 \pm 0.03$	$219 \pm 7/221 \pm 7$	$120.0 \pm 10.0/125.0 \pm 10.0$	0.19/0.27	22/22
	$K^{\pm}$	$178 \pm 7/180 \pm 6$	$0.85 \pm 0.03/0.84 \pm 0.06$	$262 \pm 20/236 \pm 10$	$14.0 \pm 1.0/10.0 \pm 1.0$	0.27/0.92	22/21
	$p/\bar{p}$	$239 \pm 7/247 \pm 10$	1.00/1.00	–	$10.0 \pm 2.0/2.0 \pm 0.2$	0.84/1.28	21/20
39	$\pi^{\pm}$	$114 \pm 10/116 \pm 10$	$0.78 \pm 0.02/0.78 \pm 0.02$	$222 \pm 7/222 \pm 6$	$129.0 \pm 12.0/128.0 \pm 10.0$	0.39/0.20	22/22
	$K^{\pm}$	$189 \pm 7/189 \pm 5$	$0.95 \pm 0.01/0.96 \pm 0.02$	$332 \pm 18/347 \pm 23$	$14.0 \pm 1.5/11.0 \pm 1.0$	0.20/0.36	22/22
	$p/\bar{p}$	$250 \pm 20/252 \pm 20$	1.00/1.00	–	$8.3 \pm 1.0/2.8 \pm 0.3$	0.81/2.42	20/21
62.4	$\pi^{\pm}$	$139 \pm 6/137 \pm 8$	1.00/1.00	–	$146.0 \pm 10.0/150.0 \pm 10.0$	1.14/1.07	8/8
	$K^{\pm}$	$210 \pm 25/214 \pm 25$	1.00/1.00	–	$15.8 \pm 1.0/13.6 \pm 0.8$	0.02/0.37	8/8
	$p/\bar{p}$	$335 \pm 20/346 \pm 40$	1.00/1.00	–	$8.3 \pm 0.4/3.8 \pm 0.3$	1.97/1.60	13/14
130	$\pi^{\pm}$	$136 \pm 8/137 \pm 7$	1.00/1.00	–	$181.0 \pm 13.0/185.0 \pm 11.0$	1.98/3.68	8/8
	$K^{\pm}$	$204 \pm 12/210 \pm 13$	1.00/1.00	–	$18.5 \pm 1.4/17.3 \pm 1.3$	0.39/0.20	11/11
	$p/\bar{p}$	$373 \pm 15/385 \pm 22$	1.00/1.00	–	$7.5 \pm 0.7/5.5 \pm 0.3$	1.73/0.47	11/11
200	$\pi^{\pm}$	$116 \pm 8/115 \pm 5$	$0.76 \pm 0.03/0.76 \pm 0.03$	$263 \pm 23/262 \pm 25$	$209.0 \pm 14.0/215.0 \pm 12.0$	0.21/0.17	7/7
	$K^{\pm}$	$235 \pm 30/239 \pm 30$	1.00/1.00	–	$19.3 \pm 1.1/18.5 \pm 1.0$	0.06/0.06	8/8
	$p/\bar{p}$	$382 \pm 40/393 \pm 35$	1.00/1.00	–	$8.3 \pm 0.9/6.4 \pm 0.6$	0.10/0.25	14/15

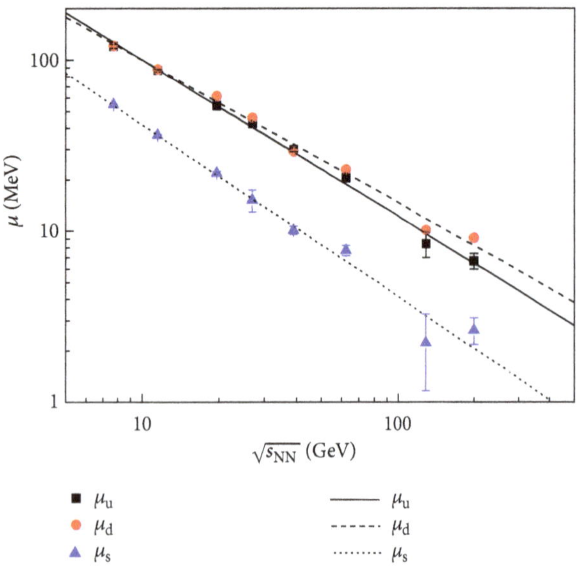

FIGURE 2: Correlations between μ_q and $\sqrt{s_{NN}}$ for central Au+Au collisions at RHIC. The symbols represent μ_q obtained from the ratios by integrating the yield over the given p_T ranges available in experiments in Figure 1. The solid, dashed, and dotted curves are fitted results corresponding to μ_u, μ_d, and μ_s, respectively.

the lowest BES energy the difference between the chemical potentials is dozens of MeV, while at the highest RHIC energy these quantities are around a few MeV. The decrease in μ_q is obvious, which indicates the change of mean free path of produced quarks in the middle state. If the produced quarks at the lowest BES energy have a small mean free path which looks as if a liquid-like middle state is formed, the produced quarks at the highest RHIC energy should have a large mean free path which looks as if a gas-like middle state is formed. The main difference at different energies is different mean free paths of the produced quarks. To search for the critical energy at which the change from a liquid-like middle state to a gas-like middle state had happenned is beyond the focus of the present work.

From (9) we can obtain a linear relation between $\ln \mu_q$ and $\ln \sqrt{s_{NN}}$,

$$\ln \mu_q = a - |b| \cdot \ln \sqrt{s_{NN}}, \qquad (10)$$

where the intercept a and slope $-|b|$ can be obtained from the parameters in (9). In particular, $-|b|$ is close to -1. The large negative slope shows an obvious anticorrelation between $\ln \mu_q$ and $\ln \sqrt{s_{NN}}$. It is expected that $\ln \mu_q$ will be smaller at higher energy or larger at lower energy. In particular, at the LHC energies, $\ln \mu_q$ will be negative since μ_q will be less than 1 MeV. The limiting value of μ_q is close to 0 at the LHC [35], which results in an obvious negative $\ln \mu_q$.

The main conclusion observed from Figure 2 is that μ_q is high (from dozens of MeV to ~100 MeV) at the BES and close to 0 at the LHC [35]. This is consistent with the trend of μ_{baryon} (~100–300 MeV at the BES and ~1 MeV at the LHC) obtained from other works [1, 29–32, 36]. This is natural due to the fact that baryon consists of valence quarks. If we regard $\mu_{baryon} = \Sigma\mu_q$, where Σ denotes the sum over all valence quarks in baryon, the present work is consistent with the models which study μ_{baryon} [1, 29–32, 36].

We would like to point out that although we have used the (two-component) standard distribution in the fits of p_T spectra and T_1 (T_2) has been used, the values of μ_q obtained by us are independent of models and parameters. In fact, μ_q is only related to k_j if T_{ch} is known. We can use directly the yield ratios of data to obtain μ_q. The reason why we use the function form instead of data is to extend p_T spectrum in intermediate region to low and high regions where the data are not available. In our opinion, the function that fitted the data in intermediate p_T region can predict approximately the trends in low and high p_T regions.

4. Conclusions

In summary, we found a good fit of the transverse momentum spectra of charged particles produced in central Au+Au collisions at the RHIC at its BES energies. It is shown that the (two-component) standard distribution successfully fitted the data measured at midrapidity by the STAR Collaboration, though other distributions are also acceptable. The effective temperature parameter increases with the increase of the particle mass and the collision energy.

At BES energies, the chemical potentials of light flavor quarks were obtained from the yield ratios of negatively to positively charged particles in given transverse momentum ranges available in experiments. At low energy, the chemical potentials of up and down quarks are consistent but differ from that of strange quark. At high energy, the three chemical potentials seem to deviate from each other, and they finally approach zero at very high energy.

From the lowest BES energy to the highest RHIC energy, with the increase of logarithmic collision energy, an exponential decrease of the chemical potentials of light flavor quarks is observed. The similarity in up and down quark masses renders the similarity in their chemical potentials. The difference between the chemical potentials of up (or down) and strange quarks is caused by their different masses. The difference between the chemical potentials changes from dozens of MeV to a few MeV. The decrease in chemical potential indicates that the mean free path of produced quarks changes from a small value to a large one.

Conflicts of Interest

The authors declare that there are no conflicts of interest regarding the publication of this paper.

Acknowledgments

Communications from Edward K. Sarkisyan-Grinbaum are highly acknowledged. This work was supported by the

National Natural Science Foundation of China under Grant Nos. 11747063, 11575103, and 11747319, the Doctoral Scientific Research Foundation of Taiyuan University of Science and Technology under Grant No. 20152043, the Shanxi Provincial Natural Science Foundation under Grant No. 201701D121005, and the Fund for Shanxi "1331 Project" Key Subjects Construction.

References

[1] P. Braun-Munzinger and J. Wambach, "Colloquium: Phase diagram of strongly interacting matter," *Reviews of Modern Physics*, vol. 81, no. 3, pp. 1031–1050, 2009.

[2] Y. Aoki, G. Endrődi, Z. Fodor, S. D. Katz, and K. K. Szabó, "The order of the quantum chromodynamics transition predicted by the standard model of particle physics," *Nature*, vol. 443, no. 7112, pp. 675–678, 2006.

[3] T. Bhattacharya, M. I. Buchoff, N. H. Christ et al., "QCD phase transition with chiral quarks and physical quark masses," *Physical Review Letters*, vol. 113, Article ID 082001, 2014.

[4] M. Cheng, N. H. Christ, and S. Datta, "QCD equation of state with almost physical quark masses," *Physical Review D*, vol. 77, no. 1, Article ID 014511, 2008.

[5] S. Ejiri, "Canonical partition function and finite density phase transition in lattice QCD," *Physical Review D*, vol. 78, no. 7, Article ID 074507, 2008.

[6] M. Asakawa and K. Yazaki, "Chiral restoration at finite density and temperature," *Nuclear Physics A*, vol. 504, no. 4, pp. 668–684, 1989.

[7] A. Barducci, R. Casalbuoni, S. De Curtis, R. Gatto, and G. Pettini, "Chiral-symmetry breaking in QCD at finite temperature and density," *Physics Letters B*, vol. 231, no. 4, pp. 463–470, 1989.

[8] A. Barducci, R. Casalbuoni, S. de Curtis, R. Gatto, and G. Pettini, "Chiral phase transitions in QCD for finite temperature and density," *Physical Review D*, vol. 41, no. 5, pp. 1610–1619, 1990.

[9] M. A. Stephanov, "QCD phase diagram and the critical point," *Progress of Theoretical Physics Supplement*, vol. 153, pp. 139–156, 2004.

[10] Z. Fodor and S. D. Katz, "Critical point of QCD at finite T and μ, lattice results for physical quark masses," *Journal of High Energy Physics*, vol. 2004, Article ID 050, 2004.

[11] A. Li, A. Alexandru, and K.-F. Liu, "Critical point of $N_f = 3$ QCD from lattice simulations in the canonical ensemble," *Physical Review D*, vol. 84, Article ID 071503, 2011.

[12] K. Nagata, K. Kashiwa, A. Nakamura, and S. M. Nishigaki, "Lee-Yang zero distribution of high temperature QCD and the Roberge-Weiss phase transition," *Physical Review D*, vol. 91, no. 9, Article ID 094507, 2015.

[13] P. de Forcrand, J. Langelage, O. Philipsen, and W. Unger, "Lattice QCD phase diagram in and away from the strong coupling limit," *Physical Review Letters*, vol. 113, no. 15, Article ID 152002, 2014.

[14] M. Stephanov, K. Rajagopal, and E. Shuryak, "Signatures of the tricritical point in QCD," *Physical Review Letters*, vol. 81, no. 22, pp. 4816–4819, 1998.

[15] B. I. Abelev et al., "Identified particle production, azimuthal anisotropy, and interferometry measurements in Au + Au collisions at $\sqrt{s_{NN}}$ = 9.2 GeV," *Physical Review C*, vol. 81, no. 2, Article ID 024911, 2010.

[16] L. Kumar et al., "STAR results from the rhic beam energy scan-I," *Nuclear Physics A*, vol. 904-905, pp. 256c–263c, 2013.

[17] L. Kumar, "Review of recent results from the RHIC beam energy scan," *Modern Physics Letters A*, vol. 28, no. 36, Article ID 1330033, 2013.

[18] https://drupal.star.bnl.gov/STAR/starnotes/public/sn0670.

[19] L. Adamczyk et al., "Bulk properties of the medium produced in relativistic heavy-ion collisions from the beam energy scan program," *Physical Review C*, vol. 96, no. 4, Article ID 044904, 2017.

[20] P. Braun-Munzinger, I. Heppe, and J. Stachel, "Chemical equilibration in Pb+Pb collisions at the SPS," *Physics Letters B*, vol. 465, no. 1–4, pp. 15–20, 1999.

[21] B. I. Abelev, M. M. Aggarwal, and Z. Ahammed, "Systematic measurements of identified particle spectra in *pp*, *d* + Au and Au + Au collisions at the STAR detector," *Physical Review C*, vol. 79, no. 3, Article ID 034909, 2009.

[22] F.-H. Liu, "Unified description of multiplicity distributions of final-state particles produced in collisions at high energies," *Nuclear Physics A*, vol. 810, no. 1, pp. 159–172, 2008.

[23] J. L. Synge, *The Relativistic Gas*, North-Holland, Amsterdam, The Netherlands, 1957.

[24] P. Z. Ning, L. Li, and D. F. Min, *Foundation of Nuclear Physics: Nucleons and Nuclei*, Higher Education Press, Beijing, China, 2003.

[25] C. D. Dermer, "The production spectrum of a relativistic Maxwell-Boltzmann gas," *The Astrophysical Journal*, vol. 280, no. 1, pp. 328–333, 1984.

[26] J. Cleymans and D. Worku, "Relativistic thermodynamics: transverse momentum distributions in high-energy physics," *The European Physical Journal A*, vol. 48, article 160, 2012.

[27] P. Braun-Munzinger, D. Magestro, K. Redlich, and J. Stachel, "Hadron production in Au-Au collisions at RHIC," *Physics Letters B*, vol. 518, no. 1-2, pp. 41–46, 2001.

[28] S. S. Adler, S. Afanasiev, and C. Aidala, "Identified charged particle spectra and yields in Au + Au collisions $\sqrt{s_{NN}}$ = 200 GeV," *Physical Review C*, vol. 69, no. 3, Article ID 034909, 2004.

[29] A. Andronic, P. Braun-Munzinger, and J. Stachel, "Thermal hadron production in relativistic nuclear collisions," *Acta Physica Polonica B*, vol. 40, no. 4, pp. 1005–1012, 2009.

[30] A. Andronic, P. Braun-Munzinger, and J. Stachel, "The Horn, the hadron mass spectrum and the QCD phase diagram: the statistical model of hadron production in central nucleus-nucleus collisions," *Nuclear Physics A*, vol. 834, pp. 237c–240c, 2010.

[31] A. Andronic, P. Braun-Munzinger, and J. Stachel, "Hadron production in central nucleus-nucleus collisions at chemical freeze-out," *Nuclear Physics A*, vol. 772, no. 3-4, pp. 167–199, 2006.

[32] J. Cleymans, H. Oeschler, K. Redlich, and S. Wheaton, "Comparison of chemical freeze-out criteria in heavy-ion collisions," *Physical Review C*, vol. 73, Article ID 034905, 2006.

[33] I. Arsene, I. G. Bearden, and D. Beavis, "Quark-gluon plasma and color glass condensate at RHIC? The perspective from the BRAHMS experiment," *Nuclear Physics A*, vol. 757, no. 1-2, pp. 1–27, 2005.

[34] H. Zhao and F.-H. Liu, "On extraction of chemical potentials of quarks from particle transverse momentum spectra in high energy collisions," *Advances in High Energy Physics*, vol. 2015, Article ID 137058, 9 pages, 2015.

[35] H.-L. Lao, Y.-Q. Gao, and F.-H. Liu, "Energy dependent chemical potentials of light particles and quarks from yield ratios of negative to positive particles in high energy collisions," 2018, https://arxiv.org/abs/1806.04309.

[36] J. Rafelski, "Melting hadrons, boiling quarks," *The European Physical Journal A*, vol. 51, no. 9, article 114, 2015.

16

The Third Five-Parametric Hypergeometric Quantum-Mechanical Potential

T. A. Ishkhanyan[1,2,3] **and A. M. Ishkhanyan**[1,2]

[1] *Russian-Armenian University, Yerevan 0051, Armenia*
[2] *Institute for Physical Research, Ashtarak 0203, Armenia*
[3] *Moscow Institute of Physics and Technology, Dolgoprudny 141700, Russia*

Correspondence should be addressed to A. M. Ishkhanyan; aishkhanyan@gmail.com

Guest Editor: Andrzej Okniński

We introduce the third *five-parametric* ordinary hypergeometric energy-independent quantum-mechanical potential, after the Eckart and Pöschl-Teller potentials, which is proportional to an arbitrary variable parameter and has a shape that is independent of that parameter. Depending on an involved parameter, the potential presents either a short-range singular well (which behaves as inverse square root at the origin and vanishes exponentially at infinity) or a smooth asymmetric step-barrier (with variable height and steepness). The general solution of the Schrödinger equation for this potential, which is a member of a general Heun family of potentials, is written through fundamental solutions each of which presents an irreducible linear combination of two Gauss ordinary hypergeometric functions.

1. Introduction

The solutions of the Schrödinger equation in terms of special mathematical functions for energy-independent potentials which are proportional to an arbitrary variable parameter and have a shape independent of that parameter are very rare [1–10] (see the discussion in [11]). It is a common convention to refer to such potentials as *exactly* solvable in order to distinguish them from the *conditionally* integrable ones for which a condition is imposed on the potential parameters such that the shape of the potential is not independent of the potential strength (e.g., a parameter is fixed to a constant or different term-strengths are not varied independently). While there is a relatively large set of potentials of the latter type (see, e.g., [12–20] for some examples discussed in the past and [21–25] for some recent examples), the list of the known exactly integrable potentials is rather limited even for the potentials of the most flexible *hypergeometric* class. The list of the exactly solvable hypergeometric potentials currently involves only ten items [1–10]. Six of these potentials are solved in terms of the confluent hypergeometric functions [1–6]. These are the classical Coulomb [1], harmonic oscillator [2], and Morse [3] potentials and the three recently derived potentials, which are the inverse square root [4], the Lambert-W step [5], and Lambert-W singular [6] potentials. The remaining four exactly integrable potentials which are solved in terms of the Gauss ordinary hypergeometric functions are the classical Eckart [7] and Pöschl-Teller [8] potentials and the two new potentials that we have introduced recently [9, 10].

An observation worth mentioning here is that all five classical hypergeometric potentials, both confluent and ordinary, involve *five* arbitrary variable parameters, while all new potentials are four-parametric. In this communication we show that the two four-parametric ordinary hypergeometric potentials [9, 10] are in fact particular cases of a more general five-parametric potential which is solved in terms of the hypergeometric functions. This generalization thus suggests the third five-parametric ordinary hypergeometric quantum-mechanical potential after the ones by Eckart [7] and Pöschl-Teller [8].

The potential we introduce belongs to one of the eleven independent eight-parametric general Heun families [25] (see also [26]). From the mathematical point of view, a peculiarity of the potential is that this is the only known case when the location of a singularity of the equation to which the Schrödinger equation is reduced is not fixed to a particular point but stands for a variable potential-parameter. Precisely, in our case the third finite singularity of the Heun equation, located at a point $z = a$ of the complex z-plane (that is, the singularity which is additional if compared with the ordinary hypergeometric equation), is not fixed but is variable; it stands for the fifth free parameter of the potential.

The potential is in general defined parametrically as a pair of functions $V(z), x(z)$. However, in several cases the coordinate transformation $x(z)$ is inverted thus producing explicitly written potentials given as $V = V(z(x))$ through an elementary function $z = z(x)$. All these cases are achieved by fixing the parameter a to a particular value; hence, all these particular potentials are four-parametric. The mentioned two recently presented four-parametric ordinary hypergeometric potentials [9, 10] are just such cases.

The potential we present is either a singular well (which behaves as the inverse square root in the vicinity of the origin and exponentially vanishes at infinity) or a smooth asymmetric step-barrier (with variable height, steepness, and asymmetry). The general solution of the Schrödinger equation for this potential is written through fundamental solutions each of which presents an irreducible linear combination of two ordinary hypergeometric functions ${}_2F_1$. The singular version of the potential describes a short-range interaction and for this reason supports only a finite number of bound states. We derive the exact equation for energy spectrum and estimate the number of bound states.

2. The Potential

The potential is given parametrically as

$$V(z) = V_0 + \frac{V_1}{z}, \tag{1}$$

$$x(z) = x_0 + \sigma(a\ln(z-a) - \ln(z-1)), \tag{2}$$

where $a \neq 0, 1$ and x_0, σ, V_0, V_1 are arbitrary (real or complex) constants. Rewriting the coordinate transformation as

$$\frac{(z-a)^a}{z-1} = e^{(x-x_0)/\sigma}, \tag{3}$$

it is seen that for real rational a the transformation is rewritten as a polynomial equation for z; hence, in several cases it can be inverted.

Since $a \neq 0, 1$, the possible simplest case is when the polynomial equation is quadratic. This is achieved for $a = -1, 1/2, 2$. It is checked, however, that these three cases lead to four-parametric subpotentials which are equivalent in the sense that each is derived from another by specifications of the involved parameters. For $a = -1$ the potential reads [9]

$$V(x) = V_0 + \frac{V_1}{\sqrt{1 + e^{(x-x_0)/\sigma}}}, \tag{4}$$

where we have changed $\sigma \longrightarrow -\sigma$.

The next are the cubic polynomial reductions which are achieved in six cases: $a = -2, -1/2, 1/3, 2/3, 3/2, 3$. It is again checked, however, that these choices produce only one independent potential. This is the four-parametric potential presented in [10]:

$$V = V_0 + \frac{V_1}{z},$$
$$z = -1 + \frac{1}{\left(e^{x/(2\sigma)} + \sqrt{1+e^{x/\sigma}}\right)^{2/3}} + \left(e^{x/(2\sigma)} + \sqrt{1+e^{x/\sigma}}\right)^{2/3}, \tag{5}$$

where one should replace x by $x - x_0$. Similar potentials in terms of elementary functions through quartic and quintic reductions of (3) are rather cumbersome; we omit those.

For arbitrary real $a \neq 0, 1$, assuming $z \in (0, 1)$ and shifting

$$x_0 \longrightarrow x_0 - \sigma a \ln(-a) + i\pi\sigma, \tag{6}$$

the potential (1), (2) presents a singular well. In the vicinity of the origin it behaves as $x^{-1/2}$,

$$V|_{x\longrightarrow 0} \sim \sqrt{\frac{(a-1)\sigma}{2a}}\frac{V_1}{\sqrt{x}}, \tag{7}$$

and exponentially approaches a constant, $V_0 + V_1$, at infinity,

$$V|_{x\longrightarrow +\infty} \sim \left(\frac{a-1}{a}\right)^a V_1 e^{-x/\sigma}. \tag{8}$$

The potential and the two asymptotes are shown in Figure 1.

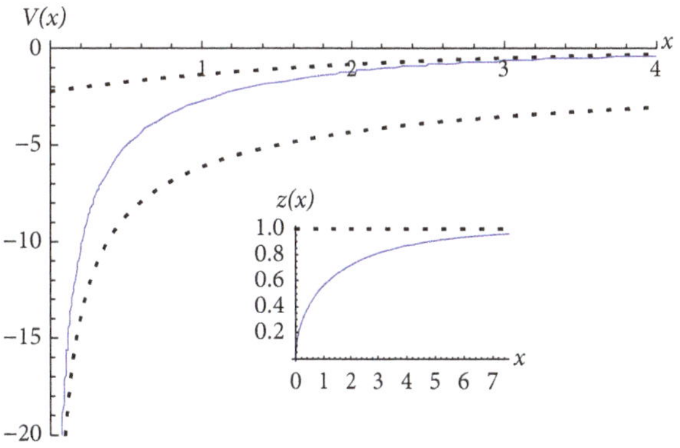

Figure 1: Potential (1), (2) for $a = -2$ and $(\sigma, x_0, V_0, V_1) = (2, 0, 5, -5)$. The inset presents the coordinate transformation $z(x) \in (0, 1)$ for $x \in (0, \infty)$.

A potential of a different type is constructed if one allows the parameterization variable z to vary within the interval $z \in (1, \infty)$ for $a < 1$ or within the interval $z \in (1, a)$ for $a > 1$. This time, shifting (compare with (6))

$$x_0 \longrightarrow x_0 - \sigma a \ln(1-a), \tag{9}$$

we derive an asymmetric step-barrier the height of which depends on V_0 and V_1, while the asymmetry and steepness

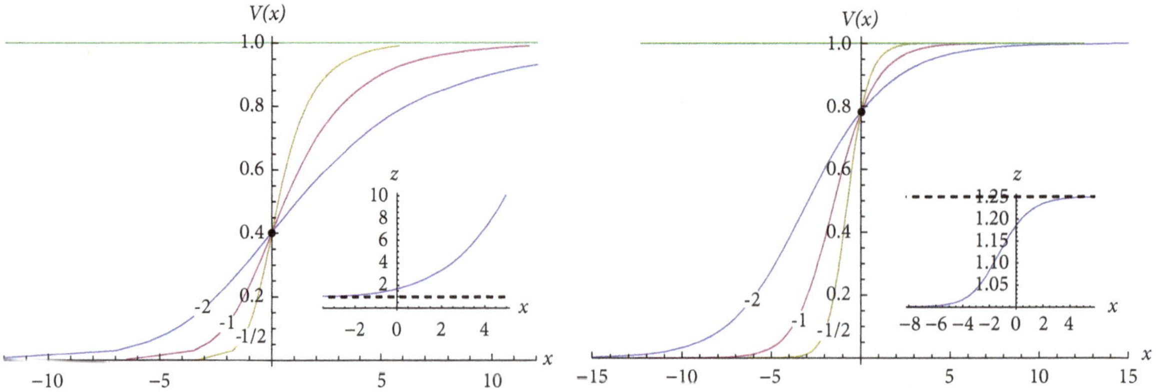

FIGURE 2: Potential (1), (2) for $a = -2$ and $(x_0, V_0, V_1) = (0, 1, -1)$ (left figure) and for $a = 1.25$ and $(x_0, V_0, V_1) = (0, 5, -5)$ (right figure); $\sigma = -2, -1, -1/2$. The fixed points are marked by filled circles. The insets present the coordinate transformation $z(x)$ for $\sigma = -1$.

are controlled by the parameters a and σ. The shape of the potential is shown in Figure 2 for $a = -2$ and $a = 1.25$. We note that in the limit $\sigma \longrightarrow 0$ the potential turns into the abrupt-step potential and that the subfamily of barriers generated by variation of σ at constant V_0 and V_1 has a σ-independent fixed point located at $x = x_0$ (marked in Figure 2 by filled circles).

3. Reduction to the General Heun Equation

The solution of the one-dimensional Schrödinger equation for potential (1), (2),

$$\frac{d^2\psi}{dx^2} + \frac{2m}{\hbar^2}(E - V(x))\psi = 0, \tag{10}$$

is constructed via reduction to the general Heun equation [27–29]

$$\frac{d^2u}{dz^2} + \left(\frac{\gamma}{z-a_1} + \frac{\delta}{z-a_2} + \frac{\varepsilon}{z-a_3}\right)\frac{du}{dz} + \frac{\alpha\beta z - q}{(z-a_1)(z-a_2)(z-a_3)}u = 0. \tag{11}$$

The details of the technique are presented in [11, 25]. It has been shown that the energy-independent general-Heun potentials, which are proportional to an arbitrary variable parameter and have shapes which are independent of that parameter, are constructed by the coordinate transformation $z = z(x)$ of the Manning form [30] given as

$$\frac{dz}{dx} = \frac{(z-a_1)^{m_1}(z-a_2)^{m_2}(z-a_3)^{m_3}}{\sigma}, \tag{12}$$

where $m_{1,2,3}$ are integers or half-integers and σ is an arbitrary scaling constant. As it is seen, the coordinate transformation is solely defined by the singularities $a_{1,2,3}$ of the general Heun equation. The canonical form of the Heun equation assumes two of the three finite singularities at 0 and 1, and the third one at a point a, so that $a_{1,2,3} = (0, 1, a)$ [27–29]. However, it may be convenient for practical purposes to apply a different specification of the singularities, so for a moment we keep the parameters $a_{1,2,3}$ unspecified.

The coordinate transformation is followed by the change of the dependent variable

$$\psi = (z-a_1)^{\alpha_1}(z-a_2)^{\alpha_2}(z-a_3)^{\alpha_3}u(z) \tag{13}$$

and application of the ansatz

$$V(z) = \frac{v_0 + v_1 z + v_2 z^2 + v_3 z^3 + v_4 z^4}{(z-a_1)^2(z-a_2)^2(z-a_3)^2}\left(\frac{dz}{dx}\right)^2, \quad v_{0,1,2,3,4} = \text{const}. \tag{14}$$

The form of this ansatz and the permissible sets of the parameters $m_{1,2,3}$ are revealed through the analysis of the behavior of the solution in the vicinity of the finite singularities of the general Heun equation [11]. This is a crucial point which warrants that all the parameters involved in the resulting potentials can be varied independently.

It has been shown that there exist in total thirty-five permissible choices for the coordinate transformation each being defined by a triad (m_1, m_2, m_3) satisfying the inequalities $-1 \le m_{1,2,3} \le 1$ and $1 \le m_1 + m_2 + m_3 \le 3$ [25]. However, because of the symmetry of the general Heun equation with respect to the transpositions of its singularities, only eleven of the resultant potentials are independent [25]. The potential (1), (2) belongs to the fifth independent family with $m_{1,2,3} = (1, 1, -1)$ for which from (14) we have

$$V(z) = \frac{V_4 + V_3 z + V_2 z^2 + V_1 z^3 + V_0 z^4}{(z-a_3)^4} \tag{15}$$

with arbitrary $V_{0,1,2,3,4} = \text{const}$, and, from (12),

$$\frac{x - x_0}{\sigma} = \frac{a_1 - a_3}{a_1 - a_2}\ln(z - a_1) + \frac{a_3 - a_2}{a_1 - a_2}\ln(z - a_2). \tag{16}$$

It is now convenient to have a potential which does not explicitly involve the singularities. Hence, we put $a_3 = 0$ and apply the specification $a_{1,2,3} = (a, 1, 0)$ to derive the potential

$$V(z) = V_0 + \frac{V_1}{z} + \frac{V_2}{z^2} + \frac{V_3}{z^3} + \frac{V_4}{z^4} \quad (17)$$

$$\text{with} \quad \frac{(x - x_0)}{\sigma/(a-1)} = a \ln(z - a) - \ln(z - 1). \quad (18)$$

The solution of the Schrödinger equation (10) for this potential is written in terms of the general Heun function H_G as

$$\psi = (z-a)^{\alpha_1} z^{\alpha_2} (z-1)^{\alpha_3} \cdot H_G(a_1, a_2, a_3; q; \alpha, \beta, \gamma, \delta, \varepsilon; z), \quad (19)$$

where the involved parameters $\alpha, \beta, \gamma, \delta, \varepsilon$, and q are given through the parameters $V_{0,1,2,3,4}$ of potential (17) and the exponents $\alpha_{1,2,3}$ of the prefactor by the equations [25]

$$(\gamma, \delta, \varepsilon) = (1 + 2\alpha_1, 1 + 2\alpha_2, -1 + 2\alpha_3), \quad (20)$$

$$1 + \alpha + \beta = \gamma + \delta + \varepsilon, \quad \alpha\beta = (\alpha_1 + \alpha_2 + \alpha_3)^2 + \frac{2m\sigma^2 (E - V_0)}{\hbar^2}, \quad (21)$$

$$q = \frac{2m\sigma^2}{\hbar^2}(V_1 - (1+a)(E - V_0)) + \left(-\alpha_2^2 + (-1 + \alpha_1 + \alpha_3)(\alpha_1 + \alpha_3)\right) + a\left(-\alpha_1^2 + (-1 + \alpha_2 + \alpha_3)(\alpha_2 + \alpha_3)\right); \quad (22)$$

the exponents $\alpha_{1,2,3}$ of the prefactor are defined by the equations

$$\alpha_1^2 = \frac{2m\sigma^2}{a^2 (a-1)^2 \hbar^2}\left(V_4 + aV_3 + a^2 V_2 + a^3 V_1 + a^4 (V_0 - E)\right), \quad (23)$$

$$\alpha_2^2 = -\frac{2m\sigma^2}{(a-1)^2 \hbar^2}(E - V_0 - V_1 - V_2 - V_3 - V_4), \quad (24)$$

$$\alpha_3(\alpha_3 - 2) = \frac{2m\sigma^2 V_4}{a^2 \hbar^2}. \quad (25)$$

4. The Solution of the Schrödinger Equation in Terms of the Gauss Functions

Having determined the parameters of the Heun equation, the next step is to examine the cases when the general Heun function H_G is written in terms of the Gauss hypergeometric functions ${}_2F_1$. An observation here is that the direct one-term Heun-to-hypergeometric reductions discussed by many authors (see, e.g., [27, 28, 31–34]) are achieved by such restrictions and imposed on the involved parameters (three or more conditions), which are either not satisfied by the Heun potentials or produce very restrictive potentials. It is checked that the less restrictive reductions reproduce the classical Eckart and Pöschl-Teller potentials, while the other reductions result in conditionally integrable potentials.

More advanced are the finite-sum solutions achieved by termination of the series expansions of the general Heun function in terms of the hypergeometric functions [35–39]. For such reductions, only two restrictions are imposed on the involved parameters and, notably, these restrictions are such that in many cases they are satisfied. The solutions for the above-mentioned four-parametric subpotentials [9, 10] have been constructed right in this way. Other examples achieved by termination of the hypergeometric series expansions of the functions of the Heun class include the recently reported inverse square root [4], Lambert-W step [5], and Lambert-W singular [6] potentials.

The series expansions of the general Heun function in terms of the Gauss ordinary hypergeometric functions are governed by three-term recurrence relations for the coefficients of the successive terms of the expansion. A useful particular expansion in terms of the functions of the form ${}_2F_1(\alpha, \beta; \gamma_0 - n; z)$ which leads to simpler coefficients of the recurrence relation is presented in [25, 39]. If the expansion functions are assumed irreducible to simpler functions, the termination of this series occurs if $\varepsilon = -N$, $n = 0, 1, 2, \ldots$, and a $(N+1)$th degree polynomial equation for the accessory parameter q is satisfied. For $\varepsilon = 0$ the latter equation is $q = a\alpha\beta$, which corresponds to the trivial direct reduction of the general Heun equation to the Gauss hypergeometric equation. This case reproduces the classical Eckart and Pöschl-Teller potentials [25]. For the first nontrivial case $\varepsilon = -1$ the termination condition for singularities $a_{1,2,3} = (a, 1, 0)$ takes a particularly simple form:

$$q^2 + q(\gamma - 1 + a(\delta - 1)) + a\alpha\beta = 0. \quad (26)$$

The solution of the Heun equation for a root of this equation is written as [39]

$$u = {}_2F_1\left(\alpha, \beta; \gamma; \frac{a-z}{a-1}\right) + \frac{\gamma - 1}{q + a(\delta - 1)} \cdot {}_2F_1\left(\alpha, \beta; \gamma - 1; \frac{a-z}{a-1}\right), \quad (27)$$

This solution has a representation through Clausen's generalized hypergeometric function ${}_3F_2$ [40, 41].

Consider if the termination condition (26) for $\varepsilon = -1$ is satisfied for the parameters given by (20)-(25). From (20) we find that for $\varepsilon = -1$ holds $\alpha_3 = 0$. It then follows from (25) that $V_4 = 0$. With this, (26) is reduced to

$$V_2 + V_3\left(\frac{1+a}{a} - \frac{2m\sigma^2}{a^2\hbar^2}V_3\right) = 0. \quad (28)$$

This equation generally defines a conditionally integrable potential in that the potential parameters V_2 and V_3 are not varied independently. Alternatively, if the potential parameters are assumed independent, the equation is satisfied only if $V_2 = V_3 = 0$. Thus, we put $V_{2,3,4} = 0$ and potential

(17) is reduced to that given by (1). Furthermore, since σ is arbitrary, in order for (18) to exactly reproduce the coordinate transformation (2), we replace $\sigma/(1-a) \longrightarrow \sigma$.

With this, the solution of the Schrödinger equation (10) for potential (1) is written as

$$\psi = (z-a)^{\alpha_1}(z-1)^{\alpha_2}\left({}_2F_1\left(\alpha,\beta;\gamma;\frac{a-z}{a-1}\right) + \frac{2\alpha_1}{a\alpha_2-\alpha_1}\cdot {}_2F_1\left(\alpha,\beta;\gamma-1;\frac{a-z}{a-1}\right)\right) \quad (29)$$

$$\text{with } (\alpha,\beta,\gamma) = (\alpha_1+\alpha_2+\alpha_0, \alpha_1+\alpha_2-\alpha_0, 1+2\alpha_1), \quad (30)$$

$$\alpha_{0,1,2} = \left(\pm\sqrt{\frac{2m\sigma^2(a-1)^2}{\hbar^2}(V_0-E)}, \pm\sqrt{\frac{2m\sigma^2a^2}{\hbar^2}\left(V_0-E+\frac{V_1}{a}\right)}, \pm\sqrt{\frac{2m\sigma^2}{\hbar^2}(V_0-E+V_1)}\right). \quad (31)$$

This solution applies for any real or complex set of the involved parameters. Furthermore, we note that any combination for the signs of $\alpha_{1,2}$ is applicable. Hence, by choosing different combinations, one can construct different independent fundamental solutions. Thus, this solution supports the general solution of the Schrödinger equation.

A final remark is that using the contiguous functions relations for the hypergeometric functions one can replace the second hypergeometric function in (29) by a linear combination of the first hypergeometric function and its derivative. In this way we arrive at the following representation of the general solution of the Schrödinger equation:

$$\psi = (z-a)^{\alpha_1}(z-1)^{\alpha_2}\left(F + \frac{z-a}{\alpha_1+a\alpha_2}\frac{dF}{dz}\right), \quad (32)$$

$$\text{where } F = c_1\cdot {}_2F_1\left(\alpha,\beta;\gamma;\frac{a-z}{a-1}\right) + c_2 \cdot {}_2F_1\left(\alpha,\beta;1+\alpha+\beta-\gamma;\frac{z-1}{a-1}\right). \quad (33)$$

5. Bound States

Consider the bound states supported by the singular version of potential (1), (2), achieved by shifting $x_0 \longrightarrow x_0 - \sigma a\ln(-a) + i\pi\sigma$ in (2). Since the potential vanishes at infinity exponentially, it is understood that this is a short-range potential. The integral of the function $xV(x)$ over the semiaxis $x \in (0,+\infty)$ is finite, hence, according to the general criterion [42–46], the potential supports only a finite number of bound states. These states are derived by demanding the wave function to vanish both at infinity and in the origin (see the discussion in [47]). We recall that for this potential the coordinate transformation maps the interval $x \in (0,+\infty)$ onto the interval $z \in (0,1)$. Thus, we demand $\psi(z=0) = \psi(z=1) = 0$.

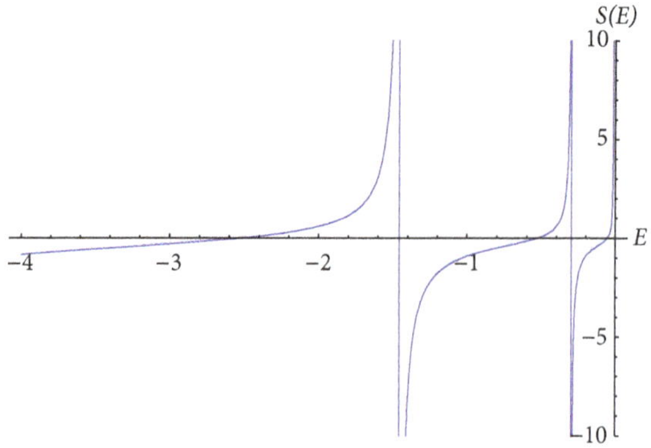

FIGURE 3: Graphical representation of the spectrum equation (35) for $m, \hbar, V_0, \sigma, a = 1, 1, 5, 2, -2$.

The condition $V(+\infty) = 0$ assumes $V_0 + V_1 = 0$; hence, α_2 is real for negative energy. Choosing, for definiteness, the plus signs in (31), we have $\alpha_2 > 0$. Then, examining the equation $\psi(z=1) = 0$, we find that

$$\psi|_{z\longrightarrow 1} \sim c_1A_1(1-z)^{-\alpha_2} + c_2A_2(1-z)^{\alpha_2} \quad (34)$$

with some constants $A_{1,2}$. Since for positive α_2 the first term diverges, we conclude $c_1 = 0$. The condition $\psi(z=0) = 0$ then gives the following exact equation for the spectrum:

$$S(E) \equiv 1 + \frac{\alpha_1+a\alpha_2}{2(1-a)\alpha_2} \cdot \frac{{}_2F_1(\alpha+1,\beta+1;1+2\alpha_2;1/(1-a))}{{}_2F_1(\alpha,\beta;2\alpha_2;1/(1-a))} = 0. \quad (35)$$

The graphical representation of this equation is shown in Figure 3. The function $S(E)$ has a finite number of zeros. For the parameters $m, \hbar, V_0, \sigma, a = 1, 1, 5, 2, -2$ applied in the figure there are just three bound states.

According to the general theory, the number of bound states is equal to the number of zeros (not counting $x = 0$) of the zero-energy solution, which vanishes at the origin [42–46]. We note that for $E = 0$ the lower parameter of the second hypergeometric function in (33) vanishes: $1+\alpha+\beta-\gamma = 0$. Hence, a different second independent solution should be applied. This solution is constructed by using the first hypergeometric function with α_1 everywhere replaced by $-\alpha_1$. The result is rather cumbersome. It is more conveniently written in terms of the Clausen functions as

$$\psi_{E=0} = c_1(z-a)^{\alpha_1}\, {}_3F_2\left(-\sqrt{\frac{a-1}{a}}\alpha_1+\alpha_1, \sqrt{\frac{a-1}{a}}\alpha_1+\alpha_1, 1+\alpha_1; \alpha_1, 1+2\alpha_1; \frac{a-z}{a-1}\right) + c_2(z-a)^{-\alpha_1} \cdot {}_3F_2\left(-\sqrt{\frac{a-1}{a}}\alpha_1-\alpha_1, \sqrt{\frac{a-1}{a}}\alpha_1-\alpha_1, 1-\frac{\alpha_1}{a}; -\frac{\alpha_1}{a}, 1; \frac{z-1}{a-1}\right), \quad (36)$$

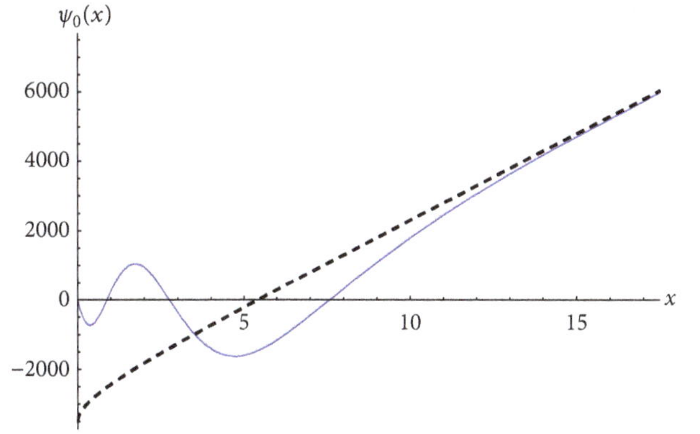

FIGURE 4: The zero-energy solution for $m, \hbar, V_0, \sigma, a = 1, 1, 5, 2, -2$. The dashed line shows the logarithmic asymptote at infinity: $\psi_0|_{x\to\infty} \sim A + B\ln(1-z)$.

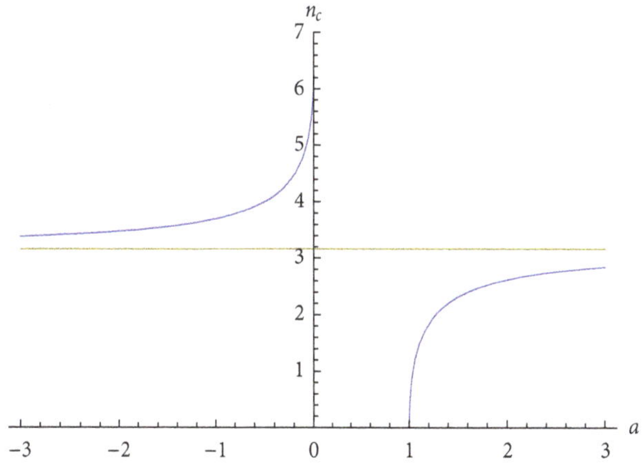

FIGURE 5: The dependence of Chadan's estimate $n_c = I_C/2$ for the number of bound states on the parameter a ($m, \hbar, V_0, \sigma = 1, 1, 5, 2$).

where $\alpha_1 = \sqrt{2a(a-1)m\sigma^2 V_0/\hbar^2}$ and the relation between c_1 and c_2 is readily derived from the condition $\psi_{E=0}(0) = 0$. This solution is shown in Figure 4. It is seen that for parameters $m, \hbar, V_0, \sigma, a = 1, 1, 5, 2, -2$ used in Figure 3 the number of zeros (excluded the origin) is indeed 3.

For practical purposes, it is useful to have an estimate for the number of bound states. The absolute upper limit for this number is given by the integral [42, 43]

$$I_B = \int_0^{\infty} r \left| V\left(x \longrightarrow \frac{r\hbar}{\sqrt{2m}}\right)\right| dr = (1-a) \cdot \left(Li_2\left(\frac{1}{1-a}\right) + 2a\coth^{-1}(1-2a)^2\right)\frac{2m\sigma^2 V_0}{\hbar^2}. \quad (37)$$

where Li_2 is Jonquière's polylogarithm function of order 2 [48, 49]. Though of general importance, however, in many cases this is a rather overestimating limit. Indeed, for the parameters applied in Figure 3 it gives $n \le I_B \approx 24$.

More stringent are the estimates by Calogero [44] and Chadan [45] which are specialized for everywhere monotonically nondecreasing attractive central potentials. Calogero's estimate reads $n \le I_C$ with [44]

$$I_C = \frac{2/\pi}{\hbar/\sqrt{2m}} \int_0^{\infty} \sqrt{-V(x)}dx = \left(1 + \left(\sqrt{1-a} - \sqrt{-a}\right)^2\right)\sqrt{\frac{2m\sigma^2 V_0}{\hbar^2}}, \quad (38)$$

We note that $I_C \approx \sqrt{2I_B}$. The result by Chadan further tunes the upper limit for the number of bound states to the half of that by Calogero; that is $n \le I_C/2$ [45]. For the parameters applied in Figure 3 this gives $n \le 3.48$, which is, indeed, an accurate estimate. The dependence of the function $n_c = I_C/2$ on the parameter a for $a \in (-\infty, 0) \cup (1, +\infty)$ is shown in Figure 5. It is seen that more bound states are available for a close to zero. The maximum number achieved in the limit $a \longrightarrow 0$ is $\sqrt{2m\sigma^2 V_0/\hbar^2}$; hence, for sufficiently small V_0 or σ such that $2m\sigma^2 V_0 < \hbar^2$, bound states are not possible at all.

6. Discussion

Thus, we have presented the third five-parametric quantum-mechanical potential for which the solution of the Schrödinger equation is written in terms of the Gauss ordinary hypergeometric functions. The potential involves five (generally complex) parameters which are varied independently. Depending on the particular specifications of these parameters, the potential suggests two different appearances. In one version we have a smooth step-barrier with variable height, steepness, and asymmetry, while in the other version this is a singular potential-well which behaves as the inverse square root in the vicinity of the origin and exponentially vanishes at infinity.

The potential is in general given parametrically; however, in several cases the involved coordinate transformation allows inversion thus leading to particular potentials which are explicitly written in terms of elementary functions. These reductions are achieved by particular specifications of a parameter standing for the third finite singularity of the general Heun equation. The resultant subpotentials all are four-parametric (see, e.g., [9, 10]). These particular cases are defined by coordinate transformations which are roots of polynomial equations. It turns out that different polynomial equations of the same degree produce the same potential (with altered parameters). The reason for this is well understood in the case of quadratic equations. In that case the third singularity of the general Heun equation, to which the Schrödinger equation is reduced, is specified as $a = -1, 1/2$ or 2. We then note that the form-preserving transformations of the independent variable map the four singularities of the Heun equation, $z = 0, 1, a, \infty$, onto the points $z = 0, 1, a_1, \infty$ with a_1 adopting one of the six possible values $a, 1/a, 1-a, 1/(1-a), a/(1-a), (a-1)/a$ [27–29]. It is seen that the triad $(-1, 1/2, 2)$ is a specific set which remains invariant at form-preserving transformations of the independent variable.

The potential belongs to the general Heun family $m_{1,2,3} = (1, 1, -1)$. This family allows several conditionally integrable reductions too [25]. A peculiarity of the exactly integrable potential that we have presented here is that the location of a finite singularity of the general Heun equation is not fixed to a particular point of the complex z-plane but serves as a variable potential-parameter. In the step-barrier version of the potential, this parameter stands for the asymmetry of the potential.

The solution of the Schrödinger equation for the potential we have presented is constructed via termination of a series expansion of the general Heun equation in terms of the Gauss ordinary hypergeometric functions. The general solution of the problem is composed of fundamental solutions each of which is an *irreducible* combination of two hypergeometric functions. Several other potentials allowing solutions of this type have been reported recently [4–6, 9, 10, 23–25]. Further cases involve the solutions for supersymmetric partner potentials much discussed in the past [15, 50, 51] and for several nonanalytic potentials discussed recently [52–54]. One should distinguish these solutions from the case of reducible hypergeometric functions [55–59] when the solutions eventually reduce to quasi-polynomials, e.g., discussed in the context of quasi-exactly solvability [57–59]. We note that, owing to the contiguous functions relations [60], the two-term structure of the solution is a general property of all finite-sum hypergeometric reductions of the general Heun functions achieved via termination of series solutions. It is checked that in our case the linear combination of the involved Gauss functions is expressed through a single generalized hypergeometric function ${}_3F_2$ [40, 41].

We have presented the explicit solution of the problem and discussed the bound states supported by the singular version of the potential. We have derived the exact equation for the energy spectrum and estimated the number of bound states. The exact number of bound states is given by the number of zeros of the zero-energy solution which we have also presented.

Conflicts of Interest

The authors declare that they have no conflicts of interest.

Acknowledgments

The research was supported by the Russian-Armenian (Slavonic) University at the expense of the Ministry of Education and Science of the Russian Federation, the Armenian Science Committee (SC Grants no. 18RF-139 and no. 18T-1C276), and the Armenian National Science and Education Fund (ANSEF Grant no. PS-4986). T. A. Ishkhanyan acknowledges the support from SPIE through a 2017 Optics and Photonics Education Scholarship and thanks the French Embassy in Armenia for a doctoral grant as well as the Agence universitaire de la Francophonie with Armenian Science Committee for a Scientific Mobility grant.

References

[1] E. Schrödinger, "Quantisierung als eigenwertproblem," *Annalen der Physik*, vol. 384, no. 4, pp. 361–376, 1926.

[2] E. Schrödinger, "Quantisierung als Eigenwertproblem. Zweite Mitteilung," *Annalen der Physik*, vol. 384, no. 6, pp. 489–527, 1926.

[3] P. M. Morse, "Diatomic molecules according to the wave mechanics. II. Vibrational levels," *Physical Review A: Atomic, Molecular and Optical Physics*, vol. 34, no. 1, pp. 57–64, 1929.

[4] A. M. Ishkhanyan, "Exact solution of the Schrödinger equation for the inverse square root potential V0/√x," *EPL (Europhysics Letters)*, vol. 112, no. 1, 2015.

[5] A. M. Ishkhanyan, "The Lambert-*W* step-potential – an exactly solvable confluent hypergeometric potential," *Physics Letters A*, vol. 380, no. 5-6, pp. 640–644, 2016.

[6] A. M. Ishkhanyan, "A singular Lambert-W Schrödinger potential exactly solvable in terms of the confluent hypergeometric functions," *Modern Physics Letters A*, vol. 31, no. 33, 1650177, 11 pages, 2016.

[7] C. Eckart, "The penetration of a potential barrier by electrons," *Physical Review A: Atomic, Molecular and Optical Physics*, vol. 35, no. 11, article 1303, 1930.

[8] G. Pöschl and E. Teller, "Bemerkungen zur Quantenmechanik des anharmonischen Oszillators," *Zeitschrift für Physik*, vol. 83, no. 3-4, pp. 143–151, 1933.

[9] A. Ishkhanyan, "The third exactly solvable hypergeometric quantum-mechanical potential," *EPL (Europhysics Letters)*, vol. 115, no. 2, 2016.

[10] T. A. Ishkhanyan, V. A. Manukyan, A. H. Harutyunyan, and A. M. Ishkhanyan, "A new exactly integrable hypergeometric potential for the Schrödinger equation," *AIP Advances*, vol. 8, no. 3, 2018.

[11] A. Ishkhanyan and V. Krainov, "Discretization of Natanzon potentials," *The European Physical Journal Plus*, vol. 131, no. 9, 2016.

[12] F. H. Stillinger, "Solution of a quantum mechanical eigenvalue problem with long range potentials," *Journal of Mathematical Physics*, vol. 20, no. 9, pp. 1891–1895, 1979.

[13] G. P. Flessas and A. Watt, "An exact solution of the Schrödinger equation for a multiterm potential," *Journal of Physics A: Mathematical and General*, vol. 14, no. 9, pp. L315–L318, 1981.

[14] J. N. Ginocchio, "A class of exactly solvable potentials. I. One-dimensional Schrödinger equation," *Annals of Physics*, vol. 152, no. 1, pp. 203–219, 1984.

[15] F. Cooper, J. N. Ginocchio, and A. Khare, "Relationship between supersymmetry and solvable potentials," *Physical Review D: Particles, Fields, Gravitation and Cosmology*, vol. 36, no. 8, pp. 2458–2473, 1987.

[16] A. De Souza Dutra, "Conditionally exactly soluble class of quantum potentials," *Physical Review A: Atomic, Molecular and Optical Physics*, vol. 47, no. 4, pp. R2435–R2437, 1993.

[17] R. Dutt, A. Khare, and Y. P. Varshni, "New class of conditionally exactly solvable potentials in quantum mechanics," *Journal of Physics A: Mathematical and General*, vol. 28, no. 3, pp. L107–L113, 1995.

[18] C. Grosche, "Conditionally solvable path integral problems," *Journal of Physics A: Mathematical and General*, vol. 28, no. 20, pp. 5889–5902, 1995.

[19] H. Exton, "The exact solution of two new types of Schrödinger equation," *Journal of Physics A: Mathematical and General*, vol. 28, no. 23, pp. 6739–6741, 1995.

[20] G. Junker and P. Roy, "Conditionally exactly solvable problems and non-linear algebras," *Physics Letters A*, vol. 232, no. 3-4, pp. 155–161, 1997.

[21] B. W. Williams, "Exact solutions of a Schrödinger equation based on the Lambert function," *Physics Letters A*, vol. 334, no. 2-3, pp. 117–122, 2005.

[22] A. López-Ortega, "New conditionally exactly solvable inverse power law potentials," *Physica Scripta*, vol. 90, no. 8, p. 085202, 2015.

[23] G. Junker and P. Roy, "Conditionally exactly solvable potentials: a supersymmetric construction method," *Annals of Physics*, vol. 270, no. 1, pp. 155–177, 1998.

[24] A. M. Ishkhanyan, "A conditionally exactly solvable generalization of the inverse square root potential," *Physics Letters A*, vol. 380, pp. 3786–3790, 2016.

[25] A. M. Ishkhanyan, "Schrödinger potentials solvable in terms of the general Heun functions," *Annals of Physics*, vol. 388, pp. 456–471, 2018.

[26] A. Lemieux and A. K. Bose, *Construction de Potentiels Pour Lesquels L'équation De Schrödinger Est Soluble*, vol. 10, Annales de l'Institut Henri Poincaré, 1969.

[27] A. Ronveaux, *Heun's Differential Equations*, Oxford University Press, Oxford, UK, 1995.

[28] S. Y. Slavyanov and W. Lay, *Special Functions, A Unified Theory Based on Singularities*, Oxford Mathematical Monographs, Oxford, 2000.

[29] F. W. J. Olver, D. W. Lozier, R. F. Boisvert, and C. W. Clark, *NIST Handbook of Mathematical Functions*, Cambridge University Press, 2010.

[30] M. F. Manning, "Exact solutions of the schrödinger equation," *Physical Review A: Atomic, Molecular and Optical Physics*, vol. 48, no. 2, pp. 161–164, 1935.

[31] R. S. Maier, "On reducing the Heun equation to the hypergeometric equation," *Journal of Differential Equations*, vol. 213, no. 1, pp. 171–203, 2005.

[32] R. Vidunas and G. Filipuk, "Parametric transformations between the Heun and Gauss hypergeometric functions," *Funkcialaj Ekvacioj. Serio Internacia*, vol. 56, no. 2, pp. 271–321, 2013.

[33] R. Vidunas and G. Filipuk, "A classification of coverings yielding Heun-to-hypergeometric reductions," *Osaka Journal of Mathematics*, vol. 51, no. 4, pp. 867–903, 2014.

[34] M. van Hoeij and R. Vidunas, "Belyi functions for hyperbolic hypergeometric-to-Heun transformations," *Journal of Algebra*, vol. 441, pp. 609–659, 2015.

[35] N. Svartholm, "Die Lösung der Fuchs'schen Differentialgleichung zweiter Ordnung durch Hypergeometrische Polynome," *Mathematische Annalen*, vol. 116, no. 1, pp. 413–421, 1939.

[36] A. Erdélyi, "The Fuchsian equation of second order with four singularities," *Duke Mathematical Journal*, vol. 9, pp. 48–58, 1942.

[37] A. Erdélyi, "Certain expansions of solutions of the Heun equation," *Quarterly Journal of Mathematics*, vol. 15, pp. 62–69, 1944.

[38] E. G. Kalnins and J. Miller, "Hypergeometric expansions of Heun polynomials," *SIAM Journal on Mathematical Analysis*, vol. 22, no. 5, pp. 1450–1459, 1991.

[39] T. A. Ishkhanyan, T. A. Shahverdyan, and A. M. Ishkhanyan, "Expansions of the Solutions of the General Heun Equation Governed by Two-Term Recurrence Relations for Coefficients," *Advances in High Energy Physics*, vol. 2018, Article ID 4263678, 9 pages, 2018.

[40] J. Letessier, G. Valent, and J. Wimp, "Some differential equations satisfied by hypergeometric functions," in *Approximation and computation (West Lafayette, IN, 1993)*, vol. 119 of *Internat. Ser. Numer. Math.*, pp. 371–381, Birkhauser Boston, Boston, Mass, USA, 1994.

[41] R. S. Maier, "P-symbols, Heun identities, and 3 F2 identities," in *Special Functions and Orthogonal Polynomials*, vol. 471 of *Contemp. Math.*, pp. 139–159, Amer. Math. Soc., Providence, RI, 2008.

[42] V. Bargmann, "On the number of bound states in a central field of force," *Proceedings of the National Acadamy of Sciences of the United States of America*, vol. 38, pp. 961–966, 1952.

[43] J. Schwinger, "On the bound states of a given potential," *Proceedings of the National Acadamy of Sciences of the United States of America*, vol. 47, pp. 122–129, 1961.

[44] F. Calogero, "Upper and lower limits for the number of bound states in a given central potential," *Communications in Mathematical Physics*, vol. 1, pp. 80–88, 1965.

[45] K. Chadan, "The asymptotic behaviour of the number of bound states of a given potential in the limit of large coupling," *Il Nuovo Cimento A*, vol. 58, no. 1, pp. 191–204, 1968.

[46] F. Brau, "Limits on the number of bound states and conditions for their existence," in *Studies in mathematical physics research*, pp. 1–54, Nova Sci. Publ., New York, 2004.

[47] M. Znojil, "Comment on "Conditionally exactly soluble class of quantum potentials"," *Physical Review A: Atomic, Molecular and Optical Physics*, vol. 61, no. 6, 2000.

[48] L. Euler, "Institutiones Calculi Integralis," *Opera Omnia*, vol. 11, pp. 110–113, 1768.

[49] A. Jonquière, "Note sur la série," *Bulletin de la Société Mathématique de France*, vol. 17, pp. 142–152, 1889.

[50] F. Cooper, A. Khare, and U. Sukhatme, "Supersymmetry and quantum mechanics," *Physics Reports*, vol. 251, no. 5-6, pp. 267–385, 1995.

[51] B. K. Bagchi, *Supersymmetry in Quantum And Classical Mechanics*, Chapman & Hall/CRC, 2000.

[52] R. Sasaki and M. Znojil, "One-dimensional Schrödinger equation with non-analytic potential and its exact Bessel-function solvability," *Journal of Physics A: Mathematical and General*, vol. 49, no. 44, Article ID 445303, 2016.

[53] M. Znojil, "Symmetrized exponential oscillator," *Modern Physics Letters A*, vol. 31, no. 34, 1650195, 11 pages, 2016.

[54] M. Znojil, "Morse potential, symmetric Morse potential and bracketed bound-state energies," *Modern Physics Letters A*, vol. 31, no. 14, p. 1650088, 2016.

[55] S. K. Bose, "Exact bound states for the central fraction power singular potential," *Nuovo Cimento*, vol. 109, no. 11, pp. 1217–1220, 1994.

[56] J. Karwowski and H. A. Witek, "Biconfluent Heun equation in quantum chemistry: Harmonium and related systems," *Theoretical Chemistry Accounts*, vol. 133, no. 7, 2014.

[57] A. G. Ushveridze, *Quasi-Exactly Solvable Models in Quantum Mechanics*, IOP, Bristol, UK, 1994.

[58] C. M. Bender and M. Monou, "New quasi-exactly solvable sextic polynomial potentials," *Journal of Physics A: Mathematical and General*, vol. 38, no. 10, pp. 2179–2187, 2005.

[59] A. V. Turbiner, "One-dimensional quasi-exactly solvable Schrödinger equations," *Physics Reports*, vol. 642, pp. 1–71, 2016.

[60] M. Abramowitz and I. A. Stegun, *Handbook of Mathematical Functions, with Formulas, Graphs, and Mathematical Tables*, Dover, New York, NY, USA, 1972.

17

Mass Spectra and Decay Constants of Heavy-Light Mesons: A Case Study of QCD Sum Rules and Quark Model

Halil Mutuk

Physics Department, Ondokuz Mayis University, 55139 Samsun, Turkey

Correspondence should be addressed to Halil Mutuk; halilmutuk@gmail.com

Guest Editor: Xian-Wei Kang

We visited mass spectra and decay constants of pseudoscalar and vector heavy-light mesons (B, B_s, D, and D_S) in the framework of QCD sum rule and quark model. The harmonic oscillator wave function was used in quark model while a simple interpolating current was used in QCD sum rule calculation. We obtained good results in accordance with the available experimental data and theoretical studies.

1. Introduction

The ultimate objective of particle physics is to investigate and examine the structure and the origin of matter. For this purpose many theoretical and experimental endeavors are made, and a resulting model was theorized, which we call the Standard Model of particle physics. Quark model which was proposed by Gell-Mann and Zweig in 1964 [1] is a part of the Standard Model, and interprets hadrons fairly compatible with the experimental data. According to the quark model, mesons are made of quark-antiquark pairs ($q\overline{q}$) and baryons are made of three quarks qqq or antiquarks $\overline{q}\overline{q}\overline{q}$. These quarks interact with each other via emitting and/or absorbing gluons. The resulting theory which explains these interactions is the Quantum Chromodynamics (QCD).

The interaction of quarks is described by QCD, which is part of the Standard Model of particle physics. QCD is thought to be the *true* theory of strong interactions. QCD is a SU(3) gauge theory describing the interactions of six quarks which transform under the fundamental representation of SU(3) group via the exchange of gluons that transform under the adjoint representation. Although it has been more than 50 years that QCD has been proposed, a solution has been evaded. Contrary to electroweak theory, where it is possible to obtain precise results using perturbation theory, the order of precision obtained in QCD has been lower by orders of magnitude. The main reason for this is that the coupling constant (which should be the perturbation parameter) of QCD is of the order one in low energies; hence the truncation of the perturbative expansion cannot be carried out. However, it is an important subject to study the spectrum of particles predicted by QCD.

Since perturbation theory is not applicable, a nonperturbative approach has to be used to study systems that involve strong interactions. Some of the nonperturbative approaches to strongly interacting systems are the QCD sum rules, quark models, and lattice QCD. The advantage of QCD sum rules and lattice QCD is that they are based on QCD itself, whereas, in quark models, one assumes a potential energy between the quarks and solves a Schrödinger-like equation. The advantage of quark models, on the other hand, is that it allows one to study also the excited states, whereas, in QCD sum rules and lattice QCD, only the ground state or in some exceptional cases the first excited state can be studied.

In quark models one assumes a potential interaction among quarks which makes model as a nonrelativistic approach. Therefore, the systems that are best suited for study in quark models are the heavy quark system which contain c or b quarks. The bare masses of u, d, and s quarks are 2 MeV, 4 MeV, and 96 MeV, respectively [2]. At a first look, quark model seems rather difficult to apply to light quarks. Capstick et al. presented reasonable explanations to link quark models including a minimal amount of relativity to the basics of QCD [3]. Although the pole masses of u, d and s quarks are very low and hence they are relativistic, in constituent quark models, instead of treating the physical u, d, and s quarks, one treats the so-called constituent quarks, which are nothing else than quarks dressed by gluons and other sea quarks inside the hadron. The masses of constituent quarks are around 300 MeV and hence they can also be treated in nonrelativistic quark models. Such an approach has been applied to light quark systems with a surprising success [4–6], leading to that model so-called Constituent Quark Model (CQM), which, based on the Gell Mann-Zweig idea, explains meson and baryon bound systems.

A different situation is for heavy-light quark systems ($Q\overline{q}$). For example an electron is more relativistic in the hydrogen atom (p, e^-) than in the positronium atom (e^+e^-) [7]. Positronium can be taken as a naive model for quarkonium. The binding energy of the positronium is half of the hydrogen atom and is small compared to the electron mass. For this reason the positronium bound state can be described by nonrelativistic quantum mechanics. But the decay of the positronium resonance is a purely relativistic phenomenon. Nevertheless, we can attempt to apply the quark model to heavy-light mesons. The outcome of this attempt is not directly using of Heavy Quark Symmetry (HQS), but one aspect of it. Mesons are two particle systems and the reduced mass is dominated by the light quark mass, $1/\mu = 1/m + 1/M \simeq 1/M$ if $M \gg m$. The spectra for $(c\overline{q})$ and $(b\overline{q})$ should be very similar under this assumption [7]. Indeed reasonable spectroscopy of D and B mesons can be obtained. There is a rich literature for the spectrum and dynamics of the heavy-light mesons, for example [8–25]. In [26], they studied semileptonic D and D_s decays based on the predictions of the relevant form factors from the covariant light-front quark model. In [27], the authors studied the Cabibbo-Kobayashi-Maskawa matrix element $|V_{ub}|$ which is not determined up to now in inclusive or exclusive B decays.

Light quark physics is a key topic to understand the nature of QCD. They can be thought of a probe of the strong interactions by means of nonperturbative effects [28]. Heavy-light meson systems ($Q\overline{q}$) is also central to enlighten the nature of QCD and strong interactions. Heavy-light meson spectroscopy has been the subject of both theoretical and experimental studies since the 2000s. Especially in the charm sector, new excited states were observed in D and D_s mesons [28–32].

An important feature of B meson physics is that it is sensitive to New Physics (NP) Beyond the Standard Model (BSM) via rare decays. Furthermore hadronic decay channels of B mesons might have more systematic uncertainties due to the model indetermination, compared to the lepton/photon decay channels. Thus studying $B \longrightarrow lepton/photon$ decays present a play field for the search of NP. Besides that, B factory experiments BaBar and Belle were built to test the description of quark mixing in the Standard Model. The first theoretical description of quark mixing was proposed by Cabibo in 1963 [33]. One year later in 1964, Christenson et al. discovered CP violation in neutral kaon decays with a tiny friction [34]. This phenomenon is referred to as conclusion that matter and antimatter might behave differently. Kobayashi and Maskawa generalized Cabibbo's idea by adjusting new quarks to the model [35]. In the framework of Standard Model, CP violation can be accommodated by introducing a complex phase in the 3×3 unitary Cabibo-Kobayashi-Maskawa (CKM) matrix. Indeed this phase can be measured in experiments. The cost of adding a parameter is to use a third generation of quarks. CP violation also occurs in B decays. The B factories were built to test for this purpose. B factories gave a substantial contribution to particle physics such as first observation of CP violation apart from the kaon, measurements of CKM matrix elements, measurements of purely leptonic B meson decays, and searches for new physics.

In this work, we obtained mass spectrum and decay constants of the D and B mesons via QCD Sum Rule and a Quark Model potential. We also predicted decay constant for the B_s meson where there are no specific experimental data. Harmonic oscillator wave function is used in the quark model and a sufficiently trivial interpolating current is used in QCD Sum Rule calculations. We studied ground states since they are accessible in the framework of QCD Sum Rules.

2. QCD Sum Rule Formalism

In perturbation theory we assume that the eigenvalues and eigenfunctions can be expanded in a power series as follows:

$$\begin{aligned} E_n &= E_n^0 + \lambda E_n^1 + \lambda^2 E_n^2 + \cdots \\ |n\rangle &= |n^0\rangle + \lambda |n^1\rangle + \lambda^2 |n^2\rangle + \cdots, \end{aligned} \tag{1}$$

where n is the principal quantum number and λ is a parameter. These series are in principle divergent, but they are asymptotic. This means that when the perturbation parameter is small, the first two or three terms are convergent so that the rest of the series can be ignored. In the case of QCD, due to the largeness of the parameter in lower energies, such a truncation, cannot be performed. The nonperturbative aspect of QCD makes it almost impossible to study bound states in terms of perturbation theory. For this reason, there is a need of nonperturbative methods to overwhelm this situation and study bound states. Among others such as Effective Field Theory and Lattice QCD, QCD Sum Rule is maybe the most popular nonperturbative method.

QCD Sum Rule is first formulated by Shifman, Vansthein, and Zakharov for mesons in [36] and generalized to baryons by Iofe in [37]. The basic idea of the this formalism is to study bound state phenomena in QCD from the asymptotic freedom side, *i.e.*, to start evaluation of correlation function at short distances, where quark-gluon dynamics are perturbative and move to larger distances where hadronization occurs, including nonperturbative effects and using some approximate procedure to get information on hadronic properties [38].

To obtain physical observables from QCD sum rules, a correlator of two hadronic currents which is defined as follows

$$\Pi = i \int d^4 x e^{ipx} \left\langle 0 \left| \mathcal{T} j(x) j^\dagger(0) \right| 0 \right\rangle, \tag{2}$$

is studied. Here p is momentum and $j(x)$ is a current composed of quarks and gluon fields with the hadron's quantum numbers. When this operator is applied to vacuum, it can create the hadron that we study. Equation (2) is known as correlation function. The fundamental assumption of the QCD sum rules is that there is a region of p where correlation function can be equivalently described at both quark and hadron sector. The former is known as QCD or OPE (Operator Product Expansion) side, and the latter is known as the phenomenological side. Matching these two sides of the sum rule, one can obtain information about hadron properties [38].

For $p^2 > 0$, resolution of identity operator of hadron states can be written between the operators. This results in correlation function as follows:

$$\Pi = \sum_h \langle 0 | j | h(p) \rangle \frac{1}{p^2 - m_h^2} \langle h(p) | j^\dagger | 0 \rangle + \text{higher states}. \tag{3}$$

It can be seen from (3) the poles in the correlation function, which indicates the presence of hadrons, created by operator $j(x)$.

For $-p^2 \gg \Lambda_{QCD}^2 (p^2 < 0)$, major contribution to the correlation function will come from the $x \sim 0$ region [39]. In this case the product of two operators can be written in terms of OPE:

$$\mathcal{T} j(x) j^\dagger(0) = \sum_d C_d(x) O_d. \tag{4}$$

Here $C_d(x)$ are the coefficients, which can be calculated by the perturbation theory, and $O_d' s$ are the operators with the mass dimension d. If Fourier transformation applies to (4), correlation function can be written as follows:

$$\Pi = \sum_d C_d^f(p) \frac{\langle O_d \rangle}{p^d}, \tag{5}$$

where $\langle O_d \rangle$ are the vacuum condensates that cannot be calculated by perturbation theory except $d = 0$. $d = 0$ corresponds to unitary operator and can be calculated via perturbation theory. Other operators can be written as $\langle \bar{q}q \rangle$ $(d = 3)$; $m_q \langle \bar{q}q \rangle$ $(d = 4)$, $\langle G_{\mu\nu} G^{\mu\nu} \rangle$ $(d = 4)$, $\langle \bar{q} g \sigma G q \rangle$ $(d = 5)$. For $d = 1$ and $d = 2$ there exists no operator. As a result of this, the expansion converges quickly although it is an infinite summation.

In order to get sum rules we must equate (3) and (5). But these two expressions are obtained in different regions of p. By using spectral density representation of correlation function, this matching can be made:

$$\Pi(p^2) = \int_0^\infty \frac{\rho(s)}{s - p^2} + \text{polynomials of } p^2. \tag{6}$$

Spectral density $\rho(s)$ can be acquired from (3). Inserting $\rho(s)$ into (6), an expression of correlation function can be obtained from (3) for $p^2 < 0$ region. If we denote $\rho^{phen}(s)$ as spectral density from (3) and $\rho^{QCD}(s)$ from (5), we get

$$\int_0^\infty ds \frac{\rho^{phen}(s)}{s - p^2} + \text{polynomials} = \int_0^\infty ds \frac{\rho^{QCD}(s)}{s - p^2} + \text{polynomials}. \tag{7}$$

In order to extract physical properties from this expression, one must eliminate the polynomial terms, for example, by using derivatives. In principle, no one knows the polynomial degree and how many polynomials are. The correct procedure is then to use the Borel transformation, which contains infinite derivative:

$$\mathcal{B}_M^2 [\Pi(q^2)] = \lim_{-q^2, n \longrightarrow \infty, -q^2/n = M^2} \frac{-(q^2)^{n+1}}{n!} \left(\frac{d}{dq^2} \right)^n \Pi(q^2) \cdot \tag{8}$$

Here M^2 is defined as the Borel parameter [36]. This transformation effectively removes the polynomials and makes

$$\frac{1}{s - p^2} \longrightarrow e^{-s/M^2}. \tag{9}$$

Then,

$$\sum_h |\langle 0 | j | h(p) \rangle|^2 e^{-m_h^2/M^2} + \text{higher states} = \int_0^\infty \rho^{QCD}(s) e^{-s/M^2}, \tag{10}$$

which resembles QCD parameters and hadronic properties. This equation still shows presence of unknown parameters. The $e^{-m_h^2/M^2}$ factor makes the contribution of small masses dominant. To parameterize contributions of higher states, quark-hadron duality approximation is used. According to quark-hadron duality, for $s > s_0$, $\rho^{phen}(s) \simeq \rho^{QCD}(s)$. $\rho^{phen}(s)$ has contribution of higher states and heavier hadrons when $s > s_0$. s_0 is called as continuum threshold and is related mass of the hadron that is studied in sum rules. So, we can write (10) as follows:

$$|\langle 0 | j | m_h(p) \rangle|^2 e^{-m_h^2/M^2} = \int_0^{s_0} \rho^{QCD}(s) e^{-s/M^2}. \tag{11}$$

In this equation m_h is the hadron of the smallest mass which can be created by j.

Physical properties extracted from the sum rules must be independent of Borel parameter, (M^2). Here we assume that there exist a range of M^2, called Borel window, in which two sides have a good overlap and information on the lowest state can be extracted. Minimum and maximum values of Borel window can be extracted in a way that QCD side convergence gives the minimum value, and the condition that pole contribution should be bigger than the continuum contribution gives the maximum value of Borel window [38].

2.1. Mass Sum Rule. The mass sum rule can be obtained by matching QCD and phenomenological sides of correlation function [36, 38, 39]. Here we will give the following formula:

$$m^2 = \frac{\int_{s_{min}}^{s_0} dse^{-s/M^2} s\rho^{QCD}(s)}{\int_{s_{min}}^{s_0} dse^{-s/M^2} \rho^{QCD}(s)}. \quad (12)$$

2.2. Decay Constant. The decay constant can be obtained from the formula [40] as follows:

$$f_{m_h}^2 = e^{m_h/M^2} \frac{1}{m_h^2} \int_{s_{min}}^{s_0} dse^{-s/M^2} \rho^{QCD}(s), \quad (13)$$

where m_h is the hadron mass extracted from sum rules.

3. Quark Model

Also known as potential model or quark potential model, quark model considers one or more interacting particles under a given potential. In the early 60s quarks were modelled and experimental evidences were found subsequently. This approach provided a reliable basis to study and investigate particle physics and gave compatible results with the experiments.

The most important part of the quark model is the potential. After the November revolution of particle physics in 1974, the year in which charmonium states were observed, new models were proposed to calculate spectrum and radiative transitions [41–43]. The so-called Cornell potential, proposed in [42], reads as

$$V(r) = -\frac{a}{r} + br + c, \quad (14)$$

where a, b, and c are some parameters to extract from fit to the experimental data. This potential is still used with some modifications to account, for example, for hyperfine splittings in the energy levels. The other potentials such as power law potential [44], logarithmic potential [45], Richardson potential [46], Buchmüller-Tye potential [47], and Song-Lin potential [48] were used to fit quarkonium spectra and gave good results in agreement with experiments. These were phenomenological spin-independent potentials and not directly QCD motivated. The interquark potential was not derived from first principles of QCD in the early quarkonium phenomenology. This means, in terms of QCD, that potential is universal (flavour independent) and since quarks are colorless particles, it was reasonable to assume the universality as valid, despite the fact that gluons couple to color charge. These spin-independent potential models performed good but not complete explanation of the energy level splittings. If we want to accommodate these splittings in the theory, we have to take care, *i.e.*, of *spin-spin* and *spin-orbit* interactions in the model. Reference [49] reports an example of a QCD-motivated, spin- and velocity-dependent potential. These potentials deliver reliable results.

TABLE 1: Mass spectra of heavy-light mesons in MeV. QM denotes quark model and SR denotes sum rule calculations. The parameters are $\kappa = 0.471$, $a = 0.192\ GeV^2$, $m_c = 1.320\ GeV$, $m_b = 4.740\ GeV$ [52], $\langle \bar{q}q \rangle = 0.241\ GeV^3$, $m_u = m_d = 0.340\ GeV$, and $m_s = 0.600\ GeV$.

Meson	Exp. [2]	QM	SR	[9]	[14]
D^0/D^+	1869.3 ± 0.4	1859	1972 ± 94	1870.82	1854.7
D_s^+	1968.2 ± 0.4	2056	2118 ± 75	1966.62	1974.5
B^+/B^0	5279.0 ± 0.5	5260	5259 ± 109	5273.50	5277.2
B_s^0	5367.7 ± 1.8	5442	5488 ± 76	5365.99	5384.8

4. Elaboration of the Problem

4.1. QCD Sum Rules. In QCD sum rules, the choice of the $j(x)$ current is important, since it creates hadrons from vacuum. We used the following current:

$$j(x) = i\overline{Q}_a(x)\gamma_5 q_a(x), \quad (15)$$

where Q is heavy quark, q is light quark, a is the color index, and γ_5 is the Dirac matrix. We take care of $m_q \longrightarrow 0$ limit. In the limit of $m_q \longrightarrow 0$, there appears a flavor symmetry between b and c quarks. By this symmetry it is possible to extract information about c and b sector with the same current. b and c quarks are heavy quarks so that it cannot be expected to be in the vacuum by themselves. So it is possible to ignore such condensate terms like $\langle \alpha_s(GG/\pi) \rangle$ and $\langle \bar{q}g_s\sigma Gq \rangle$. By introducing the current term into (2), one can obtain the following spectral density:

$$\begin{aligned} F(s_0, M^2) &= -\langle q\bar{q} \rangle e^{-m_Q^2/M^2} m_Q \\ &+ 6e^{-s_0/M^2} e^{-s_u/M^2} (u_1 - u_2) \\ &\times \left[e^{-s_0/M^2} M^2 \left(m_Q^2 + s(u) + M^2 \right) \right. \\ &\left. - e^{-s_u/M^2} \left(m_Q^4 + M^2 \left(s_0 + M^2 \right) \right) \right] \end{aligned} \quad (16)$$

where $\langle q\bar{q} \rangle$ is the condensate, and u_1 and u_2 are solutions of $s(u) = m_Q^2/(1-u) + m_q^2/u = s_0$.

The mass sum rule can be obtained by taking derivative with respect to $1/M^2$ and dividing the result by (16):

$$m_h^2 = M^4 \frac{1}{F(s_0, M^2)} \frac{dF(s_0, M^2)}{dM^2}. \quad (17)$$

The decay constant sum rule can be obtained as follows:

$$f_{m_h}^2 = e^{m_h^2/M^2} \frac{1}{m_h^2} F(s_0, M^2). \quad (18)$$

The mass values and decays constants for heavy-light mesons are presented in Tables 1 and 2 and Figures 1–8.

4.2. Quark Model. Energy eigenvalues can be obtained by solving the Schrödinger equation in the quark model. The Schrödinger equation reads as follows:

$$H|\Psi_n\rangle = E_n|\Psi_n\rangle, \quad (19)$$

TABLE 2: Pseudoscalar and vector decay constants of heavy-light mesons in MeV. QM denotes quark model and SR denotes sum rule calculations.

Meson	Exp.	QM	SR	[9]	[16]	[53]
D^0/D^+	206 ± 8.9	199	210.25 ± 11.60	205.14	$206.2 \pm 7.3 \pm 5.1$	207.53
D_s^+	249	253	245.70 ± 7.46	241.84	$245.3 \pm 15.7 \pm 4.5$	262.56
B^+/B^0	204 ± 31	209	223.45 ± 12.4	201.09	$193.4 \pm 12.3 \pm 4.3$	208.13
B_s^0		275	277.22 ± 11	292.04	$232.5 \pm 18.6 \pm 2.4$	262.39

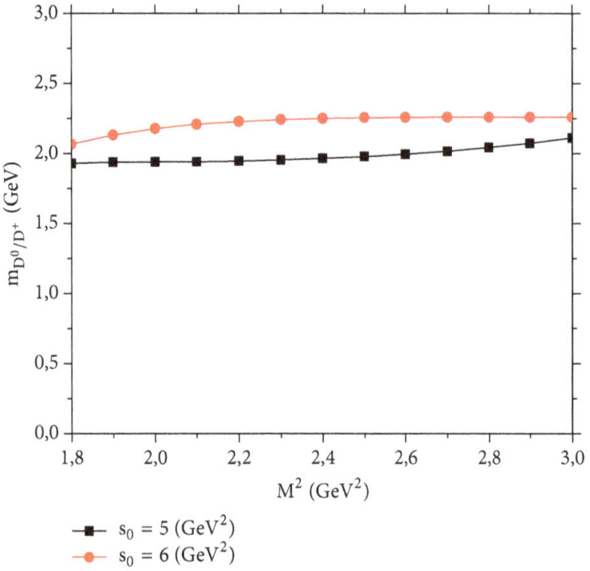

FIGURE 1: Borel parameter dependence of the D^0/D^+ masses.

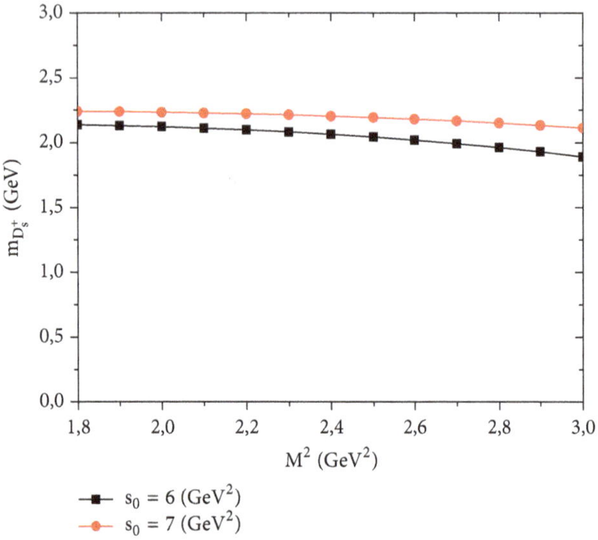

FIGURE 2: Borel parameter dependence of the D_s^+ mass.

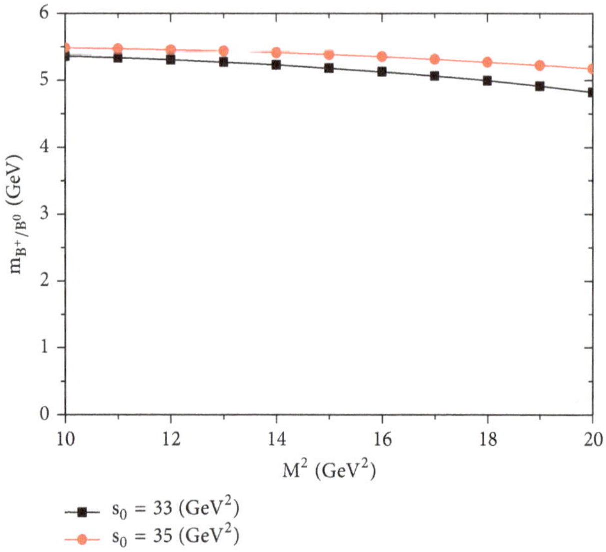

FIGURE 3: Borel parameter dependence of the B^+/B^0 masses.

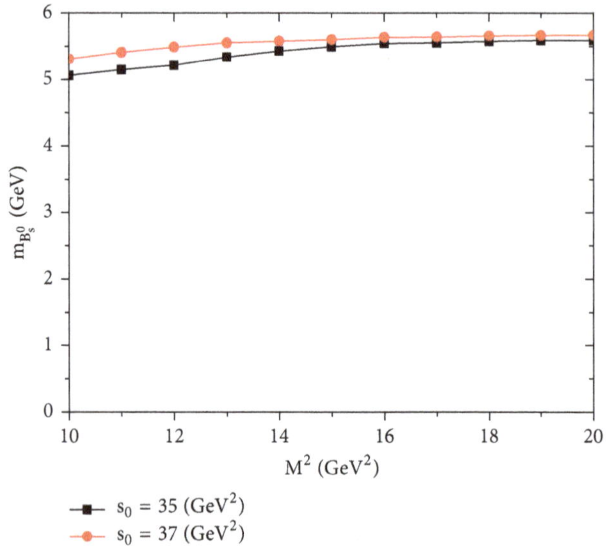

FIGURE 4: Borel parameter dependence of the B_s^0 mass.

where n denotes the principal quantum number. We can separate the wave function into radial R_{nl} and angular parts $Y_{lm}(\theta, \phi)$ as follows:

$$\Psi_{nlm}(r, \theta, \phi) = R_{nl}(r) Y_{lm}(\theta, \phi). \quad (20)$$

R_{nl} is the radial wave function given as follows:

$$R_{nl} = N_{nl} r^l e^{-\nu r^2} L_{(n-l)/2}^{l+1/2}(2\nu r^2), \quad (21)$$

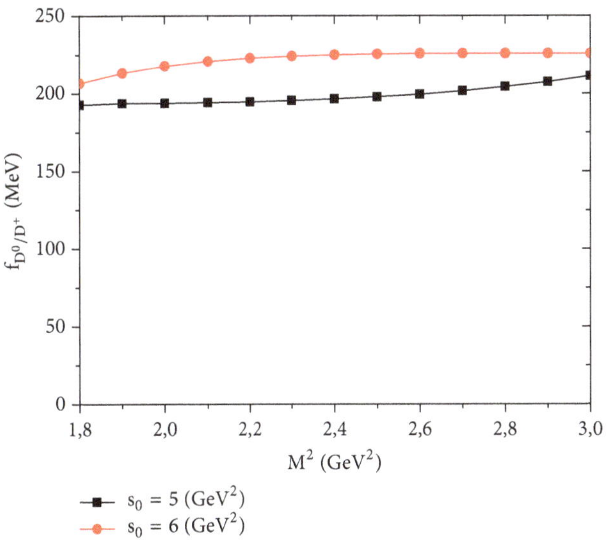

FIGURE 5: Borel parameter dependence of the D^0/D^+ decay constants.

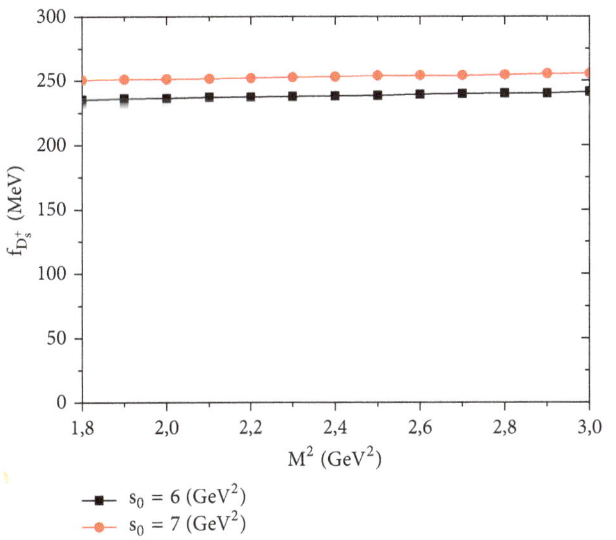

FIGURE 6: Borel parameter dependence of the D_s^+ decay constant.

with the associated Laguerre polynomials $L_{(n-l)/2}^{l+1/2}$ and the normalization constant:

$$N_{nl} = \sqrt{\sqrt{\frac{2\nu^3}{\pi}} \frac{2\left(\left(n-l\right)/2\right)! \nu^l}{\left(\left(n+l\right)/2+1\right)!!}}. \quad (22)$$

With the wave function in hand one can obtain masses as well as decay constants for heavy and light mesons. The mass spectra can be obtained by solving (19). For the decay constants we employ the following formulas:

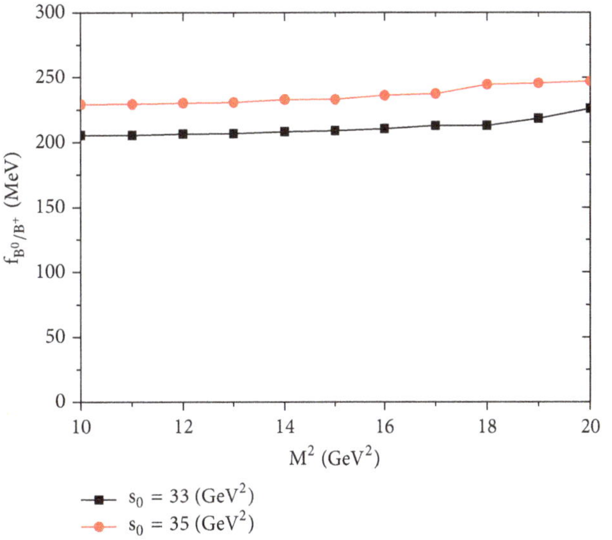

FIGURE 7: Borel parameter dependence of the B^+/B^0 decay constants.

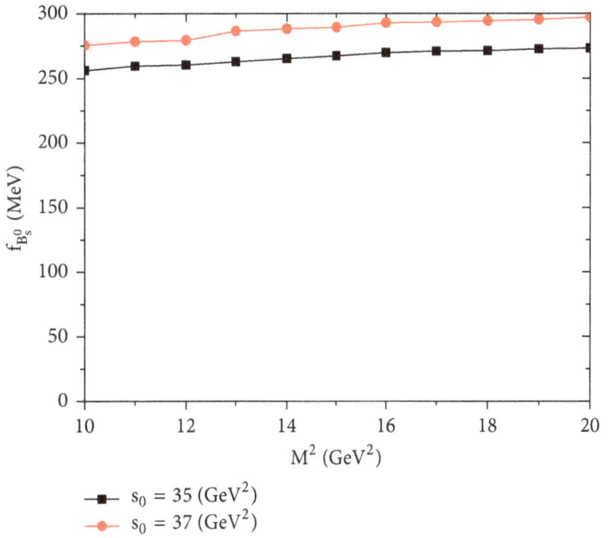

FIGURE 8: Borel parameter dependence of the B_s^0 decay constant.

The results are as follows:

$$f_p = \sqrt{\frac{3}{m_p}} \times \int \frac{d^3k}{(2\pi)^3} \sqrt{1+\frac{m_q}{E_k}} \times \sqrt{1+\frac{m_{\bar{q}}}{E_{\bar{k}}}} \times \left(1 - \frac{k^2}{\left(E_k+m_q\right)\left(E_{\bar{k}}+m_{\bar{q}}\right)}\right) \phi\left(\vec{k}\right), \quad (23)$$

for pseudoscalar mesons and

$$f_v = \sqrt{\frac{3}{m_v}} \times \int \frac{d^3k}{(2\pi)^3} \sqrt{1+\frac{m_q}{E_k}} \times \sqrt{1+\frac{m_{\bar{q}}}{E_{\bar{k}}}} \times \left(1 + \frac{k^2}{3\left(E_k+m_q\right)\left(E_{\bar{k}}+m_{\bar{q}}\right)}\right) \phi\left(\vec{k}\right), \quad (24)$$

for the vector mesons [50].

In the nonrelativistic limit, these two equations take a simple form, which is known to be Van Royen and Weisskopf relation [51]. For the meson decay constants,

$$f_{p/v}^2 = \frac{12\left|\Psi_{p/v}(0)\right|^2}{m_{p/v}}. \quad (25)$$

Here $m_{p/v}$ denotes the pseudoscalar and vector mass of the related meson.

The results are shown in Tables 1 and 2.

5. Summary and Conclusions

In this paper, we calculated mass spectra and decay constants of pseudoscalar and vector heavy-light mesons (B, B_s, D, and D_S) in the framework of QCD sum rule and quark model. Obtained results for masses of B and D mesons are in good agreement with the available experimental data. In the mass spectra, the extrapolation via quark model gave close results to experimental data compared to the QCD sum rule consideration. The QCD sum rules approach gives reasonable but not very good-matching results compared to the experimental values, because of the adopted approximation when evaluating the current, whereas the higher dimension of that operator could improve the estimates. Other potentials and further studies should be taken into consideration for a better understanding.

The heavy-light mesons under study in this paper are well established indeed, and any prediction or reproduction of mass spectrum does not directly guarantee the validity of the model, but shows a possible path to follow for a further investigation. Therefore other physical observables such as decay constants should be experimentally investigated to give more inputs to the theory. For example the only precise value of decay constant is known for D mesons, and systematics are evaluated. The other mesons in this study need more experimental data. For B_s there is no available experimental data. We predicted for the first time decay constant value for B_s in this manner.

Decay constants give information about short distance structure of hadrons. The obtained results for decay constants are in agreement with the other studies and available data. We did not consider in this work relativistic corrections.

In QCD Sum Rule calculations, physical observables must be independent of the Borel parameter. In Figures 1–8 the smoothness of the graphs is compatible with existing data. It is worthy to note that in Figures 1 and 5 the 'slope' of the two curves of D^0/D^+ is not in the same range. The reason for that could be the smallness of the Borel parameter and continuum threshold energy, since correlation function receives main contribution at $s \neq M^2$. On the other hand, the smallness of Borel parameter can blow up the corrections to the perturbative part of the correlation function.

In summary, we obtained good results in accordance with the available data and theoretical studies. As mentioned before, other potential models and interpolating currents can be used to study mass spectra and decay constants. Heavy-light systems in view of the quark model are important to study hadronic interactions. In particular, Heavy Quark Spin Symmetry can play an essential role in heavy-light systems. The higher dimensions of the operators in interpolating currents would deliver more accurate results.

Conflicts of Interest

The author declares that there are no conflicts of interest.

References

[1] M. Gell-Mann, "A schematic model of baryons and mesons," *Physics Letters*, vol. 8, no. 3, pp. 214-215, 1964.

[2] C. Patrignani et al., "Review of Particle Physics," *Chinese Physics C*, vol. 40, article 100001, 2016.

[3] S. Capstick, S. Godfrey, N. Isgur, and J. Paton, "Taking the "naive" and "non-relativistic" out of the quark potential model," *Physics Letters B*, vol. 175, no. 4, pp. 457–461, 1986.

[4] S. Gasiorowicz and J. L. Rosner, "Hadron spectra and quarks," *American Journal of Physics*, vol. 49, no. 10, pp. 954–984, 1981.

[5] B. Juliá-Díaz and D. O. Riska, "Baryon magnetic moments in relativistic quark models," *Nuclear Physics A*, vol. 739, no. 1-2, pp. 69–88, 2004.

[6] K. B. Vijaya Kumar, B. Hanumaiah, and S. Pepin, "Meson spectrum in a relativistic harmonic model with instanton-induced interaction," *The European Physical Journal A*, vol. 19, no. 2, pp. 247–250, 2004.

[7] J.-M. Richard, "An Introduction to Quark Model," 2012, https://arxiv.org/abs/1205.4326.

[8] D. Ebert, R. N. Faustov, and V. O. Galkin, "Radiative M1-decays of heavy–light mesons in the relativistic quark model," *Physics Letters B*, vol. 537, no. 3-4, pp. 241–248, 2002.

[9] K. K. Pathak, D. K. Choudhury, and N. S. Bordoloi, "Leptonic decay of Heavy-light Mesons in a QCD Potential," *International Journal of Modern Physics A*, vol. 28, no. 02, p. 1350010, 2013.

[10] D. Ebert, R. N. Faustov, and V. O. Galkin, "Decay constants of heavy-light mesons in the relativistic quark model," *Modern Physics Letters A*, vol. 17, no. 13, pp. 803–807, 2002.

[11] T. A. Lähde, C. J. Nyfält, and D. O. Riska, "Spectra and M1 decay widths of heavy-light mesons," *Nuclear Physics A*, vol. 674, no. 1-2, pp. 141–167, 2000.

[12] D. U. Matrasulov, F. C. Khanna, and H. Yusupov, "Spectra of heavy-light mesons," *Journal of Physics G: Nuclear and Particle Physics*, vol. 29, no. 3, pp. 475–483, 2003.

[13] B. L. G. Bakker, "Spectrum and decay donstants of heavy-light mesons," *Few-Body Systems*, vol. 44, pp. 91–93, 2008.

[14] B. H. Yazarloo and H. Mehraban, "Mass spectrum and decay properties of heavy-light mesons: D, D_s, B and B_s mesons," *The European Physical Journal Plus*, vol. 132, no. 2, p. 80, 2017.

[15] J. B. Liu and M. Z. Yang, "Heavy-light mesons in a relativistic model," *Chinese Physics C*, vol. 40, no. 7, p. 073101, 2016.

[16] W. Lucha, D. Melikhov, and S. Simula, "Decay constants of heavy pseudoscalar mesons from QCD sum rules," *Journal of Physics G: Nuclear and Particle Physics*, vol. 38, no. 10, Article ID 105002, 2011.

[17] J. Liu and C. Lü, "Spectra of heavy–light mesons in a relativistic model," *The European Physical Journal C*, vol. 77, no. 5, p. 312, 2017.

[18] T. Lesiak, "B Meson Spectroscopy," *Acta Physica Polonica*, vol. 29, pp. 3379–3386, 1998.

[19] H. A. Alhendi, T. M. Aliev, and M. Savcı, "Strong decay constants of heavy tensor mesons in light cone QCD sum rules," *Journal of High Energy Physics*, vol. 2016, no. 4, 2016.

[20] Z. Wang, "Analysis of the masses and decay constants of the heavy-light mesons with QCD sum rules," *The European Physical Journal C*, vol. 75, no. 9, p. 427, 2015.

[21] A. K. Rai, R. H. Parmar, and P. C. Vinodkumar, "Masses and decay constants of heavy-light flavour mesons in a variational scheme," *Journal of Physics G: Nuclear and Particle Physics*, vol. 28, no. 8, pp. 2275–2282, 2002.

[22] T. Huang and C. Luo, "Decay constants of heavy-light mesons in heavy quark effective theory," *Physical Review D: Particles, Fields, Gravitation and Cosmology*, vol. 53, no. 9, pp. 5042–5050, 1996.

[23] A. Duncan, E. Eichten, J. M. Flynn, B. R. Hill, and H. Thacker, "Masses and decay constants of heavy-light mesons using the multistate smearing technique," *Nuclear Physics B (Proceedings Supplements)*, vol. 34, no. C, pp. 445–452, 1994.

[24] P. Gelhausen, A. Khodjamirian, A. A. Pivovarov, and D. Rosenthal, "Erratum: Decay constants of heavy-light vector mesons from QCD sum rules (Phys. Rev. D (2013) 88 (014015))," *Physical Review D: Particles, Fields, Gravitation and Cosmology*, vol. 91, no. 9, 2015.

[25] S. Narison, "Decay Constants of Heavy-Light Mesons from QCD," *Nuclear and Particle Physics Proceedings*, vol. 270-272, pp. 143–153, 2016.

[26] H.-y. Cheng and X.-W. Kang, "Branching fractions of semileptonic D and Ds decays from the covariant light-front quark model," *European Physical Journal C*, vol. 77, no. 9, p. 587, 2017.

[27] X. Kang, B. Kubis, C. Hanhart, and U. Meissner, "Bl4 decays and the extraction of —Vub—," *Physical Review D*, vol. 89, no. 5, Article ID 053015, 2014.

[28] N. Brambilla, S. Eidelman, P. Foka et al., "QCD and strongly coupled gauge theories: challenges and perspectives," *The European Physical Journal C*, vol. 74, no. 10, 2014.

[29] B. Aubert, R. Barate, M. Bona et al., "Observation of a New Ds Meson Decaying to DK at a Mass of 2.86GeV/c2," *Physical Review Letters*, vol. 97, Article ID 222001, 2006.

[30] J. Brodzicka, H. Palka, I. Adachi et al., "Observation of a New D_{sJ} Meson in $B^+ \longrightarrow D^0 D^0 K^+$ Decays," *Physical Review Letters*, vol. 100, Article ID 092001, 2008.

[31] R. Aaij, B. Adeva, M. Adinolfi et al., "Dalitz plot analysis of $B_s^0 \longrightarrow D^0 K^- \pi^+$ decays," *Physical Review*, vol. 90, Article ID 072003, 2014.

[32] R. Aaij, B. Adeva, M. Adinolfi et al., "Observation of Overlapping Spin-1 and Spin-3 D0K- Resonances at Mass 2.86GeV/c2," *Physical Review Letters*, vol. 113, Article ID 162001, 2014.

[33] N. Cabibbo, "Unitary symmetry and leptonic decays," *Physical Review Letters*, vol. 10, no. 12, pp. 531–533, 1963.

[34] J. H. Christenson, J. W. Cronin, V. L. Fitch, and R. Turlay, "Evidence for the 2π Decay of the K02 Meson," *Physical Review Letters*, vol. 13, no. 4, pp. 138–140, 1964.

[35] M. Kobayashi and T. Maskawa, "CP violation in the renormalizable theory of weak interaction," *Progress of Theoretical and Experimental Physics*, vol. 49, pp. 652–657, 1973.

[36] M. A. Shifman, A. I. Vainshtein, and V. I. Zakharov, "QCD and resonance physics applications," *Nuclear Physics B*, vol. 147, no. 5, pp. 448–518, 1979.

[37] B. L. Ioffe, "Calculation of baryon masses in quantum chromodynamics," *Nuclear Physics B*, vol. 188, no. 2, pp. 317–341, 1981.

[38] M. Nielsen, F. S. Navarra, and S. H. Lee, "New charmonium states in QCD sum rules: a concise review," *Physics Reports*, vol. 497, no. 2-3, pp. 41–83, 2010.

[39] P. Colangelo and A. Khodjamirian, "QCD sum rules, a modern perspective," in *at The Frontier of Particle Physics*, M. Shifman, Ed., vol. 3, pp. 1495–1576, 2001.

[40] H. Sundu, "The Mass and Current-Meson Coupling Constant of the Exotic X(3872) State from QCD Sum Rules," *Süleyman Demirel University Journal of Natural and Applied Sciences*, vol. 20, no. 3, p. 448, 2016.

[41] A. De Rújula and S. L. Glashow, "Is bound charm found?" *Physical Review Letters*, vol. 34, no. 1, pp. 46–49, 1975.

[42] E. Eichten, K. Gottfried, T. Kinoshita, J. Kogut, K. D. Lane, and T.-M. Yan, "Spectrum of charmed quark-antiquark bound states," *Physical Review Letters*, vol. 34, no. 6, pp. 369–372, 1975.

[43] T. Appelquist, A. De Rújula, H. D. Politzer, and S. L. Glashow, "Spectroscopy of the new mesons," *Physical Review Letters*, vol. 34, no. 6, pp. 365–369, 1975.

[44] A. Martin, "A fit of upsilon and charmonium spectra," *Physics Letters B*, vol. 93, no. 3, pp. 338–342, 1980.

[45] C. Quigg and J. L. Rosner, "Quarkonium level spacings," *Physics Letters B*, vol. 71, no. 1, pp. 153–157, 1977.

[46] J. L. Richardson, "The heavy quark potential and the Υ, J/ψ systems," *Physics Letters B*, vol. 82, no. 2, pp. 272–274, 1979.

[47] W. Buchmüller and S.-H. H. Tye, "Quarkonia and quantum chromodynamics," *Physical Review D: Particles, Fields, Gravitation and Cosmology*, vol. 24, no. 1, pp. 132–156, 1981.

[48] S. Xiaotong and L. Hefen, "A new phenomenological potential for heavy quarkonium," *Zeitschrift für Physik C Particles and Fields*, vol. 34, no. 2, pp. 223–231, 1987.

[49] E. Eichten and F. Feinberg, "Spin-dependent forces in quantum chromodynamics," *Physical Review D: Particles, Fields, Gravitation and Cosmology*, vol. 23, no. 11, pp. 2724–2744, 1981.

[50] O. Lakhina and E. S. Swanson, "Dynamic properties of charmonium," *Physical Review D: Particles, Fields, Gravitation and Cosmology*, vol. 74, no. 1, 2006.

[51] R. Van Royen and V. F. Weisskopf, "Protsyessy raspada adronov i modyelcyrillic small soft sign kvarkov," *Il Nuovo Cimento A*, vol. 50, no. 3, pp. 617–645, 1967.

[52] S. Jacobs, M. G. Olsson, and C. Suchyta, "Comparing the Schrödinger and spinless Salpeter equations for heavy-quark bound states," *Physical Review D: Particles, Fields, Gravitation and Cosmology*, vol. 33, no. 11, pp. 3338–3348, 1996.

[53] S. Nam, "Extended nonlocal chiral-quark model for the D- and B- meson weak-decay constants," *Physical Review D*, vol. 85, no. 3, Article ID 034019, 2012.

18

Remark on Remnant and Residue Entropy with GUP

Hui-Ling Li ,[1] Wei Li,[1] and Yi-Wen Han[2]

[1]*College of Physics Science and Technology, Shenyang Normal University, Shenyang 110034, China*
[2]*School of Computer Science and Information Engineering, Chongqing Technology and Business University, Chongqing 400070, China*

Correspondence should be addressed to Hui-Ling Li; lhl51759@126.com

Academic Editor: Edward Sarkisyan-Grinbaum

In this article, close to the Planck scale, we discuss the remnant and residue entropy from a Rutz-Schwarzschild black hole in the frame of Finsler geometry. Employing the corrected Hamilton-Jacobi equation, the tunneling radiation of a scalar particle is presented, and the revised tunneling temperature and revised entropy are also found. Taking into account generalized uncertainty principle (GUP), we analyze the remnant stability and residue entropy based on thermodynamic phase transition. In addition, the effects of the Finsler perturbation parameter, GUP parameter, and angular momentum parameter on remnant and residual entropy are also discussed.

1. Introduction

Although a complete self-consistent theory of quantum gravity has not been established, it is an effective way to understand the behavior of gravity by combining various models of quantum gravitational effects. In various quantum gravitation models, such as string theory, loop quantum gravity, and non-commutative geometry, it is believed that there exists a minimum observable length, and this minimum observable length should have the order of the Planck scale. Considering a minimum observable length, a Hilbert space representation of quantum mechanics has been formulated by Kempf et al. in [1], which plays an important role on modifications of general relativity and black hole physics for research programs in recent years. Furthermore, Nozari and Etemad generalized the seminal work of Kempf et al. to the case that there is also a maximal particles' momentum, which resolved some shortcomings such as an infinite amount of energy for a free test particle and provided several novel and interesting features [2]. Especially, very recently, Nozari et al. have addressed the origin of natural cutoff, including a minimal measurable length, and suggested that quantum gravity cutoffs are global (topological) properties of the symplectic manifolds [3]. Now, one of the research directions of quantum gravity is to construct new theoretical models from the minimum observable scale, including Double Special Relativity (DSR) [4], Gravity's Rainbow [5, 6], Modified Dispersion Relation (MDR) [7], and Generalized Uncertainty Principle (GUP) [1, 8–10]. Due to the problems of semiclassical tunneling radiation, the quantum gravitational effect caused by GUP is considered to study the quantum tunneling radiation of black holeswhich leading to a very interesting result: GUP can prevent a black hole from evaporating completely, leaving remnant.

Recently, based on GUP, people began to study the quantum tunneling and remnant of black hole. Combining the GUP with the minimum observed length into the tunneling method of Parikh and Wilczek, Nozari and Mehdipour [11] have successfully derived the quantum tunneling radiation of scalar particles from a Schwarzschild black hole. They pointed out that GUP has an important correction to the final state of black hole evaporation, and the black hole cannot evaporate completely. Following that, the possible effects of natural cutoffs as a minimal length, a maximal momentum, and a minimal momentum on the quantum tunneling have been completely discussed [12]. Later, by utilizing the deformed Hamilton-Jacobi equations, Benrong et al. pointed out that breaking covariance in quantum gravity effective models is a key for a black hole to have the remnant left in the evaporation [13]. In addition, in the frame of improved exponential GUP [14], the remnant of a Schwarzschild black hole has been given at the end of the evaporation process [15]. On the other hand, considering the quantum gravitational effect under the influence of GUP, adopting the fermion tunneling method and the modified Dirac equation, Chen and Wang etc. [16–18] elaborated the fermion tunneling of black holes and black rings. It is shown that GUP may slow down the temperature of the black hole increase and prevent the black hole from evaporating completely, which leads to the existence of the minimum non-zero mass, that is, the black hole remnant. Subsequently, the quantum tunneling radiation and remnant of Gödel black hole and high dimensional Myers-Perry black hole have also been deeply studied [19–21]. Above research proves that the black hole remnant can exist, but related issues of the stability of remnant and residue entropy have been comparatively less discussed for a Finsler black hole.

Although Einstein's general relativity described by Riemannian geometry is one of the most successful gravitational theories, it still has some problems in explaining the accelerating expansion of the universe and establishing a complete theory of quantum gravity. One has considered that the difficulties caused by general relativity may have something to do with the mathematical tools it uses. So, people try to establish a modified gravitational theory described by Finsler geometry. Finsler geometry is the most general differential geometry, which regards Riemannian geometry as its special case, and it is just Riemannian geometry without quadratic restriction [22]. In recent years, the application of Finsler geometry in black hole physics has gradually aroused people's interest. People began to construct field equations of all kinds of Finsler spacetime. With the study of Finsler field equation, people try to construct various solutions of Finsler black hole [22–25]. The study of these black hole solutions makes us understand the physical properties of Finsler spacetime more deeply. In this paper, based on GUP, we take a simple Finsler black hole as an example and investigate the remnant and residue entropy from a Rutz-Schwarzschild black hole.

2. Tunneling Radiation of a Scalar Particle Based on GUP

In this section, considering the effect of GUP, applying with the corrected Hamilton-Jacobi equation, we will focus on investigating the scalar particle's tunneling radiation from a Rutz-Schwarzschild black hole. Under the frame of Finsler geometry, Rutz constructed the generalized Einstein field equation and derived a Finsler black hole solution. The metric (Rutz-Schwarzschild black hole) is given by [24]

$$ds^2 = -\left(1-\frac{2M}{r}\right)\left(1-\frac{\varepsilon d\Omega}{dt}\right)dt^2 + \left(1-\frac{2M}{r}\right)^{-1}dr^2 + r^2 d\Omega^2. \quad (1)$$

Here $\varepsilon \ll 1$ is the Finsler perturbation parameter, and the line element reduces to the Schwarzschild metric when $\varepsilon = 0$. The solution is a non-Riemannian solution, and the time component of the metric depends on the tangent vector $d\Omega/dt$. It is obvious that the correction term $\varepsilon d\Omega/dt$ remains while the mass vanishes and still exists when $r \longrightarrow \infty$. The metric is very different from the Schwarzschild black hole, and it is very interesting to discuss on the scalar particle's tunneling radiation based on GUP.

By taking into account the effect of GUP, in a curved spacetime, the revised Hamilton-Jacobi equation for the motion of scalar particles can be expressed as [26]

$$g^{00}\left(\partial_0 S + eA_0\right)^2 + \left[g^{kk}\left(\partial_k S + eA_k\right)^2 + m^2\right] \times \left\{1 - 2\beta\left[g^{jj}\left(\partial_j S\right)^2 + m^2\right]\right\} = 0. \quad (2)$$

Here $\beta = \beta_0 l_p^2/\hbar^2 = \beta_0/M_p^2c^2$, $\beta_0(\le 10^{34})$ is a dimensionless constant and l_P and M_P are Planck length and Planck mass. Adopting rational approximation, according to the line element and the modified Hamilton-Jacobian equation, we can get the following motion equation of a scalar particle:

$$\begin{aligned}&-\left[\left(1-\frac{2M}{r}\right)\left(1-\frac{\varepsilon d\Omega}{dt}\right)\right]^{-1}\left(\partial_t S\right)^2 - \left[\left(1-\frac{2M}{r}\right)\left(\partial_r S\right)^2 + \frac{1}{r^2}\left(\partial_\theta S\right)^2\right.\\&\left. + \frac{1}{r^2\sin^2\theta}\left(\partial_\varphi S\right)^2 + m^2\right]\left\{1 - 2\beta\left[\left(1-\frac{2M}{r}\right)\left(\partial_r S\right)^2\right.\right.\\&\left.\left. + \frac{1}{r^2}\left(\partial_\theta S\right)^2 + \frac{1}{r^2\sin^2\theta}\left(\partial_\varphi S\right)^2 + m^2\right]\right\} = 0.\end{aligned} \quad (3)$$

Setting $S = -\omega t + W(r) + \Theta(\theta, \varphi)$, with ω standing for the energy of a scalar particle and inserting the action S into (3), we have

$$\begin{aligned}&2\beta\left(1-\frac{2M}{r}\right)^2\left(\partial_r W(r)\right)^4 - \left(1-\frac{2M}{r}\right)\left(\partial_r W(r)\right)^2\\&+\left[\left(1-\frac{2M}{r}\right)\left(\partial_r W(r)\right)^2 + \frac{1}{r^2}\left(\partial_\theta \Theta(\theta,\varphi)\right)^2\right.\\&\left.+\frac{1}{r^2\sin^2\theta}\left(\partial_\varphi \Theta(\theta,\varphi)\right)^2\right] \times \left[4\beta\left(1-\frac{2M}{r}\right)\right.\\&\left.\cdot\left(\partial_r W(r)\right)^2 - 1\right] + \omega^2\left[\left(1-\frac{2M}{r}\right)\left(1-\frac{\varepsilon d\Omega}{dt}\right)\right]^{-1}\\&= -\lambda\end{aligned} \quad (4)$$

and

$$2\beta\left[\frac{1}{r^2}\left(\partial_\theta \Theta(\theta,\varphi)\right)^2 + \frac{1}{r^2\sin^2\theta}\left(\partial_\varphi \Theta(\theta,\varphi)\right)^2\right] = \lambda, \quad (5)$$

For (5), the mode of angular momentum of a tunneling particle is associated with its components $\partial_\theta\Theta$ and $\partial_\varphi\Theta$ as follows [27]:

$$\frac{1}{r^2}\left(\partial_\theta S\right)^2 + \frac{1}{r^2\sin^2\theta}\left(\partial_\varphi S\right)^2 = L^2. \quad (6)$$

Equations (5) and (6) yield

$$2\left(L^2\right)^2 = \frac{\lambda}{\beta}. \quad (7)$$

Formula (7) shows that λ is related to the angular momentum L of the scalar particle, and it is pointed out that the angular part will have an effect on the tunneling radiation of the scalar particle. Thus, (4) can change to

$$2\beta\left(1-\frac{2M}{r}\right)^2(\partial_r W(r))^4+\left(4m^2\beta+\sqrt{8\beta\lambda}-1\right)\times\left(1-\frac{2M}{r}\right)(\partial_r W(r))^2+\left(2m^2\beta+\sqrt{8\beta\lambda}-1\right)m^2+\omega^2\left[\left(1-\frac{2M}{r}\right)\left(1-\frac{\varepsilon d\Omega}{dt}\right)\right]^{-1}-\sqrt{\frac{\lambda}{2\beta}}+\lambda=0. \tag{8}$$

Equation (8) is solved and the higher-order term of β is ignored; then $W(r)$ is taken as

$$W(r)_\pm=\pm\int\frac{\sqrt{(1-2M/r)(1-\varepsilon d\Omega/dt)\left(m^2(1-2\beta m^2)-\lambda+\sqrt{\lambda/2\beta}\right)+\omega^2}}{\sqrt{(1-\varepsilon d\Omega/dt)}(1-2M/r)}\times\left\{1+\beta\left[m^2+\omega^2\left[\left(1-\frac{2M}{r}\right)\left(1-\frac{\varepsilon d\Omega}{dt}\right)\right]^{-1}\right]+\frac{\beta\lambda}{2}\right\}dr, \tag{9}$$

where $+(-)$ represents the solution of the outgoing (ingoing) wave. The Laurent series is expanded at the event horizon $r=r_+$, and the solution of (9) is obtained by using the contour integral

$$W(r_+)_\pm=\pm i\frac{2\pi M\omega}{\sqrt{1-\varepsilon d\Omega/dt}}\left[1+\frac{1}{2}\beta\left(3m^2+\lambda+\frac{4\omega^2}{1-\varepsilon d\Omega/dt}\right)+\sqrt{\frac{\beta\lambda}{8}}\right]+\texttt{real part}. \tag{10}$$

Applying the invariant tunneling rate under the canonical transformation [28, 29], considering the influence of the time part on the tunneling rate, through the Kruskal coordinate (T,R), that is, $T=e^{\kappa r^*}\sinh(\kappa t)$ and $R=e^{\kappa r^*}\cosh(\kappa t)$, we get the contribution of the extra imaginary part of the time segment and have $\operatorname{Im}\omega t_{out(in)}=-\pi\omega/2\kappa$. As a result, the total tunneling rate of a scalar particle passing through the event horizon of Rutz-Schwarzschild black hole is as follows:

$$\Gamma\propto\exp\left\{\left[\operatorname{Im}(\omega t_{out})+\operatorname{Im}(\omega t_{in})-\operatorname{Im}\oint P dr\right]\right\}=\exp\left\{-4\pi\frac{2M\omega}{\sqrt{1-\varepsilon d\Omega/dt}}\left[1+\frac{1}{2}\beta\left(3m^2+\lambda+\frac{4\omega^2}{1-\varepsilon d\Omega/dt}\right)+\sqrt{\frac{\beta\lambda}{8}}\right]\right\}. \tag{11}$$

Compared with the Boltzmann factor $\Gamma=\exp(-\omega/T)$, the corrected tunneling temperature of the black hole is

$$T=\frac{\sqrt{1-\varepsilon d\Omega/dt}}{8\pi M}\left[1+\frac{1}{2}\beta\left(3m^2+\lambda+\frac{4\omega^2}{1-\varepsilon d\Omega/dt}\right)+\sqrt{\frac{\beta\lambda}{8}}\right]^{-1}=T_H\left[1+\frac{1}{2}\beta\left(3m^2+\lambda+\frac{4\omega^2}{1-\varepsilon d\Omega/dt}\right)+\sqrt{\frac{\beta\lambda}{8}}\right]^{-1}. \tag{12}$$

Here $T_H=\sqrt{1-\varepsilon d\Omega/dt}/8\pi M$ is the Hawking temperature. According to the first law of black hole thermodynamics, the entropy of Rutz-Schwarzschild black hole is calculated as

$$S=\int\frac{dM}{T}=\int\frac{8\pi M}{\sqrt{1-\varepsilon d\Omega/dt}}\left[1+\frac{1}{2}\beta\left(3m^2+\lambda+\frac{4\omega^2}{1-\varepsilon d\Omega/dt}\right)+\sqrt{\frac{\beta\lambda}{8}}\right]dM. \tag{13}$$

From (11)–(13), we obviously find that the scalar particle's tunneling rate, tunneling temperature, and the entropy are all not only dependent on the black hole mass M, tunneling scalar particle's mass m, and energy ω, but also dependent on Finsler perturbation parameter ε, GUP parameter β, and angular momentum parameter λ.

3. Remnant and Entropy Based on Thermodynamics Phase Transition

On the basis of the above scalar particle's tunneling radiation, considering the GUP, now we focus on discussing on the remnant and entropy at the end of evaporation. Since all the tunneling particles at the event horizon can be regarded as massless, the mass of scalar particles is no longer considered in the following process. According to the uncertainty relation $\Delta p \geq \hbar/\Delta x$ and the lower limit of tunneling particle energy [30, 31] $\omega \geq \hbar/\Delta x$, near the event horizon, the uncertainty of the position can be taken as the radius of the black hole [30, 31]; that is, $\Delta x \approx r_{BH} = r_+$. Consequently, the tunneling temperature of the black hole evaporating to Planck scale is

$$T = \frac{\sqrt{1-\varepsilon d\Omega/dt}}{8\pi M}\left\{1+\frac{1}{2}\beta\left(\lambda+\frac{4\omega^2}{1-\varepsilon d\Omega/dt}\right)+\sqrt{\frac{\beta\lambda}{8}}\right\}^{-1} = T_H\left\{1-\frac{1}{2}\beta\left(\lambda+\frac{4\hbar^2}{(1-\varepsilon d\Omega/dt)\,r_+^2}\right)-\sqrt{\frac{\beta\lambda}{8}}\right\}. \tag{14}$$

It can be seen that the modified tunneling temperature near Planck scale is related to the properties of the background spacetime of Finsler black hole, the energy of a tunneling particle, and the parameter of quantum gravitational effect. When the radius of the black hole satisfies

$$r_+ < \sqrt{\frac{8\beta\hbar^2}{(1-\varepsilon d\Omega/dt)\left(4-2\beta\lambda-\sqrt{2\beta\lambda}\right)}}, \tag{15}$$

the revised tunneling temperature $T < 0$ violates the third law of thermodynamics. This means that, by considering the effect of GUP, the evaporation will stop when the tunneling temperature is infinitely close to absolute zero, which leads to a minimum radius; namely,

$$r_i = \sqrt{\frac{8\beta\hbar^2}{(1-\varepsilon d\Omega/dt)\left(4-2\beta\lambda-\sqrt{2\beta\lambda}\right)}} = \ell_p\sqrt{\frac{8\hbar^2\beta_0}{(1-\varepsilon d\Omega/dt)\left(4\hbar^2-2\lambda\beta_0\ell_p^2-\ell_p\hbar\sqrt{2\lambda\beta_0}\right)}}. \tag{16}$$

Expression (16) is represented by the mass. By using the relation between event horizon and mass $r_+ = 2M$, near the Planck scale, the tunneling temperature can be expressed as

$$T = \frac{\sqrt{1-\varepsilon d\Omega/dt}}{8\pi GM}\left\{1-\frac{1}{2}\beta\left(\lambda+\frac{\hbar^2}{(1-\varepsilon d\Omega/dt)\,G^2M^2}\right)-\sqrt{\frac{\beta\lambda}{8}}\right\}. \tag{17}$$

In order to ensure the temperature $T \geq 0$, the mass of the black hole satisfies

$$M \geq \sqrt{\frac{2\beta\hbar^2}{(1-\varepsilon d\Omega/dt)\left(4-2\beta\lambda-\sqrt{2\beta\lambda}\right)G^2}}, \tag{18}$$

which implies the minimum mass of a black hole

$$M_{min} = \sqrt{\frac{2\beta\hbar^2}{(1-\varepsilon d\Omega/dt)\left(4-2\beta\lambda-\sqrt{2\beta\lambda}\right)G^2}} = M_p\sqrt{\frac{2\beta_0\hbar^2}{c^2(1-\varepsilon d\Omega/dt)\left(4-2\lambda\beta_0/M_p^2c^2-\sqrt{2\lambda\beta_0/M_p^2c^2}\right)}}. \tag{19}$$

The value is the Rutz-Schwarzschild black hole's remnant; that is, $M_{res} = M_{min}$. In order to explain the problem better, we can analyze heat capacity C, which is

$$C = T\left(\frac{\partial S}{\partial T}\right) = T\frac{\partial S}{\partial M}\frac{\partial M}{\partial T} = T\frac{\partial S}{\partial M}\left(\frac{\partial T}{\partial M}\right)^{-1} = \frac{N}{F}. \tag{20}$$

Here S is the modified black hole entropy; at the Planck scale it can be rewritten as

$$S = \int\frac{dM}{T} = \int\frac{8\pi GM}{\sqrt{1-\varepsilon d\Omega/dt}}\left\{\left(1+\frac{1}{2}\beta\lambda+\sqrt{\frac{\beta\lambda}{8}}\right)+\left(\frac{\beta\hbar^2}{2(1-\varepsilon d\Omega/dt)\,G^2M^2}\right)\right\}dM = \frac{4\pi GM^2}{\sqrt{1-\varepsilon d\Omega/dt}}\left(1+\frac{1}{2}\beta\lambda+\sqrt{\frac{\beta\lambda}{8}}\right)+\frac{4\pi\beta\hbar^2}{G(1-\varepsilon d\Omega/dt)^{3/2}}\ln M, \tag{21}$$

and the coefficients N and F are, respectively,

$$N = \left[2G^2M^2(1-\varepsilon d\Omega/dt)\left(1-\frac{1}{2}\beta\lambda-\sqrt{\frac{\beta\lambda}{8}}\right)-\beta\hbar^2\right]\times\left[16\pi G^3M^4\sqrt{1-\varepsilon d\Omega/dt}\right] \tag{22}$$

and

$$F = \left[2G^2M^2(1-\varepsilon d\Omega/dt)\left(1+\frac{1}{2}\beta\lambda+\sqrt{\frac{\beta\lambda}{8}}\right)+\beta\hbar^2\right]\times\left[3\beta\hbar^2-2G^2M^2\left(1-\frac{1}{2}\beta\lambda-\sqrt{\frac{\beta\lambda}{8}}\right)(1-\varepsilon d\Omega/dt)\right]. \tag{23}$$

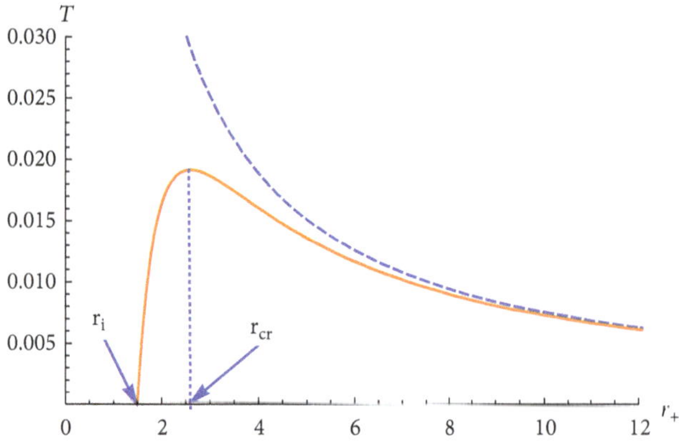

FIGURE 1: The figure shows the temperature T versus the radius of event horizon r_+ for varying β. Above blue dashed curve is $\gamma = 0.1$ and $\beta = 0$. Below orange curve is $\gamma = 0.1$ and $\beta = 1$.

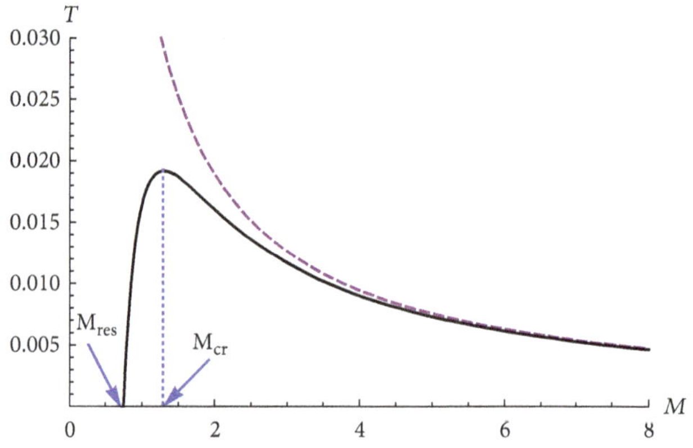

FIGURE 2: The figure shows the temperature T versus the mass M for varying β. Above purple dashed curve is $\gamma = 0.1$ and $\beta = 0$. Below black curve is $\gamma = 0.1$ and $\beta = 1$.

From expression (20), we find

$$M = M_{Cr} = M_p \sqrt{\frac{6\beta_0 \hbar^2}{c^2 (1 - \varepsilon d\Omega/dt) \left(4 - 2\lambda\beta_0/M_p^2c^2 - \sqrt{2\lambda\beta_0/M_p^2c^2}\right)}}, \quad (24)$$

$$C \longrightarrow 0,$$

$$M \longrightarrow M_{min} = M_{res} = M_p \sqrt{\frac{2\beta_0 \hbar^2}{c^2 (1 - \varepsilon d\Omega/dt) \left(4 - 2\lambda\beta_0/M_p^2c^2 - \sqrt{2\lambda\beta_0/M_p^2c^2}\right)}}, \quad (25)$$

$$C \longrightarrow 0.$$

Expression (25) is consistent with expression (19). From the above equations we can see that, for the Rutz-Schwarzschild black hole, considering the tunneling of scalar particles, when the black hole evaporates to the Planck scale, the phase transition takes place, leaving a stable remnant. Then, based on expressions (14), (17), (20), and (21), we draw the following thermodynamic curves 1—5 and further analyze the entropy and remnant in detail. We set $\gamma = \varepsilon d\Omega/dt$ and $M_p = c = \hbar = 1$ in all of the following drawings for research convenience.

As Figures 1 and 2 show, in a large range of radius and mass, for the cases $\beta = 0$ and $\beta = 1$, the tunneling temperatures tend to be consistent. But as the radius (or mass) closes to the critical value r_{cr} (or M_{cr}), these curves become markedly different. The tunneling temperature reaches the maximum value at r_{cr} (or M_{cr}) for the case of $\beta = 1$; then it decreases with the decrease of radius (or mass). Finally, the tunneling temperature tends to zero when the radius (or mass) attains to the minimum value r_i (or M_{res}) for the case of $\beta = 1$, which leads to remnant. We can also explain it through Figure 3.

In Figure 3, the heat capacity monotonically increases with the mass increasing for the case of $\beta = 0$, and it is always negative. However, for the case of $\beta = 1$, we find that, at $M = M_{cr}$, heat capacity is divergent, which means the existence of phase transition. That is the black hole undergoes

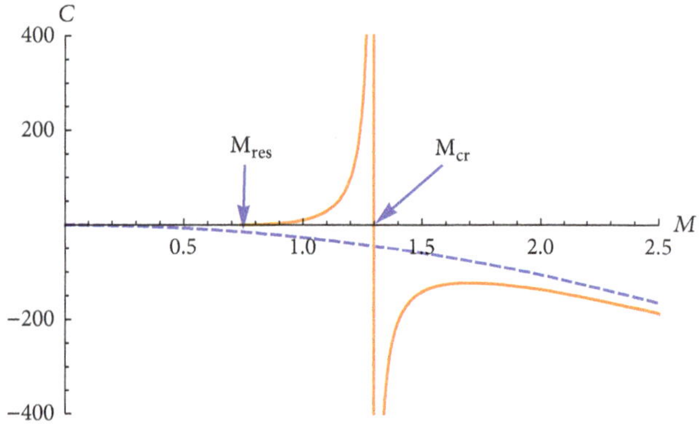

FIGURE 3: The figure shows the heat capacity C versus the mass M for varying β. The blue dashed curve is $\gamma = 0.1$ and $\beta = 0$. The orange curve is $\gamma = 0.1$ and $\beta = 1$.

a phase transition from $C < 0$ (unstable hole) to $C > 0$ (stable hole), which also corresponds to the phase transitions in Figures 1 and 2 at the critical values r_{cr} and M_{cr}. Furthermore, we can see that the mass tends to minimum value (namely $M_{min} = M_{res}$) when $C \longrightarrow 0$. In addition, we also note that the same problem is treated in noncommutative geometry in [32] (see Figure 7). The result shows that the heat capacity tends to zero when the mass approaches a certain value, but some unusual feature is also presented.

In Figures 4 and 5, the crucial difference is shown for the two curves. To clearly see the difference between them, we can refer to Figure 5. The traditional entropy is zero as the mass becomes zero for the case of $\beta = 0$. However, by considering GUP, the entropy is also revised for the case of $\beta = 1$, and there exits minimum S_{min} (namely, residual entropy $S_{res} = S_{min}$) at the remnant, which means the entropy is no longer zero in the final stages of black hole evaporation.

To see clearly the Finsler perturbation parameter and angular momentum parameter effect on remnant and residual entropy, we present Tables 1 and 2 for different γ and λ when $\beta = 1$. From Tables 1 and 2, we can find that remnant and residual entropy increase as Finsler perturbation

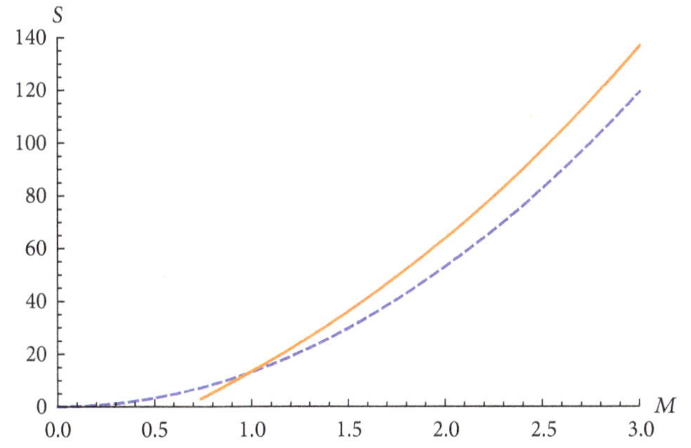

FIGURE 4: The figure shows the entropy S versus the mass M for varying β on a large scale of entropy. The blue dashed curve is $\gamma = 0.1$ and $\beta = 0$. The orange curve is $\gamma = 0.1$ and $\beta = 1$.

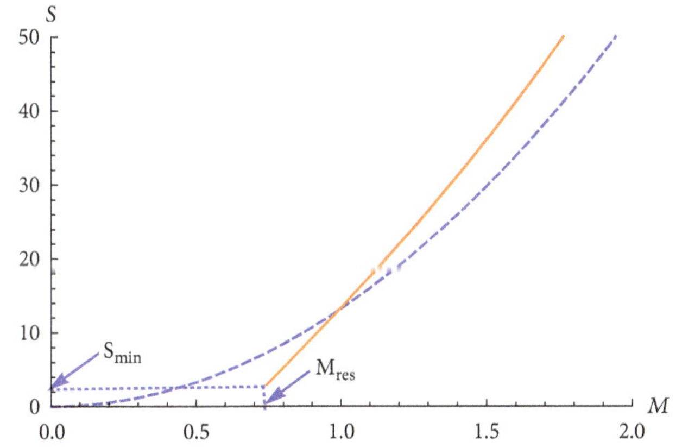

FIGURE 5: The figure shows the entropy S versus the mass M for varying β within a small scale of entropy. The blue dashed curve is $\gamma = 0.1$ and $\beta = 0$. The orange curve is $\gamma = 0.1$ and $\beta = 1$.

TABLE 1: Remnant and residual entropy with $\lambda = 0.001$ for different γ.

γ	M_{res}	S_{res}
0.1	0.749748	3.29388
0.01	0.714856	1.90766
0.0001	0.711309	1.76666

TABLE 2: Remnant and residual entropy with $\gamma = 0.1$ for different λ.

λ	M_{res}	S_{res}
0.5	1.054090	22.8521
0.1	0.814124	7.17324
0.01	0.760867	3.95553

parameter increases, and they are also increase as angular momentum parameter increases.

In addition, when $\beta = 0$ and $\varepsilon = 0$ (namely, $\gamma = 0$), the results reduce to that of Schwarzschild black hole without GUP. In this case, the black hole can radiate continuously until the mass and entropy decrease to zero. In the case of $\beta = 1$ and $\varepsilon = 0$, the results reduce to that of Schwarzschild black hole with GUP. From (21) and (25), we can see that the remnant and the residue entropy in the frame of Finsler geometry are larger than those of Schwarzschild black hole. Moreover, we also see the influence of Finsler parameter on remnant and residue entropy from Table 1.

4. Discussion on Remnant Based on Parikh-Wilczek Tunneling

Based on GUP, using Parikh-Wilczek tunneling method, we continue to explore the quantum tunneling and remnant of a Rutz-Schwarzschild black hole in the frame of Finsler geometry. The Rutz-Schwarzschild metric in the Painleve coordinate system can be rewritten as

$$ds^2 = -\left(1 - \frac{2M}{r}\right)\left(1 - \frac{\varepsilon d\Omega}{dt}\right)dt'^2 + 2\sqrt{\frac{2M}{r}\left(1 - \frac{\varepsilon d\Omega}{dt}\right)}dt'dr + dr^2 + r^2 d\Omega^2, \tag{26}$$

which is obtained by the coordinate transformation

$$dt \longrightarrow dt' - \frac{\sqrt{2M/r}}{1 - 2M\sqrt{1 - \varepsilon d\Omega/dt}/r}dr. \tag{27}$$

We can derive the radial null geodesics

$$\dot{r} \equiv \frac{dr}{dt'} = \sqrt{1 - \frac{\varepsilon d\Omega}{dt}}\left(\pm 1 - \sqrt{\frac{2M}{r}}\right), \tag{28}$$

where $+(-)$ represents the solution of the outgoing (ingoing) geodesics. Now, we consider the influence of GUP with a minimal length on quantum tunneling. The expression GUP with a minimal length is [12]

$$\Delta x \Delta p \geq \hbar\left[1 + \beta(\Delta p)^2\right], \tag{29}$$

and the generalized energy [11] is

$$\omega = E\left(1 + \beta E^2\right), \tag{30}$$

where $\beta = \alpha^2 l_p^2$. From the GUP expression, the revised commutation relation between the radial coordinate and the conjugate momentum becomes

$$[r, p_r] = i\hbar\left(1 + \beta p_r^2\right). \tag{31}$$

In the classical limit, the above commutation relation can be replaced by the following poisson bracket:

$$\{r, p_r\} = 1 + \beta p_r^2. \tag{32}$$

In order to obtain the imaginary part of the action, we use the following deformed Hamiltonian equation:

$$\dot{r} = \{r, H\} = \{r, p_r\}\left.\frac{dH}{dp_r}\right|_r. \tag{33}$$

In the framework of Parikh and Wilczek's tunneling radiation, the imaginary part of the action can be given by

$$\operatorname{Im} I = \operatorname{Im} \int_{r_{in}}^{r_{out}} p_r dr = \operatorname{Im} \int_{r_{in}}^{r_{out}} \int_0^{p_r} p'_r dr. \quad (34)$$

According to the work in [12], the Hamiltonian is $H = M - \omega'$, $p_r^2 \approx \omega'^2$, $p_r \approx \omega'$. Thus, by eliminating the momentum in the favor of the energy in (34), we have

$$\begin{aligned} \operatorname{Im} I &= \operatorname{Im} \int_M^{M-\omega} \int_{r_{in}}^{r_{out}} \frac{1+\beta\omega'^2}{\dot{r}} dr dH' \\ &= \operatorname{Im} \int_M^{M-\omega} \int_{r_{in}}^{r_{out}} \frac{1+\beta\omega'^2}{(1-\varepsilon d\Omega/dt)(1-2(M-\omega')/r)} dr d(M-\omega') \quad (35) \\ &= \frac{4\pi M\omega}{(1-\varepsilon d\Omega/dt)} \left[1 - \frac{\omega}{2M} + \beta\omega^2 \left(\frac{1}{3} - \frac{\omega}{4M}\right)\right]. \end{aligned}$$

The tunneling rate with effect of GUP with a minimal length is therefore

$$\begin{aligned} \Gamma \sim \exp(-2 \operatorname{Im} S) = \exp\Bigg[&-\frac{8\pi ME}{(1-\varepsilon d\Omega/dt)} \\ &+ \frac{4\pi E^2}{(1-\varepsilon d\Omega/dt)} - \frac{32\pi ME^3}{3(1-\varepsilon d\Omega/dt)} \quad (36) \\ &+ \frac{10\pi\beta E^4}{(1-\varepsilon d\Omega/dt)} + O(\beta^2)\Bigg] = \exp(\Delta S), \end{aligned}$$

where ΔS is the difference in black hole entropies before and after emission [11, 12, 33]. It is shown that Parikh-Wilczek's procedure is also valid for the Rutz-Schwarzschild metric. When $\beta = 0$ and $\varepsilon = 0$, the result reduces to the result of Schwarzschild black hole, which is the same with that derived by Parikh and Wilczek [33]. Note that, for a Schwarzschild black hole, with Parikh-Wilczek tunneling mechanism, the quantum tunneling of massless particles from black hole horizon has been studied based on GUP [11, 12]. It is obvious that our result coincides with the one in [12] when $\varepsilon = 0$. We find that (36) gives tunneling rate's correction which caused by the varied background spacetime and GUP. When $\beta = 0$, the tunneling temperature

$$T = \left(1 + \frac{E}{2M}\right) T_H \quad (37)$$

is higher than the original temperature T_H. This predicts the varied spacetime can accelerate the evaporation and the remnant does not arise. when GUP is taken into account, the tunneling temperature

$$T = \left(1 + \frac{E}{2M} - \frac{4}{3}\beta E^2 + \frac{5}{4M}\beta E^3\right) T_H \quad (38)$$

is lower than the original Hawking temperature T_H, which implies that GUP can slow down the increase of the Hawking temperature and stop the black hole from completely evaporating. As a result, a remnant of the black hole is left at the end of evaporation.

5. Conclusion

In conclusion, employing the tunneling radiation of a scalar particle and generalized uncertainty principle, we study the remnant and residue entropy from a Rutz-Schwarzschild black hole based on black hole thermodynamic. Firstly, based on the Hamilton-Jacobi equation revised by GUP, we present the modified tunneling temperature-uncertainty relation and modified entropy-uncertainty relation by using quantum tunneling method. Then, using the generalized uncertainty principle, we calculate the remnant and residual entropy after evaporation when the black hole reaches the Planck scale. Finally, based on black hole thermodynamic phase transition, a detailed analysis of whether there is a stable remnant and residual entropy in the final stage of evaporation is given; that is, the thermodynamic stability of a black hole is related to its thermal capacity and temperature. In order to ensure the thermal stability of a black hole, when the black hole evaporates to the Planck scale, the black hole with negative heat capacity must be transformed into a black hole with positive heat capacity. As a result, at remnant, the modified tunneling temperature and heat capacity tend to zero and modified entropy reaches the minimum value, which imply the black holes are in thermal equilibrium with the outside environment. In addition, the effects of the Finsler perturbation parameter and angular momentum parameter on remnant and residual entropy are also discussed.

The emergence of Finsler black hole solution infuses new vitality into general relativity and puts forward a new research and development idea for black hole physics theory. In this paper, we consider a simple Finsler black hole. For the Rutz-Schwarzschild black hole, we investigate the remnant and residue entropy based on the scalar particles' tunneling radiation via semiclassical Hamilton-Jacobi method. In view of the size of the Finsler perturbation parameter $\varepsilon < 3.15 \times 10^{-4} s$ in geometrical units [24], the small physical quantity is ignored and we approximately adopt Hamilton-Jacobi equation based on GUP to study tunneling radiation from horizon. Strictly speaking, we should apply new Hamilton-Jacobi equation in the framework of complicated Finsler geometry to discuss the related problems. Up to now, Finsler gravity theory has not been established completely. Here, adopting the approximate method we only to understand simply some quantum results of Finsler gravity. We hope that our research can provide some enlightenments for the thermodynamic evolution of black hole physics.

We also note that, the expression of GUP is not unique, which gives rise to different correction on tunneling radiation. Therefore, the remnant and residue entropy should also be distinct from that in the frameworks of other GUPS. We are going to discuss the related issues in further.

Conflicts of Interest

The authors declare that they have no conflicts of interest.

Acknowledgments

This work is supported in part by the National Natural Science Foundation of China (Grant no. 11703018) and Natural Science Foundation of Liaoning Province, China (Grant no. 20180550275).

References

[1] A. Kempf, G. Mangano, and R. B. Mann, “Hilbert space representation of the minimal length uncertainty relation,” *Physical Review D: Particles, Fields, Gravitation and Cosmology*, vol. 52, no. 2, pp. 1108–1118, 1995.

[2] K. Nozari and A. Etemadi, “Minimal length, maximal momentum, and Hilbert space representation of quantum mechanics,” *Physical Review D: Particles, Fields, Gravitation and Cosmology*, vol. 85, Article ID 104029, 2012.

[3] K. Nozari, M. A. Gorji, V. Hosseinzadeh, and B. Vakili, “Natural cutoffs via compact symplectic manifolds,” *Classical and Quantum Gravity*, vol. 33, no. 2, Article ID 025009, 21 pages, 2016.

[4] G. Amelino-Camelia, G. Mandanici, A. Procaccini, and J. Kowalski-Glikman, “Phenomenology of doubly special relativity,” *International Journal of Modern Physics A*, vol. 20, no. 26, pp. 6007–6037, 2005.

[5] J. Magueijo and L. Smolin, “Gravity's rainbow,” *Classical and Quantum Gravity*, vol. 21, pp. 1725–1736, 2004.

[6] A. F. Ali, S. Das, and E. C. Vagenas, “Discreteness of space from the generalized uncertainty principle,” *Physics Letters B*, vol. 678, no. 5, pp. 497–499, 2009.

[7] S. Hossenfelder, M. Bleicher, S. Hofmann, J. Ruppert, S. Scherer, and H. Stöcker, “Signatures in the Planck regime,” *Physics Letters B*, vol. 575, no. 1-2, pp. 85–99, 2003.

[8] A. Kempf and G. Mangano, “Minimal length uncertainty relation and ultraviolet regularization,” *Physical Review D: Particles, Fields, Gravitation and Cosmology*, vol. 55, no. 12, article 7909, 1997.

[9] S. Hossenfelder, “Interpretation of quantum field theories with a minimal length scale,” *Physical Review D*, vol. 73, no. 10, pp. 381–385, 2006.

[10] S. Das, E. C. Vagenas, and A. F. Ali, “Discreteness of space from GUP II: relativistic wave equations,” *Physics Letters B*, vol. 690, no. 4, pp. 407–412, 2010.

[11] K. Nozari and S. H. Mehdipour, “Quantum gravity and recovery of information in black hole evaporation,” *EPL (Europhysics Letters)*, vol. 84, no. 2, Article ID 20008, 2008.

[12] K. Nozari and S. Saghafi, “Natural cutoffs and quantum tunneling from black hole horizon,” *Journal of High Energy Physics*, vol. 2012, no. 11, article 005, 2012.

[13] M. Benrong, P. Wang, and H. Yang, “Covariant GUP Deformed Hamilton-Jacobi Method,” *Advances in High Energy Physics*, vol. 2017, Article ID 3191839, 8 pages, 2017.

[14] Y.-G. Miao and Y.-J. Zhao, “Interpretation of the cosmological constant problem within the framework of generalized uncertainty principle,” *International Journal of Modern Physics D*, vol. 23, no. 7, Article ID 1450062, 2014.

[15] Y.-G. Miao, Y.-J. Zhao, and S.-J. Zhang, “Maximally localized states and quantum corrections of black hole thermodynamics in the framework of a new generalized uncertainty principle,” *Advances in High Energy Physics*, vol. 2015, Article ID 627264, 15 pages, 2015.

[16] D. Y. Chen, Q. Q. Jiang, P. Wang, and H. Yang, “Remnants, fermions' tunnelling and effects of quantum gravity,” *Journal of High Energy Physics*, vol. 2013, no. 11, article 176, 2013.

[17] D. Chen, H. Wu, and H. Yang, “Fermion's tunnelling with effects of quantum gravity,” *Advances in High Energy Physics*, vol. 2013, Article ID 432412, 6 pages, 2013.

[18] D. Chen, H. Wu, and H. Yang, “Observing remnants by fermions' tunneling,” *Journal of Cosmology and Astroparticle Physics*, vol. 2014, no. 3, article 36, 2014.

[19] H.-L. Li and X.-T. Zu, “Black hole remnant and quantum tunnelling in three-dimensional Gödel spacetime,” *Astrophysics and Space Science*, vol. 357, no. 1, article 6, 2015.

[20] H.-L. Li and R. Lin, “Black hole remnant and quantum tunneling corrections to the five-dimensional Myers-Perry black hole based on generalized uncertainty principle,” *Canadian Journal of Physics*, vol. 94, no. 11, pp. 1153–1157, 2016.

[21] H.-L. Li, Z.-W. Feng, and X.-T. Zu, “Quantum tunneling from a high dimensional Gödel black hole,” *General Relativity and Gravitation*, vol. 48, no. 2, article 18, 2016.

[22] X. Li and Z. Chang, “Exact solution of vacuum field equation in Finsler spacetime,” *Physical Review D: Particles, Fields, Gravitation and Cosmology*, vol. 90, no. 6, 2014.

[23] I. P. Lobo, N. Loret, and F. Nettel, “Investigation of Finsler geometry as a generalization to curved spacetime of Planck-scale-deformed relativity in the de Sitter case,” *Physical Review D: Particles, Fields, Gravitation and Cosmology*, vol. 95, no. 4, 046015, 16 pages, 2017.

[24] S. F. Rutz, “A Finsler generalisation of Einstein's vacuum field equations,” *General Relativity and Gravitation*, vol. 25, no. 11, pp. 1139–1158, 1993.

[25] S. I. Vacaru, “Finsler black holes induced by noncommutative anholonomic distributions in Einstein gravity,” *Classical and Quantum Gravity*, vol. 27, no. 10, 105003, 19 pages, 2010.

[26] Z. W. Feng, H. L. Li, X. T. Zu, and S. Z. Yang, “Quantum corrections to the thermodynamics of Schwarzschild-Tangherlini black hole and the generalized uncertainty principle,” *The European Physical Journal C*, vol. 76, article 212, 2016.

[27] M. Benrong, P. Wang, and H. Yang, “Minimal length effects on tunnelling from spherically symmetric black holes,” *Advances in High Energy Physics*, vol. 2015, Article ID 898916, 8 pages, 2015.

[28] V. Akhmedova, T. Pilling, A. de Gill, and D. Singleton, “Temporal contribution to gravitational WKB-like calculations,” *Physics Letters B*, vol. 666, no. 3, pp. 269–271, 2008.

[29] E. T. Akhmedov, V. Akhmedova, and D. Singleton, “Hawking temperature in the tunneling picture,” *Physics Letters B*, vol. 642, no. 1-2, pp. 124–128, 2006.

[30] G. Amelino-Camelia, M. Arzano, and A. Procaccini, “Severe constraints on the loop-quantum-gravity energy-momentum dispersion relation from the black-hole area-entropy law,” *Physical Review D: Particles, Fields, Gravitation and Cosmology*, vol. 70, no. 10, Article ID 107501, 2004.

[31] R. J. Adler, P. Chen, and D. I. Santiago, “The generalized uncertainty principle and black hole remnants,” *General Relativity and Gravitation*, vol. 33, no. 12, pp. 2101–2108, 2001.

[32] K. Nozari and S. H. Mehdipour, “Failure of standard thermodynamics in planck scale black hole system,” *Chaos, Solitons & Fractals*, vol. 39, no. 2, pp. 956–970, 2009.

[33] M. K. Parikh and F. Wilczek, “Hawking radiation as tunneling,” *Physical Review Letters*, vol. 85, no. 24, pp. 5042–5045, 2000.

19

Absorption Cross Section and Decay Rate of Dilatonic Black Strings

Huriye Gürsel and İzzet Sakallı

Physics Department, Faculty of Arts and Sciences, Eastern Mediterranean University, Famagusta, North Cyprus, Mersin 10, Turkey

Correspondence should be addressed to İzzet Sakallı; izzet.sakalli@emu.edu.tr

Academic Editor: Piero Nicolini

We studied in detail the propagation of a massive tachyonic scalar field in the background of a five-dimensional (5D) Einstein–Yang–Mills–Born–Infeld–dilaton black string: the massive Klein–Gordon equation was solved, exactly. Next we obtained complete analytical expressions for the greybody factor, absorption cross section, and decay rate for the tachyonic scalar field in the geometry under consideration. The behaviors of the obtained results are graphically represented for different values of the theory's free parameters. We also discuss why tachyons should be used instead of ordinary particles for the analytical derivation of the greybody factor of the dilatonic 5D black string.

1. Introduction

A wealth of information about quantum gravity can be obtained by studying the unique and fascinating objects known as black holes (BHs). In BH physics, greybody factors (GFs) modify black-body radiation, or predicted Hawking radiation [1, 2], within the limits of geometrical optics [3]. In other words, GFs modify the Hawking radiation spectrum observed at spatial infinity (SI), so that the radiation is not purely Planckian [4].

GF, absorption cross section (ACS), and decay rate (DR) are quantities dependent upon both the frequency of radiation and the geometry of spacetime. Currently, although there are many studies of GF, ACS, and DR (see, for example, [5–10] and the references therein), the number of analytical studies of GFs that consider modified black-body radiation of higher-dimensional ($D > 4$) spacetimes, like the BHs in string theory and black strings [11–13], is rather limited (see, for instance, [6, 7, 14–18]). This paucity of studies has arisen from the mathematical difficulty of obtaining an analytical solution to the wave equation of the stringent geometry being considered; in fact, analytical GF computations apply to spacetimes in which the metric components are independent of time. It is also worth noting that although BSs are defined as a higher-dimensional generalization of a BH, in which the event horizon is topologically equivalent to $S^2 \times S^1$ and spacetime is asymptotically $M^{D-1} \times S^1$, four-dimensional (4D) BSs are also derived. Lemos and Santos [19–21] showed that cylindrically symmetric static solutions, with a negative cosmological constant, of the Einstein–Maxwell equations admit charged 4D BSs. A rotating version of the charged 4D BSs [22, 23] exhibits features similar to the Kerr-Newman BH in spherical topology. The problem of analyzing GFs of scalar fields from charged 4D BSs has recently been discussed by Ahmed and Saifullah [24]. An interesting point about GFs has been reported by [25]: BH energy loss during Hawking radiation depends, crucially, on the GF and the particles' degrees of freedom.

As mentioned above, further study of the GFs of BSs is required. To fill this literature gap, in the current study, we considered dilatonic 5D BS [26], which is a solution to the Einstein–Yang–Mills–Born–Infeld–dilaton (EYMBID) theory. We analytically studied its GF, ACS, and DR for massive scalar fields; however, we considered tachyonic scalar particles instead of ordinary ones. The main reason for this consideration is that using ordinary mass in the

Klein–Gordon equation (KGE) of the dilatonic $5D$ BS (as will be explained in detail later) leads to the diverging of GFs. Roughly speaking, this is due to the flux of the propagating waves of the ordinary massive scalar fields. Namely, once the scalar fields to be considered belong to the massive ordinary particles, the incoming SI flux becomes zero. The latter remark implies that detectable radiation emitted from a dilatonic $5D$ BS spacetime belongs to the massive tachyonic scalar fields. Therefore, the current study focuses on the wave dynamics of tachyonic particles moving in dilatonic $5D$ BS spacetime. However, using tachyonic modes in $5D$ geometry should not be seen as nonphysical; instead they should be considered as the imaginary mass fields rather than faster-than-light particles [27]. First, Feinberg [28, 29] proposed that tachyonic particles could be quanta of a quantum field with imaginary mass. It was soon realized that excitations of *such imaginary mass fields do not in fact propagate faster than light* [30].

Following the idea of Kaluza–Klein [31], any $4D$ physical trajectory is the projection of higher-dimensional worldlines. Effective $4D$ worldlines associated with massive particles are causality constrained to be timelike. However, the corresponding higher-dimensional worldlines need not be exclusively timelike, which gives rise to a topological classification of physical objects. In particular, elementary particles in a $5D$ geometry should be viewed as tachyonic modes. The existence of tachyons in higher dimensions has been thoroughly studied by Davidson and Owen [32]. Furthermore, the reader may refer to [33] to understand tachyon condensation in the evaporation process of a BS. To find the analytical GF, ACS, and DR, we have shown how to obtain the complete analytical solution to the massive KGE in the geometry of a dilatonic $5D$ BS.

Our work is organized as follows. Following this introduction, a brief overview of the geometry of the dilatonic $5D$ BS is provided in Section 2. Section 3 describes the KGE of the tachyonic fields in the dilatonic $5D$ BS geometry; we present the exact solution of the radial equation in terms of hypergeometric functions. In Section 4, we compute the GF and consequently the ACS and DR of the dilatonic $5D$ BS, respectively. We then graphically exhibit the results of the ACS and DR. Section 5 concludes with the final remarks drawn from our study.

2. Dilatonic $5D$ BS in EYMBID Theory

D $(= d+1)$-dimensional action in the EYMBID theory is given by [26]

$$I_{EYMBID} = -\frac{1}{16\pi G_{(D)}}\int_{\mathcal{M}} d^d x\sqrt{-g} \cdot \left[\mathcal{R} - \frac{4(\nabla\psi)^2}{D-2} + 4\chi^2 e^{-b\psi}\left(1-\sqrt{1+\frac{Fe^{2b}}{2\chi^2}}\right)\right], \tag{1}$$

where ψ is the dilaton field, χ denotes the Born–Infeld parameter [34], and $b = -(4/(d-2))\alpha$ with the dilaton parameter $\alpha = 1/\sqrt{d-1}$. $G_{(D)}$ represents the D-dimensional Newtonian constant and its relation to its $4D$ form ($G_{(4)}$) is given by

$$G_{(D)} = G_{(4)}L^{D-4}, \tag{2}$$

where L is the upper limit of the compact coordinate ($\int_0^L dz = L$). Furthermore, $\mathcal{R}$ stands for the Ricci scalar and $F = F^{(\overline{a})}_{\lambda\rho}F^{(\overline{a})\lambda\rho}$ where the 2-form Yang–Mills field is given by

$$F^{(a)} = dA^{(\overline{a})} + \frac{1}{2\sigma}C^{(\overline{a})}_{(\overline{b})(\overline{c})}\left(A^{(\overline{b})}\wedge A^{(\overline{c})}\right), \tag{3}$$

with $C^{(\overline{a})}_{(\overline{b})(\overline{c})}$ and σ being structure and coupling constants, respectively. The Yang–Mills potential $A^{(a)}$ is defined by following the Wu-Yang ansatz [35]

$$A^{(\overline{a})} = \frac{Q}{r^2}\left(x_i dx_j - x_j dx_i\right), \tag{4}$$

$$r^2 = \sum_{i=1}^{d-1} x_i^2, \qquad 2 \le j+1 \le i \le d-1, \quad 1 \le \overline{a} \le \frac{(d-1)(d-2)}{2}, \tag{5}$$

where Q is the Yang–Mills charge. The solution for the dilaton is as follows:

$$\psi = -\frac{(d-2)}{2}\frac{\alpha \ln r}{\alpha^2+1}. \tag{6}$$

On the other hand, the line-element of the dilatonic $5D$ BS is given by [26]

$$ds_0^2 = -\frac{f(r)}{\beta}d\tilde{t}^2 + \frac{\beta dr^2}{rf(r)} + rd\tilde{z}^2 + \beta\left(d\theta^2 + \sin^2\theta d\phi^2\right), \tag{7}$$

where $f(r) = r - r_+$ and $\beta = 4Q^2/3$. r_+ represents the outer event horizon having the following $(d+1)$–dimensional form:

$$\frac{32}{L^{d-4}}\left(\frac{Q^2 d}{d-1}\right)^{(d-2)/2} = r_+^{(d(d-2)+2)/d}. \tag{8}$$

Because, in our case, $D = 5$ (i.e., $d = 4$), the horizon becomes

$$r_+ = 4\beta^{2/5} = 4.488Q^{4/5}. \tag{9}$$

After rescaling the metric (7)

$$ds^2 = \frac{ds_0^2}{\beta} = -\frac{f(r)}{\beta^2}d\tilde{t}^2 + \frac{dr^2}{rf(r)} + \frac{r}{\beta}d\tilde{z}^2 + d\theta^2 + \sin^2\theta d\phi^2, \tag{10}$$

and in sequel assigning $\tilde{t}$ and $\tilde{z}$ coordinates to the new coordinates

$$\tilde{t} \longrightarrow \beta t, \qquad \tilde{z} \longrightarrow \beta z, \tag{11}$$

we get the metric that will be used in our computations:

$$ds^2 = -f(r)\,dt^2 + \frac{dr^2}{rf(r)} + \beta r dz^2 + d\theta^2 + \sin^2\theta d\phi^2. \quad (12)$$

It is worth noting that the surface gravity [36] of the dilatonic $5D$ BS can be evaluated by

$$\kappa^2 = -\frac{1}{2}\nabla^\rho \Upsilon^\sigma \nabla_\rho \Upsilon_\sigma\Big|_{r=r_+}, \quad (13)$$

in which Υ^μ represents the timelike Killing vector:

$$\Upsilon^\mu = [1,0,0,0,0]. \quad (14)$$

Then, (13) results in

$$\kappa = \frac{\sqrt{r}f'}{2}\Big|_{r=r_+} = \frac{\sqrt{r_+}}{2}, \quad (15)$$

where the prime denotes the derivative with respect to r. Furthermore, the associated Hawking temperature is expressed by

$$T_H = \frac{\kappa}{2\pi} = \frac{\sqrt{r_+}}{4\pi}. \quad (16)$$

It is important to remark that the Hawking temperature of the dilatonic BS given in (40) of [26] is incorrect. The authors of [26] computed the Hawking temperature of the dilatonic $5D$ BS considering the metric to be symmetric, which is not the case since $g_{tt} \neq 1/g_{rr}$. Meanwhile, it is obvious that dilatonic $5D$ BS (12) has a nonasymptotically flat structure. Therefore, it possesses a quasilocal mass [37–39], which can be computed as follows:

$$M_{QL} = \frac{1}{6}\beta r_h^{3/2} L = \frac{4}{3}\beta^{8/5} L \cong 2.113 Q^{16/5} L. \quad (17)$$

Thus, the first law of thermodynamics is satisfied:

$$dM_{QL} = T_H dS_{BH}, \quad (18)$$

where S_{BH} denotes the Bekenstein-Hawking entropy [36], which takes the following form for the dilatonic $5D$ BS:

$$S_{BH} = \frac{A_H}{4} = \frac{1}{4}\beta r_h \int_0^\pi \sin\theta d\theta \int_0^{2\pi} d\phi \int_0^L dz = \pi\beta L r_h. \quad (19)$$

3. Wave Equation of a Massive Scalar Tachyonic Field in Dilatonic $5D$ BS

As the scalar waves being studied belong to the massive scalar tachyons, the corresponding KGE is given by

$$\left[\square - (i\mu)^2\right]\Psi(t,\mathbf{r}) = \left[\square + \mu^2\right]\Psi(t,\mathbf{r}) = 0. \quad (20)$$

We chose the ansatz as follows:

$$\Psi(t,\mathbf{r}) = R(r)\,Y_l^m(\theta,\phi)\,e^{ikz}e^{-i\omega t}, \quad (21)$$

where $Y_l^m(\theta,\phi)$ is the usual spherical harmonics and k is a constant. After making straightforward calculations, we obtained the radial equation as follows:

$$rf\frac{\ddot{R}}{R} + \frac{\dot{R}}{R}\left(f + r\dot{f}\right) + \frac{\omega^2}{f} - \frac{k^2}{\beta r} + \mu^2 - \lambda = 0, \quad (22)$$

where $\lambda = l(l+1)$ and a dot mark denotes a derivative with respect to r. Multiplying each term by $r\beta f(r)R(r)$ and using the ansatz $y = (r_+ - r)/r_+$, which in turn implies $r = r_+ - yr_+$, one gets

$$y(1-y)R'' + (1-2y)R' + \left[\frac{\omega^2}{yr_+} + \frac{k^2}{\beta(1-y)r_+} - \mu^2 + \lambda\right]R = 0, \quad (23)$$

where prime denotes derivative with respect to y. Setting

$$\left[\frac{\omega^2}{yr_+} + \frac{k^2}{\beta(1-y)r_+} - \mu^2 + \lambda\right] = \frac{A^2}{y} - \frac{B^2}{1-y} + C, \quad (24)$$

one can obtain

$$A = -\frac{\omega}{2\kappa}, \quad B = \frac{ik}{2\kappa\sqrt{\beta}}, \quad C = \lambda - \mu^2. \quad (25)$$

Equation (23) can be solved by comparing it with the standard hypergeometric differential equation [40] which admits the following solution:

$$R = \xi_1(-y)^{iA}(1-y)^{-B}F(a,b;c;y) + \xi_2(-y)^{-iA}(1-y)^{-B}F(\alpha,\varsigma;\eta;y), \quad (26)$$

where

$$a = \frac{1}{2}\left(1 + \sqrt{1+4C}\right) + iA - B, \quad (27)$$

$$b = \frac{1}{2}\left(1 - \sqrt{1+4C}\right) + iA - B, \quad (28)$$

$$c = 1 + 2iA. \quad (29)$$

And

$$\alpha = a - c + 1, \quad (30)$$

$$\varsigma = b - c + 1, \quad (31)$$

$$\eta = 2 - c. \quad (32)$$

According to our calculations, to have a nonzero SI incoming flux (see (53)) or nondivergent GF (see (55)), $\sqrt{1+4C}$ must be imaginary. To this end, we must impose the following condition:

$$4\mu^2 > 4\lambda + 1, \quad (33)$$

such that

$$\sqrt{1+4C} = i\tau, \tag{34}$$

where

$$\tau = \sqrt{4\mu^2 - 4\lambda - 1}, \quad \tau \in \mathbb{R}. \tag{35}$$

To obtain a physically acceptable solution, we must terminate the outgoing solution at the horizon, which can be simply done by imposing $\xi_1 = 0$. Thus, the physical radial solution reduces to

$$R = \xi_2 (-y)^{-iA} (1-y)^{-B} F(\alpha, \varsigma; \gamma; y). \tag{36}$$

It should be noted that checking the forms of (36) both at the horizon and at SI is essential. Section 3 shows that both are needed for the evaluation of the GF. For the near horizon (NH) where $y \longrightarrow 0$, one can state

$$R_{NH} = \xi_2 (-y)^{-iA}, \tag{37}$$

which implies that the purely ingoing plane wave reads

$$\psi_{NH} = \xi_2 e^{-i\omega(\hat{r}_* + t)} e^{ikz}, \tag{38}$$

where

$$\begin{aligned} r^* &= \int \frac{dr}{\sqrt{r} f} \longrightarrow \\ \hat{r}_* &= \lim_{r \longrightarrow r_+} r^* \simeq \frac{\ln(-y)}{\sqrt{r_+}}, \\ \hat{r}_* &\simeq \frac{1}{2\kappa} \ln(-y) \Longrightarrow \\ y &= -e^{2\kappa \hat{r}_*}. \end{aligned} \tag{39}$$

On the other hand, for $y \longrightarrow \infty$, the inverse transformation of the hypergeometric function is given by [41]

$$\begin{aligned} F(\alpha, \varsigma; \eta; y) &= (-y)^{-\alpha} \frac{\Gamma(\eta)\Gamma(\varsigma-\alpha)}{\Gamma(\varsigma)\Gamma(\eta-\alpha)} \\ &\quad \times F\left(\alpha, \alpha+1-\eta; \alpha+1-\varsigma; \frac{1}{y}\right) \\ &\quad + (-y)^{-\varsigma} \frac{\Gamma(\eta)\Gamma(\alpha-\varsigma)}{\Gamma(\alpha)\Gamma(\eta-\varsigma)} \\ &\quad \times F\left(\varsigma, \varsigma+1-\eta; \varsigma+1-\alpha; \frac{1}{y}\right), \end{aligned} \tag{40}$$

which yields the following asymptotic solution:

$$\begin{aligned} R_{SI} &\simeq \xi_2 (-y)^{-iA-B-\alpha} \frac{\Gamma(\eta)\Gamma(\varsigma-\alpha)}{\Gamma(\varsigma)\Gamma(\eta-\alpha)} \\ &\quad + \xi_2 (-y)^{-iA-B-\varsigma} \frac{\Gamma(\eta)\Gamma(\alpha-\varsigma)}{\Gamma(\alpha)\Gamma(\eta-\varsigma)}. \end{aligned} \tag{41}$$

To express (41) in a more compact form, let us perform the following simplifications. Considering

$$-iA - B - \alpha = -\frac{1}{2}(1 + i\tau), \tag{42}$$

together with

$$-iA - B - \varsigma = -\frac{1}{2}(1 - i\tau), \tag{43}$$

and letting $x = -y$, the radial equation for $r \longrightarrow \infty$ takes the form

$$\begin{aligned} R_{SI} &= \frac{1}{\sqrt{x}} \left[\xi_2 x^{-i\tau/2} \frac{\Gamma(\eta)\Gamma(\varsigma-\alpha)}{\Gamma(\varsigma)\Gamma(\eta-\alpha)} \right. \\ &\quad \left. + \xi_2 x^{i\tau/2} \frac{\Gamma(\eta)\Gamma(\alpha-\varsigma)}{\Gamma(\alpha)\Gamma(\eta-\varsigma)} \right]. \end{aligned} \tag{44}$$

One can express x in terms of the tortoise coordinate at SI as

$$\begin{aligned} r^* &= \int \frac{dr}{\sqrt{r} f} \longrightarrow \\ \hat{r}_* &= \lim_{r \longrightarrow \infty} r^* \simeq -\frac{2}{\sqrt{r}}, \end{aligned} \tag{45}$$

such that

$$\begin{aligned} x &= r - r_+ \\ x|_{r \longrightarrow \infty} &\simeq r = 4e^{-2\hat{r}_*}, \end{aligned} \tag{46}$$

where $\hat{r}_* = \ln r^*$. Therefore, we have

$$R_{SI} = \frac{1}{\sqrt{r}} \left[\Lambda_1 e^{i\hat{r}_* \tau} + \Lambda_2 e^{-i\hat{r}_* \tau} \right], \tag{47}$$

where

$$\Lambda_1 = 2^{-i\tau} \xi_2 \frac{\Gamma(\eta)\Gamma(\varsigma-\alpha)}{\Gamma(\varsigma)\Gamma(\eta-\alpha)}, \tag{48}$$

$$\Lambda_2 = 2^{-i\tau} \xi_2 \frac{\Gamma(\eta)\Gamma(\alpha-\varsigma)}{\Gamma(\alpha)\Gamma(\eta-\varsigma)}. \tag{49}$$

Thus, the asymptotic wave solution becomes

$$\psi_{SI} = \frac{e^{ikz}}{\sqrt{r}} \left[\Lambda_1 e^{i(\hat{r}_* \tau - \omega t)} + \Lambda_2 e^{-i(\hat{r}_* \tau + \omega t)} \right]. \tag{50}$$

4. Radiation of Dilatonic 5*D* BS

4.1. The Flux Computation. In this section, we compute the ingoing flux at the horizon ($r \longrightarrow r_+$) and the asymptotic flux for the SI region ($r \longrightarrow \infty$). The evaluation of these flux values will enable us to calculate the GF and, subsequently, the ACS and DR.

The NH-flux can be calculated via [42, 43]

$$\digamma_{NH} = \frac{A_{BH}}{2i} \left(\overline{\psi}_{NH} \partial r_* \psi_{NH} - \psi_{NH} \partial r_* \overline{\psi}_{NH} \right), \tag{51}$$

which, after a few manipulations, can be written as

$$F_{NH} = -4\pi\beta \left|\xi_2\right|^2 r_+. \tag{52}$$

The incoming flux at SI is computed via

$$F_{SI} = \frac{A_{BH}}{2i} \left(\overline{\psi}_{SI}\partial r_* \psi_{SI} - \psi_{SI}\partial r_* \overline{\psi}_{SI}\right). \tag{53}$$

Having performed the steps to evaluate the derivatives with respect to the tortoise coordinate, the incoming flux at SI takes the form

$$F_{SI} = -4\pi\beta \left|\Lambda_2\right|^2 \tau. \tag{54}$$

It is important to remark that if we were dealing with the standard particles rather than tachyons, τ (35) would be imaginary, i.e., $\tau \longrightarrow i\tau$, and therefore SI incoming wave would lead this flux evaluation (53) to be zero. This would indicate the existence of a divergent GF.

4.2. ACS of Dilatonic 5D BS. The GF of the dilatonic 5D BS is obtained by the following expression [6, 8]:

$$\gamma^{l,k} = \frac{F_{NH}}{F_{SI}} = \frac{-4\pi\beta \left|\xi_2\right|^2 r_+}{-4\pi\beta \left|\Lambda_2\right|^2 \tau}, \tag{55}$$

which is nothing but

$$\gamma^{l,k} = \frac{\left|\xi_2\right|^2 r_+}{\left|\Lambda_2\right|^2 \tau}. \tag{56}$$

After a few manipulations, with (see [40])

$$\left|\Gamma(iy)\right|^2 = \frac{\pi}{y\sinh(\pi y)}, \tag{57}$$

$$\left|\Gamma(1+iy)\right|^2 = \frac{\pi y}{\sinh(\pi y)}, \tag{58}$$

$$\left|\Gamma\left(\frac{1}{2}+iy\right)\right|^2 = \frac{\pi}{\cosh(\pi y)}, \tag{59}$$

(56) can be presented as

$$\gamma^{l,k} = \frac{\kappa r_+}{\omega}\left(e^{2\pi\omega/\kappa} - 1\right)\Xi, \tag{60}$$

where

$$\Xi = \frac{e^{2\pi\tau} - 1}{\left[e^{\pi(\tau+\omega/\kappa-k/\kappa\sqrt{\beta})} + 1\right]\left[e^{\pi(\tau+\omega/\kappa+k/\kappa\sqrt{\beta})} + 1\right]}. \tag{61}$$

To evaluate the ACS of the dilatonic 5D BS concerned, we follow the study of [44]. Thus, one can get the ACS expression in 5D as follows:

$$\sigma^{l,k} = \frac{4\pi(l+1)^2}{\omega^3}\gamma^{l,k}, \tag{62}$$

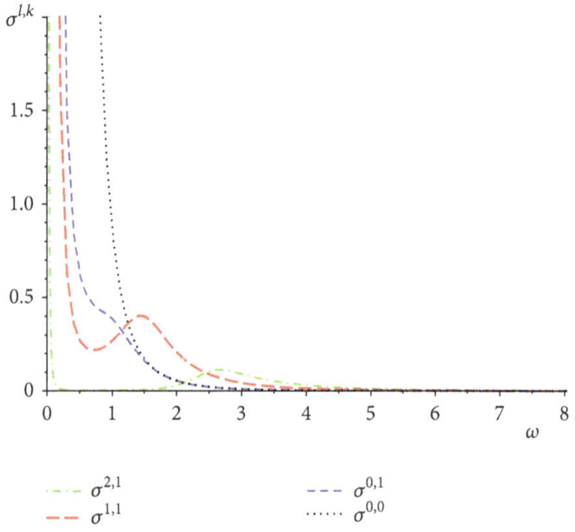

FIGURE 1: Plots of the ACS ($\sigma^{l,k}$) versus frequency ω. The plots are governed by (63). The configuration of the dilatonic 5D BS is as follows: $\mu = 3$ and $Q = 0.2$.

which, in our case, becomes

$$\sigma^{l,k} = \frac{4\pi(l+1)^2 \kappa r_+}{\omega^4}\left(e^{2\pi\omega/\kappa} - 1\right)\Xi. \tag{63}$$

Furthermore, one can also get the total ACS as follows [45]:

$$\sigma_{abs}^{Total} = \sum_{l=0}^{\infty}\sigma^{l,k}. \tag{64}$$

In Figure 1, the relationship between absorption ACS and frequency is examined; the figure is drawn based on (63). In the high frequency regime, all ACSs tend to vanish by following the same curve. Unlike the high frequency regime, ACSs diverge in the low frequency regime as $\omega \longrightarrow 0$. As a final remark, negative $\sigma^{l,k}$ behavior has not been observed in our graphical analyses, which means that superradiance does not occur [46], as expected (as the dilatonic 5D BS (12) does not rotate).

4.3. DR of Dilatonic 5D BS. The final step follows from the ACS evaluation. The DR of the dilatonic 5D BS can be computed via [8]

$$\Gamma_{DR}^{l,k} = \frac{\sigma^{l,k}}{e^{2\pi\omega/\kappa} - 1} = \frac{4\pi(l+1)^2 \kappa r_+}{\omega^4}\Xi. \tag{65}$$

Figure 2 shows how the DR behaves with respect to the frequency. By taking (65) as the reference, the plots for increasing l are illustrated. In the high frequency regime, all DRs fade in the same way. In the low frequency regime, DRs tend to diverge. However, it can be observed that when l has larger values, the corresponding DR diverges when ω is much closer to zero.

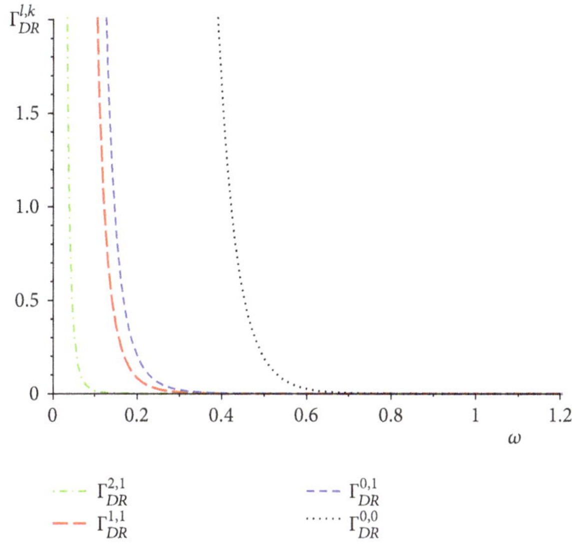

FIGURE 2: Plots of the DR ($\Gamma_{DR}^{l,k}$) versus frequency ω. The plots are governed by (65). The configuration of the dilatonic 5D BS is as follows: $\mu = 3$ and $Q = 0.2$.

5. Conclusion

This article evaluated the GF, ACS, and DR for the dilatonic 5D BS geometry arising from the EYMBID theory. As a result of the analytical method we followed, it was shown that the radiation of the dilatonic 5D BS spacetime can only be caused by tachyons. The crucial point here is that if standard scalar particles had been used rather than tachyonic ones, zero incoming flux at SI would have been obtained, which would lead to the diverging of the GF. Therefore, in a way, we were forced to use tachyons to solve this problem, and this carries great importance as it implies that the fifth dimension could be directly linked to tachyons. In short, according to our analytical method, we obtained results (compatible with boundary conditions) when the radiation of the dilatonic 5D BS was provided by the tachyons.

In future study, we want to extend our analysis to the Dirac equation for the geometry of the dilatonic 5D BS. Hence, we are planning to undertake similar analysis for fermions and compare the results with scalar ones.

Conflicts of Interest

The authors declare that they have no conflicts of interest.

Acknowledgments

We wish to thank Prof. Dr. Mustafa Halilsoy for drawing our attention to this problem and for his helpful comments and suggestions. This work is supported by Eastern Mediterranean University through the project BAPC-04-18-01.

References

[1] S. W. Hawking, "Black hole explosions?" *Nature*, vol. 248, no. 5443, pp. 30-31, 1974.

[2] S. W. Hawking, "Particle creation by black holes," *Communications in Mathematical Physics*, vol. 43, no. 3, pp. 199–220, 1975, erratum: ibid, vol. 46, p. 206, 1976.

[3] S. Creek, O. Efthimiou, P. Kanti, and K. Tamvakis, "Greybody factors for brane scalar fields in a rotating black hole background," *Physical Review D: Particles, Fields, Gravitation and Cosmology*, vol. 75, no. 8, 084043, 17 pages, 2007.

[4] P. Boonserm, "Rigorous bounds on Transmission, Reflection, and Bogoliubov coefficients," 2009, https://arxiv.org/abs/0907.0045.

[5] C. Hoyos and P. Koroteev, "On the null energy condition and causality in lifshitz holography," *Physical Review D: Particles, Fields, Gravitation and Cosmology*, vol. 82, Article ID 109905, 2010.

[6] T. Harmark, J. Natario, and R. Schiappa, "Greybody Factors for d-Dimensional Black Holes," *Advances in Theoretical and Mathematical Physics*, vol. 14, no. 3, pp. 727–793, 2010.

[7] R. Kallosh, A. Linde, and B. Vercnocke, "Natural inflation in supergravity and beyond," *Physical Review D: Particles, Fields, Gravitation and Cosmology*, vol. 90, Article ID 041303(R), 2014.

[8] A. Sánchez-Bretaño, M. Callejo, M. Montero, Á. L. Alonso-Gómez, M. J. Delgado, and E. Isorna, "Performing a hepatic timing signal: glucocorticoids induce gper1a and gper1b expression and repress gclock1a and gbmal1a in the liver of goldfish," *Journal of Comparative Physiology B: Biochemical, Systemic, and Environmental Physiology*, vol. 186, no. 1, pp. 73–82, 2016.

[9] G. Panotopoulos and A. Rincon, "Greybody factors for a minimally coupled massless scalar field in Einstein-Born-Infeld dilaton spacetime," *Physical Review D: Particles, Fields, Gravitation and Cosmology*, vol. 96, no. 2, 025009, 7 pages, 2017.

[10] Y. Zhang, C. Li, and B. Chen, "Greybody factors for a spherically symmetric Einstein-Gauss-Bonnet–de Sitter black hole," *Physical Review D: Particles, Fields, Gravitation and Cosmology*, vol. 97, Article ID 044013, 2018.

[11] J. Kunz, "Black holes in higher dimensions (black strings and black rings)," in *Proceedings of the MG13 Meeting on General Relativity*, pp. 568–581, Stockholm University, Sweden, 2015.

[12] R. Gregory and R. Laflamme, "The instability of charged black strings and p-branes," *Nuclear Physics. B. Theoretical, Phenomenological, and Experimental High Energy Physics. Quantum Field Theory and Statistical Systems*, vol. 428, no. 1-2, pp. 399–434, 1994.

[13] R. Gregory and R. Laflamme, "Black strings and p-branes are unstable," *Physical Review Letters*, vol. 70, no. 19, pp. 2837–2840, 1993.

[14] I. R. Klebanov and S. D. Mathur, "Black hole greybody factors and absorption of scalars by effective strings," *Nuclear Physics. B. Theoretical, Phenomenological, and Experimental High Energy Physics. Quantum Field Theory and Statistical Systems*, vol. 500, no. 1-3, pp. 115–132, 1997.

[15] J. Callan, S. S. Gubser, I. R. Klebanov, and A. A. Tseytlin, "Absorption of fixed scalars and the D-brane approach to black holes," *Nuclear Physics. B. Theoretical, Phenomenological, and Experimental High Energy Physics. Quantum Field Theory and Statistical Systems*, vol. 489, no. 1-2, pp. 65–94, 1997.

[16] M. Cvetic and F. Larsen, "General rotating black holes in string theory: greybody factors and event horizons," *Physical Review D: Particles, Fields, Gravitation and Cosmology*, vol. 56, no. 8, pp. 4994–5007, 1997.

[17] I. R. Klebanov and M. Krasnitz, "Fixed scalar greybody factors in five and four dimensions," *Physical Review D: Particles, Fields, Gravitation and Cosmology*, vol. 55, no. 6, pp. R3250–R3254, 1997.

[18] S. S. Gubser and I. R. Klebanov, "Emission of charged particles from four- and five-dimensional black holes," *Nuclear Physics. B. Theoretical, Phenomenological, and Experimental High Energy Physics. Quantum Field Theory and Statistical Systems*, vol. 482, no. 1-2, pp. 173–186, 1996.

[19] J. P. Lemos, "Two-dimensional black holes and planar general relativity," *Classical and Quantum Gravity*, vol. 12, no. 4, pp. 1081–1086, 1995.

[20] J. P. S. Lemos, "Three dimensional black holes and cylindrical general relativity," *Physics Letters B*, vol. 353, p. 46, 1995.

[21] N. O. Santos, "Solution of the vacuum Einstein equations with nonzero cosmological constant for a stationary cylindrically symmetric spacetime," *Classical and Quantum Gravity*, vol. 10, no. 11, pp. 2401–2406, 1993.

[22] J. P. S. Lemos and V. T. Zanchin, "Rotating charged black strings and three-dimensional black holes," *Physical Review D: Particles, Fields, Gravitation and Cosmology*, vol. 54, no. 6, pp. 3840–3853, 1996.

[23] R. G. Cai and Y. Z. Zhang, "Black plane solutions in four-dimensional spacetimes," *Physical Review D*, vol. 54, p. 4891, 1996.

[24] J. Ahmed and K. Saifullah, "Greybody factor of scalar fields from black strings," *The European Physical Journal C*, vol. 77, p. 885, 2017.

[25] E. K. Akhmedov and J. Pulido, "Solar neutrino oscillations and bounds on neutrino magnetic moment and solar magnetic field," *Physics Letters B*, vol. 553, no. 1-2, pp. 7–17, 2003.

[26] S. H. Mazharimousavia and M. Halilsoy, "Non-Abelian magnetic black strings versus black holes," *The European Physical Journal Plus*, vol. 131, p. 138, 2016.

[27] A. Sen, "Rolling Tachyon," *Journal of High Energy Physics*, vol. 2002, no. 04, p. 048, 2002.

[28] G. Feinberg, "Possibility of faster-than-light particles," *Physical Review A: Atomic, Molecular and Optical Physics*, vol. 159, no. 5, pp. 1089–1105, 1967.

[29] L. Wolfenstein, "Neutrino oscillations in matter," *Physical Review D: Particles, Fields, Gravitation and Cosmology*, vol. 17, no. 9, pp. 2369–2374, 1978.

[30] Y. Aharonov, A. Komar, and L. Susskind, "Superluminal Behavior, Causality, and Instability," *Physical Review A: Atomic, Molecular and Optical Physics*, vol. 182, no. 5, pp. 1400–1403, 1969.

[31] O. Klein, "Quantentheorie und fünfdimensionale relativitätstheorie," *Zeitschrift für Physik*, vol. 37, no. 12, pp. 895–906, 1926.

[32] A. Davidson and D. A. Owen, "Elementary particles as higher-dimensional tachyons," *Physics Letters B*, vol. 177, no. 1, pp. 77–81, 1986.

[33] G. Horowitz, "Tachyon condensation and black strings," *Journal of High Energy Physics*, vol. 2005, no. 08, p. 091, 2005.

[34] M. Born and L. Infeld, "Foundations of the new field theory," *Proceedings of the Royal Society A Mathematical, Physical and Engineering Sciences*, vol. 144, no. 852, pp. 425–451, 1934.

[35] T. T. Wu and C. N. Yang, *in Properties of Matter Under Unusual Conditions*, New York, NY, USA, Interscience, 1969, edited by H. Mark and S. Fernbach.

[36] R. M. Wald, *General Relativity*, University of Chicago Press, London, UK, 1984.

[37] J. D. Brown and J. W. York Jr., "Quasilocal energy and conserved charges derived from the gravitational action," *Physical Review D: Particles, Fields, Gravitation and Cosmology*, vol. 47, no. 4, pp. 1407–1419, 1993.

[38] S. Bose and N. Dadhich, "Brown-York quasilocal energy, gravitational charge, and black hole horizons," *Physical Review D: Particles, Fields, Gravitation and Cosmology*, vol. 60, no. 6, Article ID 064010, 7 pages, 1999.

[39] S. Chakraborty and N. Dadhich, "Brown-York quasilocal energy in Lanczos-Lovelock gravity and black hole horizon," *Journal of High Energy Physics*, vol. 12, p. 003, 2015.

[40] M. Abramowitz and I. A. Stegun, *Handbook of Mathematical Functions, with Formulas, Graphs, and Mathematical Tables*, Dover, New York, NY, USA, 1972.

[41] S. Y. Slavyanov and W. Lay, *Special Functions, A Unified Theory Based on Singularities*, Oxford Mathematical Monographs, Oxford, UK, 2000.

[42] S. Fernando, "Greybody factors of charged dilaton black holes in 2 + 1 dimensions," *General Relativity and Gravitation*, vol. 37, no. 3, pp. 461–481, 2005.

[43] M. Andrews, M. Lewandowski, M. Trodden, and D. Wesley, "Distinguishing *k*-defects from their canonical twins," *Physical Review D: Particles, Fields, Gravitation and Cosmology*, vol. 82, Article ID 105006, 2010.

[44] P. Kanti, J. Grain, and A. Barrau, "Bulk and brane decay of a-dimensional Schwarzschild-de SITter black hole: scalar radiation," *Physical Review D: Particles, Fields, Gravitation and Cosmology*, vol. 71, no. 10, 104002, 18 pages, 2005.

[45] W. G. Unruh, "Absorption cross section of small black holes," *Physical Review D: Particles, Fields, Gravitation and Cosmology*, vol. 14, no. 12, p. 3251.

[46] S. Chandrasekhar, *The Mathematical Theory of Black Holes*, The Clarendon Press, New York, NY, USA, 1983.

20

Dark Matter in the Standard Model Extension with Singlet Quark

Vitaly Beylin and Vladimir Kuksa

Research Institute of Physics, Southern Federal University, 344090 Rostov-on-Don, Pr. Stachky 194, Russia

Correspondence should be addressed to Vladimir Kuksa; vkuksa47@mail.ru

Academic Editor: Jouni Suhonen

We analyze the possibility of hadron Dark Matter carriers consisting of singlet quark and the light standard one. It is shown that stable singlet quarks generate effects of new physics which do not contradict restrictions from precision electroweak data. The neutral and charged pseudoscalar low-lying states are interpreted as the Dark Matter particle and its mass-degenerated partner. We evaluate their masses and lifetime of the charged component and describe the potential asymptotes of low-energy interactions of these particles with nucleons and with each other. Some peculiarities of Sommerfeld enhancement effect in the annihilation process are also discussed.

1. Introduction

The problem of Dark Matter (DM) explanation has been in the center of fundamental physics attention for a long time. The existence of the DM is followed from astrophysical data and remains the essential phenomenological evidence of new physics' manifestations beyond the Standard Model (SM) [1, 2]. Appropriate candidates as DM carriers should be stable particles which weakly interact with ordinary matter (so called, WIMPs). Such particles usually are considered in the framework of supersymmetric, hypercolor, or other extensions of the SM (see, for instance, review [3]). The last experimental rigid restrictions on cross section of spin-independent WIMP-nucleon interaction [4] exclude many variants of WIMPs as the DM carriers. So, another scenarios are discussed in literature, such as quarks from fourth generation, hypercolor quarks, dark atoms, and axions [3]. In spite of some theoretical peculiarities, the possibility of hadronic DM is not excluded and considered, for example, in [5–11]. The possibility of new hadrons existence, which can be interpreted as carriers of the DM, was analyzed in detail within the framework of the SM chiral-symmetric extension [11].

Principal feature of the hadronic DM structure is that the strong interaction of new stable quarks with standard ones leads to the formation of neutral stable meson or baryon heavy states. Such scenario can be realized in the extensions of the SM with extra generation [5–9], in mirror and chiral-symmetric models [11, 12] or in extensions with singlet quark [13–17]. The second variant was considered in detail in [11], where the quark structure and low-energy phenomenology of new heavy hadrons were described. It was shown that the scenario does not contradict cosmochemical data, cosmological tests, and known restrictions for new physics effects. However, the explicit realization of the chiral-symmetric scenario faces some theoretical troubles, which can be eliminated with the help of artificial assumptions. The extensions of SM with fourth generation and their phenomenology were considered during last decades in spite of strong experimental restrictions which, for instance, follow from invisible Z-decay channel, unitary condition for CM-matrix, FCNC, etc. The main problem of 4th generation is the contribution of new heavy quarks to the Higgs boson decays [18]. The contribution of new heavy quarks to vector boson coupling may be compensated by the contribution of 50 GeV neutrino [19–21]; however, such assumption seems artificial. In this paper, we analyze the hypothesis of hadronic Dark Matter which follows from the SM extension with singlet quark.

The paper is organised as follows. In the second section we describe the extension of the SM with singlet quark and consider the restrictions on its phenomenology, following from precision electroweak data. Quark composition and interaction of new hadrons with the standard ones at low energies are analyzed in the third section. The masses of new hadrons, decay properties of charged partner of the DM carrier, and annihilation cross section are analyzed in the fourth section.

2. Standard Model Extension with Stable Singlet Quark

There is a wide class of high-energy extensions of the SM with singlet quarks which are discussed during many decades. Here, we consider the simplest extension of the SM with singlet quarks as the framework for description of the DM carrier. Singlet (or vector-like) quark is defined as fermion with standard $U_Y(1)$ and $SU_C(3)$ gauge interactions but it is singlet under $SU_W(2)$ transformations. The low-energy phenomenology of both down- and up-type quarks (D and U) was considered in detail in large number of works (see, for instance, [10, 22–24] and references therein). As a rule, singlet quark is supposed to be unstable due to the mixing with the ordinary ones. This mixing leads to the FCNC appearing at the tree level. As a consequence, we get additional contributions into rare processes, such as rare lepton and semileptonic decays, and mixing in the systems of neutral mesons ($M^0-\overline{M}^0$ oscillations). The current experimental data on new physics phenomena give rigid restrictions for the angles of ordinary-singlet quark mixing. In this work, we consider alternative aspect of the extensions with singlet quark Q, namely, the scenario with the absence of such mixing. As a result, we get stable singlet quark which has no decay channels due to absence of nondiagonal Q-quark currents. More exactly, due to confinement, the singlet quark forms bound states with the ordinary ones, for instance (Qq), and the lightest state is stable. In this work, we consider some properties of such particles and analyze the possibility of interpreting the stable neutral meson $M^0 = (\overline{Q}q)$ as the DM carrier.

Now, we examine the minimal variants of the SM extension with singlet quark Q_A, where subscript $A = U, D$ denotes up- or down- type with charge $q = 2/3, -1/3$. According to the definition, the field Q is singlet with respect to $SU_W(2)$ group and has standard transformations under abelian $U_Y(1)$ and color $SU_C(3)$ groups. So, the minimal additional gauge-invariant Lagrangian has the form

$$L_Q = i\overline{Q}\gamma^\mu\left(\partial_\mu - ig_1\frac{Y}{2}V_\mu - ig_s\frac{\lambda_a}{2}G^a_\mu\right)Q - M_Q\overline{Q}Q, \quad (1)$$

where $Y/2 = q$ is charge in the case of singlet Q, and M_Q denotes phenomenological mass of quark. Note, singlet quark (SQ) cannot get mass term from the standard Higgs mechanism because the Higgs doublet is fundamental representation of $SU(2)$ group. Abelian part of the interaction Lagrangian (1), which will be used in further considerations, includes the interactions with physical photon A and Z boson:

$$L_Q^{int} = g_1 q V_\mu \overline{Q}\gamma^\mu Q = qg_1\left(c_w A_\mu - s_w Z_\mu\right)\overline{Q}\gamma^\mu Q, \quad (2)$$

where $c_w = \cos\theta_w$, $s_w = \sin\theta_w$, $g_1 c_w = e$, and θ_w is Weinberg angle of mixing. Note, the left and right parts of the singlet field Q have the same transformation properties, interaction (2) has vector-like (chiral-symmetric) form, and singlet quark usually is named vector-like quark [23, 24].

First of all, we should take into account direct and indirect restrictions on new physics (NF) manifestations which follow from the precision experimental data. The additional chiral quarks, for instance from standard fourth generation, are excluded at the 5 σ level by LHC data on Higgs searches [22]. As the vector-like (nonchiral) singlet fermions do not receive their masses from a Higgs doublet, they are allowed by existing experimental data on Higgs physics. The last limits on new colored fermions follow from the jets data from the LHC [25]. The corresponding limits for effective colored factors $n_{eff} = 2, 3, 6$ are about 200 GeV, 300 GeV, and 400 GeV. Note that these limits are much less than the estimation of quark mass which follows from the DM analysis (see the fourth section). Theoretical and experimental situation for long-lived heavy quarks were considerably discussed in the review [10], where it was noted that vector-like new heavy quarks can elude experimental constraints from LHC.

Indirect limits on new fermions follow from precision electroweak measurements of the effects, such as flavor-changing neutral currents (FCNC) and vector boson polarization, which take place at the loop level in the SM. Because we consider the case of stable singlet quark, there is no mixing with ordinary quarks and, consequently, FCNC effects are absent. The NF manifestations in polarization effects of gauge bosons γ, Z, W are usually described by oblique Peskin-Takeuchi parameters [26] (PT parameters). From (2), it follows that the singlet quark gives nonzero contributions into polarization of γ- and Z-bosons which are described by the values of $\Pi_{\gamma\gamma}, \Pi_{\gamma Z}, \Pi_{ZZ}$. As W-boson does not interact with the SQ, corresponding contribution into polarization operator is zero, $\Pi_{WW} = 0$. These parameters are expressed in terms of vector bosons polarization $\Pi_{ab}(p^2)$, where $a, b = W, Z, \gamma$. Here, we use the definition $\Pi_{\mu\nu}(p^2) = p_\mu p_\nu P(p^2) + g_{\mu\nu}\Pi(p^2)$ and the expressions for PT oblique parameters from [27]. In the case under consideration, $\Pi_{ab}(0) = 0$ and PT parameters can be represented by the following expressions:

$$\alpha S = 4s_w^2c_w^2\left[\frac{\Pi_{ZZ}\left(M_Z^2, M_U^2\right)}{M_Z^2} - \frac{c_w^2 - s_w^2}{s_w c_w}\Pi'_{\gamma Z}\left(0, m_U^2\right)\right.$$
$$\left. - \Pi'_{\gamma\gamma}\left(0, M_U^2\right)\right];$$

$$\alpha U = -4s_w^2\left[c_w^2\frac{\Pi_{ZZ}\left(M_Z^2, M_U^2\right)}{M_Z^2} + 2s_w c_w\Pi'_{\gamma Z}\left(0, M_U^2\right)\right.$$
$$\left. + s_w^2\Pi'_{\gamma\gamma}\left(0, M_U^2\right)\right];$$

$$\alpha T = -\frac{\Pi_{ZZ}\left(0, M_U^2\right)}{M_Z^2} = 0;$$

$$\alpha V = \Pi'_{ZZ}\left(M_Z^2, M_U^2\right) - \frac{\Pi_{ZZ}\left(M_Z^2, M_U^2\right)}{M_Z^2};$$
$$\alpha W = 0 \quad \left(W \sim \Pi_{WW} = 0\right);$$
$$\alpha X = -s_w c_w \left[\frac{\Pi_{\gamma Z}\left(M_Z^2, M_U^2\right)}{M_Z^2} - \Pi'_{\gamma Z}\left(0, M_U^2\right)\right]. \quad (3)$$

In (3) polarization $\Pi_{ab}(p^2, M_U^2)$, where $a, b = \gamma, Z$, in one-loop approach can be represented in simple form (for the case of SQ with $q = 2/3$):

$$\Pi_{ab}\left(p^2, M_U^2\right) = \frac{g_1^2}{9\pi^2} k_{ab} F\left(p^2, M_U^2\right);$$
$$k_{ZZ} = s_w^2,$$
$$k_{\gamma\gamma} = c_w^2, \quad (4)$$
$$k_{\gamma Z} = -s_w c_w;$$
$$F\left(p^2, M_U^2\right) = -\frac{1}{3}p^2 + 2M_U^2 + 2A_0\left(M_U^2\right) + \left(p^2 + 2M_U^2\right) B_0\left(p^2, M_U^2\right).$$

In (4) the function $F(p^2, M_U^2)$ contains divergent terms in the one-point, $A_0(M_U^2)$, and two-point, $B_0(p^2, M_U^2)$, Veltman functions which are exactly compensated in physical parameters (3). Using standard definitions of the functions $A_0(M_U^2)$ and $B_0(p^2, M_U^2)$ and the equality $B_0'(0, M_U^2) = M_U^2/6$, by straightforward calculations we get simple expressions for oblique parameters:

$$S = -U = \frac{16 s_w^4}{9\pi}\left[-\frac{1}{3} + 2\left(1 + 2\frac{M_Q^2}{M_Z^2}\right)\left(1 - \sqrt{\beta}\arctan\frac{1}{\sqrt{\beta}}\right)\right], \quad (5)$$

where $\beta = 4M_Q^2/M_Z^2 - 1$. We check that in the limit $M_Q^2/M_Z^2 \longrightarrow \infty$ the values of S and U go to zero as $\sim M_Z^2/M_Q^2$ in accordance with well-known results for the case of vector-like interactions [2, 27]. From (5) it follows that beginning from $M_Q = 500$ GeV the parameter $S < 10^{-2}$ and the remaining nonzero parameters have nearly the same values. These values are significantly less than the current experimental limits [28]: $S = 0.00 + 0.11(-0.10)$, $U = 0.08 \pm 0.11$, $T = 0.02 + 0.11(-0.12)$; that is, the scenario with up-type singlet quark satisfies the restrictions on indirect manifestations of heavy new fermions. Note that the parameters V, W, X describe the contributions of new fermions with masses close to the electroweak scale. In the case of down-type singlet quark, having charge $q = -1/3$, the contributions into all polarization and, consequently, into PT parameters are four times smaller.

In the quark-gluon phase (QGP) of the Universe evolution, stable SQ interacts with standard quarks through exchanges by gluons g, γ, and Z according to (1). So, we have large cross section for annihilation into gluons and quarks, $Q\overline{Q} \longrightarrow gg$ and $Q\overline{Q} \longrightarrow q\overline{q}$ correspondingly, and also small additional contributions in electroweak channels $Q\overline{Q} \longrightarrow \gamma\gamma, ZZ$. These cross sections can be simply derived from the known expressions for the processes $gg \longrightarrow Q\overline{Q}$ and $q\overline{q} \longrightarrow Q\overline{Q}$ (see review in [28]) by time inversion. Two-gluon cross section in the low-energy limit looks like

$$\sigma\left(U\overline{U} \longrightarrow gg\right) = \frac{14\pi}{3}\frac{\alpha_s^2}{v_r M_U^2}, \quad (6)$$

where M_U is mass of U-quark and $\alpha_s = \alpha_s(M_U)$ is strong coupling at the corresponding scale. Two-quark channel in the massless limit $m_q \longrightarrow 0$ is as follows:

$$\sigma\left(U\overline{U} \longrightarrow q\overline{q}\right) = \frac{2\pi}{9}\frac{\alpha_s^2}{v_r M_U^2}. \quad (7)$$

So, the two-gluon channel dominates. We should note that the cross section of SQ-annihilation is suppressed by large M_U in comparison with the annihilation of standard quarks.

After the transition from quark-gluon plasma to hadronization stage, the singlet quarks having standard strong interactions (gluon exchange) form coupled states with ordinary quarks. New heavy hadrons can be constructed as coupled states which consist of heavy stable quark Q and a light quark from the SM quark sector. Here, we consider the simplest two-quark states, neutral and charged mesons. The lightest of them, for instance, neutral meson $M = (\overline{Q}q)$, is stable and can be considered as the carrier of cold Dark Matter. Possibility of existence of heavy stable hadrons was carefully analyzed in [11], where it was shown that this hypothesis does not contradict cosmochemical data and cosmological test. This conclusion was based on the important property of new hadron, namely, repulsive strong interaction with nucleons at large distances. The effect will be qualitatively analyzed for the case of MM and MN interactions in the next section.

3. Quark Composition of New Hadrons and Their Interactions with Nucleons

At the hadronization stage, heavy SQ form the coupled states with the ordinary light quarks. Classification of these new heavy hadrons was considered in [11], where quark composition of two-quark (meson) and three-quark (fermion) states was represented for the case of up- and down-types of quark Q. Stable and long-lived new hadrons are divided into three families of particles with characteristic values of masses M, 2M and 3M, where M is the mass of Q-quark. Quantum numbers and quark content of these particles for the case of up-type quark $Q = U$ are represented in Table 1.

Some states in Table 1 were also considered in [10] for the case of long-lived vector-like heavy quark and in [29], where U-type quark belongs to the sequential 4th generation.

Table 1: Characteristics of U-type hadrons.

$J^P = 0^-$	$T = \frac{1}{2}$	$M = (M^0\ M^-)$	$M^0 = \overline{U}u,\ M^- = \overline{U}d$
$J = \frac{1}{2}$	$T = 1$	$B_1 = (B_1^{++}\ B_1^{+}\ B_1^{0})$	$B_1^{++} = Uuu,\ B_1^{+} = Uud,\ B_1^{0} = Udd$
$J = \frac{1}{2}$	$T = \frac{1}{2}$	$B_2 = (B_2^{++}\ B_2^{+})$	$B_2^{++} = UUu,\ B_2^{+} = UUd$
$J = \frac{3}{2}$	$T = 0$	(B_3^{++})	$B_3^{++} = UUU$

Table 2: Characteristics of D-type hadrons.

$J^P = 0^-$	$T = \frac{1}{2}$	$M_D = (M_D^{+}\ M_D^{0})$	$M_D^{+} = \overline{D}u,\ M_D^{0} = \overline{D}d$
$J = \frac{1}{2}$	$T = 1$	$B_{1D} = (B_{1D}^{+}\ B_{1D}^{0}\ B_{1D}^{-})$	$B_{1D}^{+-} = Duu(Ddd),\ B_{1D}^{0} = Dud$
$J = \frac{1}{2}$	$T = \frac{1}{2}$	$B_{2D} = (B_{2D}^{0}\ B_{2D}^{-})$	$B_{2D}^{0} = DDu,\ B_{2D}^{-} = DDd$
$J = \frac{3}{2}$	$T = 0$	(B_{3D}^{-})	$B_{3D}^{-} = DDD$

In [30], an important property of suppression of hadronic interaction of heavy quark systems containing three new quarks, like (UUU) states, was considered. This model has $SU(3) \times SU(2) \times SU(2) \times U(1)$ symmetry and offers a novel alternative for the DM carriers—they can be an electromagnetically bound states made of terafermions. The charged M^- and neutral M^0 particles can manifest themselves in cosmic rays and as carrier of the DM. In [7–9] a possibility is discussed that new stable charged hadrons exist but are hidden from detection, being bounded inside neutral dark atoms. For instance, stable particles with charge $Q = -2$ can be bound with primordial helium.

Interactions of the baryon-type particles B_1 and B_2 (the second and third line in Table 1) are similar to the nucleonic ones, and they may compose atomic nuclei together with nucleons. As it was demonstrated in [11], this circumstance does not prevent the B_1 and B_2 burnout in the course of cosmochemical evolution. There are no problems also with interaction of B_3 isosinglet with nucleons which proceeds mainly through exchange by mesons, η and η'. The constants of such interactions, as it follows from the quark model of the mesonic exchange (see [11]), are not a large one; i.e., B_3N interaction is suppressed in comparison with the NN interaction.

There is another type of hypothetical hadrons which possess analogous properties of strong interactions. They are constructed from stable quark of the down-type (D-quark) with $Q = -1/3$ electric charge. Quantum numbers and quark content of these particles are represented in Table 2 (see the corresponding analysis and comments in [11]).

In this table, the states $M_D^{+}, B_{1D}^{0}, B_{2D}^{0}, B_{3D}^{-}$ are stable. Particles possessing a similar quark composition appear in various high-energy generalizations of SM, in which D-quark is a singlet with respect to weak interactions group. For example, each quark-lepton generation in $E(6) \times E(6)$-model contains two singlet D-type quarks. This quark appears, also, from the Higgs sector in supersymmetric generalization of $SU(5)$ Great Unification model. As a rule, with reference to cosmological restrictions, it is assumed that new hadrons are unstable due to the mixing of singlet D-quarks with the standard quarks of the down-type. Note that the consequences for cosmochemical evolution, caused by existence of the hypothetical stable U- and D-types hadrons, are very different.

Cosmochemical evolution of new hadrons at hadronization stage was qualitatively studied both for U and D cases in [11]. A very important conclusion was arrived from this analysis: baryon asymmetry in new quark sector must exist and has a sign opposite to asymmetry in standard quark sector (quarks U disappear but antiquarks $\overline{U}$ remain). This conclusion follows from the strong cosmochemical restriction for the ratio "anomalous/natural" hydrogen $C \leqslant 10^{-28}$ for $M_Q \lesssim 1$ TeV [31] and anomalous helium $C \leqslant 10^{-12}–10^{-17}$ for $M_Q \leq 10$ TeV [32]. In our case, the state $B_1^{+} = (Uud)$ is heavy (anomalous) proton which can form anomalous hydrogen. At the stage of hadronization, B_1^{+} can be formed by direct coupling of quarks and as a result of reaction $\overline{M}^0 + N \longrightarrow B_1^{+} + X$, where X is totality of leptons and photons in the final state. The antiparticles $\overline{B}_1^{+}$ are burning out due to the reaction $\overline{B}_1^{+} + N \longrightarrow M^0 + X$. The states like (pM^0) can also manifest themselves as anomalous hydrogens, but as it was shown in [11], interaction of p and M^0 has a potential barrier at large distances. So, formation of coupled states (pM^0) at low energies is strongly suppressed. As it follows from the experimental restrictions on anomalous hydrogen and helium [31, 32] that baryon symmetry in extra sector of quarks is not excluded for the case of superheavy new quarks with masses $M_Q \gg 1$ TeV (see, also, the fourth section). Further, we consider the interaction of new hadrons with nucleons and their self-interaction in more detail.

At low energies the hadrons interactions can be approximately described by a model of meson exchange in terms of an effective Lagrangian. It was shown in [33] that low-energy baryon-meson interactions are effectively described by $U(1) \times SU(3)$ gauge theory, where $U(1)$ is the group of semistrong interaction and $SU(3)$ is group of hadronic unitary symmetry. Effective physical Lagrangian which was used for calculation of MN interaction potential is represented in [11]. By straightforward calculations, it was demonstrated there that the dominant contribution resulted from the exchanges by ρ and ω mesons. This Lagrangian at low energies can be applied for analyzing both MN and MM interactions. Here, we give the part of Lagrangian with vector meson exchange which will be used for evaluation of the potential:

$$\begin{aligned} L_{int} &= g_\omega \omega^\mu \overline{N}\gamma_\mu N + g_\rho \overline{N}\gamma_\mu \vec{\rho}^\mu N \\ &+ i g_{\omega M}\omega^\mu \left(M^\dagger \partial_\mu M - \partial_\mu M^\dagger M\right) \\ &+ i g_{\rho M}\left(M^\dagger \vec{\rho}^\mu \partial_\mu M - \partial_\mu M^\dagger \vec{\rho}^\mu M\right). \end{aligned} \tag{8}$$

In (8) $N = (p,n), M = (M^0, M^-), M^\dagger = (\overline{M}^0, M^+)$, and coupling constants are the following [11]:

$$\begin{aligned} g_\rho &= g_{\rho M} = \frac{g}{2}, \\ g_\omega &= \frac{\sqrt{3}g}{2\cos\theta}, \\ g_{\omega M} &= \frac{g}{4\sqrt{3}\cos\theta}, \\ \frac{g^2}{4\pi} &\approx 3.16, \\ \cos\theta &= 0.644. \end{aligned} \tag{9}$$

Note that the one-pion exchange, which is dominant in NN interaction, is forbidden in the $MM\pi$-vertex due to parity conservation.

In Born approximation, the potential of interaction and the nonrelativistic amplitude of scattering for the case of nonpolarized particles are connected by the relation:

$$U(\vec{r}) = -\frac{1}{4\pi^2\mu}\int f(q)\exp\left(i\vec{q}\vec{r}\right)d^3q, \tag{10}$$

where μ is the reduced mass of scattering particles. For the case of M scattering off nucleons, this potential was calculated in [11], where the relation $f(q) = -2\pi i\mu F(q)$ between nonrelativistic amplitude, $f(q)$, and Feynman amplitude, $F(q)$, was utilized. As it was shown, contributions of scalar and two-pion exchanges are suppressed by the factor $\sim m_N/m_M$. Expressions for potentials of interaction of various pairs from doublets (M^0, M^-) and (p,n) have the following form:

$$\begin{aligned} U\left(M^0, p; r\right) &= U\left(M^-, n; r\right) \approx U_\omega(r) + U_\rho(r), \\ U\left(M^0, n; r\right) &= U\left(M^-, p; r\right) \approx U_\omega(r) - U_\rho(r). \end{aligned} \tag{11}$$

In (11) the terms $U_\omega(r)$ and $U_\rho(r)$ are defined by the following expressions:

$$\begin{aligned} U_\omega &= \frac{g^2K_\omega}{16\pi\cos^2\theta}\frac{1}{r}\exp\left(-\frac{r}{r_\omega}\right), \\ U_\rho &= \frac{g^2K_\rho}{16\pi}\frac{1}{r}\exp\left(-\frac{r}{r_\rho}\right), \end{aligned} \tag{12}$$

where $K_\omega = K_\rho \approx 0.92$, $r_\omega = 1.04/m_\omega$, $r_\rho = 1.04/m_\rho$. Taking into account these values and $m_\omega \approx m_\rho$, we rewrite expressions (11) in the form

$$\begin{aligned} U\left(M^0, p; r\right) &= U\left(M^-, n; r\right) \approx 2.5\frac{1}{r}\exp\left(-\frac{r}{r_\rho}\right), \\ U\left(M^0, n; r\right) &= U\left(M^-, p; r\right) \approx 1.0\frac{1}{r}\exp\left(-\frac{r}{r_\rho}\right). \end{aligned} \tag{13}$$

Two consequences can be deduced from expressions (13). Firstly, all the four pairs of particles have repulsive potential ($U > 0$) of interaction at long distances, where Born approximation is valid. Secondly, due to potential barrier the DM particles at low energies cannot interact with nucleons; i.e., they cannot form the coupled states (pM^0) which manifest themselves as anomalous protons. So, they cannot be directly detected. To overcome the barrier, nucleons should have energy $\sim$ 1 GeV or more and this situation takes place in high-energy cosmic rays.

Potential of MM interaction can be also reconstructed with the help of the above given method. Here, we determine only the sign of potential which defines characteristic (attractive or repulsive) of interaction at long distances. This characteristic plays crucial role for low-energy collisions of the DM particles and nucleons. To determine the sign of potential we use the definition of Lagrangian in the nonrelativistic limit:

$$L = L_0 + L_{int} \longrightarrow W_k - U, \tag{14}$$

where W_k is kinetic part and U is potential. There is a relation between effective $L_{int}(q)$ and Feynman amplitude $F(q)$: $F(q) = ikL_{int}(q)$, where $k > 0$ is real coefficient depending on the type of particles. As a result, we get equality $signum(U) = signum(iF)$, where amplitude of interaction is determined by one-particle exchange diagrams for the process $M_1M_2 \longrightarrow M_1'M_2'$. Here, $M = (M^0, \overline{M^0})$ and vertexes are defined by the low-energy Lagrangian (8). With the help of this simple approach, one can check previous conclusion about repulsive character of MN interactions. First of all it should be noted that low-energy effective Lagrangians of NM^0 and $N\overline{M}^0$ have opposite sign due to different sign of vertexes ωM^0M^0 and $\omega\overline{M}^0\overline{M}^0$. This effect can be seen from the differential structure of corresponding part of Lagrangian (8) and representation of field function of the M-particle in the form:

$$M(x) = \sum_p \hat{a}_p^-(M)\exp(-ipx) + \hat{a}_p^+(\overline{M})\exp(ipx),$$
$$M^\dagger(x) = \sum_p \hat{a}_p^+(M)\exp(ipx) + \hat{a}_p^-(\overline{M})\exp(-ipx). \quad (15)$$

In (15), $a_p^{\pm}(M)$ and $a_p^{\pm}(\overline{M})$ are the operators of creation and destruction of particles M and antiparticles $\overline{M}$ with momentum p. As a result, we get the vertexes $\omega(q)M^0(p)M^0(p-q)$ and $\omega(q)\overline{M}^0(p)\overline{M}^0(p-q)$ in momentum representation with opposite signs, $L_{int} = \pm g_{\omega M}(2p-q)$, respectively. This leads to the repulsive and attractive potentials of NM and $N\overline{M}$ low-energy effective interactions via ω exchange. Thus, the absence of potential barrier in the last cases gives rise to the problem of coupled states $p\overline{M}^0$ (the problem of anomalous hydrogen). As it was noted earlier, to overcome this problem we make the suggestion that the hadronic DM is baryon asymmetric ($\overline{M}^0$ is absent at low-energy stage of hadronization) or particles $\overline{M}^0$ are superheavy. Properties of interactions of baryons B_1 and B_2 are similar to nucleonic one (the main contribution gives one-pion and vector meson exchanges) and together with nucleons they may compose an atomic nuclei. So, new baryons can form superheavy nucleons which in the process of evolution are concentrated due to gravitation in the center of massive planets or stars.

Further, we have checked that the potential of M^0M^0 and $\overline{M}^0\overline{M}^0$ interactions is attractive ($U < 0$) for the case of scalar meson exchange and repulsive for the case of vector meson exchange. Potential of $M^0\overline{M}^0$ scattering has attractive asymptotes both for scalar and vector meson exchanges. Thus, the presence of potential barrier in the processes of M^0M^0 and $\overline{M}^0\overline{M}^0$ scattering depends on the relative contribution of scalar and vector mesons. In the case of $M^0\overline{M}^0$ scattering the total potential is attractive and this property can lead to increasing of annihilation cross section in an analogy with Sommerfeld effect [34].

4. Main Properties of New Hadrons as the DM Carriers

The mass of heavy quark M_Q and the mass splitting of the charged M^- and neutral M^0 mesons, $\delta m = m^- - m^0$, are significant characteristics of these states both for their physical interpretation and for application in cosmology. In this analysis, we take into consideration standard electromagnetic and strong interactions only. So, some properties of new mesons doublet $M = (M^0, M^-)$ are analogous to properties of standard mesons consisting of pairs of heavy and light quarks. From experimental data on mass splitting in neutral-charged meson pairs $K = (K^0, K^{\pm})$, $D = (D^0, D^{\pm})$, and $B = (B^0, B^{\pm})$, it is seen that, for down-type mesons K and B, the mass splitting $\delta m < 0$, while for up-type meson D, the value of $\delta m > 0$. Such results can be explained by the fitting data on current masses of quarks, $m_d > m_u$, and binding energy of the systems $(\overline{Q}u)$ and $(\overline{Q}d)$, where Coulomb contributions have different signs. The absolute value of δm for the case of $K-$ and $D-$ mesons is O(MeV), but for $B-$ mesons it is less. Taking into account these data, for the case of up SQ we assume

$$\delta m = m(M^-) - m(M^0) > 0,$$
$$\delta m = O(MeV). \quad (16)$$

Then, we conclude that neutral state $M^0 = (\overline{U}u)$ is stable and can play the role of the DM carrier. The charged partner $M^- = (\overline{U}d)$ has only one decay channel with very small phase space:

$$M^- \longrightarrow M^0 e^- \overline{\nu}_e, \quad (\text{if } \delta m > m_e). \quad (17)$$

This semileptonic decay resulted from the weak transition $d \longrightarrow u + W^- \longrightarrow u + e^-\overline{\nu}_e$, where heavy quark $\overline{U}$ is considered as spectator. The width of decay can be calculated in a standard way and final expression for differential width as follows (see also review by R. Kowalski in [28]):

$$\frac{d\Gamma}{d\omega} = \frac{G_F^2}{48\pi^3}|U_{ud}|^2(m_- + m_0)^2 m_0^3(\omega^2-1)^{3/2}G^2(\omega). \quad (18)$$

In the case under consideration $m_- \approx m_0$, $\omega = k^0/m_0 \approx 1$ and $G(\omega) \approx 1$ (HQS approximation). Here, $G(\omega)$ is equivalent to normalized form factor $f_+(q)$, where q is the transferred momentum. In the vector dominance approach this form factor is defined as $f_+(q) = f_+(0)/(1 - q^2/m_v^2)$, where m_v is the mass of vector intermediate state. So, HQS approximation corresponds to the conditions $q^2 \ll m_v^2$ and $f_+(0) \approx 1$ for the case $\omega = k^0/m_0 \approx 1$. Using (18), for the total width we get

$$\Gamma \approx \frac{G_F^2|U_{ud}|^2 m_0^5}{12\pi^3}\int_1^{\omega_m}(\omega^2-1)^{3/2}d\omega;$$
$$\omega_m = \frac{m_0^2 + m_-^2}{2m_0 m_-}. \quad (19)$$

After integration, expression (19) can be written in the simple form:

$$\Gamma \approx \frac{G_F^2}{60\pi^3}(\delta m)^5, \quad (20)$$

where weak coupling constant is taken at a low-energy scale because of small transferred momentum in the process. From expression (20) one can see that the width crucially depends on the mass splitting, $\Gamma \sim (\delta m)^5$, and does not depend on the mass of meson M. For instance, in the interval $\delta m = (1-10)$ MeV we get the following estimations:

$$\Gamma \sim (10^{-29} - 10^{-24})\,\text{GeV};$$
$$\tau \sim (10^5 - 10^0)\,\text{s}. \quad (21)$$

Thus, charged partner of M^0, which is long-lived (metastable), can be directly detected in the processes of M^0N- collisions with energetic nucleons, N. This conclusion is in accordance with the experimental evidence of heavy charged metastable particles presence in cosmic rays (see [11] and references therein). Note also that the models of DM with a long-lived coannihilation partner are discussed in literature (see, for instance, [10, 35]).

Experimental and theoretical premises of new heavy hadron existence were discussed in [11]. With the help of low-energy model of baryon-meson interactions, it was shown that the potential of MN-interaction has repulsive asymptotics. So, the low-energy particles M do not form coupled states with nucleon and the hypothesis of their existence does not contradict the cosmochemical data.

Now, we estimate the mass of new hadrons which are interpreted as carriers of the DM. The data on Dark Matter relic concentration result in value of the cross section of annihilation at the level

$$(\sigma v_r)^{exp} \approx 10^{-10}\ GeV^{-2}. \tag{22}$$

Comparing the model annihilation cross section (which depends on the mass) to this value, we estimate the mass of the meson M^0. Note that the calculations are fulfilled for the case of hadron-symmetrical DM, that is, the relic abundance is suggested the same for M^0 and $\overline{M}^0$. To escape the contradiction with strong restriction on anomalous helium, we should expect the mass of M^0 above 10 TeV. Approximate evaluation of the model cross section $\sigma(M^0\overline{M}^0)$ can be fulfilled in spectator approach $\sigma(M^0\overline{M}^0) \sim \sigma(U\overline{U})$ considering the light u-quarks as spectators. Main contributions to this cross section result from subprocesses $U\overline{U} \longrightarrow gg$ and $U\overline{U} \longrightarrow q\overline{q}$, where g and q are standard gluon and quark. Corresponding cross sections are represented in the second section ((6) and (7)) and their sum is used for approximate evaluation of the full annihilation cross section of the processes $M^0\overline{M}^0 \longrightarrow$ hadrons. Thus, we can estimate M_U mass from the following approximate equation:

$$(\sigma v_r)^{exp} \approx \frac{44\pi}{9}\frac{\alpha_s^2}{M_U^2}. \tag{23}$$

Now, from (22) and (23) we get $m(M^0) \approx M_U \approx 20$ TeV at $\alpha_s = \alpha_s(M_U)$. Note that this value gets into the range (10–100) TeV which was declared for the case of heavy WIMPonium states in [36].

As it was noted in the previous section, attractive potential of $M^0\overline{M}^0$ interaction at long distances can increase the cross section due to the light meson exchange. This effect leads to Sommerfeld enhancement [34] of the cross section:

$$\sigma v_r = (\sigma v_r)_0\, S\left(\frac{\alpha}{v}\right), \tag{24}$$

where $(\sigma v_r)_0$ is initial cross section which results from the left side of the expression (23); $\alpha = g^2/4\pi$ is defined by the effective coupling according to (9) and $v = v_r/2$. At $m \ll M \approx M_U$, where m is mass of mesons (the light force carriers), Sommerfeld enhancement (SE) factor can be represented in the form [34]

$$S\left(\frac{\alpha}{v}\right) = \frac{\pi\alpha/v}{1 - \exp(-\pi\alpha/v)}. \tag{25}$$

In our case, the light force carriers are ω- and ρ-mesons and $\alpha \sim 1$ (see (8) and (9)), so from (25), we get $10^2 \lesssim S(\alpha/v)/\pi \lesssim 10^3$ in the interval $10^{-2} > v > 10^{-3}$. In this case, from (23)-(25) it follows that at $v \sim 10^{-2}$ the mass of new quark $M_U \sim 10^2$ TeV, which agrees with the evaluation of the mass of baryonic DM in [37] ($M \sim 100$ TeV). Thus, we get too heavy M^0 which cannot be detected in the searching for signals of anomalous hydrogen ($M_{max} \lesssim$ 1 TeV) and anomalous helium ($M_{max} \lesssim 10$ TeV). Note, however, that in these calculations we take into account the light mesons only, ($m \ll M_U$), which act at long distances $r \sim m_\rho^{-1}$. At short distance, near the radius of coupling state $M^0 = (\overline{U}u)$, i.e., at $r \sim M_U^{-1}$, the exchange by heavy mesons containing heavy quark U is possible, for instance, by vector or scalar M-mesons. In this case, expression (25) is not valid because of $M_\chi \sim M_U$, where M_χ is the mass of heavy force carriers. To evaluate SE factor in this case, we use its numerical calculation from [38], where isocontours of the SE corrections are presented as functions of $y = \alpha M/M_\chi$ and $x = \alpha/v$. Then $y \approx 1$, and from [38] (see Figure 1 there) it follows that $S \approx 10$ in the interval $10^{-1} > v > 10^{-3}$. As a result, from (23) and (25) it follows that $M_U \approx 60\ TeV$ which does not change the situation crucially. It should be noted that full description of SE requires an account of weak vector bosons Z, W which interact with light quarks only. Thus, SE effect is formed at various energy regions corresponding to various distances and has very complicated and vague nature (see, also, [39]).

5. Conclusion

We have analyzed a scenario of the hadronic DM based on the simplest extension of the SM with singlet quark. It was shown in a previous work that the existence of new heavy hadrons does not contradict cosmological constraints. Here, we demonstrate that the scenario is in accordance with the precision electroweak restrictions on manifestations of new physics. With the help of effective model Lagrangian, we describe the asymptotes of interaction potential at low energies for interactions of new hadrons with nucleons and with each other. These asymptotics, both attractive and repulsive, occur for different pairs of interacting particles N, M and their antiparticles. The cosmochemical constrictions on anomalous hydrogen and anomalous helium lead to the conclusion that abundance of particles M and antiparticles $\overline{M}$ is strongly asymmetrical, or new hadrons M are superheavy (with mass larger 10 TeV).

Approximate value of the mass splitting for charged and neutral components was evaluated and lifetime of charged metastable hadron component was calculated; it is rather large, $\tau \gg 1$ s. Using the value of the DM relic concentration and the expression for the model cross section of annihilation, mass of the hadronic DM carrier is estimated. The value of mass without account of SE effect is near 20 TeV and the SE increases it up to an order of 10^2 TeV. These results agree with the evaluations of mass of baryonic DM, which are represented in literature (see previous section). So, superheavy new hadrons cannot be generated in the LHC experiments and detected in the searching for anomalous hydrogen and helium. Some peculiarities of Sommerfeld enhancement effect in the process of annihilation are analyzed. It should be underlined that the model annihilation cross section was evaluated at the level of subprocesses. So, for the description of the hadronic Dark Matter in more detail, it is necessary to clarify the mechanism of annihilation process at various energy scales.

Conflicts of Interest

The authors declare that they have no conflicts of interest.

Acknowledgments

The work was supported by Russian Scientific Foundation (RSCF) [Grant No.: 18-12-00213].

References

[1] F. Sannino, "Conformal dynamics for TeV physics and cosmology," *Acta Physica Polonica B*, vol. 40, no. 12, pp. 3533–3743, 2009.

[2] R. Pasechnik, V. Beylin, V. Kuksa, and G. Vereshkov, "Vector-like technineutron dark matter: is a QCD-type Technicolor ruled out by XENON100?" *The European Physical Journal C*, vol. 74, no. 2, article no. 2728, 2014.

[3] M. Khlopov, "Cosmological reflection of particle symmetry," *Symmetry*, vol. 8, p. 81, 2016.

[4] E. Aprile et al., "First Dark Matter Search Results from the XENON1 Experiment," *Physical Review Letters*, vol. 119, 2017.

[5] M. Maltoni, V. A. Novikov, L. B. Okun, A. N. Rozanov, and M. I. Vysotsky, "Extra quark-lepton generations and precision measurements," *Physics Letters B*, vol. 476, no. 1-2, pp. 107–115, 2000.

[6] K. M. Belotsky, D. Fargion, M. Y. Khlopov et al., "Heavy hadrons of 4th family hidden in our Universe and close to detection," *Gravitation and Cosmology*, vol. 11, pp. 3–15, 2005.

[7] M. Y. Khlopov, "Physics of dark matter in the light of dark atoms," *Modern Physics Letters A*, vol. 26, no. 38, pp. 2823–2839, 2011.

[8] M. Y. Khlopov, "Introduction to the special issue on "indirect dark matter searches"," *Modern Physics Letters A*, vol. 29, 2014.

[9] J. R. Cudell and M. Khlopov, "Dark atoms with nuclear shell: A status review," *International Journal of Modern Physics D*, vol. 24, no. 13, 2015.

[10] M. Buchkremer and A. Schmidt, "Long-Lived Heavy Quarks: A Review," *Advances in High Energy Physics*, vol. 2013, Article ID 690254, 17 pages, 2013.

[11] Y. N. Bazhutov, G. M. Vereshkov, and V. I. Kuksa, "Experimental and theoretical premises of new stable hadron existence," *International Journal of Modern Physics A*, vol. 2, 2017.

[12] J. C. Pati and A. Salam, "Lepton number as the fourth colour," *Physics Review D*, vol. 10, no. 1, pp. 275–289, 1974.

[13] V. Barger, N. G. Deshpande, R. J. Phillips, and K. Whisnant, "Extra fermions in E6 superstring models," *Physical Review D*, vol. 33, no. 7, pp. 1912–1924, 1986.

[14] V. D. Angelopoulos, J. Ellis, H. Kowalski, D. V. Nanopoulos, N. D. Tracas, and F. Zwirner, "Search for new quarks suggested by superstring," *Nuclear Physics B*, vol. 292, pp. 59–92, 1987.

[15] P. Langacker and D. London, "Mixing between ordinary and exotic fermions," *Physical Review D*, vol. 38, no. 3, pp. 886–906, 1988.

[16] V. A. Beylin, G. M. Vereshkov, and V. I. Kuksa, "Mixing of singlet quark with standard ones and the properties of new mesons," *Physics of Atomic Nuclei*, vol. 55, no. 8, pp. 2186–2192, 1992.

[17] R. Rattazzi, "Phenomenological implications of a heavy isosinglet up-type quark," *Nuclear Physics B*, vol. 335, no. 2, pp. 301–310, 1990.

[18] M. Y. Khlopov and R. M. Shibaev, "Probes for 4th generation constituents of dark atoms in higgs boson studies at the LHC," *Advances in High Energy Physics*, vol. 2014, Article ID 406458, 7 pages, 2014.

[19] M. Maltoni, V. Novikov, L. Okun, A. Rozanov, and M. Vysotsky, "Extra quark-lepton generations and precision measurements," *Physics Letters B*, vol. 476, no. 1-2, pp. 107–115, 2000.

[20] V. Ilyin, M. Maltoni, V. Novikov, L. Okun, A. Rozanov, and M. Vysotsky, "On the search for 50 GeV neutrinos," *Physics Letters B*, vol. 503, no. 1-2, pp. 126–132, 2001.

[21] V. A. Novikov, L. B. Okun, A. N. Rozanov, and M. I. Vysotsky, "Extra generations and discrepancies of electroweak precision data," *Physics Letters B*, vol. 529, no. 1-2, pp. 111–116, 2002.

[22] O. Eberhardt, G. Herbert, H. Lacker et al., "Impact of a Higgs Boson at a Mass of 126 GeV on the Standard Model with Three and Four Fermion Generations," *Physical Review Letters*, vol. 109, no. 24, 2012.

[23] F. J. Botella, G. C. Branco, and M. Nebot, "The hunt for New Physics in the Flavour Sector with up vector-like quarks," *Journal of High Energy Physics*, vol. 12, 2012.

[24] A. Kumar Alok, S. Banerjee, D. Kumar, S. U. Sankar, and D. London, "New-physics signals of a model with vector singlet up-type quark," *Physical Review D*, vol. 92, no. 1, Article ID 013002, 2015.

[25] J. Llorente and B. Nachman, "Limites on new coloured fermions using precision data from Large Hadron Collider," *Nuclear Physics B*, vol. 936, pp. 106–117, 2018.

[26] M. E. Peskin and T. Takeuchi, "Estimations of oblique electroweak corrections," *Physical Review D*, vol. 46, no. 1, pp. 381–409, 1992.

[27] C. P. Burgess, S. Godfrey, H. König, D. London, and I. Maksymyk, "A global fit to extended oblique parameters," *Physics Letters B*, vol. 326, no. 3-4, pp. 276–281, 1994.

[28] M. Tanabashi et al., "The Review of Particle Physics (2018)," *Physical Review D*, vol. 98, 2018.

[29] K. Belotsky, M. Khlopov, and K. Shibaev, "Stable quarks of the 4th family?" in *The Physics of Quarks: New Research*, Nova Science Publishers, Hauppauge, NY, USA, 2008.

[30] S. G. Glashow, "A Sinister Extension of the Standard Model to SU(3)×SU(2)×SU(2)×U(1)," in *Proceedings of the 7th International Workshop on Neutrino Telescopes*, 9 pages, Venezia, Italy, 2005.

[31] P. Smith, J. Bennett, G. Homer, J. Lewin, H. Walford, and W. Smith, "A search for anomalous hydrogen in enriched D2O, using a time-of-flight spectrometer," *Nuclear Physics B*, vol. 206, no. 3, pp. 333–348, 1982.

[32] P. Mueller, L. Wang, R. J. Holt, Z. Lu, T. P. O'Connor, and J. P. Schiffer, "Search for Anomalously Heavy Isotopes of Helium in the Earth's Atmosphere," *Physical Review Letters*, vol. 92, no. 2, 2004.

[33] G. M. Vereshkov and V. I. Kuksa, "U(1)SU(3)-gauge model of baryon-meson interactions," *Physics of Atomic Nuclear*, vol. 54, no. 12, pp. 1700–1704, 1991.

[34] R. Iengo, "Sommerfeld enhancement: general results from field theory diagrams," *Journal of High Energy Physics. A SISSA Journal*, no. 5, 024, 15 pages, 2009.

[35] V. V. Khoze, A. D. Plascencia, and K. Sakurai, "Simplified models of dark matter with a long-lived co-annihilation partner," *Journal of High Energy Physics*, vol. 2017, no. 6, 2017.

[36] P. Asadi, M. Baumgart, and P. J. Fitzpatric, "Capture and decay of EW WIMPonium," *Journal of Cosmology and Astroparticle Physics*, 2017.

[37] R. Huo, S. Matsumoto, Y. S. Tsai, and T. T. Yanagida, "A scenario of heavy but visible baryonic dark matter," *Journal of High Energy Physics*, p. 162, 2016.

[38] M. Cirelli, A. Strumia, and M. Tamburini, "Cosmology and astrophysics of minimal dark matter," *Nuclear Physics B*, vol. 787, no. 1-2, pp. 152–175, 2007.

[39] K. Blum, R. Sato, and T. R. Slatyer, "Self-consistent calculation of the Sommerfeld enhancement," *Journal of Cosmology and Astroparticle Physics*, vol. 6, 2016.

21

Neutral-Current Neutrino-Nucleus Scattering off Xe Isotopes

P. Pirinen[1] **J. Suhonen**[1] **and E. Ydrefors**[2]

[1] *University of Jyvaskyla, Department of Physics, P.O. Box 35, 40014, Finland*
[2] *Instituto Tecnológico de Aeronáutica, DCTA, 12228-900 São José dos Campos, Brazil*

Correspondence should be addressed to P. Pirinen; pekka.a.pirinen@student.jyu.fi

Academic Editor: Athanasios Hatzikoutelis

Large liquid xenon detectors aiming for dark matter direct detection will soon become viable tools also for investigating neutrino physics. Information on the effects of nuclear structure in neutrino-nucleus scattering can be important in distinguishing neutrino backgrounds in such detectors. We perform calculations for differential and total cross sections of neutral-current neutrino scattering off the most abundant xenon isotopes. The nuclear-structure calculations are made in the nuclear shell model for elastic scattering and also in the quasiparticle random-phase approximation (QRPA) and microscopic quasiparticle-phonon model (MQPM) for both elastic and inelastic scattering. Using suitable neutrino energy distributions, we compute estimates of total averaged cross sections for ^{8}B solar neutrinos and supernova neutrinos.

1. Introduction

When the idea of neutrinos was first suggested by Pauli in 1930, it was thought that they would never be observed experimentally. Only two decades later interaction of neutrinos with matter was detected in the famous Cowan-Reines experiment [1]. More recently, detection and research of neutrinos have become more and more of an everyday commodity, and various more versatile ways to examine interactions of the little neutral one have emerged and are being tested in laboratories all over the world.

Coherent elastic neutrino-nucleus scattering (CEνNS) is a process where the neutrino interacts with the target nucleus as a whole instead of a single nucleon. Although CEνNS has been predicted since the 1970s [2], it was discovered only very recently by the COHERENT collaboration [3]. Due to the coherent enhancement, this experiment had the remarkable feature of detecting neutrinos with a compact 14.6 kg detector instead of a massive detector volume which is used in conventional neutrino experiments. Coherent neutrino-nucleus scattering is on one hand an important potential source of information for beyond-standard-model physics [4–11], but on the other hand it may also hinder new discoveries as it will start disturbing dark matter detectors in the near future.

A great experimental effort has been put into directly detecting dark matter in the past few decades (see [12] for a review). The next-generation detectors are expected to be sensitive enough to probe cross sections low enough to start observing CEνNS as an irreducible background [13, 14]. Solar neutrinos, atmospheric neutrinos, and diffuse supernova background neutrinos provide a natural source of background neutrinos, which for obvious reasons cannot be shielded against. As there are uncertainties in the fluxes of each of the aforementioned neutrino types, the sensitivity of WIMP (weakly interacting massive particle) detection is basically limited to the magnitude of this uncertainty. To make matters worse, it has been shown that for some specific WIMP masses and cross sections the recoil spectra of CEνNS very closely mimic that of scattering WIMPs [14].

It is therefore of utmost importance to devise a way to go through this neutrino floor. One potential way of achieving this is having directional sensitivity in the detector [15, 16]. As solar and atmospheric neutrinos have a distinct source within the solar system, it is expected that their recoil direction would be different to that of WIMPs, which are typically assumed to be gravitationally bound in a halo spanning the galaxy. Also arising from the different origin of neutrinos and WIMPs is the idea of using timing information to discriminate between neutrino and WIMP induced events in a detector [17]. Due to the motion of the Earth around the Sun, it is expected that the solar neutrino flux peaks around January, but the WIMP flux peaks in June when the velocities of the Sun and Earth are the most in phase. The recoil spectra of WIMPs and neutrinos could also be distinguished if the WIMP-nucleus interaction happens via a nonstandard operator emerging in the effective field theory framework [18, 19].

Some of the leading dark matter experiments use a liquid xenon target [20–24], which allows for easy scalability to larger detector volumes. It is expected that the xenon detectors are the first to hit the neutrino floor. In this article we compute cross sections for elastic and inelastic neutrino-nucleus scattering for the most abundant xenon isotopes. For the coherent scattering we use the quasiparticle random-phase approximation (QRPA) framework and the nuclear shell model to model the nuclear structure and we compare the results between the two models. The wave functions of the states of odd-mass xenon isotopes are obtained by using the microscopic quasiparticle-phonon model (MQPM) on top of a QRPA calculation. Inelastic scattering is computed in the QRPA/MQPM formalism. In our calculations we consider ^{8}B solar neutrinos and supernova neutrinos.

A similar QRPA calculation has been made in [25] for ^{136}Xe, where both charged-current and neutral-current inelastic scattering was examined. Similar computations of neutral-current neutrino-nucleus scattering cross sections have been made before for the stable cadmium isotopes in [26] and for molybdenum isotopes in [27]. Both calculations used the QRPA/MQPM approach. To our knowledge this article presents the first calculation of neutral-current neutrino-nucleus scattering within a complete microscopic nuclear framework for Xe isotopes other than ^{136}Xe.

This article is organized as follows. In Section 2 we outline the formalism used to compute neutral-current neutrino-nucleus scattering. In Section 3 we summarize the nuclear-structure calculations made for the target xenon isotopes. In Section 4 we discuss the results of our cross-section calculations and in Section 5 conclusions are drawn.

2. Neutral-Current Neutrino-Nucleus Scattering

In this section we summarize the formalism used to compute neutral-current neutrino-nucleus scattering processes. We examine standard-model reactions mediated by the neutral Z^0 boson, namely, the processes

$$\nu + (A, Z) \longrightarrow \nu + (A, Z), \quad (1)$$

$$\nu + (A, Z) \longrightarrow \nu + (A, Z)^{*}, \quad (2)$$

i.e., the elastic and inelastic scattering of neutrinos off a nucleus (with A nucleons and Z protons), respectively. In the elastic process the initial and final states of the target

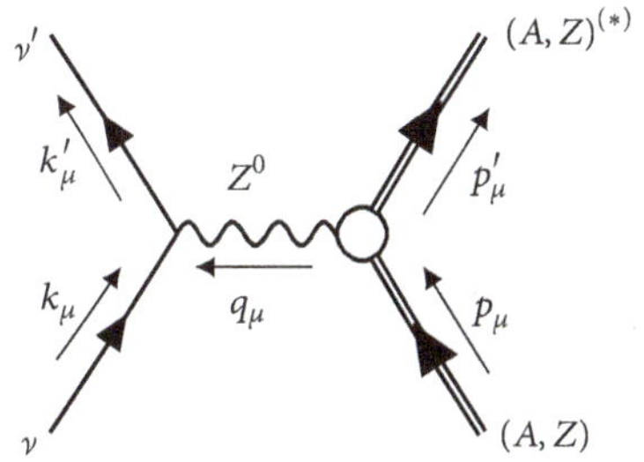

Figure 1: A diagram of the neutral-current scattering process. The four momenta of the involved particles are labeled in the figure.

nucleus are the same, while in the inelastic process excitation of the target nucleus takes place. The kinematics of the scattering process is illustrated in Figure 1. We label the four momenta of the incoming and outgoing neutrino as k_μ and k'_μ, respectively. The momenta of the target nucleus before and after interacting with the neutrino are p_μ and p'_μ. The momentum transfer to the nucleus is referred to as $q_\mu = k'_\mu - k_\mu = p_\mu - p'_\mu$. The neutrino kinetic energy before and after scattering is E_k and $E_{k'}$.

The neutral-current neutrino-nucleus scattering differential cross section to an excited state of energy E_{ex} can be written as [28]

$$\frac{d^2\sigma}{d\Omega dE_{\text{ex}}} = \frac{G_F^2 \left|\mathbf{k}'\right| E_{k'}}{\pi (2J_i + 1)} \left(\sum_{J\geq 0} \sigma_{\text{CL}}^{J} + \sum_{J\geq 1} \sigma_{\text{T}}^{J} \right), \quad (3)$$

which comprises the Coulomb-longitudinal (σ_{CL}^{J}) and transverse (σ_{T}^{J}) parts. They are defined as

$$\begin{aligned}\sigma_{\text{CL}}^{J} &= (1+\cos\theta) \left|\left\langle J_f \left\| \mathcal{M}_J(q) \right\| J_i \right\rangle\right|^2 \\ &+ \left(1+\cos\theta - 2b\sin^2\theta\right) \left|\left\langle J_f \left\| \mathcal{L}_J(q) \right\| J_i \right\rangle\right|^2 \\ &+ \frac{E_{\text{ex}}}{q}(1+\cos\theta) \\ &\times 2\,\text{Re}\left\{ \left\langle J_f \left\| \mathcal{M}_J(q) \right\| J_i \right\rangle^{*} \left\langle J_f \left\| \mathcal{L}_J(q) \right\| J_i \right\rangle \right\},\end{aligned} \quad (4)$$

and

$$\begin{aligned}\sigma_{\text{T}}^{J} &= \left(1-\cos\theta + b\sin^2\theta\right) \\ &\cdot \left[\left|\left\langle J_f \left\| \mathcal{T}_J^{\text{mag}}(q) \right\| J_i \right\rangle\right|^2 + \left|\left\langle J_f \left\| \mathcal{T}_J^{\text{el}}(q) \right\| J_i \right\rangle\right|^2 \right] \\ &\mp \frac{E_k + E_{k'}}{q}(1-\cos\theta) \times 2 \\ &\cdot \text{Re}\left\{ \left\langle J_f \left\| \mathcal{T}_J^{\text{mag}}(q) \right\| J_i \right\rangle \left\langle J_f \left\| \mathcal{T}_J^{\text{el}} \right\| J_i \right\rangle^{*} \right\},\end{aligned} \quad (5)$$

where the minus sign is taken for neutrino scattering and the plus sign for antineutrino scattering. J_i and J_f are the initial and final state angular momenta of the nucleus. We use the abbreviation

$$b = \frac{E_{\mathbf{k}} E_{\mathbf{k}'}}{q^2}, \quad (6)$$

and q is the magnitude of the three-momentum transfer. The formalism and various different operators involved are discussed in detail in [28, 29].

To compute the averaged cross section $\langle\sigma\rangle$, we need to fold the computed cross sections with the energy distribution of the incoming neutrinos. We take the supernova neutrino spectrum to be of a two-parameter Fermi-Dirac character

$$f_{\mathrm{FD}}(E_k) = \frac{1}{F_2(\alpha_\nu)T_\nu}\frac{(E_k/T_\nu)^2}{1+e^{E_k/(T_\nu-\alpha_\nu)}}, \quad (7)$$

where α_ν is the so-called pinching parameter and T_ν is the neutrino temperature. The normalization factor $F_2(\alpha_\nu)$ is defined by the formula

$$F_k(\alpha_\nu) = \int \frac{x^k dx}{1+e^{x-\alpha_\nu}}, \quad (8)$$

and the temperature and mean energy of neutrinos are related by

$$\frac{\langle E_\nu\rangle}{T_\nu} = \frac{F_3(\alpha_\nu)}{F_2(\alpha_\nu)}. \quad (9)$$

We also examine solar neutrinos from ^{8}B beta decay. We use an ^{8}B neutrino energy spectrum from [30].

3. Nuclear Structure of the Target Nuclei

In this section we outline the nuclear-structure calculations performed for the investigated nuclei ^{128,129,130,131,132,134,136}Xe. We have performed computations in the quasiparticle random-phase approximation (QRPA), microscopic quasiparticle-phonon model (MQPM), and the nuclear shell model.

3.1. QRPA/MQPM Calculations. The nuclear structure of even-even Xe isotopes was computed by using the charge-conserving QRPA framework. The QRPA is based on a BCS calculation [31], where quasiparticle creation and annihilation operators are defined via the Bogoliubov-Valatin transformation as

$$\begin{aligned} a_\alpha^\dagger &= u_a c_\alpha^\dagger + v_a \tilde{c}_\alpha, \\ \tilde{a}_\alpha &= u_a \tilde{c}_\alpha - v_a c_\alpha^\dagger, \end{aligned} \quad (10)$$

with the regular particle creation and annihilation operators $c_\alpha^\dagger$ and $\tilde{c}$ defined in [32]. Here α contains the quantum numbers (a, m_α) with $a = (n_a, l_a, j_a)$. The excited states with respect to the QRPA vacuum are created with the phonon creation operator

$$Q_\omega^\dagger = \sum_{ab} \mathcal{N}_{ab}(J_\omega)\left(X_{ab}^\omega \left[a_a^\dagger a_b^\dagger\right]_{J_\omega M_\omega} + Y_{ab}^\omega \left[\tilde{a}_a \tilde{a}_b\right]_{J_\omega M_\omega}\right) \quad (11)$$

for an excited state $\omega = (J_\omega, M_\omega, \pi_\omega, k_\omega)$, where k_ω is a number labeling the excited states of given J^π. In the above equation

$$\mathcal{N}_{ab}(J_\omega) = \frac{\sqrt{1+\delta_{ab}(-1)^{J_\omega}}}{1+\delta_{ab}}, \quad (12)$$

and X_{ab}^ω and Y_{ab}^ω are amplitudes describing the wave function that are solved from the QRPA equation

$$\begin{bmatrix} \mathrm{A} & \mathrm{B} \\ -\mathrm{B}^* & -\mathrm{A}^* \end{bmatrix}\begin{bmatrix} \mathrm{X}^\omega \\ \mathrm{Y}^\omega \end{bmatrix} = E_\omega \begin{bmatrix} \mathrm{X}^\omega \\ \mathrm{Y}^\omega \end{bmatrix}, \quad (13)$$

where the matrix A is the basic Tamm-Dankoff matrix and B is the so-called correlation matrix, both defined in detail in [32].

We perform the QRPA calculations using large model spaces consisting of the entire $0s$–$0d$, $1p$–$0f$–$0g$, $2s$–$1d$–$0h$, and $1f$–$2p$ major shells, adding also the $0i_{13/2}$ and $0i_{11/2}$ orbitals. The single-particle bases are constructed by solving the Schrödinger equation for a Coulomb-corrected Woods-Saxon potential. We use the Woods-Saxon parameters given in [33]. We make an exception for ^{136}Xe, adopting the set of adjusted values of single-particle energies from [25]. Due to the neutron-magic nature of ^{136}Xe, adjusted single-particle energies are necessary to get agreement with experimental energy levels. The Bonn one-boson exchange potential [34] was used to estimate the residual two-body interaction.

The QRPA formalism involves several parameters that have to be fixed by fitting observables to experimental data. In the BCS calculation we fit the proton and neutron pairing strengths $G_{\mathrm{pair}}^{\mathrm{p}}$ and $G_{\mathrm{pair}}^{\mathrm{n}}$ so that the lowest quasiparticle energy matches the empirical pairing gap given by the three-point formula [35]:

$$\begin{aligned} \Delta_{\mathrm{p}}(A,Z) &= \frac{1}{4}(-1)^{Z+1}\left[S_{\mathrm{p}}(A+1,Z+1) - 2S_{\mathrm{p}}(A,Z) + S_{\mathrm{p}}(A-1,Z-1)\right], \\ \Delta_{\mathrm{n}}(A,Z) &= \frac{1}{4}(-1)^{A-Z+1}\left[S_{\mathrm{n}}(A+1,Z) - 2S_{\mathrm{n}}(A,Z) + S_{\mathrm{n}}(A-1,Z)\right]. \end{aligned} \quad (14)$$

It should be noted that for the neutron-magic ^{136}Xe this procedure cannot be done for the neutron pairing strength. We have instead used a bare value of $G_{\mathrm{pair}} = 1.0$ for ^{136}Xe.

The particle-particle and particle-hole terms of the two-body matrix elements are scaled by strength parameters G_{pp} and G_{ph}, respectively. The energies of the computed QRPA states are quite sensitive to these model parameters. We fit the lowest excited states of each J^π separately to experimental values from [36] by altering the values of G_{pp} and G_{ph}. The values used for the model parameters are given in Table 1.

The QRPA process is known to produce states that are spurious, namely, the first excited 0^+ state and the first 1^- state. The first 0^+ state has been deemed spurious in [26, 37]. The first 1^- state is spurious due to center-of-mass motion as described in [32]. We have fitted the energies of these states to zero, if possible, by using the model parameters G_{pp} and G_{ph}, and subsequently the states have been omitted from calculations for the even-mass isotopes and also from the MQPM calculations for the odd-mass isotopes. The contributions of these spurious states to the total neutrino-nucleus scattering cross section would be tiny in any case.

TABLE 1: Model parameters used in the BCS and QRPA calculations. For each nucleus (column 1) the values of G_{pp} and G_{ph} (column 2) are given for the important J^{π} phonons in columns 3 to 9.

Nucleus	G	0^+	1^-	2^+	3^-	4^+	5^-	6^+
^{128}Xe	pp	0.796	1.000	1.000	1.000	1.000	1.000	1.000
	ph	0.298	0.500	0.527	0.500	0.652	0.883	0.934
^{130}Xe	pp	0.730	1.000	1.000	1.000	1.000	1.000	1.000
	ph	0.303	0.500	0.531	0.500	0.581	0.833	0.788
^{132}Xe	pp	0.653	1.000	1.000	1.000	1.000	1.000	1.000
	ph	0.319	0.500	0.533	0.500	0.436	0.933	1.000
^{134}Xe	pp	0.500	1.000	1.000	1.000	1.000	1.000	1.000
	ph	0.370	0.500	0.511	0.500	0.596	1.000	0.891
^{136}Xe	pp	0.843	1.000	1.000	1.000	1.000	1.000	1.000
	ph	0.100	0.500	0.583	0.500	0.700	0.747	0.891

TABLE 2: The valence-space truncations made in the shell-model calculations. The first column labels the Xe isotope; the following five columns give the minimum/maximum number of neutrons on the single-particle orbitals $0g_{7/2}$, $1d_{5/2}$, $1d_{3/2}$, $2s_{1/2}$, and $1h_{11/2}$, respectively.

Nucleus	$0g_{7/2}$	$1d_{5/2}$	$1d_{3/2}$	$2s_{1/2}$	$1h_{11/2}$
^{128}Xe	8/8	6/6	0/4	0/2	4/12
^{129}Xe	8/8	6/6	0/4	0/2	4/12
^{130}Xe	8/8	4/6	0/4	0/2	0/12
^{131}Xe	8/8	6/6	0/4	0/2	0/12
^{132}Xe	0/8	0/8	0/4	0/2	0/12
^{134}Xe	0/8	0/8	0/4	0/2	0/12
^{136}Xe	0/8	0/8	0/4	0/2	0/12

Odd-mass xenon isotopes ^{129,131}Xe are then computed by using the MQPM formalism, in which we use a combination of one- and three-quasiparticle states by coupling a quasiparticle with a QRPA phonon to form the three-quasiparticle configurations. The MQPM basic excitation can be written in terms of quasiparticle and QRPA-phonon creation operators as [38]

$$\Gamma_k^{\dagger}(jm) = \sum_n C_n^k a_{njm}^{\dagger} + \sum_{a,\omega} D_{a\omega}^k \left[a_a^{\dagger} Q_{\omega}^{\dagger}\right]_{jm}. \quad (15)$$

The amplitudes C and D are computed by solving the MQPM equations of motion. The detailed description of the process can be found in [38]. No additional model parameters are required for the MQPM calculation aside for the parameters fitted for the BCS/QRPA calculation described above. We do the MQPM calculations of ^{129}Xe and ^{131}Xe using ^{130}Xe and ^{132}Xe as reference nuclei, respectively. We select all QRPA phonons of $J \leq 6$ with an energy less than 10 MeV to be used in the calculation.

3.2. Shell-Model Calculations. We perform shell-model calculations for Xe isotopes using the shell-model code NuShellX@MSU [39]. We use the $0g_{7/2}$, $1d_{5/2}$, $1d_{3/2}$, $2s_{1/2}$, and $0h_{11/2}$ valence space and the SN100PN interaction [40]. The single-particle energies associated with the aforementioned orbitals in the SN100PN interaction are 0.8072, 1.5623, 3.3160, 3.2238, and 3.6051 MeV, respectively, for protons, and −10.6089, −10.2893, −8.7167, −8.6944, and −8.8152 MeV for neutrons.

The matrix dimension in the shell-model calculation increases rapidly when moving away from the $N = 82$ shell closure of ^{136}Xe. For ^{132,134,136}Xe we were able to do a full calculation with no truncations, but for $^{128-131}$Xe we had to put restrictions on the neutron valence space. The truncations made for each isotope are shown in detail in Table 2. For the isotopes $^{128-131}$Xe we assume a completely filled $0g_{7/2}$ orbital and for ^{128,129,131}Xe we also assume the $1d_{5/2}$ orbital to be full. These should be reasonable approximations when aiming to describe the ground state and low-lying excited states in the xenon nuclei. The orbitals $0g_{7/2}$ and $1d_{5/2}$ have the lowest single-particle energies and the excitations are likely to take place from higher orbitals when the neutron number of the nuclei is quite large.

The computed energy levels of the even-mass xenon isotopes are given in Figure 2 and the odd-mass isotopes in Figure 3. For the even-mass isotopes the experimental energy spectra are very well reproduced by the shell-model calculations. The accuracy is somewhat diminished when moving to lower masses from the closed neutron major shell of ^{136}Xe, but a decent correspondence between experimental and theoretical levels can be found. For the odd-mass isotopes the situation is more complex, but the positive-parity states are well reproduced by the calculations. However, the negative-parity states $11/2^-$ and $9/2^-$ are computed to be much lower than in the experimental spectrum. This effect has been observed in earlier calculations using the SN100PN interaction in this mass region [41]. The experimental data for the xenon isotopes was obtained from [36].

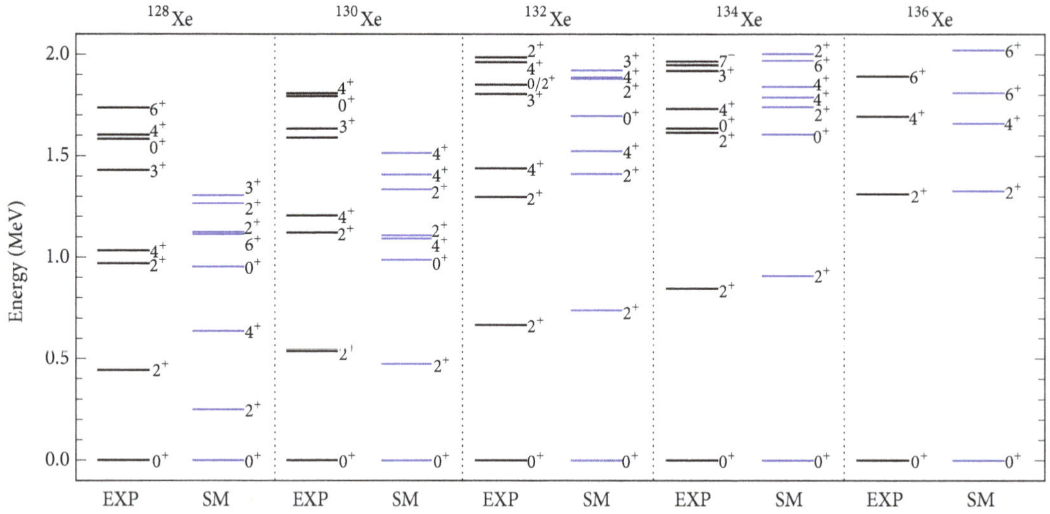

FIGURE 2: Experimental and shell-model energy spectra of even-mass xenon isotopes. A maximum of eight lowest energy levels are shown for each isotope. From left to right: ^{128}Xe, ^{130}Xe, ^{132}Xe, ^{134}Xe, and ^{136}Xe.

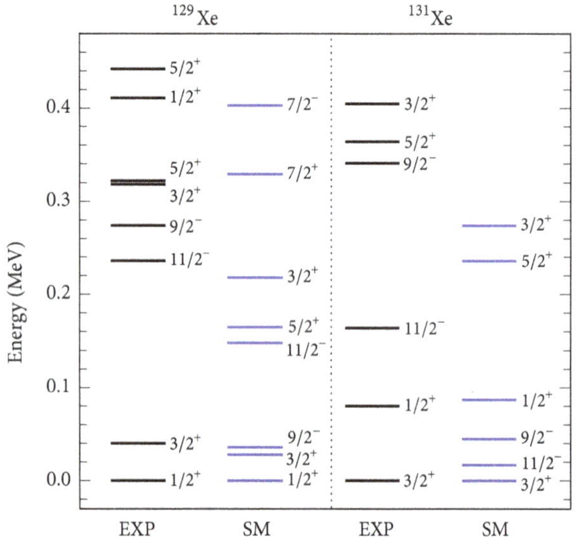

FIGURE 3: Experimental and shell-model energy spectra of odd-mass xenon isotopes ^{129}Xe (left) and ^{131}Xe (right).

To give a further measure of accuracy of our calculation, we computed the ground state magnetic moments for ^{129}Xe and ^{131}Xe. For ^{129}Xe the experimental magnetic moment of the $1/2^+$ ground state is $\mu_{\text{exp}} = -0.7779763(84)\mu_{\text{N}}$ while the shell-model calculated value is $\mu_{\text{sm}} = -1.360\mu_{\text{N}}$. For ^{131}Xe $3/2^+$ ground state the numbers are $\mu_{\text{exp}} = +0.691862(4)\mu_{\text{N}}$ and $\mu_{\text{sm}} = +1.059\mu_{\text{N}}$ for experiment and shell model, respectively. The sign of the magnetic moment in both cases is correct, but the magnitude of both of our calculated values is somewhat larger than that of the experimental ones.

4. Neutrino Scattering Results

In this section we present the results of our calculations for neutrino-nucleus scattering cross sections by methods described in Section 2. We have computed total cross sections for coherent and inelastic neutrino-nucleus scattering as a function of the neutrino energy and also averaged total cross sections for solar ^{8}B neutrinos and supernova neutrinos scattering off the most abundant xenon isotopes. In the following calculations of averaged supernova neutrino cross sections we have used two different neutrino temperatures corresponding to different neutrino flavors. We follow the choices of [26, 37] and have the electron neutrinos described by parameters $\alpha = 3.0$, $\langle E_\nu \rangle = 11.5$ MeV, and $T_\nu = 2.88$ MeV, and the muon and tau neutrinos by $\alpha = 3.0$, $\langle E_\nu \rangle = 16.3$ MeV, and $T_\nu = 4.08$ MeV. Whenever we refer to supernova neutrinos in the following text these parameter values are used in the calculations.

4.1. Coherent Elastic Scattering. In Table 3 we present the total cross section for coherent neutrino-nucleus scattering off the target xenon isotopes as a function of neutrino energy. In Table 3 we only show calculations in the nuclear shell model, but the values for the QRPA/MQPM formalism are very similar, which is reflected on the total averaged cross sections shown later. The cross sections rise rapidly for small neutrino energies and start to saturate when approaching 100 MeV. The cross sections are larger for the higher-A isotopes, following the N^2 coherent enhancement.

We present the total averaged cross section for supernova neutrinos as well as solar ^{8}B neutrinos in Table 4. Results for coherent scattering are shown for the shell model and QRPA/MQPM calculations. The results between the shell model and quasiparticle approaches are very similar. Some small differences can be observed in the results for the odd-mass isotopes, but those are still not very significant. The cross sections for the supernova neutrinos are larger than for ^{8}B neutrinos by roughly a factor of 3 or 5 depending on the neutrino flavor. This is due to the average energy of the supernova neutrinos being larger at 11.5 MeV or 16.3 MeV, while the ^{8}B spectrum peaks at around 7 MeV.

TABLE 3: Coherent elastic neutral-current scattering cross section for neutrinos scattering off xenon targets as a function of neutrino energy. The cross sections for each isotope are given in units of cm^2 in columns 2-8 as a function of the neutrino energy (column 1). The computations were made in the nuclear shell model.

E_ν (MeV)	σ (cm^2) ^{128}Xe	^{129}Xe	^{130}Xe	^{131}Xe	^{132}Xe	^{134}Xe	^{136}Xe
5	5.16×10^{-40}	5.31×10^{-40}	5.46×10^{-40}	5.61×10^{-40}	5.76×10^{-40}	6.08×10^{-40}	6.40×10^{-40}
10	2.02×10^{-39}	2.08×10^{-39}	2.14×10^{-39}	2.20×10^{-39}	2.26×10^{-39}	2.38×10^{-39}	2.50×10^{-39}
20	7.44×10^{-39}	7.65×10^{-39}	7.86×10^{-39}	8.07×10^{-39}	8.29×10^{-39}	8.73×10^{-39}	9.19×10^{-39}
30	1.47×10^{-38}	1.51×10^{-38}	1.55×10^{-38}	1.59×10^{-38}	1.63×10^{-38}	1.71×10^{-38}	1.80×10^{-38}
40	2.19×10^{-38}	2.25×10^{-38}	2.31×10^{-38}	2.37×10^{-38}	2.43×10^{-38}	2.55×10^{-38}	2.67×10^{-38}
50	2.80×10^{-38}	2.88×10^{-38}	2.94×10^{-38}	3.02×10^{-38}	3.09×10^{-38}	3.24×10^{-38}	3.39×10^{-38}
60	3.25×10^{-38}	3.33×10^{-38}	3.40×10^{-38}	3.49×10^{-38}	3.57×10^{-38}	3.73×10^{-38}	3.91×10^{-38}
70	3.55×10^{-38}	3.64×10^{-38}	3.72×10^{-38}	3.81×10^{-38}	3.89×10^{-38}	4.07×10^{-38}	4.25×10^{-38}
80	3.75×10^{-38}	3.84×10^{-38}	3.92×10^{-38}	4.02×10^{-38}	4.10×10^{-38}	4.28×10^{-38}	4.48×10^{-38}

TABLE 4: Total averaged cross section for ^{8}B solar neutrinos and electron and muon/tau supernova neutrinos (SNν_e/SNν_x) scattering off xenon targets. The results are shown for calculations in the nuclear shell model (SM) and the QRPA/MQPM formalisms. Cross sections for coherent scattering are given in units of 10^{-39} cm^2 and for inelastic scattering in 10^{-43} cm^2.

Nucleus	Model	$\langle\sigma\rangle_{\mathrm{coh},^8\mathrm{B}}$ (10^{-39} cm^2)	$\langle\sigma\rangle_{\mathrm{coh,SN}\nu_e}$ (10^{-39} cm^2)	$\langle\sigma\rangle_{\mathrm{coh,SN}\nu_x}$ (10^{-39} cm^2)	$\langle\sigma\rangle_{\mathrm{inel},^8\mathrm{B}}$ (10^{-43} cm^2)	$\langle\sigma\rangle_{\mathrm{inel,SN}\nu_e}$ (10^{-43} cm^2)	$\langle\sigma\rangle_{\mathrm{inel,SN}\nu_x}$ (10^{-43} cm^2)
^{128}Xe	SM	1.064	3.051	5.692	-	-	-
	QRPA	1.065	3.052	5.696	1.567	38.10	152.0
^{129}Xe	SM	1.095	3.138	5.853	-	-	-
	MQPM	1.105	3.166	5.903	2.208	45.11	173.4
^{130}Xe	SM	1.125	3.223	6.008	-	-	-
	QRPA	1.126	3.225	6.013	1.564	40.94	161.0
^{131}Xe	SM	1.157	3.313	6.173	-	-	-
	MQPM	1.167	3.336	6.215	3.699	54.14	195.4
^{132}Xe	SM	1.188	3.401	6.335	-	-	-
	QRPA	1.189	3.403	6.339	2.341	48.21	180.4
^{134}Xe	SM	1.253	3.585	6.671	-	-	-
	QRPA	1.253	3.585	6.673	3.107	56.10	201.7
^{136}Xe	SM	1.320	3.773	7.016	-	-	-
	QRPA	1.320	3.773	7.016	2.102	53.43	200.5

4.2. Inelastic Scattering. Due to the limitations of the shell model in describing high-lying excited states, we compute inelastic scattering properties using only the QRPA/MQPM formalism, which is known to depict well the collective properties of excited nuclear states. The total cross section as a function of neutrino energy is given in Table 5 for each xenon isotope. For smaller neutrino energies, 0 to 30 MeV, the cross sections are slightly larger for the odd-mass isotopes than for their neighboring isotopes. The energies of solar neutrinos fit completely into this range, which leads to the averaged cross sections for solar neutrinos to be larger for the odd-mass isotopes.

The total averaged inelastic cross sections are listed in Table 4. The inelastic scattering cross sections are some orders of magnitude smaller than the coherent cross sections, as expected. Here the cross sections of the supernova neutrinos are an order of magnitude or two larger than of ^{8}B solar neutrinos, again due to the supernova neutrinos having on average a higher energy. The effect of neutrino energy appears more pronounced in inelastic scattering than in coherent scattering, however. The cross sections of the odd-mass isotopes are again slightly larger than those of the neighboring isotopes.

We can compare our inelastic scattering results with those calculated in [26] for Cd isotopes using the same supernova neutrino parameters. The results for Cd isotopes in [26] in the case of electron neutrino range from 4.38×10^{-42} cm^2 for ^{106}Cd to 4.96×10^{-42} cm^2 for ^{111}Cd, with a general decreasing trend with increasing mass number for even-mass nuclei. Our results for Xe isotopes in Table 4 are very similar in magnitude, but the trend is rather rising than decreasing with increasing mass number. This could be a shell effect, as adding neutrons to Cd isotopes takes the nucleus further away from a closed major shell, but for the xenon nuclei it gets closer to a shell closure. Same conclusions can be made for the other neutrino flavors.

We show the contributions from different multipole channels to the total averaged cross sections in Figure 4

TABLE 5: Inelastic neutral-current scattering cross section for neutrinos scattering off xenon targets as a function of neutrino energy. The cross sections are given in units of cm^2. The computations were made in the QRPA/MQPM formalism.

E_ν	σ (cm^2)						
(MeV)	^{128}Xe	^{129}Xe	^{130}Xe	^{131}Xe	^{132}Xe	^{134}Xe	^{136}Xe
5	1.71×10^{-45}	2.10×10^{-45}	1.29×10^{-46}	2.74×10^{-44}	7.74×10^{-45}	1.28×10^{-44}	2.00×10^{-47}
10	3.56×10^{-43}	5.27×10^{-43}	3.54×10^{-43}	8.49×10^{-43}	5.49×10^{-43}	7.57×10^{-43}	4.75×10^{-43}
20	1.41×10^{-41}	1.71×10^{-41}	1.53×10^{-41}	2.00×10^{-41}	1.78×10^{-41}	2.06×10^{-41}	2.02×10^{-41}
30	6.50×10^{-41}	7.45×10^{-41}	6.85×10^{-41}	8.18×10^{-41}	7.53×10^{-41}	8.29×10^{-41}	8.41×10^{-41}
40	1.85×10^{-40}	1.94×10^{-40}	1.91×10^{-40}	2.05×10^{-40}	2.02×10^{-40}	2.16×10^{-40}	2.20×10^{-40}
50	3.99×10^{-40}	3.85×10^{-40}	4.05×10^{-40}	3.95×10^{-40}	4.20×10^{-40}	4.38×10^{-40}	4.47×10^{-40}
60	7.17×10^{-40}	6.41×10^{-40}	7.20×10^{-40}	6.45×10^{-40}	7.35×10^{-40}	7.55×10^{-40}	7.66×10^{-40}
70	1.14×10^{-39}	9.51×10^{-40}	1.14×10^{-39}	9.44×10^{-40}	1.15×10^{-39}	1.16×10^{-39}	1.18×10^{-39}
80	1.66×10^{-39}	1.30×10^{-39}	1.64×10^{-39}	1.28×10^{-39}	1.65×10^{-39}	1.66×10^{-39}	1.67×10^{-39}

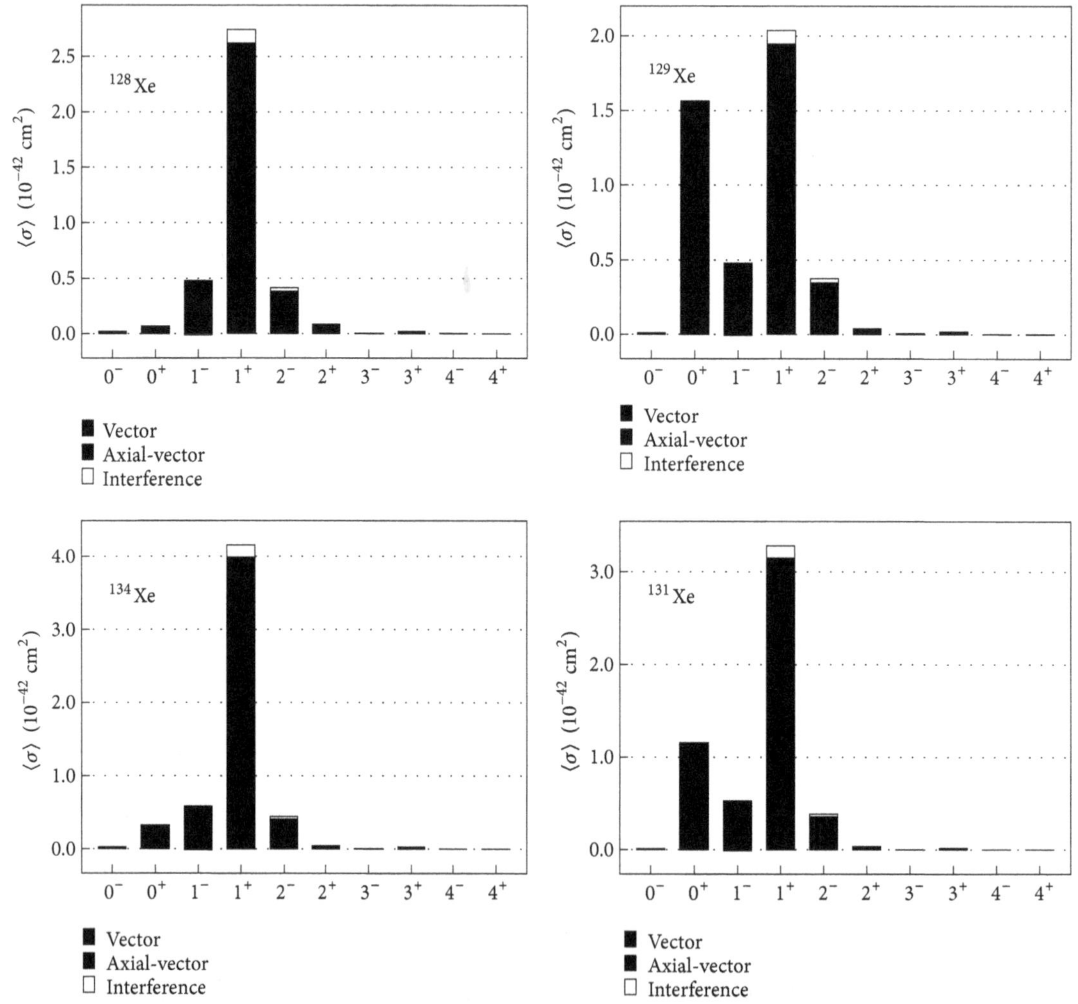

FIGURE 4: The contributions of multipole channels $J \leq 4$ to the total averaged cross section for inelastic scattering of supernova electron neutrinos. Bar plots are shown for a representative sample of ^{128}Xe (top left), ^{129}Xe (top right), ^{134}Xe (bottom left), and ^{131}Xe (bottom right). A division to vector, axial-vector, and interference parts of the interaction is shown. Cross sections are given in units of 10^{-42} cm^2.

for supernova electron neutrinos and Figure 5 for solar neutrinos. It is evident that the most dominant contribution comes from an axial-vector 1^+ multipole transition in all cases but one. Smaller, yet still important contributions arise from the axial-vector 1^- and 2^- channels for higher neutrino energies. This is characteristic behavior for neutral-current scattering, which has been observed in [26] for Cd isotopes and in [27] for Mo isotopes. The contributions get more evenly distributed among the different multipoles with increasing neutrino energy.

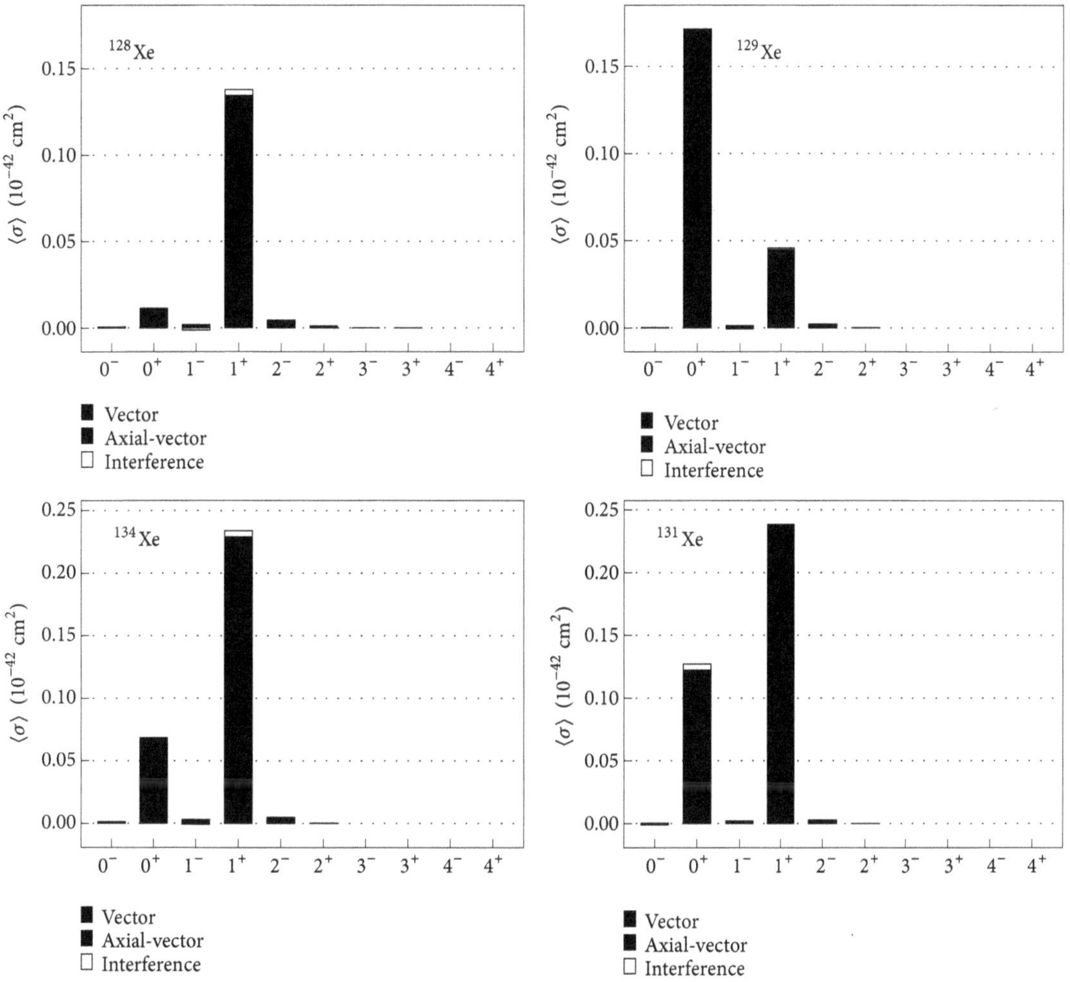

FIGURE 5: The contributions of multipole channels $J \leq 4$ to the total averaged cross section for inelastic scattering of solar ^{8}B neutrinos. Bar plots are shown for a representative sample of ^{128}Xe (top left), ^{129}Xe (top right), ^{134}Xe (bottom left), and ^{131}Xe (bottom right). A division to vector, axial-vector, and interference parts of the interaction is shown. Cross sections are given in units of 10^{-42} cm^2.

For the odd-mass nuclei our calculations also show a significant contribution from a vector 0^+ channel, and for solar neutrinos scattering off ^{129}Xe this channel in fact becomes the strongest. For the even-mass isotopes this channel is more suppressed, but it becomes more significant for the lower energy solar neutrinos. Similar large 0^+ contributions were observed in [26] for Cd isotopes. This is problematic as, in principle, the 0^+ contribution is expected to be small because it vanishes at the limit $q \longrightarrow 0$. The particle-number violation of the quasiparticle framework can be an explanation for the large computed 0^+ contribution. A detailed examination on the origins of the 0^+ anomaly will be conducted in a later study. At this time one should regard the 0^+ contributions with caution as they are probably at least partially spurious.

In Figures 6 and 7 we show the dominating contributions to the inelastic scattering cross section from various final states of ^{128}Xe and ^{131}Xe, respectively. We notice that the major contributions are very similar for the solar and supernova electron neutrinos for the even-mass ^{128}Xe, where the leading contributions come from 1^+ states at 8.4 MeV, 5.0 MeV, and 6.7 MeV. For solar neutrinos there is also a notable contribution from a 0^+ state at 2.4 MeV. The situation is very much different for the odd-mass ^{131}Xe, where for supernova neutrinos there is a pile-up of $5/2^+$, $3/2^+$, and $1/2^+$ states at roughly 8 MeV giving large contributions to the total cross section in addition to the large contributions from lower-lying $5/2^+$ and $3/2^+$ states. However, for solar ^{8}B neutrinos this peak at 8 MeV is much smaller, and the leading contributions are more localized to the $5/2^+$ state at 1.8 MeV and the $3/2^+$ state at 2.9 MeV. It is interesting that a relatively small change in the average neutrino energy can lead to the higher-lying states to give much larger contributions to the total cross section.

Following the discussion on the anomalously large 0^+ multipole contribution in ^{129}Xe we show the dominant final states for neutrinos scattering off ^{129}Xe in Figure 8. As expected from the large 0^+ multipole, the largest contributions here come from $1/2^+$ states at energies of roughly 2 – 3 MeV. Something in the nuclear-structure calculation seems to favor the 0^+ multipole transition to $1/2^+$ final states over the 1^+ multipole transition to $3/2^+$ states. Otherwise similar conclusions can be made for ^{129}Xe as for ^{131}Xe above about

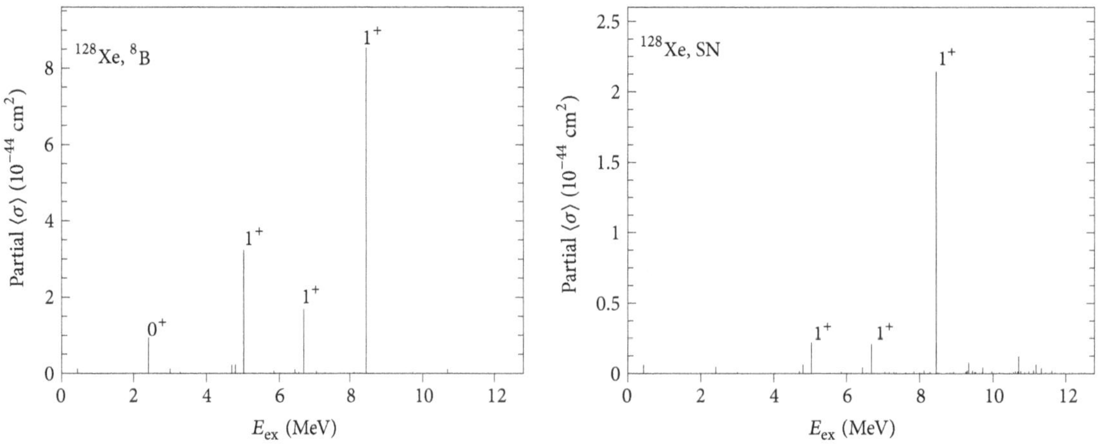

FIGURE 6: Contributions to the inelastic scattering averaged cross section arising from various final states of ^{128}Xe at energies E_{ex}. Results are shown for ^{8}B solar neutrinos (left panel) and supernova electron neutrinos (right panel). Cross sections are given in units of 10^{-44} cm^2 for solar neutrinos and 10^{-42} cm^2 for supernova neutrinos.

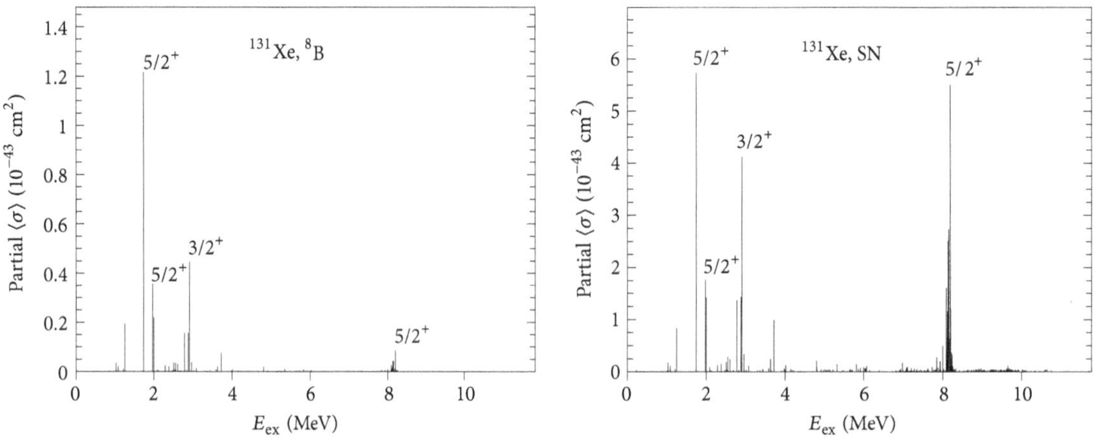

FIGURE 7: Contributions to the inelastic scattering averaged cross section arising from various final states of ^{131}Xe at energies E_{ex}. Results are shown for ^{8}B solar neutrinos (left panel) and supernova electron neutrinos (right panel). In both panels cross sections are given in units of 10^{-43} cm^2.

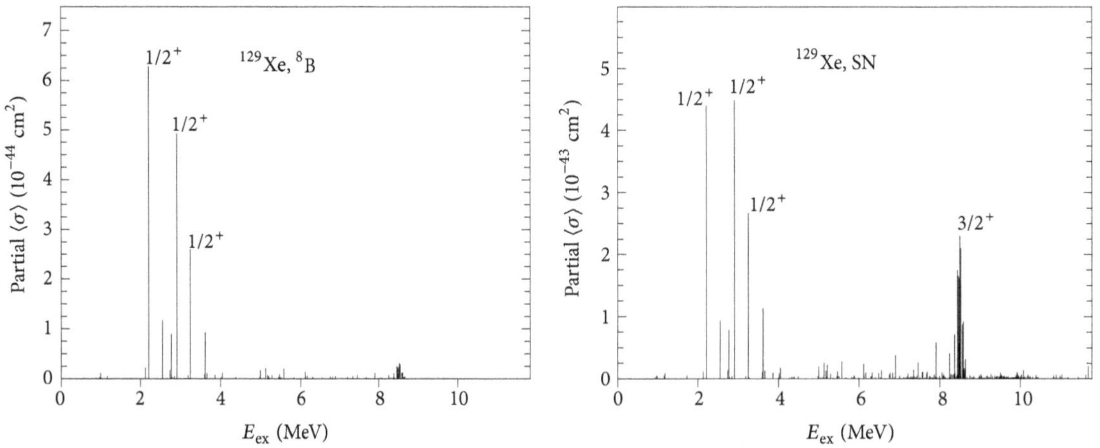

FIGURE 8: Contributions to the inelastic scattering averaged cross section arising from various final states of ^{129}Xe at energies E_{ex}. Results are shown for ^{8}B solar neutrinos (left panel) and supernova electron neutrinos (right panel). Cross sections are given in units of 10^{-44} cm^2 for solar neutrinos and 10^{-43} cm^2 for supernova neutrinos.

the location of the peaks in energy and differences between solar and supernova neutrinos.

5. Conclusions

We have computed various properties of cross sections of neutral-current neutrino-nucleus scattering off the most abundant Xe isotopes. The nuclear structure of our target Xe nuclei was computed in the nuclear shell model for elastic scattering and in the QRPA framework for both elastic and inelastic scattering. For the odd-mass nuclei ^{129}Xe and ^{131}Xe an MQPM calculation was performed based on the QRPA calculation for ^{130}Xe and ^{132}Xe, respectively. We used realistic neutrino energy distributions for solar neutrinos from ^{8}B beta decay and supernova neutrinos to compute the averaged cross sections for each neutrino scenario.

The total averaged cross sections for supernova neutrinos are dependent on the values of the parameters α_ν and $\langle E_\nu \rangle$. We have shown results of only one set of parameters for electron neutrinos and one for muon/tau neutrinos. The dependence of the cross sections on the parameter α_ν is typically quite mild, unless the change is large [25, 26]. The values $\alpha_\nu = 3.0$, $\langle E_{\nu_e} \rangle = 11.5$ MeV, and $\langle E_{\nu_x} \rangle = 16.3$ MeV used in this work are reasonable estimates and allow comparison of results with the ^{8}B solar neutrinos, for which the energy distribution is better known. A mapping of cross sections for various supernova neutrino parameters is out of scope of this work. However, we have tabulated total cross section as a function of neutrino energy, which can be used to obtain estimates of total averaged cross sections for any neutrino energy profile.

The scattering process in even-even nuclei is dominated by transitions to high-lying 1^+ states and for odd-mass nuclei typically by states differing from the initial state by one unit of angular momentum. We found that in even-mass nuclei the leading contributions from various final states are quite similar between solar neutrinos and supernova neutrinos. In odd-mass nuclei, however, the smaller energy of the solar neutrinos does not allow large contributions to the total cross section to arise from high-lying states. We also noted that the smaller energies of solar neutrinos lead into an enhancement in the vector 0^+ multipole channel in comparison to the otherwise dominating 1^+ axial-vector channel, especially in the odd-mass Xe nuclei. However, the large contribution from the 0^+ multipole can be mostly spurious, possibly due to the particle-number violation of the quasiparticle framework. This matter will be investigated further and subsequently reported elsewhere.

Conflicts of Interest

The authors declare that there are no conflicts of interest regarding the publication of this paper.

Acknowledgments

This work has been partially supported by the Academy of Finland under the Finnish Centre of Excellence Programme 2012-2017 (Nuclear and Accelerator Based Programme at JYFL). P. Pirinen was supported by a graduate student stipend from the Magnus Ehrnrooth Foundation. E. Ydrefors acknowledges the financial support of Grant no. 2016/25143-7 from the Sao Paulo Research Foundation (FAPESP).

References

[1] C. L. Cowan Jr, F. Reines, F. B. Harrison, H. W. Kruse, and A. D. McGuire, "Detection of the Free Neutrino: a Confirmation," *Science*, vol. 124, p. 103, 1956.

[2] D. Z. Freedman, "Coherent effects of a weak neutral current," *Physical Review D: Particles, Fields, Gravitation and Cosmology*, vol. 9, no. 5, pp. 1389–1392, 1974.

[3] D. Akimov, J. B. Albert, P. Awe et al., "Observation of coherent elastic neutrino-nucleus scattering," *Science*, vol. 357, p. 1123, 2017.

[4] A. J. Anderson, J. M. Conrad, E. Figueroa-Feliciano et al., "Measuring active-to-sterile neutrino oscillations with neutral current coherent neutrino-nucleus scattering," *Physical Review D: Particles, Fields, Gravitation and Cosmology*, vol. 86, no. 1, Article ID 013004, 2012.

[5] B. Dutta, Y. Gao, A. Kubik et al., "Sensitivity to oscillation with a sterile fourth generation neutrino from ultralow threshold neutrino-nucleus coherent scattering," *Physical Review D*, vol. 94, Article ID 093002, 2016.

[6] T. Kosmas, D. Papoulias, M. Tórtola, and J. Valle, "Probing light sterile neutrino signatures at reactor and Spallation Neutron Source neutrino experiments," *Physical Review D: Particles, Fields, Gravitation and Cosmology*, vol. 96, no. 6, Article ID 063013, 2017.

[7] T. S. Kosmas, O. G. Miranda, D. K. Papoulias, M. Tórtola, and J. W. F. Valle, "Probing neutrino magnetic moments at Spallation Neutron Source facility," *Physical Review D*, vol. 92, no. 1, Article ID 013011, 2015.

[8] J. Barranco, O. G. Miranda, and T. I. Rashba, "Sensitivity of low energy neutrino experiments to physics beyond the standard model," *Physical Review D: Particles, Fields, Gravitation and Cosmology*, vol. 76, Article ID 073008, 2007.

[9] P. deNiverville, M. Pospelov, and A. Ritz, "Light new physics in coherent neutrino-nucleus scattering experiments," *Physical Review D*, vol. 92, Article ID 095005, 7 pages, 2015.

[10] B. Dutta, R. Mahapatra, L. E. Strigari, and J. W. Walker, "Sensitivity to Z-prime and nonstandard neutrino interactions from ultralow threshold neutrino-nucleus coherent scattering," *Physical Review D*, vol. 93, Article ID 013015, 2016.

[11] M. Lindner, W. Rodejohann, and X. Xu, "Coherent neutrino-nucleus scattering and new neutrino interactions," *Journal of High Energy Physics*, vol. 2017, no. 3, article no. 97, 2017.

[12] T. M. Undagoitia and L. Rauch, "Dark matter direct-detection experiments," *Journal of Physics G: Nuclear and Particle Physics*, vol. 43, Article ID 013001, 78 pages, 2016.

[13] J. Monroe and P. Fisher, "Neutrino Backgrounds to Dark Matter Searches," *Phys. Rev. D - Physical Review Journals - American Physical Society*, vol. 76, Article ID 033007, 6 pages, 2007.

[14] J. Billard and E. Figueroa-Feliciano, "Implication of neutrino backgrounds on the reach of next generation dark matter direct

detection experiments," *Physical Review D*, vol. 89, Article ID 023524, 2014.

[15] C. A. J. OHare, A. M. Green, J. Billard, E. Figueroa-Feliciano, and L. E. Strigari, "Readout strategies for directional dark matter detection beyond the neutrino background," *Physical Review D*, vol. 92, Article ID 063518, 2015.

[16] P. Grothaus and M. Fairbairn, "Aspects of WIMP Dark Matter Searches at Colliders and Other Probes," *Physical Review D*, vol. 90, Article ID 055018, 2014.

[17] J. H. Davis, "Dark matter vs. neutrinos: the effect of astrophysical uncertainties and timing information on the neutrino floor," *Journal of Cosmology and Astroparticle Physics*, vol. 2015, no. 3, article no. 12, 2015.

[18] J. B. Dent, B. Dutta, J. L. Newstead, and L. E. Strigari, "Effective field theory treatment of the neutrino background in direct dark matter detection experiments," *Physical Review D*, vol. 93, Article ID 075018, 2016.

[19] J. B. Dent, B. Dutta, J. L. Newstead, and L. E. Strigari, "Dark matter, light mediators, and the neutrino floor," *Physical Review D: Particles, Fields, Gravitation and Cosmology*, vol. 95, Article ID 051701(R), 20 pages, 2017.

[20] E. Aprile, "First Dark Matter Search Results from the XENON1T Experiment," *Physical Review Letters*, vol. 119, Article ID 181301, 2017.

[21] K. Abe, K. Hieda, K. Hiraide, S. Hirano et al., "XMASS detector," *Nuclear Instruments and Methods in Physics Research*, vol. 716, no. 11, pp. 78–85, 2013.

[22] V. A. Kudryavtsev and LZ Collaboration, "Expected background in the LZ experiment," in *Proceedings of the AIP Conference*, vol. 1672, Article ID 060003, 2015.

[23] D. S. Akerib and LUX Collaboration, "Results from a Search for Dark Matter in the Complete LUX Exposure," *Physical Review Letters*, vol. 118, Article ID 021303, 2017.

[24] J. Aalbers, F. Agostini, M. Alfonsi et al., "DARWIN: towards the ultimate dark matter detector," *Journal of Cosmology and Astroparticle Physics*, vol. 2016, no. 11, article no. 017, 2016.

[25] E. Ydrefors, J. Suhonen, and Y. M. Zhao, "Neutrino-nucleus scattering off ^{136}Xe," *Physical Review C*, vol. 91, Article ID 014307, 2015.

[26] W. Almosly, E. Ydrefors, and J. Suhonen, "Neutrino scattering off the stable cadmium isotopes: neutral-current processes," *Journal of Physics G: Nuclear and Particle Physics*, vol. 42, no. 2, Article ID 025106, 2014.

[27] E. Ydrefors, K. G. Balasi, T. S. Kosmas, and J. Suhonen, "Detailed study of the neutral-current neutrino–nucleus scattering off the stable Mo isotopes," *Nuclear Physics A*, vol. 896, pp. 1–23, 2012.

[28] E. Ydrefors, K. G. Balasi, J. Suhonen, and T. S. Kosmas, "Nuclear responses to supernova neutrinos for the stable molybdenum isotopes," in *Neutrinos: Properties, Sources and Detection, Physics Research and Technology*, pp. 151–175, Nova Science Publishers, 2011.

[29] T. W. Donnelly and R. D. Peccei, "Neutral current effects in nuclei," *Physics Reports*, vol. 50, pp. 1–85, 1979.

[30] J. N. Bahcall, E. Lisi, D. E. Alburger, L. De Braeckeleer, S. J. Freedman, and J. Napolitano, "Standard neutrino spectrum from B-8 Decay," *Physical Review C: Nuclear Physics*, vol. 54, no. 1, pp. 411–422, 1996.

[31] A. Bohr, B. R. Mottelson, and D. Pines, "Possible Analogy between the Excitation Spectra of Nuclei and Those of the Superconducting Metallic State," *Physical Review*, vol. 110, no. 4, pp. 936–938, 1958.

[32] J. Suhonen, *From Nucleons to Nucleus: Concepts of Microscopic Nuclear Theory*, Theoretical and Mathematical Physics, Springer, Berlin, Germany, 2007.

[33] A. Bohr and B. R. Mottelson, "Nuclear Structure: Volume 1," in *Nuclear Structure*, Benjamin, New York, NY, USA, 1969.

[34] K. Holinde, "Two-nucleon forces and nuclear matter," *Physics Reports*, vol. 68, no. 3, pp. 121–188, 1981.

[35] A. H. Wapstra and G. Audi, "The 1983 atomic mass evaluation: (I). Atomic mass table," *Nuclear Physics A*, vol. 432, pp. 1–54, 1985.

[36] "National Nuclear Data Center: Evaluated Nuclear Structure Data File," http://www.nndc.bnl.gov/ensdf/.

[37] W. Almosly, E. Ydrefors, and J. Suhonen, "Neutral- and charged-current supernova-neutrino scattering off ^{116}Cd," *Journal of Physics G: Nuclear and Particle Physics*, vol. 40, no. 9, Article ID 095201, 2013.

[38] J. Toivanen and J. Suhonen, "Microscopic quasiparticle-phonon description of odd-mass $^{127-133}$Xe isotopes and their β decay," *Physical Review C: Nuclear Physics*, vol. 57, no. 3, pp. 1237–1245, 1998.

[39] B. A. Brown and W. D. M. Rae, "The Shell-Model Code NuShellX@MSU," *Nuclear Data Sheets*, vol. 120, pp. 115–118, 2014.

[40] B. A. Brown, N. J. Stone, J. R. Stone, I. S. Towner, and M. Hjorth-Jensen, "Magnetic moments of the 2_1^+ states around ^{132}Sn," *Physical Review C*, vol. 71, no. 4, Article ID 044317, 2005.

[41] P. Pirinen, P. C. Srivastava, J. Suhonen, and M. Kortelainen, "Shell-model study on event rates of lightest supersymmetric particles scattering off ^{83}Kr and ^{125}Te," *Physical Review D*, vol. 93, Article ID 095012, 2016.

22

Feasibility Study of the Time Reversal Symmetry Tests in Decay of Metastable Positronium Atoms with the J-PET Detector

A. Gajos,[1] C. Curceanu,[2] E. Czerwiński,[1] K. Dulski,[1] M. Gorgol,[3] N. Gupta-Sharma,[1] B. C. Hiesmayr,[4] B. Jasińska,[3] K. Kacprzak,[1] Ł. Kapłon,[1,5] D. Kisielewska,[1] G. Korcyl,[1] P. Kowalski,[6] T. Kozik,[1] W. Krzemień,[7] E. Kubicz,[1] M. Mohammed,[1,8] Sz. Niedźwiecki,[1] M. Pałka,[1] M. Pawlik-Niedźwiecka,[1] L. Raczyński,[6] J. Raj,[1] Z. Rudy,[1] S. Sharma,[1] Shivani,[1] R. Shopa,[6] M. Silarski,[1] M. Skurzok,[1] W. Wiślicki,[6] B. Zgardzińska,[3] M. Zieliński,[1] and P. Moskal[1]

[1] *Faculty of Physics, Astronomy and Applied Computer Science, Jagiellonian University, 30-348 Cracow, Poland*
[2] *INFN, Laboratori Nazionali di Frascati, 00044 Frascati, Italy*
[3] *Department of Nuclear Methods, Institute of Physics, Maria Curie-Skłodowska University, 20-031 Lublin, Poland*
[4] *Faculty of Physics, University of Vienna, 1090 Vienna, Austria*
[5] *Institute of Metallurgy and Materials Science of Polish Academy of Sciences, Cracow, Poland*
[6] *Świerk Department of Complex Systems, National Centre for Nuclear Research, 05-400 Otwock-Świerk, Poland*
[7] *High Energy Department, National Centre for Nuclear Research, 05-400 Otwock-Świerk, Poland*
[8] *Department of Physics, College of Education for Pure Sciences, University of Mosul, Mosul, Iraq*

Correspondence should be addressed to A. Gajos; aleksander.gajos@uj.edu.pl

Academic Editor: Krzysztof Urbanowski

This article reports on the feasibility of testing of the symmetry under reversal in time in a purely leptonic system constituted by positronium atoms using the J-PET detector. The present state of $\mathcal{T}$ symmetry tests is discussed with an emphasis on the scarcely explored sector of leptonic systems. Two possible strategies of searching for manifestations of $\mathcal{T}$ violation in nonvanishing angular correlations of final state observables in the decay of metastable triplet states of positronium available with J-PET are proposed and discussed. Results of a pilot measurement with J-PET and assessment of its performance in reconstruction of three-photon decays are shown along with an analysis of its impact on the sensitivity of the detector for the determination of $\mathcal{T}$-violation sensitive observables.

1. Introduction

The concept of symmetry of Nature under discrete transformations has been exposed to numerous experimental tests ever since its introduction by E. Wigner in 1931 [1]. The first evidence of violation of the supposed symmetries under spatial ($\mathcal{P}$) and charge ($\mathcal{C}$) parity transformations in the weak interactions has been found already in 1956 and 1958, respectively [2, 3]. However, observation of noninvariance of a physical system under reversal in time required over 50 years more and was finally performed in the system of entangled neutral B mesons in 2012 [4]. Although many experiments proved violation of the combined $\mathcal{CP}$ symmetry, leading to $\mathcal{T}$ violation expected on the ground of the $\mathcal{CPT}$ theorem, experimental evidence for noninvariance under time reversal remains scarce to date.

The Jagiellonian PET (J-PET) experiment aims at performing a test of the symmetry under reversal in time in a purely leptonic system constituted by orthopositronium (o-Ps) with a precision unprecedented in this sector. The increased sensitivity of J-PET with respect to previous discrete symmetry tests with o-Ps⟶3γ is achieved by a large geometrical acceptance and angular resolution of the detector as well as by improved control of the positronium atoms polarization. In this work, we report on the results of feasibility studies for the planned $\mathscr{T}$ violation searches by determination of angular correlations in the o-Ps⟶3γ decay based on a test run of the J-PET detector.

This article is structured as follows: next section briefly discusses the properties of time and time reversal in quantum systems. Subsequently, Section 3 provides an overview of the present status and available techniques of testing of the symmetry under reversal in time and points out the goals of the J-PET experiment in this field. A brief description of the detector and details of the setup used for a test measurement are given in Section 4. Section 5 discusses possible strategies to test the time reversal symmetry with J-PET. Results of the feasibility studies are presented in Section 6 and their impact on the perspectives for a $\mathscr{T}$ test with J-PET is discussed in Section 7.

2. Time and Reversal of Physical Systems in Time

Although the advent of special relativity made it common equate time with spatial coordinates, time remains a distinct concept. Its treatment as an external parameter used in classical mechanics still cannot be consistently avoided in today's quantum theories [5]. As opposed to position and momentum, time lacks a corresponding operator in standard quantum mechanics and thus, countering the intuition, cannot be an observable. Moreover, a careful insight into the time evolution of unstable quantum systems reveals a number of surprising phenomena such as deviations from exponential decay law [6, 7] or emission of electromagnetic radiation at late times [8]. The decay process, inevitably involved in measurements of unstable systems is also a factor restricting possible studies of the symmetry under time reversal [9].

While efforts are taken to define a time operator, observation of $\mathscr{CP}$ violation in the decaying meson systems disproves certain approaches [10]. Alternatively, concepts of time intervals not defined through an external parameter may be considered using tunneling and dwell times [11, 12]. However, also in this case invariance under time reversal is an important factor [13].

It is important to stress that all considerations made herein are only valid if gravitational effects are not considered. In the framework of general relativity with a generic curved spacetime, the concept of inversion of time (as well as the $\mathscr{P}$ transformation) loses its interpretation specific only to the linear affine structure of spacetime [14].

The peculiar properties of time extend as well to the operation of reversing physical systems in time (the T operator), which results in grave experimental challenges limiting the possibilities of $\mathscr{T}$ violation measurements. In contrast to the unitary P and C operators, T can be shown to be antiunitary. As a consequence, no conserved quantities may be attributed to the T operation [15] excluding symmetry tests by means of, e.g., testing selection rules.

Feasibility of $\mathscr{T}$ tests based on a comparison between time evolution of a physical system in two directions, i.e., $|\psi(t)\rangle \longrightarrow |\psi(t+\delta t)\rangle$ and $|\psi(t+\delta t)\rangle \longrightarrow |\psi(t)\rangle$, is also limited as most of the processes which could be used involve a decaying state making it impractical to obtain a reverse process with the same conditions in an experiment. The only exception exploited to date is constituted by transitions of neutral mesons between their flavour-definite states and CP eigenstates [16, 17]. A comparison of such reversible transitions in a neutral B meson system with quantum entanglement of $B^0\overline{B^0}$ pairs produced in a decay of $\Upsilon(4s)$ yielded the only direct experimental evidence of violation of the symmetry under reversal in time obtained to date [4]. While a similar concept of $\mathscr{T}$ violation searches is currently pursued with the neutral kaon system [17–19], no direct tests of this symmetry have been proposed outside the systems of neutral mesons.

In the absence of conserved quantities and with the difficulties of comparing mutually reverse time evolution processes in decaying systems, manifestations of $\mathscr{T}$ violation may still be sought in nonvanishing expectation values of certain operators odd under the T transformation [20]. It follows from the antiunitarity of the T operator that for any operator $\mathscr{O}$

$$\langle\phi|\mathscr{O}|\psi\rangle = \langle\phi|T^{\dagger}T\mathscr{O}T^{\dagger}T|\psi\rangle = \langle\phi_T|\mathscr{O}_T|\psi_T\rangle^{*}, \quad (1)$$

where the T subscript denotes states and operators transformed by the operator of reversal in time. Therefore, an operator odd with respect to the T transformation (i.e., $\mathscr{O}_T = -\mathscr{O}$) must satisfy

$$\langle\phi|O|\psi\rangle = -\langle\phi_T|O|\psi_T\rangle^{*}. \quad (2)$$

For stationary states or in systems where conditions on interaction dynamics such as absence of significant final state interactions are satisfied [21], the mean value of a T-odd and Hermitian operator must therefore vanish in case of $\mathscr{T}$ invariance:

$$\langle O\rangle_T = -\langle O\rangle, \quad (3)$$

and violation of the $\mathscr{T}$ symmetry may thus be manifested as a nonzero expectation value of such an operator.

3. Status and Strategies of $\mathscr{T}$ Symmetry Testing

A number of experiments based on the property of T operator demonstrated in (1)-(3) have been conducted to date. The electric dipole moment of elementary systems, constituting a convenient T-odd operator, has been sought for neutrons and electrons in experiments reaching a precision of 10^{-26} and 10^{-28}, respectively [22, 23]. However, none of such experiments has observed $\mathcal{T}$ violation to date despite their excellent sensitivity. In another class of experiments, a T-odd operator is constructed out of final state observables in a decay process, such as the weak decay $K^+ \longrightarrow \pi^0\mu^+\nu$ studied by the KEK-E246 experiment [24] in which the muon polarization transverse to the decay plane ($\mathcal{P}_T = \mathcal{P}_K \cdot (\mathbf{p}_\pi \times \mathbf{p}_\mu)/|\mathbf{p}_\pi \times \mathbf{p}_\mu|$) was determined as an observable whose nonzero mean value would manifest $\mathcal{T}$ violation. However, neither this measurement nor similar studies using decay of polarized ^{8}Li nuclei [25] and of free neutrons [26] have observed significant mean values of T-odd final state observables.

Notably, although the property of reversal in time shown in (1)-(3) is not limited to any particular system nor interaction, it has been mostly exploited to test the $\mathcal{T}$ symmetry in weak interactions. Whereas the latter is the most promising candidate due to well proven $\mathcal{CP}$ violation, evidence for $\mathcal{T}$ noninvariance may be sought in other physical systems and phenomena using the same scheme of a symmetry test. Systems constituted by purely leptonic matter are an example of a sector where experimental results related to the time reversal symmetry—and to discrete symmetries in general—remain rare. Several measurements of neutrino oscillations are being conducted by the NOνA and T2K experiments searching for $\mathcal{CP}$ violation in the $\nu_\mu \longrightarrow \nu_e$ and $\bar{\nu}_\mu \longrightarrow \bar{\nu}_e$ channels [27, 28], which may provide indirect information on the $\mathcal{T}$ symmetry. Another notable test of discrete symmetries in the leptonic sector is the search for the violation of Lorentz and $\mathcal{CPT}$ invariance based on the Standard Model Extension framework [29] and anti-CPT theorem [30] which has also been performed by T2K [31, 32]. Other possible tests of these symmetries with the positronium system include spectroscopy of the 1S-2S transition [33] and measuring the free fall acceleration of positronium [34]. However, the question of the $\mathcal{T}$, $\mathcal{CP}$, and $\mathcal{CPT}$ symmetries in the leptonic systems remains open as the aforementioned experiments have not observed a significant signal of a violation.

Few systems exist which allow for discrete symmetry tests in a purely leptonic sector. However, a candidate competitive with respect to neutrino oscillations is constituted by the electromagnetic decay of positronium atoms, exotic bound states of an electron and a positron. With a reduced mass only twice smaller than that of a hydrogen atom, positronium is characterized by a similar energy level structure. At the same time, it is a metastable state with a lifetime strongly dependent on the spin configuration. The singlet state referred to as parapositronium, may only decay into an even number of photons due to charge parity conservation, and has a lifetime (in vacuum) of 0.125 ns. The triplet state (orthopositronium, o-Ps) is limited to decay into an odd number of photons and lives in vacuum over three orders of magnitude longer than the singlet state (τ_{o-Ps} = 142 ns) [35–37].

Being an eigenstate of the parity operator alike atoms, positronium is also characterized by symmetry under charge conjugation typical for particle-antiparticle systems. Positronium atoms are thus a useful system for discrete symmetry studies. Moreover, they may be copiously produced in laboratory conditions using typical sources of β^+ radiation [38], giving positronium-based experiments a technical advantage over those using, e.g., aforementioned neutrino oscillations. However, few results on the discrete symmetries in the positronium system have been reported to date. The most precise measurements studied the angular correlation operators in the decay of orthopositronium states into three photons and determined mean values of final state operators odd under the $\mathcal{CP}$ and $\mathcal{CPT}$ conjugations, finding no violation signal at the sensitivity level of 10^{-3} [39, 40]. Although the aforementioned studies sought for violation of $\mathcal{CP}$ and $\mathcal{CPT}$, it should be emphasized that the operators used therein were odd under the T operation as well, leading to an implicit probe also for the symmetry under reversal in time.

The results obtained to date, showing no sign of violation, were limited in precision by technical factors such as detector geometrical acceptance and resolution, uncertainty of positronium polarization, and data sample size. In terms of physical restrictions, sensitivity of such discrete symmetry tests with orthopositronium decay is only limited by possible false asymmetries arising from photon-photon final state interactions at the precision level of 10^{-9} [41, 42]. The J-PET experiment thus sets its goal to explore the $\mathcal{T}$-violating observables at precision beyond the presently established 10^{-3} level [43].

4. The J-PET Detector

The J-PET (Jagiellonian Positron Emission Tomograph) is a photon detector constructed entirely with plastic scintillators. Along with constituting the first prototype of plastic scintillator-based cost-effective PET scanner with a large field of view [44, 45], it may be used to detect photons in the sub-MeV range such as products of annihilation of positronium atoms, thus allowing for a range of studies related to discrete symmetries and quantum entanglement [43].

J-PET consists of three concentric cylindrical layers of axially arranged γ detection modules based on strips of EJ-230 plastic scintillator as shown schematically in Figure 1. Each scintillator strip is 50 cm long with a rectangular cross-section of 7×19 mm^2. Within a detection module, both ends of a scintillator strip are optically coupled to photomultiplier tubes. Due to low atomic number of the elements constituting plastic scintillators, γ quanta interact mostly through Compton scattering in the strips, depositing a part of their energy dependent on the scattering angle. The lack of exact photon energy determination in J-PET is compensated by fast decay time of plastic scintillators resulting in high time resolution and allowing for use of radioactive sources with activity as high as 10 MBq. The energy deposited by photons scattered in a scintillator is converted to optical photons which travel to both ends of a strip undergoing multiple internal reflections. Consequently, the position of γ interaction along a detection module is determined using the difference between effective light propagation times to the two photomultiplier tubes

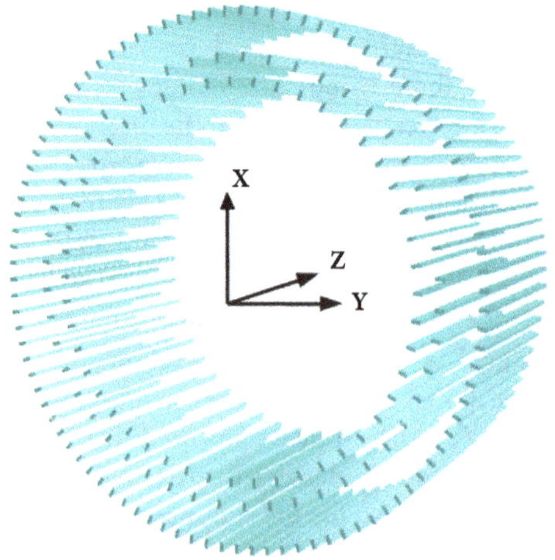

FIGURE 1: Schematic view of the J-PET detector consisting of 192 plastic scintillator strips arranged in three concentric layers with radii ranging from 42.5 cm to 57.5 cm. The strips are oriented along the Z axis of the detector barrel.

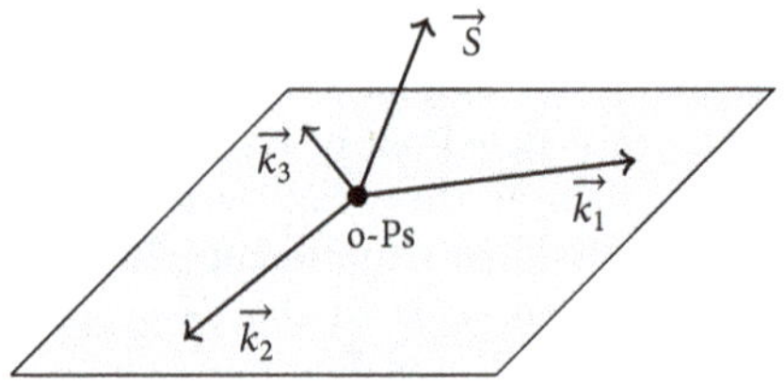

FIGURE 2: Vectors describing the final state of an o-Ps⟶3γ annihilation in the o-Ps frame of reference. $\vec{S}$ denotes orthopositronium spin and $\vec{k}_{1,2,3}$ are the momentum vectors of the annihilation photons, lying in a single plane. The operator defined in (4) is a measure of angular correlation between the positronium spin and the decay plane normal vector.

attached to a scintillator strip [46]. In the transverse plane of the detector, γ interactions are localized up to the position of a single module, resulting in an azimuthal angle resolution of about 1°.

Although the J-PET γ detection modules do not allow for a direct measurement of total photon energy, recording interactions of all photons from a 3γ annihilation allows for an indirect reconstruction of photons' momenta based on event geometry and 4-momentum conservation [47].

As J-PET is intended for a broad range of studies from medical imaging [48] through quantum entanglement [49, 50] to tests of discrete symmetries [43], its data acquisition is operating in a triggerless mode [51] in order to avoid any bias in the recorded sample of events. Electric signals produced by the photomultipliers are sampled in the time domain at four predefined voltage thresholds allowing for an estimation of the deposited energy using the time over threshold technique [52]. Further reconstruction of photon interactions as well as data preselection and handling is performed with dedicated software [53, 54]. Several extensions of the detector are presently in preparation such as improvements of the J-PET geometrical acceptance by inclusion of additional detector layers [47] as well as enhanced scintillator readout with silicon photomultipliers [55] and new front-end electronics [52].

5. Discrete Symmetry Tests with the J-PET Detector

5.1. Measurements Involving Orthopositronium Spin. The symmetry under reversal in time can be put to test in the o-Ps⟶3γ decay by using the properties of T conjugation demonstrated by (1)-(3). Spin $\vec{S}$ of the decaying orthopositronium atom and momenta of the three photons produced in the decay $\vec{k}_{1,2,3}$ (ordered according to their descending magnitude, i.e., $|\vec{k}_1| > |\vec{k}_2| > |\vec{k}_3|$) allow for construction of an angular correlation operator odd under reversal in time:

$$C_T = \vec{S} \cdot \left(\vec{k}_1 \times \vec{k}_2\right), \tag{4}$$

which corresponds to an angular correlation between the positronium spin direction and the decay plane as illustrated in Figure 2.

Such an approach which requires estimation of the spin direction of decaying positronia was used by both previous discrete symmetry tests conducted with orthopositronium decay [39, 40]. These two experiments, however, adopted different techniques to control the o-Ps spin polarization. In the $\mathscr{CP}$ violation search, positronium atoms were produced in strong external magnetic field resulting in their polarization along a thus imposed direction [39]. A setup required to provide the magnetic field, however, was associated with a limitation of the geometrical acceptance of the detectors used. The second measurement, testing the $\mathscr{CPT}$ symmetry using the Gammasphere detector which covered almost a full solid angle, did not therefore rely on external magnetic field. Instead, positronium polarization was evaluated statistically by allowing for o-Ps atoms formation only in a single hemisphere around a point-like positron source, resulting in estimation of the polarization along a fixed quantization axis with an accuracy limited by a geometrical factor of 0.5 [40]. Neither of the previous experiments attempted to reconstruct the position of o-Ps⟶3γ decay, instead limiting the volume of o-Ps creation and assuming the same origin point for all annihilations.

The J-PET experiment attempts to improve on the latter approach which does not require the use of external magnetic field. The statistical knowledge of spin polarization of the positrons forming o-Ps atoms can be significantly increased with a positronium production setup depicted in Figure 3, where polarization is estimated on an event-by-event basis instead of assuming a fixed quantization axis throughout the measurement. A trilateration-based technique of recon-

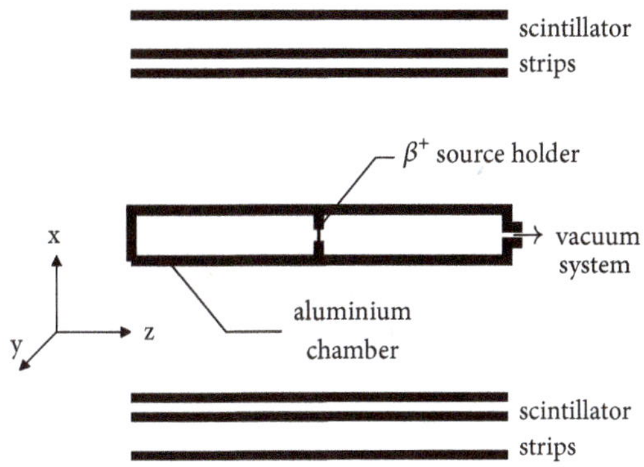

FIGURE 3: Scheme of the positronium production setup devised for positron polarization determination in J-PET experiment. Positrons are produced in a β^+ source mounted in the center of a cylindrical vacuum chamber coaxial with the detector. Positronium atoms are formed by the interaction of positrons in a porous medium covering the chamber walls. Determination of an o-Ps⟶3γ annihilation position in the cylinder provides an estimate of positron momentum direction.

structing the position of o-Ps⟶3γ decay created for J-PET allows for estimation of the direction of positron propagation in a single event with a vector spanned by a point-like β^+ source location and the reconstructed orthopositronium annihilation point [56].

The dependence of average spin polarization of positrons (largely preserved during formation of orthopositronium [57]) on the angular accuracy of the polarization axis determination is given by $(1/2)(1 + \cos\alpha)$ where α is the opening angle of a cone representing the uncertainty of polarization axis direction [58]. This uncertainty in J-PET results predominantly from the resolution of determination of the 3γ annihilation point as depicted in Figure 4 and amounts to about 15° [56], resulting in a polarization decrease smaller than 2%. By contrast, in the previous measurement with Gammasphere [40] where the polarization axis was fixed, the same geometrical factor accounted for a 50% polarization loss.

5.2. Measurements Using Polarization of Photons. The scheme of measurement without external magnetic field for positronium polarization may be further simplified with modified choice of the measured T-odd operator. This novel approach of testing the $\mathcal{T}$ symmetry may be pursued by J-PET with a spin-independent operator constructed for the o-Ps⟶3γ annihilations if the polarization vector of one of the final state photons is included [43]:

$$C_T' = \vec{k}_2 \cdot \vec{\varepsilon}_1, \quad (5)$$

where $\vec{\varepsilon}_1$ denotes the electric polarization vector of the most energetic γ quantum and $\vec{k}_2$ is the momentum of the second most energetic one. Such angular correlation operators involving photon electric polarization have never been studied in the decay of orthopositronium. Geometry of

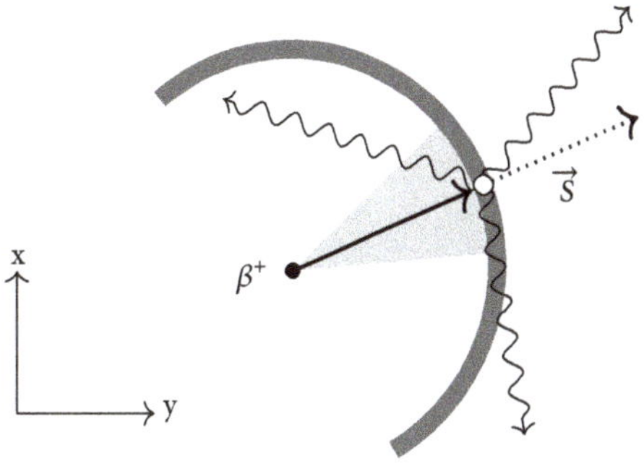

FIGURE 4: Determination of positron polarization axis using its momentum direction (black arrow) estimated using the β^+ source position and reconstructed origin of the 3γ annihilation of orthopositronium in the chamber wall (dark gray band). The shaded region represents the angular uncertainty of positron flight direction resulting from achievable resolution of the 3γ annihilation point.

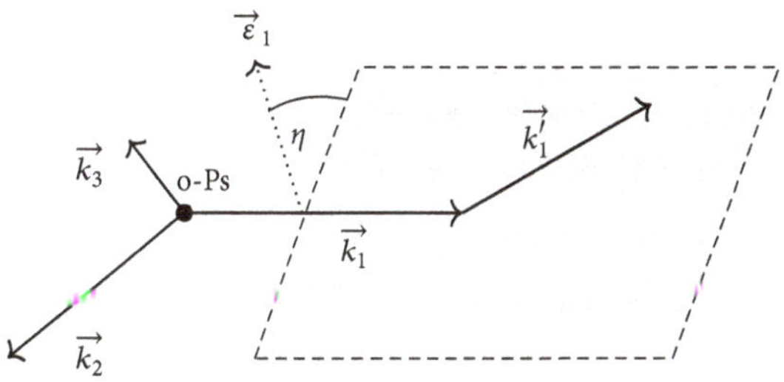

FIGURE 5: Scheme of estimation of polarization vector for a photon produced in o-Ps⟶3γ at J-PET. Photon of momentum $\vec{k}_1$ is scattered in one of the detection modules and a secondary interaction of the scattering product $\vec{k}_1'$ is recorded in a different scintillator strip. The most probable angle η between the polarization vector $\vec{\varepsilon}_1$ and the scattering plane spanned by $\vec{k}_1$ and $\vec{k}_1'$ amounts to 90°.

the J-PET detector enables a measurement of $\langle C_T' \rangle$ thanks to the ability to record secondary interactions of once scattered photons from the o-Ps⟶3γ annihilation as depicted in Figure 5.

6. Test Measurement with the J-PET Detector

The setup presented in Figure 3 was constructed and fully commissioned in 2017 [59]. One of the first test measurements was dedicated to evaluation of the feasibility of identification and reconstruction of three-photon events. A ^{22}Na β^+ source was mounted inside a cylindrical vacuum chamber of 14 cm radius. The positronium formation-enhancing medium, presently under elaboration, was not included in the measurement. Therefore, the test of 3γ event reconstruction was based on direct 3γ annihilation of positrons with electrons of the aluminium chamber walls, with a yield smaller by more than an order of magnitude than the rate of o-Ps⟶3γ annihilations expected in the

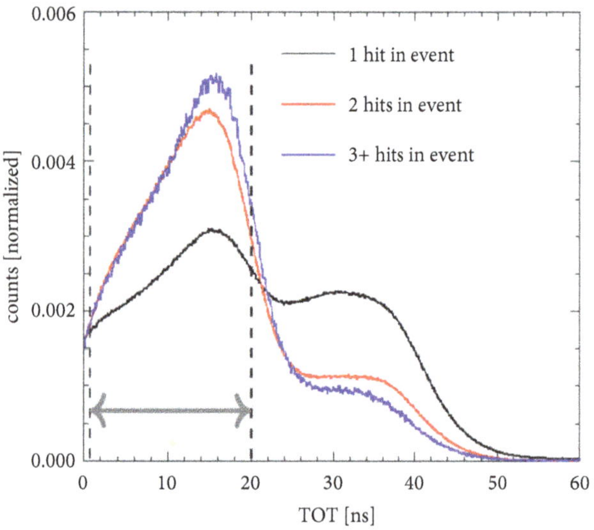

FIGURE 6: Distributions of time over threshold (TOT) values, for γ interactions observed in groups of 1, 2 and more within a coincidence time window of 2.5 ns. In the samples with increasing number of coincident photons, the contribution of γ quanta from 3γ annihilations increases with respect to other processes. Gray lines and arrow denote the region used to identify 3γ annihilation photon candidates.

final measurements with a porous medium for positronium production.

The capabilities of J-PET to select 3γ events and discriminate background arising from two-photon e^+e^- annihilations as well as from accidental coincidences are based primarily on two factors:

(i) a measure of energy deposited by a photon in Compton scattering, provided by the time over threshold (TOT) values determined by the J-PET front-end electronics,

(ii) angular dependencies between relative azimuthal angles of recorded γ interaction points, specific to topology of the event [60].

Distributions of the TOT values, after equalization of responses of each detection module, are presented in Figure 6. Separate study of TOT distributions for γ quanta observed in groups of 1, 2 and more recorded γ hits in scintillators within a short time window reveals the different composition of photons from 3γ annihilations with respect to those originating from background processes such as deexcitation of the β^+ decay products from a ^{22}Na source (1270 keV) and cosmic radiation. Two Compton edges corresponding to 511 keV and 1270 keV photons are clearly discernible in TOT distributions, allowing identifying candidates for interactions of 3γ annihilation products by TOT values located below the 511 keV Compton edge as marked with dashed lines in Figure 6.

The second event selection criterion is based on the correlations between relative azimuthal angles of photon interactions recorded in the detector in cases of three interactions observed in close time coincidence. The tests performed with Monte Carlo simulations have shown that annihilations into two and three photons can be well separated using such correlations [60]. An exemplary relative distribution of values constructed using these correlations, obtained with the test measurement, is presented in Figure 7(a). For a comparison, the same distribution obtained with a point-like annihilation medium used in another test measurement of J-PET is presented in Figure 7(b).

A sharp vertical band at $\delta\theta_2 + \delta\theta_3 \approx 180^\circ$ seen in Figure 7(b) originates from events corresponding to annihilations into two back-to-back photons. Broadening of this 2γ band in case of the extensive chamber is a result of the increased discrepancy between relative azimuthal angles of detection module locations used for the calculation and the actual relative angles in events originating in the walls of the cylindrical chamber as depicted schematically in Figure 8.

The distributions presented in Figure 7 are in good agreement with the simulation-based expectations [60]. Selection of events with values of $\delta\theta_2+\delta\theta_3$ significantly larger than 180° allows for identification of three-photon annihilations.

The aforementioned event selection techniques allowed to extract 1164 3γ event candidates from the two-day test measurement with a β^+ source activity of about 10 MBq placed in the center of the aluminium cylinder as depicted in Figure 3. Therefore, a quantitative estimation of the achievable resolution of three-photon event origin points and its impact on the positronium polarization control capabilities requires a measurement including a medium enhancing the positronium production.

Resolution of the detector and its field of view was validated with a benchmark analysis of the test data performed using the abundant 2γ annihilation events. Figure 9 presents the images of the annihilation chamber obtained using 2γ events whose selection and reconstruction was performed with the same techniques as applied to medical imaging tests performed with J-PET [48]. Although a large part of recorded annihilations originate already in the setup holding the β^+ source, a considerable fraction of positrons reach the chamber walls. The effective longitudinal field of view of J-PET for 2γ events which can be directly extended to 3γ annihilations due to similar geometrical constraints spans the range of approximately $|z| < 8$ cm.

7. Summary and Perspectives

The J-PET group attempts to perform the first search for signs of violation of the symmetry under reversal in time in the decay of positronium atoms. One of the available techniques is based on evaluation of mean values of final state observables constructed from photons' momenta and positronium spin in an o-Ps$\longrightarrow 3\gamma$ annihilation with a precision enhanced with respect to the previous realization of similar measurements by determination of positronium spin distinctly for each recorded event. Moreover, the J-PET detector enables a novel test by determination of a T-odd

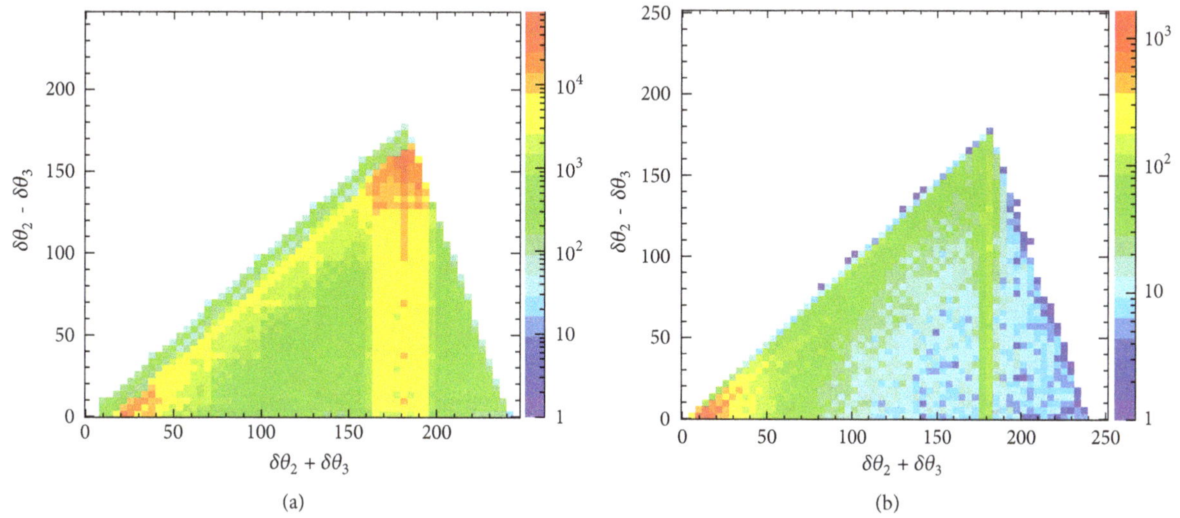

FIGURE 7: Relations between the sum and difference of two smallest relative azimuthal angles ($\delta\theta_2$ and $\delta\theta_3$, respectively) between γ interaction points in events with three recorded interactions. (a) Distribution obtained in the test measurement with the aluminium chamber presented in Figure 3. For reference, the same spectrum obtained with a point-like annihilation medium located in the detector center [59] is displayed in (b). The vertical band around $\delta\theta_2 + \delta\theta_3 \approx 180^\circ$ arises from two-photon annihilations and is broadened in the first case due to extensive dimensions of the annihilation chamber used. 3γ annihilation events are expected in the region located at the right side of the 2γ band [60].

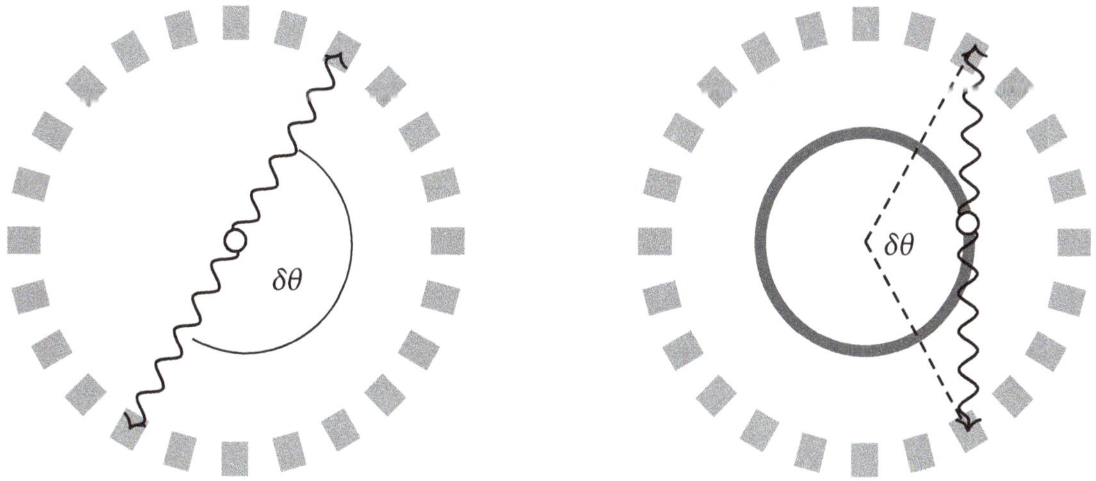

FIGURE 8: Explanation of the broadening of the 2γ annihilation band present in Figures 7(a) and 7(b) at $\delta\theta_2 + \delta\theta_3 \approx 180^\circ$. Left: when 2γ annihilations originate in a small region in the detector center, the calculated relative azimuthal angles of detection modules which registered the photons correspond closely to actual relative angles between photons' momenta. Right: with 2γ annihilations taking place in the walls of an extensive-size annihilation chamber (gray band), the broadening of the band at 180° is caused by a discrepancy between the calculated and actual relative angles. The detector scheme and proportions are not preserved for clarity.

observable constructed using the momenta and polarization of photons from annihilation.

The pilot measurement conducted with the J-PET detector demonstrated the possibility of identifying candidates of annihilation photons interactions in the plastic scintillator strips by means of the time over threshold measure of deposited energy and angular dependencies between relative azimuthal angles of γ interaction points specific to event spatial topology. A preliminary selection of three-photon annihilation events yielded 1164 event candidates from a two-day test measurement with a yield reduced by more than an order of magnitude with respect to the planned experiments with a porous positronium production target and a centrally located 10 MBq source. The annihilation reconstruction resolution and performance of the setup proposed for positron spin determination was validated with a benchmark reconstruction of two-photon annihilations. Results obtained from the test measurement confirm the feasibility of a test of symmetry under reversal in time by measurement of the angular correlation operator defined in (4) without external magnetic field once a positronium production medium is used.

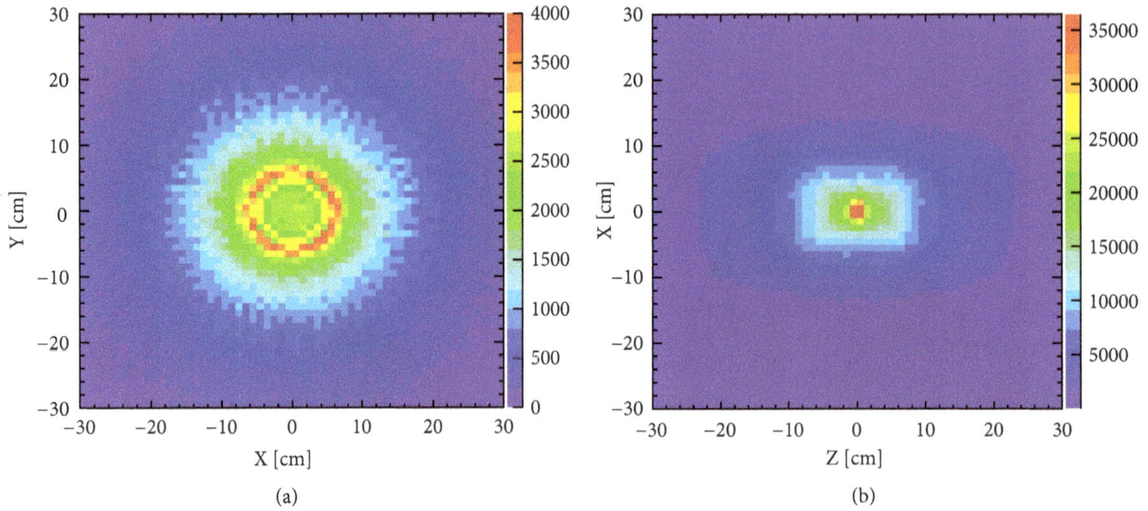

FIGURE 9: Tomographic images of the cylindrical chamber used in the test run of J-PET, obtained using reconstructed $e^+e^- \longrightarrow 2\gamma$ annihilation events. (a) Transverse view of the chamber (the central longitudinal region of $|z| < 4$ cm was excluded where the image is dominated by annihilation events originating in the setup of the β^+ source.). (b) Longitudinal view of the imaged chamber. In the central region, a strong image of the positron source and its mounting setup is visible.

Disclosure

The opinions expressed in this publication are those of the authors and do not necessarily reflect the views of the John Templeton Foundation.

Conflicts of Interest

The authors declare that there are no conflicts of interest regarding the publication of this paper.

Acknowledgments

The authors acknowledge technical and administrative support of A. Heczko, M. Kajetanowicz, and W. Migdał. This work was supported by the Polish National Center for Research and Development through Grant INNOTECH-K1/IN1/64/159174/NCBR/12, the Foundation for Polish Science through the MPD and TEAM/2017-4/39 programmes and Grants nos. 2016/21/B/ST2/01222 and 2017/25/N/NZ1/00861, the Ministry for Science and Higher Education through Grants nos. 6673/IA/SP/2016, 7150/E-338/SPUB/2017/1, and 7150/E-338/M/2017, and the EU and MSHE Grant no. POIG.02.03.00-161 00013/09. B. C. Hiesmayr acknowledges support by the Austrian Science Fund (FWF-P26783). C. Curceanu acknowledges a grant from the John Templeton Foundation (ID 58158).

References

[1] E. P. Wigner, *Group theory and its application to the quantum mechanics of atomic spectra*, Academic Press, New York, NY, USA, 1959.

[2] C. S. Wu, E. Ambler, R. W. Hayward, D. D. Hoppes, and R. P. Hudson, "Experimental test of parity conservation in beta decay," *Physical Review A: Atomic, Molecular and Optical Physics*, vol. 105, no. 4, pp. 1413–1415, 1957.

[3] P. C. Macq, K. M. Crowe, and R. P. Haddock, "Helicity of the electron and positron in muon decay," *Physical Review A: Atomic, Molecular and Optical Physics*, vol. 112, no. 6, pp. 2061–2071, 1958.

[4] J. P. Lees, V. Poireau, V. Tisserand et al., "Observation of Time Reversal Violation in the B0 Meson System," *Physical Review Letters*, vol. 109, p. 211801, 2012.

[5] J. G. Muga, R. Sala Mayato, and I. L. Egusquiza, *Time in Quantum Mechanics*, Springer-Verlag, Berlin, Germany, 2nd edition, 2008.

[6] K. Urbanowski, "Decay law of relativistic particles: quantum theory meets special relativity," *Physics Letters B*, vol. 737, pp. 346–351, 2014.

[7] F. Giacosa, "Time evolution of an unstable quantum system," *Acta Physica Polonica B*, vol. 48, no. 10, pp. 1831–1836, 2017.

[8] K. Urbanowski and K. Raczyńska, "Possible emission of cosmic X- and γ-rays by unstable particles at late times," *Physics Letters B*, vol. 731, pp. 236–241, 2014.

[9] J. Bernabeu and F. Martinez-Vidal, "Colloquium: Time-reversal violation with quantum-entangled B mesons," *Reviews of Modern Physics*, vol. 87, no. 1, pp. 165–182, 2015.

[10] T. Durt, A. Di Domenico, and B. Hiesmayr, *Falsification of a Time Operator Model based on Charge-Conjugation-Parity violation of neutral K-mesons*, 2015.

[11] N. G. Kelkar, "Electron tunneling times," *Acta Physica Polonica B*, vol. 48, no. 10, pp. 1825–1830, 2017.

[12] N. G. Kelkar, "Quantum Reflection and Dwell Times of Metastable States," *Physical Review Letters*, vol. 99, no. 21, 2007.

[13] N. G. Kelkar and M. Nowakowski, "Analysis of averaged multichannel delay times," *Physical Review A: Atomic, Molecular and Optical Physics*, vol. 78, no. 1, 2008.

[14] L. M. Sokołowski, "Discrete Lorentz symmetries in gravitational fields," *Acta Physica Polonica B*, vol. 48, no. 10, pp. 1947–1953, 2017.

[15] R. G. Sachs, *The physics of time reversal*, University of Chicago Press, Chicago, IL, USA, 1987.

[16] J. Bernabéu, F. Martínez-Vidal, and P. Villanueva-Pérez, "Time Reversal Violation from the entangled B0-B0 system," *Journal of High Energy Physics*, vol. 2012, no. 8, 2012.

[17] J. Bernabeu, A. Di Domenico, and P. Villanueva-Perez, "Direct test of time reversal symmetry in the entangled neutral kaon system at a ϕ-factory," *Nuclear Physics B*, vol. 868, no. 1, pp. 102–119, 2013.

[18] A. Gajos, "A direct test of *T* symmetry in the neutral K meson system at KLOE-2," *Journal of Physics: Conference Series*, vol. 631, p. 012018, 2015.

[19] A. Gajos, "Tests of discrete symmetries and quantum coherence with neutral kaons at the KLOE-2 experiment," *Acta Physica Polonica B*, vol. 48, no. 10, pp. 1975–1981, 2017.

[20] L. Wolfenstein, "The search for direct evidence for time reversal violation," *International Journal of Modern Physics E*, vol. 8, no. 6, pp. 501–511, 1999.

[21] H. E. Conzett, "Tests of time-reversal invariance in nuclear and particle physics," *Physical Review C: Nuclear Physics*, vol. 52, no. 2, pp. 1041–1046, 1995.

[22] J. M. Pendlebury, S. Afach, N. J. Ayres et al., "Revised experimental upper limit on the electric dipole moment of the neutron," *Physical Review D*, vol. 92, p. 092003, 2015.

[23] J. Baron, W. C. Campbell, and D. DeMille, "Order of magnitude smaller limit on the electric dipole moment of the electron," *Science*, vol. 343, no. 6168, pp. 269–272, 2014.

[24] M. Abe, M. Aoki, I. Ara et al., "Search for T-violating transverse muon polarization in the K+⟶ $\pi 0 \mu + \nu$ decay," *Physical Review Letters*, vol. D73, p. 072005, 2006.

[25] R. Huber, J. Lang, S. Navert et al., "Search for time-reversal violation in the β decay of polarized 8Li nuclei," *Physical Review Letters*, vol. 90, no. 20, p. 202301/4, 2003.

[26] A. Kozela, G. Ban, A. Białek et al., "Measurement of the transverse polarization of electrons emitted in free neutron decay," *Physical Review C nuclear physics*, vol. 85, no. 4, 2012.

[27] J. Bian, "The NOvA Experiment: Overview and Status," in *Proceedings of the Meeting of the APS Division of Particles and Fields (DPF 2013)*, Santa Cruz, Calif, USA, https://arxiv.org/abs/1309.7898.

[28] K. Abe, J. Amey, C. Andreopoulos et al., "T2K, Combined Analysis of Neutrino and Antineutrino Oscillations at T2K," *Physical Review Letters*, vol. 11, p. 151801, 2017.

[29] V. A. Kostelecký, "Formalism for CPT, T, and Lorentz violation in neutral meson oscillations," *Physical Review D: Particles, Fields, Gravitation and Cosmology*, vol. 64, no. 7, 2001.

[30] O. W. Greenberg, "CPT violation implies violation of Lorentz invariance," *Physical Review Letters*, vol. 89, no. 23, p. 231602/4, 2002.

[31] K. Abe, J. Amey, C. Andreopoulos et al., "Search for Lorentz and CPT violation using sidereal time dependence of neutrino flavor transitions over a short baseline," *Physical Review D*, vol. 95, p. 111101, 2017.

[32] V. A. Kostelecký and N. Russell, "Data tables for Lorentz and CPT violation," *Reviews of Modern Physics*, vol. 83, p. 11, 2011.

[33] K. Pachucki and S. G. Karshenboim, "Higher-order recoil corrections to energy levels of two-body systems," *Physical Review A: Atomic, Molecular and Optical Physics*, vol. 60, no. 4, pp. 2792–2798, 1999.

[34] V. A. Kostelecký and A. J. Vargas, "Lorentz and CPT tests with hydrogen, antihydrogen, and related systems," *Physical Review D: Particles, Fields, Gravitation and Cosmology*, vol. 92, no. 5, 2015.

[35] A. H. Al-Ramadhan and D. W. Gidley, "New precision measurement of the decay rate of singlet positronium," *Physical Review Letters*, vol. 72, no. 11, pp. 1632–1635, 1994.

[36] R. S. Vallery, P. W. Zitzewitz, and D. W. Gidley, "Resolution of the orthopositronium-lifetime puzzle," *Physical Review Letters*, vol. 90, no. 20, p. 203402/4, 2003.

[37] O. Jinnouchi, S. Asai, and T. Kobayashi, "Precision measurement of orthopositronium decay rate using SiO2 powder," *Physics Letters B*, vol. 572, no. 3-4, pp. 117–126, 2003.

[38] B. Jasińska, M. Gorgol, and M. Wiertel, "Determination of the 3γ fraction from positron annihilation in mesoporous materials for symmetry violation experiment with J-PET scanner," *Acta Physica Polonica B*, vol. 47, p. 453, 2016.

[39] T. Yamazaki, T. Namba, S. Asai, and T. Kobayashi, "Search for cp violation in positronium decay," *Physical Review Letters*, vol. 104, p. 083401, 2010.

[40] P. A. Vetter and S. J. Freedman, "Search for cpt-odd decays of positronium," *Physical Review Letters*, vol. 91, no. 26, p. 263401, 2003.

[41] B. K. Arbic, S. Hatamian, M. Skalsey, J. Van House, and W. Zheng, "Angular-correlation test of CPT in polarized positronium," *Physical Review A: Atomic, Molecular and Optical Physics*, vol. 37, no. 9, pp. 3189–3194, 1988.

[42] W. Bernreuther, U. Low, J. P. Ma, and O. Nachtmann, "How to Test CP, T and CPT Invariance in the Three Photon Decay of Polarized s Wave Triplet Positronium," *Zeitschrift für Physik C*, vol. 41, p. 143, 1988.

[43] P. Moskal, D. Alfs, T. Bednarski et al., "Potential of the J-PET detector for studies of discrete symmetries in decays of positronium atom - A purely leptonic system," *Acta Physica Polonica B*, vol. 47, no. 2, pp. 509–535, 2016.

[44] P. Moskal, *Strip device and the method for the determination of the place and response time of the gamma quanta and the application of the device for the positron emission thomography, Patent number US2012112079, EP2454612, JP2012533734*, 2010.

[45] S. Niedźwiecki, P. Białas, C. Curceanu et al., "J-PET: a new technology for the whole-body PET imaging," *Acta Physica Polonica A*, vol. 48, p. 1567, 2017.

[46] P. Moskal, S. Niedźwiecki, T. Bednarski et al., "Test of a single module of the J-PET scanner based on plastic scintillators," *Nuclear Instruments and Methods in Physics Research Section A: Accelerators, Spectrometers, Detectors and Associated Equipment*, vol. 764, pp. 317–321, 2014.

[47] D. Kamińska, A. Gajos, E. Czerwiński et al., "A feasibility study of ortho-positronium decays measurement with the J-PET scanner based on plastic scintillators," *The European Physical Journal C*, vol. 76, p. 445, 2016.

[48] M. Pawlik-Niedźwiecka, S. Niedźwiecki, D. Alfs et al., "Preliminary Studies of J-PET Detector Spatial Resolution," *Acta Physica Polonica A*, vol. 132, no. 5, pp. 1645–1649, 2017.

[49] B. C. Hiesmayr and P. Moskal, "Genuine Multipartite Entanglement in the 3-Photon Decay of Positronium," *Scientific Reports*, vol. 7, no. 1, 2017.

[50] M. Nowakowski and D. B. Fierro, "Three-photon entanglement from ortho-positronium revisited," *Acta Physica Polonica B*, vol. 48, no. 10, pp. 1955–1960, 2017.

[51] G. Korcyl, D. Alfs, T. Bednarski et al., "Sampling FEE and Trigger-less DAQ for the J-PET Scanner," *Acta Physica Polonica B*, vol. 47, no. 2, p. 491, 2016.

[52] M. Pałka, P. Strzempek, G. Korcyl et al., "Multichannel FPGA based MVT system for high precision time (20 ps RMS) and charge measurement," *Journal of Instrumentation*, vol. 12, no. 08, pp. P08001–P08001, 2017.

[53] W. Krzemień, A. Gajos, A. Gruntowski et al., "Analysis Framework for the J-PET Scanner," *Acta Physica Polonica A*, vol. 127, no. 5, pp. 1491–1494, 2015.

[54] W. Krzemień, D. Alfs, P. Białas et al., "Overview of the Software Architecture and Data Flow for the J-PET Tomography Device," *Acta Physica Polonica B*, vol. 47, no. 2, p. 561, 2016.

[55] P. Moskal, O. Rundel, D. Alfs et al., "Time resolution of the plastic scintillator strips with matrix photomultiplier readout for J-PET tomograph," *Physics in Medicine and Biology*, vol. 61, no. 5, pp. 2025–2047, 2016.

[56] A. Gajos, D. Kamińska, E. Czerwiński et al., "Trilateration-based reconstruction of ortho -positronium decays into three photons with the J-PET detector," *Nuclear Instruments and Methods in Physics Research Section A: Accelerators, Spectrometers, Detectors and Associated Equipment*, vol. 819, pp. 54–59, 2016.

[57] P. W. Zitzewitz, J. C. Van House, A. Rich, and D. W. Gidley, "Spin Polarization of Low-Energy Positron Beams," *Physical Review Letters*, vol. 43, no. 18, pp. 1281–1284, 1979.

[58] P. Coleman, *Positron Beams and Their Applications*, World Scientific, 2000.

[59] E. Czerwiński, K. Dulski, P. Białas et al., "Commissioning of the J-PET Detector for Studies of Decays of Positronium Atoms," *Acta Physica Polonica B*, vol. 48, no. 10, p. 1961, 2017.

[60] P. Kowalski, W. Wiślicki, L. Raczyński et al., "Scatter Fraction of the J-PET Tomography Scanner," *Acta Physica Polonica B*, vol. 47, no. 2, p. 549, 2016.

23

First Order Framework for Gauge k-Vortices

D. Bazeia,[1] L. Losano,[1] M. A. Marques,[1] and R. Menezes[1,2]

[1]*Departamento de Física, Universidade Federal da Paraíba, 58051-970 João Pessoa, PB, Brazil*
[2]*Departamento de Ciências Exatas, Universidade Federal da Paraíba, 58297-000 Rio Tinto, PB, Brazil*

Correspondence should be addressed to D. Bazeia; dbazeia@gmail.com

Academic Editor: Diego Saez-Chillon Gomez

We study vortices in generalized Maxwell-Higgs models, with the inclusion of a quadratic kinetic term with the covariant derivative of the scalar field in the Lagrangian density. We discuss the stressless condition and show that the presence of analytical solutions helps us to define the model compatible with the existence of first order equations. A method to decouple the first order equations and to construct the model is then introduced and, as a bonus, we get the energy depending exclusively on a function of the fields calculated from the boundary conditions. We investigate some specific possibilities and find, in particular, a compact vortex configuration in which the energy density is all concentrated in a unit circle.

1. Introduction

Vortices are localized structures that appear in two spatial dimensions. They are present in many areas of nonlinear science and were firstly investigated in the context of fluid mechanics [1, 2]. These objects also appear in type II superconductors [3] when one deals with the Ginzburg-Landau theory of superconductivity [4] and may also be present as magnetic domains in magnetic materials and in many other applications in condensed matter [5, 6].

In high energy physics, in particular, vortices firstly appeared in the Nielsen-Olesen work [7], which is perhaps the simplest relativistic model that supports these structures. The model consists of a Maxwell gauge field minimally coupled to a complex scalar field under the Abelian $U(1)$ symmetry in the $(2, 1)$ Minkowski spacetime. An interesting feature of the Nielsen-Olesen vortices is that they are electrically neutral and engender quantized magnetic flux. Their equations of motion are of second order and present couplings between the fields. To simplify the problem, first order equations that solve the equations of motion were found in [8, 9]. In this case, the first and second order equations are only compatible if the potential is of the Higgs type, a $|\varphi|^4$ potential that engenders spontaneous symmetry breaking. It is worth mentioning that, even with the Bogomol'nyi procedure, the analytical solutions that describe the vortices remain unknown.

Vortices have also been investigated in generalized models with distinct motivations in several works; see, e.g., [10–27]. In particular, k-vortices, which are vortices in models with generalized kinematics, similar to the models studied before in [28–31], were investigated in [12, 13], without the presence of a first order formalism and analytical solutions, but with the search for new effects. Another motivation relies on the possibility of specifying the form of potential, imposed by the first order formalism. For instance, in [23], modifications in the magnetic permeability allowed to develop a route to make the vortex compact. Also, in [27], we have developed a method to obtain vortices and to construct a class of models that supports analytical solutions. Recently, in [32], we have found vortices with internal structure, which arise in generalized models with the magnetic permeability controlled by the addition of a neutral field, enlarging the $U(1)$ symmetry to become $U(1) \times Z_2$.

Motivated by several works that appeared with generalized dynamics, we have developed a first order formalism for these models in [26]. This investigation focused on the search for the conditions that could lead to first order equations in a case similar to the one considered before in [12], with the inclusion of a quadratic kinetic term that involves the covariant derivative of the scalar field in the Lagrangian density. In the current work we further explore the subject, extending the previous results of [26, 27] to this much harder class of models. The main results show how the presence of analytical solutions can be used to construct the model, if one imposes that its equations of motion are solved by solutions of first order differential equations compatible with the stressless condition.

Although we are working in the (2, 1) dimensional spacetime with the Minkowski metric, we think that the results of the current work are also of interest to General Relativity (RG), in particular to the case of the so-called Ricci-based theories of gravity (RBG) formulated within the metric-affine approach. For instance, in the recent work [33], the authors unveiled an interesting correspondence between the space of solutions of RBG and RG, under certain circumstances. The results show that it is sometimes possible to map complicated nonlinear models into simpler ones, and we think that the models to be explored in the current work can provide novel possibilities of current interest to the scenario explored in [33, 34].

To study the subject, the work is organized in a way such that in Section 2 we present the model and the procedure, showing the requirements to make it work in the presence of first order equations. In Section 3, we illustrate our findings with some new models that support analytical solutions. In particular, we also calculate the magnetic field, energy density, and total energy of the vortex analytically and investigate the possibility of building compact solutions. Finally, in Section 4 we end the work with some conclusions and an outlook for future investigations.

2. Model and Procedure

We consider the generalized action $S = \int d^3x\mathscr{L}$ for a complex scalar field φ coupled to a gauge field A_μ under the local $U(1)$ symmetry in a three-dimensional Minkowski spacetime with metric tensor $\eta_{\mu\nu} = \mathrm{diag}(+,-,-)$. The Lagrangian density to be investigated has the form

$$\mathscr{L} = K(|\varphi|)X - Q(|\varphi|)X^2 + P(|\varphi|)Y - V(|\varphi|). \quad (1)$$

In the above expression, $K(|\varphi|)$, $Q(|\varphi|)$, and $P(|\varphi|)$ are nonnegative functions that modify the dynamics of the model and $V(|\varphi|)$ is the potential. The minus sign in the X^2 term is to keep the vortex energy nonnegative. Also, X and Y define the kinetic terms of the scalar and gauge fields, respectively, as

$$\begin{aligned} X &= \overline{D_\mu\varphi}D^\mu\varphi, \\ Y &= -\frac{1}{4}F_{\mu\nu}F^{\mu\nu}, \end{aligned} \quad (2)$$

where $D_\mu = \partial_\mu + ieA_\mu$, $F_{\mu\nu} = \partial_\mu A_\nu - \partial_\nu A_\mu$, and the overline stands for the complex conjugation. The equations of motion for this model are

$$\begin{aligned} &D_\mu(KD^\mu\varphi) - 2D_\mu(QXD^\mu\varphi) \\ &+ + \frac{\varphi}{2|\varphi|}\left(-K_{|\varphi|}X + Q_{|\varphi|}X^2 - P_{|\varphi|}Y + V_{|\varphi|}\right) = 0, \end{aligned} \quad (3a)$$

$$\partial_\mu(PF^{\mu\nu}) = J^\nu, \quad (3b)$$

where J_μ is the conserved current, given by the expression $J_\mu = ie(K - 2QX)(\overline{\varphi}D_\mu\varphi - \varphi\overline{D_\mu\varphi})$. Also, we are using the notation $V_{|\varphi|} = \partial V/\partial|\varphi|$, etc.

The energy-momentum tensor $T_{\mu\nu}$ for the generalized model (1) is

$$\begin{aligned} T_{\mu\nu} = {} & PF_{\mu\lambda}F^{\lambda}{}_{\nu} + (K - 2QX)\left(\overline{D_\mu\varphi}D_\nu\varphi + \overline{D_\nu\varphi}D_\mu\varphi\right) \\ & - \eta_{\mu\nu}\mathscr{L}. \end{aligned} \quad (4)$$

We then consider static configurations; take $A_0 = 0$ and work with the usual ansatz for vortices

$$\varphi(r,\theta) = g(r)e^{in\theta}, \quad (5a)$$

$$A_i = \epsilon_{ij}\frac{x^j}{er^2}(n - a(r)), \quad (5b)$$

in which r and θ are the polar coordinates and $n = \pm1, \pm2, \ldots$ is the vorticity. The boundary conditions for $g(r)$ and $a(r)$ are

$$\begin{aligned} g(0) &= 0, \\ a(0) &= n, \end{aligned} \quad (6)$$

$$\begin{aligned} \lim_{r\to\infty} g(r) &= v, \\ \lim_{r\to\infty} a(r) &= 0, \end{aligned} \quad (7)$$

where v is the symmetry breaking parameter which is supposed to be present in the model under investigation. The ansatz (5a) and (5b) makes X and Y be written as

$$\begin{aligned} X &= -g'^2 - \frac{a^2g^2}{r^2}, \\ Y &= -\frac{a'^2}{2e^2r^2}, \end{aligned} \quad (8)$$

where the prime denotes the derivative with respect to r. The magnetic field is given by $B = -F^{12} = -a'/(er)$. This can be used to show that the magnetic flux $\Phi = 2\pi\int_0^\infty rdrB(r)$ is quantized; that is,

$$\Phi = \frac{2\pi n}{e}. \quad (9)$$

The ansatz (5a) and (5b) can be plugged in the equations of motion (3a) and (3b), which take the form

$$\begin{aligned} &\frac{1}{r}\left(r(K - 2QX)g'\right)' - \frac{(K - 2QX)a^2g}{r^2} \\ &- \frac{1}{2}\left(-K_gX + Q_gX^2 - P_gY + V_g\right) = 0, \end{aligned} \quad (10a)$$

$$r\left(P\frac{a'}{er}\right)' - 2e(K - 2QX)ag^2 = 0, \quad (10b)$$

where $K_g = \partial K/\partial g$, etc. The components of the energy-momentum tensor are

$$T_{00} = -KX + QX^2 - PY + V, \quad (11a)$$

$$T_{12} = (K - 2QX)\left(g'^2 - \frac{a^2 g^2}{r^2}\right)\sin(2\theta), \quad (11b)$$

$$T_{11} = P\frac{a'^2}{e^2 r^2} + 2(K - 2QX)\left(g'^2\cos^2\theta + \frac{a^2 g^2}{r^2}\sin^2\theta\right) + \mathscr{L}, \quad (11c)$$

$$T_{22} = P\frac{a'^2}{e^2 r^2} + 2(K - 2QX)\left(g'^2\sin^2\theta + \frac{a^2 g^2}{r^2}\cos^2\theta\right) + \mathscr{L}. \quad (11d)$$

Up to this point, the scenario is quite similar to the one investigated before in [12]. Here, however, we want to go further and search for a first order framework that helps us to find analytical solutions. We then follow [26] and take the stressless conditions, $T_{ij} = 0$, which ensure stability of the solution under radial rescaling. This requires the solutions to obey the following first order equations

$$g' = \pm\frac{ag}{r}, \quad (12a)$$

$$-\frac{a'}{er} = \pm\sqrt{\frac{2\left(V - QX^2\right)}{P}}. \quad (12b)$$

They allow us to write $X = -2g'^2 = -2a^2g^2/r^2$. The above equations, however, must be compatible with the equations of motion (10a) and (10b). Similarly to the case that was shown in [26], for $K(|\varphi|) = 0$ and $Q(|\varphi|)$ constant, this requirement leads to a constraint that depends on a, g, and r. Therefore, it is hard to obtain a constraint in terms of g and reconstruct the model by finding the explicit form of the potential in terms of $K(|\varphi|)$, $Q(|\varphi|)$, and $P(|\varphi|)$, as in the case $Q(|\varphi|) = 0$ that was carefully investigated in [27]. The main issue appears because X does not depend exclusively on g, but also on a and r; see (8). Nevertheless, if the analytical solutions, as well as their inverses, are known, we may write X exclusively in terms of g, which we call $X(g)$. By substituting (12a) and (12b) into (10b), the following constraint arises

$$\frac{d}{dg}\sqrt{2P\left(V - QX^2(g)\right)} = -2eg\left(K - 2QX(g)\right). \quad (13)$$

One may wonder if the compatibility of (12a) and (12b) with (10a) does not imply another constraint. Nonetheless, as it was demonstrated in [26], once the above constraint is satisfied and the solutions solve (12a), (12b), (10a) becomes an identity. In our model, the choice of the functions $P(g)$, $Q(g)$, and $K(g)$ must be done in a way that allows the symmetry breaking of the potential $V(g)$ to match with the boundary conditions in (6).

The energy density $\rho = T_{00}$ is given by (11a). By using the first order equations (12a) and (12b), it can be written as

$$\begin{aligned}\rho &= P(g)\frac{a'^2}{e^2r^2} + 2K(g)g'^2 + 8Q(g)g'^4 \\ &= 2V(g) - K(g)X(g).\end{aligned} \quad (14)$$

Here, we follow the procedure developed in [26] and introduce an additional function $W(a, g)$, defined by

$$W_a = P\frac{a'}{e^2 r}, \quad (15)$$

$$W_g = 2\left(K - 2QX(g)\right)rg', \quad (16)$$

where $W_g = \partial W/\partial g$ and $W_a = \partial W/\partial a$. By combining the first order equations (12a) and (12b) and the constraint (13), one can show that

$$W(a, g) = -\frac{a}{e}\sqrt{2P\left(V - QX^2(g)\right)}. \quad (17)$$

In this case, we can write the energy density as

$$\rho = \frac{1}{r}\frac{dW}{dr}, \quad (18)$$

which can be integrated all over the plane to provide the energy

$$\begin{aligned}E &= 2\pi\left|W(a(\infty), g(\infty)) - W(a(0), g(0))\right|, \\ &= 2\pi\left|W(n, 0)\right|.\end{aligned} \quad (19)$$

Now, we follow the route suggested in [27] and develop a procedure to build analytical solutions. This can be achieved by decoupling the first order equations (12a) and (12b), as we describe below. For simplicity, we consider dimensionless fields and take $e, v = 1$; also, we work with unity vorticity, setting $n = 1$, which means to consider only the upper signs in (12a) and (12b).

In order to decouple the first order equations, we introduce the generating function $R(g)$ such that

$$r\frac{dg}{dr} = R(g). \quad (20)$$

Therefore, for a given $R(g)$, we can solve the above equation and obtain $g(r)$ obeying the boundary conditions (6). By using this into (12a) and (12b) we obtain

$$a(r) = \frac{R(g(r))}{g(r)}. \quad (21)$$

We also introduce another function, $M(g)$, which is defined by $M(g) = -\sqrt{2(V(g) - Q(g)X^2(g))/P(g)}$. By using this and the constraint in (13), we get

$$V(g) = \frac{1}{2}P(g)M^2(g) + Q(g)X^2(g), \quad (22a)$$

$$K(g) = \frac{1}{2g}\frac{d}{dg}\left(P(g)M(g)\right) + 2Q(g)X(g). \quad (22b)$$

One can show that $M(g)$ is obtained in terms of the given function $R(g)$ from (12b):

$$M(g) = \frac{R(g)}{q^2(g)} \frac{d}{dg}\left(\frac{R(g)}{g}\right), \tag{23}$$

where $q(g)$ is the inverse of $g(r)$. This procedure is valid if X is written only as a function of g. Using the definition in (20), we find

$$X(g) = \frac{-2R(g)^2}{q^2(g)}. \tag{24}$$

We can also take advantage of the function $M(g)$ to write the magnetic field as

$$B(r) = -M(g(r)), \tag{25}$$

and (17) as $W(a, g) = aP(g)M(g)$, which leads to the total energy

$$E = -2\pi P(0)M(0). \tag{26}$$

This procedure decouples the first order equations in a manner that the solutions depend only on the generating function $R(g)$. As $M(g)$ depends only on $R(g)$ and $q(g)$, we see from (22a) and (22b) that we have two equations that constrain the functions $V(|\varphi|)$, $P(|\varphi|)$, $K(|\varphi|)$, and $Q(|\varphi|)$. This means that there are several models that support the same analytical solutions defined by (20). Therefore, to find the explicit form of the models, we need to suggest two of the aforementioned functions. Even though these functions lead to the same solutions and magnetic field, they modify the energy density in (14). Thus, one must choose functions that lead to a well defined energy.

We also highlight here that the above procedure to construct the model, described by (22a), (22b), (23), and (24), is only valid in the interval $|\varphi| \in [0, 1]$, which is the one where the solution exists, according to the boundary conditions (6). Nonetheless, it is important to suggest nonnegative functions and a potential that supports a minimum at $|\varphi| = 1$, in order to include spontaneous symmetry breaking and avoid instabilities and negative energies.

3. Specific Examples

Let us now illustrate our procedure with some examples. We firstly suggest an $R(g)$ that leads to analytical solutions and then apply the method in (22a), (22b), (23), and (24) to construct the model.

3.1. First Example. The first example arises from the generating function

$$R(g) = g\left(1 - g^2\right). \tag{27}$$

This function was previously considered in [27], but with a model in which $Q(|\varphi|) = 0$, which kills the X^2 term in the Lagrangian density. By substituting the above expression in (20) and (21) we get the solutions

$$g(r) = \frac{r}{\sqrt{1 + r^2}}, \qquad a(r) = \frac{1}{1 + r^2}, \tag{28}$$

which satisfy the boundary conditions (6). The inverse function of the solution $g(r)$ in (28), combined with (23) and (24), allows us to write

$$q(g) = \frac{g}{\sqrt{1 - g^2}} \tag{29a}$$

$$M(g) = -2\left(1 - g^2\right)^2, \tag{29b}$$

$$X(g) = -2\left(1 - g^2\right)^3. \tag{29c}$$

Notice that these equations and the solutions in (28) are exclusively determined by the function $R(g)$ given in (27). This also occurs with the magnetic field, given by (25), which leads to

$$B(r) = \frac{2}{\left(1 + r^2\right)^2}. \tag{30}$$

In Figure 1, we display the solutions (28) and the magnetic field given above. Notice that their behavior is similar to the one for the Nielsen-Olesen case [7, 9].

In order to construct a model that supports the solutions in (28), we use (22a) and (22b). Firstly, though, we need to suggest an explicit form for two of the functions among $K(|\varphi|)$, $Q(|\varphi|)$, $P(|\varphi|)$, and $V(|\varphi|)$. We consider the potential

$$V(|\varphi|) = \frac{1}{2}\left|1 - |\varphi|^2\right|^s, \tag{31}$$

where $s > 2$ is a real number. It presents a set of minima at $|\varphi| = 1$ and a local maximum at $|\varphi| = 0$ as illustrated in Figure 2. The other function that we suggest is

$$Q(|\varphi|) = \frac{\alpha}{2}\left|1 - |\varphi|^2\right|^{s-6}, \tag{32}$$

where α is a real, nonnegative parameter. The case investigated in [27] is obtained for $\alpha = 0$. By substituting the above $Q(|\varphi|)$ and the potential (31) in (22a) and (22b), we obtain

$$P(|\varphi|) = \frac{1}{4}(1 - 4\alpha)\left|1 - |\varphi|^2\right|^{s-4}, \tag{33a}$$

$$K(|\varphi|) = \frac{1}{2}(s - 2 - 4\alpha(s - 1))\left|1 - |\varphi|^2\right|^{s-3}. \tag{33b}$$

In order to avoid negative coefficients in the above functions, we impose the condition $\alpha < (s - 2)/4(s - 1)$. The functions in (31), (32), (33a), and (33b) determine model (1). We want to emphasize here that this model can only be obtained explicitly because we know the analytical solutions before its construction.

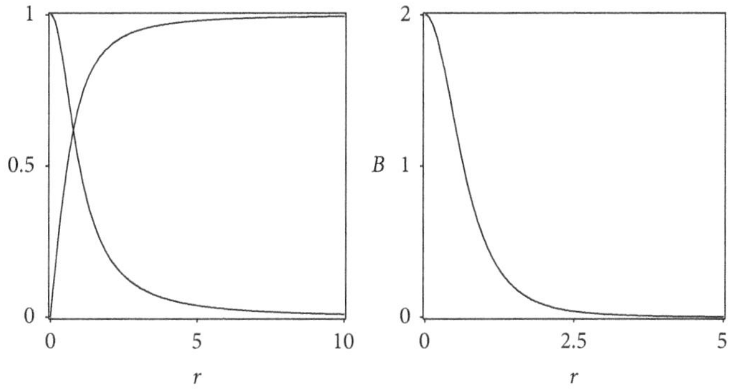

FIGURE 1: In the left panel, we display the solutions $a(r)$ (descending line) and $g(r)$ (ascending line) in (28). In the right panel, we show the magnetic field in (30).

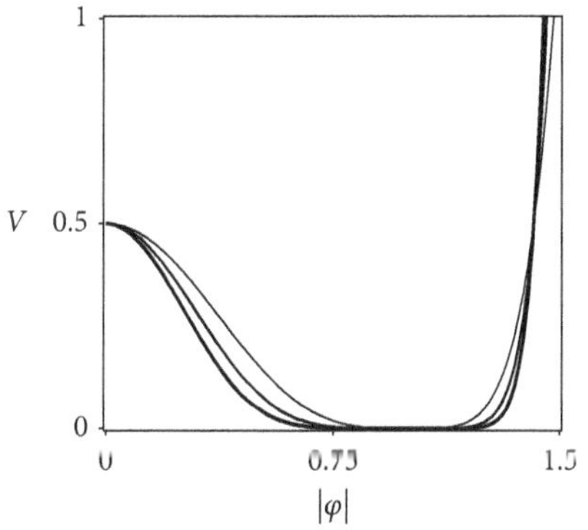

FIGURE 2: The potential in (31) for $s = 4$, $s = 6$, and $s = 8$. The thickness of the lines increases with s.

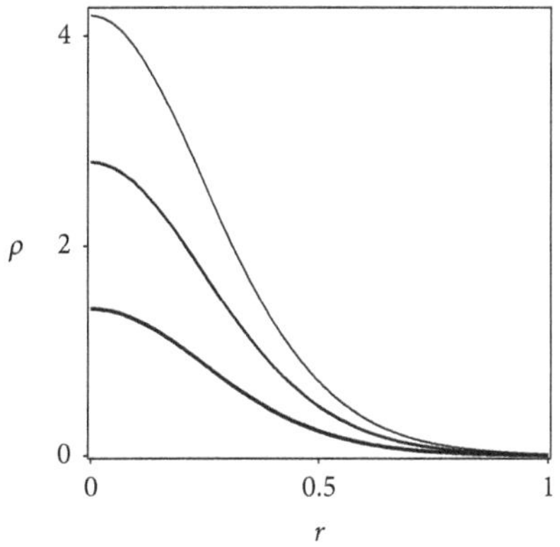

FIGURE 3: The profile of the energy density in (34) for $s = 8$ and $\alpha = 0.1, 0.15$, and 0.2. The thickness of the lines increases with α.

The energy density can be calculated from (14), which leads us to

$$\rho(r) = \frac{(1 - 4\alpha)(s - 1)}{(1 + r^2)^s}. \tag{34}$$

By a direct integration, one can show that the energy is $E = (1 - 4\alpha)\pi$, which matches with the result obtained by (26). Notice that only the parameter α modifies the energy. The above energy density can be seen in Figure 3.

Another model can be generated straightforwardly from the same choice of $R(g)$ in (27), which presents well defined $V(|\varphi|)$, $P(|\varphi|)$, $Q(|\varphi|)$, and $K(|\varphi|)$ for all φ.

3.2. Second Example. Here, we consider a generalization of the previous example by considering the generating function to be

$$R(g) = g\left(1 - g^{2l}\right), \tag{35}$$

where l is a nonnegative real parameter. This function was also investigated in [27], but with $Q(|\varphi|) = 0$. From (20) and (21), we get the analytical solutions

$$g(r) = \frac{r}{\left(1 + r^{2l}\right)^{1/2l}}, \qquad a(r) = \frac{1}{1 + r^{2l}}, \tag{36}$$

which satisfy the boundary conditions (6). From the inverse of the solution $g(r)$, combined with (23) and (24), we obtain

$$q(g) = \frac{g}{\left(1 - g^{2l}\right)^{1/2l}} \tag{37a}$$

$$M(g) = -2g^{2l-2}\left(1 - g^{2l}\right)^{1+1/l}, \tag{37b}$$

$$X(g) = -2\left(1 - g^{2l}\right)^{2+1/l}. \tag{37c}$$

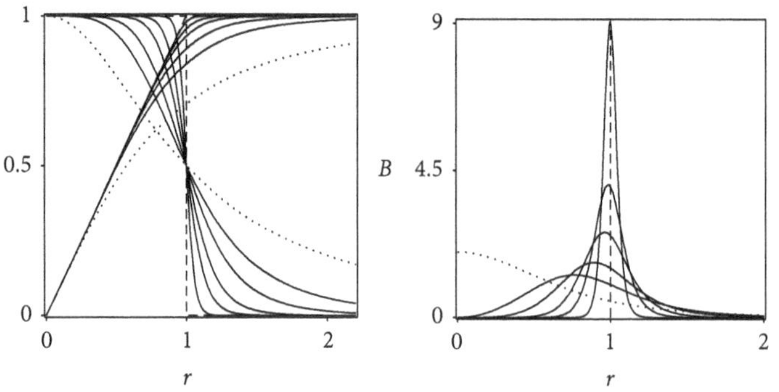

FIGURE 4: In the left panel, we display the solutions $a(r)$ (descending lines) and $g(r)$ (ascending lines) in (36). In the right panel, we show the magnetic field in (38). The dotted lines represent the case $l = 1$ and the dashed ones stand for the compact limit in (39a), (39b), and (40).

As in the previous model, these equations and the solutions in (36) are solely determined by the $R(g)$ in (27). The same is valid for the magnetic field in (25), which leads to

$$B(r) = \frac{2lr^{2l-2}}{(1+r^{2l})^2}. \tag{38}$$

One can show that, as l increases, the solutions in (36) tend to compactify

$$a_c(r) = \begin{cases} 1, & r \leq 1 \\ 0, & r > 1, \end{cases} \tag{39a}$$

$$g_c(r) = \begin{cases} r, & r \leq 1 \\ 1, & r > 1, \end{cases} \tag{39b}$$

and the same happens for the magnetic field in (38), which for very large l tends to

$$B_c(r) = \frac{\delta(r-1)}{r}, \tag{40}$$

where $\delta(z)$ is the Dirac delta function. In Figure 4, we depict the solutions (36) and the magnetic field given above for several values of l, including the compact limit in (39a) and (39b).

Again, to find the functions $K(|\varphi|)$, $Q(|\varphi|)$, $P(|\varphi|)$, and $V(|\varphi|)$ we must suggest two of them and use (22a) and (22b). We take the potential in the form

$$V(|\varphi|) = \frac{1}{2}l|\varphi|^{2l-2}\left|1-|\varphi|^{2l}\right|^{\beta l}, \tag{41}$$

where $\beta > 2$ is a real number. This potential presents minima at $|\varphi| = 1$ for any l. The point $|\varphi| = 0$ is a maximum for $l = 1$ and a minimum for $l > 1$. This behavior is shown in Figure 5. Together with the potential in (41), we keep the same lines of the previous example and suggest the X^2 term in the Lagrangian density to be modified by

$$Q(|\varphi|) = \frac{1}{2}\alpha l|\varphi|^{2l-2}\left|1-|\varphi|^{2l}\right|^{\beta l-4-2/l}, \tag{42}$$

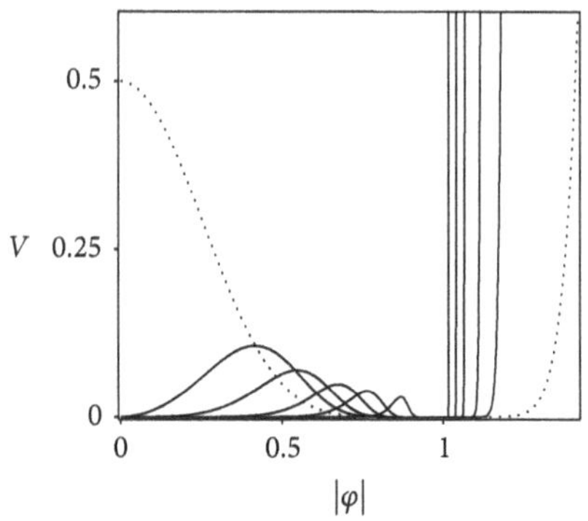

FIGURE 5: The potential in (41) for $\beta = 8$ and several values of l. The dotted line stands for the case $l = 1$.

where $\alpha > 0$ is a real parameter. Substituting $V(|\varphi|)$ and $Q(|\varphi|)$ in (22a) and (22b), we obtain

$$P(|\varphi|) = \frac{1}{4l}(1-4\alpha)|\varphi|^{2-2l}\left|1-|\varphi|^{2l}\right|^{\beta l-2-2/l}, \tag{43}$$

$$K(|\varphi|) = \frac{1}{2}\left((1-4\alpha)\left(\beta l^2-1\right)-l\right) \times |\varphi|^{2l-2}\left|1-|\varphi|^{2l}\right|^{\beta l-2-1/l}. \tag{44}$$

To avoid the presence of negative coefficients in the above expressions, we impose that $\alpha < (\beta l^2 - l - 1)/4(\beta l^2 - 1)$.

The energy density is calculated from (14), which leads to

$$\rho(r) = \frac{(1-4\alpha)\left(\beta l^2-1\right)r^{2l-2}}{(1+r^{2l})^{\beta l+1-1/l}}. \tag{45}$$

One may integrate it to get the total energy $E = (1 - 4\alpha)\pi$, which matches with the value obtained by (26). Again, only the parameter α modifies the energy of the vortices, meaning that the X^2 term in the Lagrangian density (1) plays a significant role in the model. Following a similar procedure that was done in [27], one can show that the energy density tends to compactify into a ringlike region of unit radius in the plane, described by

$$\rho(r) = \frac{1}{2}(1-4\alpha)\delta(r-1). \tag{46}$$

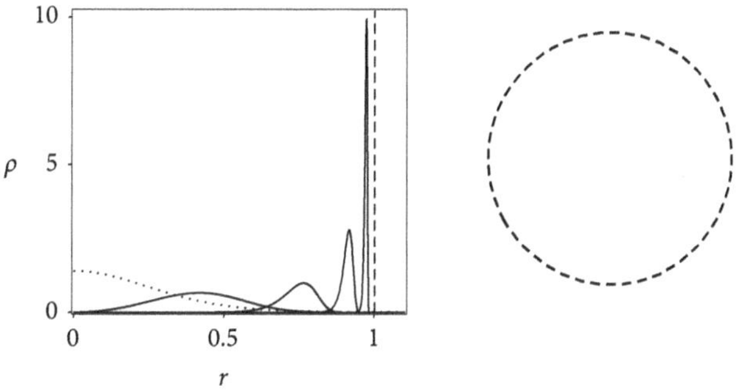

FIGURE 6: The energy density of (45) for $\beta = 8$, $\alpha = 0.2$ and several values of l (left) and its compact limit, $l \longrightarrow \infty$, in the plane (right). The dotted line represents the case $l = 1$ and the dashed ones stand for the compact limit in (46).

In Figure 6, we display the energy density for several values of α, including the compact limit given above. Its behavior, even with the presence of the parameter α, is qualitatively similar to the one found in [35] for the compactification of vortices in a generalized Chern-Simons-Higgs model.

4. Comments and Conclusions

In this work, we have developed a procedure that allows us to construct k-vortex models that support a first order framework. As we discussed above, the method is important because the constraint that dictates the form of the potential cannot be solved in general in the presence of the squared kinetic term of the scalar field, X^2, in the Lagrangian density. Thus, it seems to be very hard to start from a model with this term and find the potential that leads to the first order equations compatible with the stressless condition, vital to the stability of the system.

Nevertheless, we got inspiration from the recent works [26, 27] and noticed that, if an analytical solution is known, we can construct a model that satisfies the stressless condition and find the energy depending exclusively on a function of the fields calculated from the boundary conditions. In order to achieve this, we have introduced the generating function $R(g)$ that decouples the first order equations. It is interesting feature of this procedure that it shows there is a class of models that leads to the same analytical, stressless solutions and their respective magnetic fields, which only depend on the generating function. However, the energy density as well as the total energy depends on the model to be chosen, so we have to properly define the model, to make it behave adequately.

It is worth commenting that a similar method can be developed for the more general Lagrangian density $\mathscr{L} = f(X, |\varphi|) + P(|\varphi|)Y - V(|\varphi|)$. Thus, among the myriad of possibilities, one may develop a construction method for the kinetic term of the scalar field being of the Born-Infeld type, for instance. Other perspectives should include the possibility of considering the case in which the dynamics of the gauge field is driven by the Chern-Simons term, which cannot be multiplied by $P(|\varphi|)$ if one wants to keep gauge invariance. Since the magnetic permeability of the model is generalized, one may also investigate the presence of vortices in metamaterials; see, e.g., [36–38]. Furthermore, as the model supports the W in (17), one may seek for supersymmetric extensions, to investigate how the supersymmetry works in this scenario to lead us with first order differential equations. One may also try to extend these results to other topological structures, such as monopoles and skyrmions. We hope to report on some of the above issues in the near future.

Conflicts of Interest

The authors declare that there are no conflicts of interest regarding the publication of this paper.

Acknowledgments

We would like to acknowledge the Brazilian agency CNPq for partial financial support. D. Bazeia appreciates the support from grant 306614/2014-6, L. Losano appreciates the support from grant 303824/2017-4, M. A. Marques appreciates the support from grant 140735/2015-1, and R. Menezes appreciates the support from grant 306826/2015-1.

References

[1] H. von Helmholtz, “Über integrale der hydrodynamischen gleichungen, welche der wirbelbewegung entsprechen,” *Journal für die reine und angewandte Mathematik*, vol. 55, p. 25, 1858.

[2] P. G. Saffman, *Vortex Dynamics*, Cambridge University Press, Cambridge, UK, 1992.

[3] A. A. Abrikosov, “On the magnetic properties of superconductors of the second group,” *Soviet Physics—JETP*, vol. 5, pp. 1174–1182, 1957.

[4] V. L. Ginzburg and L. D. Landau, “On the theory of superconductivity,” *Zhurnal Eksperimental'noi i Teoreticheskoi Fiziki*, vol. 20, pp. 1064–1082, 1950.

[5] A. Hubert and R. Schäfer, *Magnetic Domains. The Analysis of Magnetic Microstructures*, Springer-Verlag, 1998.

[6] E. Fradkin, *Field Theories of Condensed Matter Physics*, Cambridge University Press, Cambridge, UK, 2013.

[7] H. B. Nielsen and P. Olesen, "Vortex-line models for dual strings," *Nuclear Physics B*, vol. 61, pp. 45–61, 1973.

[8] E. Bogomol'nyi, "The stability of classical solutions," *Soviet Journal of Nuclear Physics*, vol. 24, no. 4, pp. 449–454, 1976.

[9] H. J. de Vega and F. A. Schaposnik, "Classical vortex solution of the Abelian Higgs model," *Physical Review D: Particles, Fields, Gravitation and Cosmology*, vol. 14, no. 4, pp. 1100–1106, 1976.

[10] J. Lee and S. Nam, "Bogomol'nyi equations of Chern-Simons Higgs theory from a generalized abelian Higgs model," *Physics Letters B*, vol. 261, p. 437, 1991.

[11] D. Bazeia, "Vortices in a generalized Higgs model," *Physical Review D*, vol. 46, p. 1879, 1992.

[12] E. Babichev, "Gauge k-vortices," *Physical Review D: Particles, Fields, Gravitation and Cosmology*, vol. 77, Article ID 065021, 2008.

[13] E. Babichev, P. Brax, C. Caprini, J. Martin, and D. A. Steer, "Dirac Born Infeld (DBI) cosmic strings," *Journal of High Energy Physics*, vol. 2009, no. 03, p. 091, 2009.

[14] C. Adam, P. Klimas, J. Sánchez-Guillén, and A. Wereszczyński, "Compact gauge K vortices," *Journal of Physics A: Mathematical and Theoretical*, vol. 42, no. 13, Article ID 135401, 19 pages, 2009.

[15] D. Bazeia, E. da Hora, C. dos Santos, and R. Menezes, "BPS solutions to a generalized Maxwell-Higgs model," *The European Physical Journal C*, vol. 71, no. 12, pp. 1–9, 2011.

[16] A. N. Atmaja, H. S. Ramadhan, and E. da Hora, "More on Bogomol'nyi equations of three-dimensional generalized Maxwell-Higgs model using on-shell method," *Journal of High Energy Physics*, vol. 2016, no. 2, 2016.

[17] R. Casana, A. Cavalcante, and E. da Hora, "Self-dual configurations in Abelian Higgs models with κ-generalized gauge field dynamics," *Journal of High Energy Physics*, vol. 2016, no. 12, p. 051, 2016.

[18] A. N. Atmaja, "A method for BPS equations of vortices," *Physics Letters B*, vol. 768, pp. 351–358, 2017.

[19] E. Moreno, C. Núñez, and F. A. Schaposnik, "Electrically charged vortex solution in Born-Infeld theory," *Physical Review D: Particles, Fields, Gravitation and Cosmology*, vol. 58, no. 2, Article ID 025015, 1998.

[20] A. Alonso-Izquierdo, W. G. Fuertes, and J. M. Guilarte, "Two species of vortices in massive gauged non-linear sigma models," *Journal of High Energy Physics*, vol. 2015, no. 2, p. 139, 2015.

[21] P. Forgács, Á. Lukács, and D. Phys. Rev, "Vortices and magnetic bags in Abelian models with extended scalar sectors and some of their applications," *Phys. Rev. D*, vol. 94, Article ID 125018, 2016.

[22] J. Chagoya and G. Tasinato, "Galileon Higgs vortices," *Journal of High Energy Physics*, vol. 2016, no. 02, p. 63, 2016.

[23] D. Bazeia, L. Losano, M. A. Marques, R. Menezes, and I. Zafalan, "Compact vortices," *The European Physical Journal C*, vol. 77, no. 2, article no. 63, 2017.

[24] C. Adam, J. M. Speight, and A. Wereszczynski, "Volume of a vortex and Bradlow bound," *Physical Review D: Particles, Fields, Gravitation and Cosmology*, vol. 95, no. 11, 116007, 19 pages, 2017.

[25] D. Bazeia, M. A. Marques, and R. Menezes, "Twinlike models for kinks, vortices, and monopoles," *Physical Review D: Particles, Fields, Gravitation and Cosmology*, vol. 96, no. 2, 025010, 9 pages, 2017.

[26] D. Bazeia, L. Losano, M. A. Marques, R. Menezes, and I. Zafalan, "First order formalism for generalized vortices," *Nuclear Physics. B. Theoretical, Phenomenological, and Experimental High Energy Physics. Quantum Field Theory and Statistical Systems*, vol. 934, pp. 212–239, 2018.

[27] D. Bazeia, L. Losano, M. A. Marques, and R. Menezes, "Analytic vortex solutions in generalized models of the Maxwell-Higgs type," *Physics Letters. B. Particle Physics, Nuclear Physics and Cosmology*, vol. 778, pp. 22–29, 2018.

[28] C. Armendáriz-Picón, T. Damour, and V. Mukhanov, "k-inflation," *Physics Letters B*, vol. 458, no. 2-3, pp. 209–218, 1999.

[29] C. Armendariz-Picon, V. Mukhanov, and P. J. Steinhardt, "Dynamical solution to the problem of a small cosmological constant and late-time cosmic acceleration," *Physical Review Letters*, vol. 85, no. 21, pp. 4438–4441, 2000.

[30] C. Armendariz-Picon, V. Mukhanov, and P. J. Steinbardt, "Essentials of k-essence," *Physical Review D: Particles, Fields, Gravitation and Cosmology*, vol. 63, Article ID 103510, 2001.

[31] E. Babichev, "Global topological k-defects," *Physical Review D: Particles, Fields, Gravitation and Cosmology*, vol. 74, no. 8, Article ID 085004, 2006.

[32] D. Bazeia, M. A. Marques, and R. Menezes, "Maxwell–Higgs vortices with internal structure," *Physics Letters B*, vol. 780, p. 485, 2018.

[33] V. I. Afonso, G. J. Olmo, and D. Rubiera-Garcia, "Mapping Ricci-based theories of gravity into general relativity," *Physical Review D*, vol. 97, no. 2, Article ID 021503(R), 2018.

[34] V. I. Afonso, G. J. Olmo, E. Orazi, and D. Rubiera-Garcia, "Mapping nonlinear gravity into General Relativity with nonlinear electrodynamics," *The European Physical Journal C*, vol. 78, no. 10, p. 866, 2018.

[35] D. Bazeia, L. Losano, M. A. Marques, and R. Menezes, "Compact Chern-Simons vortices," *Physics Letters. B. Particle Physics, Nuclear Physics and Cosmology*, vol. 772, pp. 253–257, 2017.

[36] R. A. Shelby, D. R. Smith, and S. Schultz, "Experimental verification of a negative index of refraction," *Science*, vol. 292, no. 5514, pp. 77–79, 2001.

[37] S. Anantha Ramakrishna, "Physics of negative refractive index materials," *Reports on Progress in Physics*, vol. 68, p. 449, 2005.

[38] C. Caloz, "Perspectives on EM metamaterials," *Materials Today*, vol. 12, p. 12, 2009.

Semiexact Solutions of the Razavy Potential

Qian Dong,[1] F. A. Serrano,[2] Guo-Hua Sun,[3] Jian Jing,[4] and Shi-Hai Dong[1]

[1]Laboratorio de Información Cuántica, CIDETEC, Instituto Politécnico Nacional, UPALM, CDMX 07700, Mexico
[2]Escuela Superior de Ingeniera Mecánica y Eléctrica UPC, Instituto Politécnico Nacional, Av. Santa Ana 1000, México, D. F. 04430, Mexico
[3]Catedrática CONACYT, CIC, Instituto Politécnico Nacional, CDMX 07738, Mexico
[4]Department of Physics and Electronic, School of Science, Beijing University of Chemical Technology, Beijing 100029, China

Correspondence should be addressed to Shi-Hai Dong; dongsh2@yahoo.com

Academic Editor: Saber Zarrinkamar

In this work, we study the quantum system with the symmetric Razavy potential and show how to find its exact solutions. We find that the solutions are given by the confluent Heun functions. The eigenvalues have to be calculated numerically. The properties of the wave functions depending on m are illustrated graphically for a given potential parameter ξ. We find that the even and odd wave functions with definite parity are changed to odd and even wave functions when the potential parameter m increases. This arises from the fact that the parity, which is a defined symmetry for very small m, is completely violated for large m. We also notice that the energy levels ϵ_i decrease with the increasing potential parameter m.

1. Introduction

It is well-known that the exact solutions of quantum systems play an important role since the early foundation of the quantum mechanics. Generally speaking, two typical examples are studied for the hydrogen atom and harmonic oscillator in classical quantum mechanics textbooks [1, 2]. Up till now, there are a few main methods to solve the quantum soluble systems. The first is called the functional analysis method. That is to say, one solves the second-order differential equation and obtains their solutions [3], which are expressed by some well-known special functions. The second is called the algebraic method, which is realized by studying the Hamiltonian of quantum system. This method is also related to supersymmetric quantum mechanics (SUSYQM) [4], further closely with the factorization method [5]. The third is called the exact quantization rule method [6], from which we proposed proper quantization rule [7], which shows more beauty and symmetry than exact quantization rule. It should be recognized that almost all soluble potentials mentioned above belong to single well potentials. The double-well potentials have not been studied well due to their complications [8–17], in which many authors have been searching the solutions of the double-well potentials for a long history. This is because the double-well potentials could be used in the quantum theory of molecules to describe the motion of the particle in the presence of two centers of force, the heterostructures, Bose-Einstein condensates, superconducting circuits, etc.

Almost forty years ago, Razavy proposed a bistable potential [18]:

$$V(x) = \frac{\hbar^2\beta^2}{2\mu}\left[\frac{1}{8}\xi^2\cosh(4\beta x) - (m+1)\xi\cosh(2\beta x) - \frac{1}{8}\xi^2\right], \quad (1)$$

which depends on three potential parameters β, ξ, and a positive integer m. In Figure 1 we plot it as the function of the variables x with various m, in which we take $\beta = 1$ and $\xi = 3$. Choose atomic units $\hbar = \mu = 1$ and also take $\mathscr{V}(x) = 2V(x)$. Using series expansion around the origin, we have

$$\mathscr{V}(x) = (-m\xi - \xi) + x^2\left(-2m\xi + \xi^2 - 2\xi\right)$$

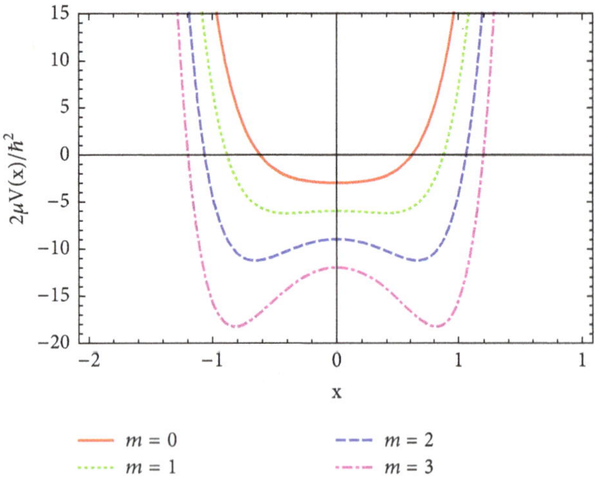

FIGURE 1: (Color online) A plot of potential as function of the variables x and m.

$$+\frac{2}{3}x^4\left(-m\xi+2\xi^2-\xi\right)+\frac{4}{45}x^6\left(-m\xi+8\xi^2-\xi\right)+O\left(x^7\right), \tag{2}$$

which shows that $\mathscr{V}(x)$ is symmetric to variable x. We find that the minimum value of the potential $\mathscr{V}_{\min}(x) = -(m+1)^2 - \xi^2/4$ at two minimum values $x = \pm(1/2)\cosh^{-1}[2(m+1)/\xi]$. For a given value $\xi = 3$, we find that the potential has a flat bottom for $n = 0$, but for $n > 1$ it takes the form of a double-well. Razavy presented the so-called exact solutions by using the "polynomial method" [18]. After studying it carefully, we find that the solutions cannot be given exactly due to the complicated three-term recurrence relation. The method presented there [18] is more like the Bethe Ansatz method as summarized in our recent book [19]. That is, the solutions cannot be expressed as one of special functions because of three-term recurrence relations. In order to obtain some so-called exact solutions, the author has to take some constraints on the coefficients in the recurrence relations as shown in [18]. Inspired by recent study of the hyperbolic type potential well [20–28], in which we have found that their solutions can be exactly expressed by the confluent Heun functions [23], in this work we attempt to study the solutions of the Razavy potential. We shall find that the solutions can be written as the confluent Heun functions but their energy levels have to be calculated numerically since the energy term is involved within the parameter η of the confluent Heun functions $H_c(\alpha,\beta,\gamma,\delta,\eta,z)$. This constraints us to use the traditional Bethe Ansatz method to get the energy levels. Even though the Heun functions have been studied well, its main topics are focused in the mathematical area. Only recent connections with the physical problems have been discovered; in particular the quantum systems for those hyperbolic type potential have been studied [20–28]. The terminology "semiexact" solutions used in [21] arise from the fact that the wave functions can be obtained analytically, but the eigenvalues cannot be written out explicitly.

This paper is organized as follows. In Section 2, we present the solutions of the Schrödinger equation with the Razavy potential. It should be recognized that the Razavy potential is single or double-well depends on the potential parameter m. In Section 3 some fundamental properties of the solutions are studied. The energy levels for different m are calculated numerically. Some concluding remarks are given in Section 4.

2. Semiexact Solutions

Let us consider the one-dimensional Schrödinger equation:

$$-\frac{\hbar^2}{2\mu}\frac{d^2}{dx^2}\psi(x)+V(x)\psi(x)=E\psi(x). \tag{3}$$

Substituting potential (1) into (3), we have

$$\frac{d^2}{dx^2}\psi(x)+\left\{\varepsilon-\left[\frac{1}{8}\xi^2\cosh(4x)-(m+1)\xi\cosh(2x)-\frac{1}{8}\xi^2\right]\right\}\cdot\psi(x)=0, \tag{4}$$

$$\epsilon = 2E.$$

Take the wave functions of the form

$$\psi(x)=e^{\xi\cosh^2(x)/2}y(x). \tag{5}$$

Substituting this into (4) allows us to obtain

$$y''(x)+\xi\sinh(2x)y'(x)+\left[(m+2)\xi\cosh(2x)+\epsilon\right]y(x)=0. \tag{6}$$

Take a new variable $z=\cosh^2(x)$. The above equation becomes

$$4(z-1)zy''(z)+\left[4z(\xi(z-1)+1)-2\right]y'(z)+((m+2)\xi(2z-1)+\epsilon)y(z)=0 \tag{7}$$

which can be rearranged as

$$y''(z)+\left[\xi+\frac{1}{2}\left(\frac{1}{z}+\frac{1}{z-1}\right)\right]y'(z)+\frac{(m+2)\xi(2z-1)+\epsilon}{4(z-1)z}y(z)=0. \tag{8}$$

When comparing this with the confluent Heun differential equation in the simplest uniform form [13]

$$\frac{d^2H(z)}{dz^2}+\left(\alpha+\frac{1+\beta}{z}+\frac{1+\gamma}{z-1}\right)\frac{dH(z)}{dz}+\left(\frac{\mu}{z}+\frac{\nu}{z-1}\right)H(z)=0, \tag{9}$$

we find the solution to (8) is given by the acceptable confluent Heun function $H_c(\alpha, \beta, \gamma, \delta, \eta; z)$ with

$$\begin{aligned}\alpha &= \xi,\\ \beta &= -\frac{1}{2},\\ \gamma &= -\frac{1}{2},\\ \mu &= \frac{\xi(m+2)-\varepsilon}{4},\\ \nu &= \frac{\xi(m+2)+\varepsilon}{4},\end{aligned} \tag{10}$$

from which we are able to calculate the parameters δ and η involved in $H_c(\alpha, \beta, \gamma, \delta, \eta; z)$ as

$$\begin{aligned}\delta &= \mu + \nu - \frac{1}{2}\alpha(\beta+\gamma+2) = \frac{1}{2}(m+1)\xi,\\ \eta &= \frac{1}{2}\alpha(\beta+1) - \mu - \frac{1}{2}(\beta+\gamma+\beta\gamma)\\ &= \frac{1}{8}\left[-2(m+1)\xi + 2\epsilon + 3\right].\end{aligned} \tag{11}$$

It is found that the parameter η related to energy levels is involved in the confluent Heun function. The wave function given by this function seems to be analytical, but the key issue is how to first get the energy levels. Otherwise, the solution becomes unsolvable. Generally, the confluent Heun function can be expressed as a series of expansions:

$$H_C(\alpha, \beta, \gamma, \delta, \eta, z) = \sum_{n=0}^{\infty} v_n(\alpha, \beta, \gamma, \delta, \eta, \xi) z^n, \quad |z| < 1. \tag{12}$$

The coefficients v_n are given by a three-term recurrence relation:

$$\begin{aligned}A_n v_n - B_n v_{n-1} - C_n v_{n-2} &= 0,\\ v_{-1} &= 0,\\ v_0 &= 1,\end{aligned} \tag{13}$$

with

$$\begin{aligned}A_n &= 1 + \frac{\beta}{n},\\ B_n &= 1 + \frac{1}{n}(\beta+\gamma-\alpha-1)\\ &\quad + \frac{1}{n^2}\left\{\eta - \frac{1}{2}(\beta+\gamma-\alpha) - \frac{\alpha\beta}{2} + \frac{\beta\gamma}{2}\right\},\\ C_n &= \frac{\alpha}{n^2}\left(\frac{\delta}{\alpha} + \frac{\beta+\gamma}{2} + n - 1\right).\end{aligned} \tag{14}$$

To make the confluent Heun functions reduce to polynomials, two termination conditions have to be satisfied [13, 14]:

$$\begin{aligned}\mu + \nu + N\alpha &= 0,\\ \Delta_{N+1}(\mu) &= 0,\end{aligned} \tag{15}$$

where

$$\begin{vmatrix} \mu - p_1 & (1+\beta) & 0 & \cdots & 0 & 0 & 0\\ N\alpha & \mu - p_2 + \alpha & 2(2+\beta) & \cdots & 0 & 0 & 0\\ 0 & (N-1)\alpha & \mu - p_3 + 2\alpha & \cdots & 0 & 0 & 0\\ \vdots & \vdots & \vdots & \ddots & \vdots & \vdots & \vdots\\ 0 & 0 & 0 & \cdots & \mu - p_{N-1} + (N-2)\alpha & (N-1)(N-1+\beta) & 0\\ 0 & 0 & 0 & \cdots & 2\alpha & \mu - p_N + (N-1)\alpha & N(N+\beta)\\ 0 & 0 & 0 & \cdots & 0 & \alpha & \mu - p_{N+1} + N\alpha \end{vmatrix} = 0 \tag{16}$$

with

$$p_N = (N-1)(N+\beta+\gamma). \tag{17}$$

For present problem, it is not difficult to see that the first condition is violated. That is, $\mu + \nu + \alpha = 0$ when $N = 1$. From this we have $m = -4$. This is contrary to the fact that m is a positive integer. Therefore, we cannot use this method to obtain the eigenvalues. On the other hand, we know that $z \in [1, \infty)$. Thus, the series expansion method is invalid. This is unlike previous study [22, 24], in which the quasiexact wave functions and eigenvalues can be obtained by studying those two constraints. The present case is similar to our previous study [20, 21], in which some constraint is violated. We have to choose other approach to study the eigenvalues as used in [20, 21].

3. Fundamental Properties

In this section we are going to study some basic properties of the wave functions as shown in Figures 2–4. We first consider the positive integer m. Since the energy spectrum

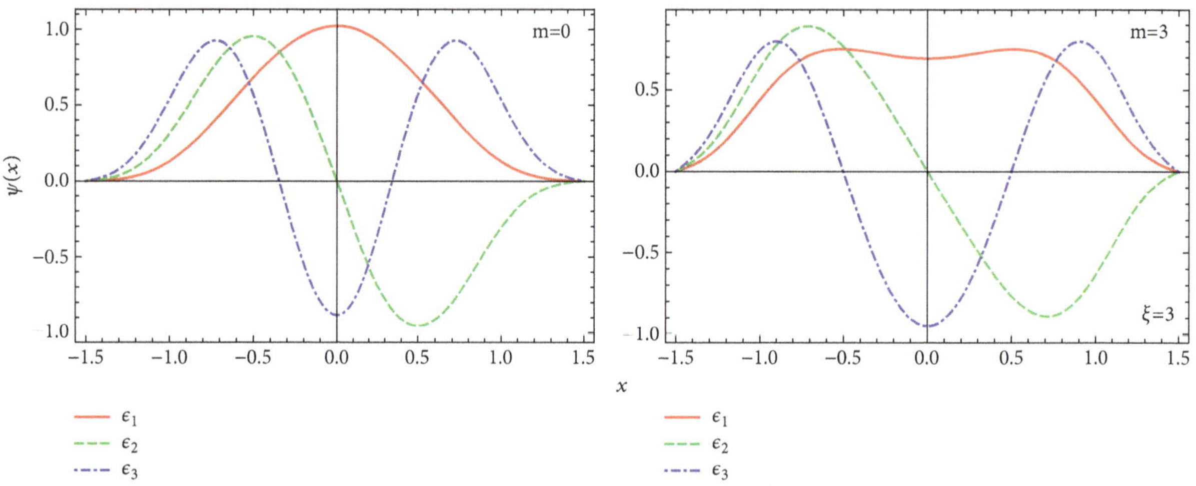

FIGURE 2: (Color online) The characteristics of the potential $V(z)$ as a function of the position z. We take $m = 0, 1$ and $\xi = 3$.

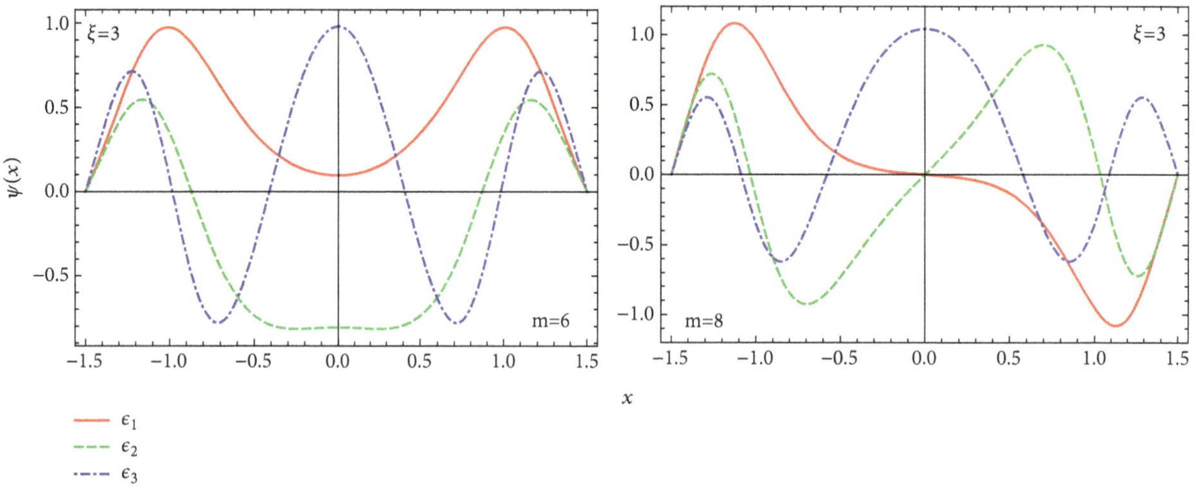

FIGURE 3: (Color online) The characteristics of the potential $V(z)$ as a function of the position z. We take $m = 6, 8$ and $\xi = 3$.

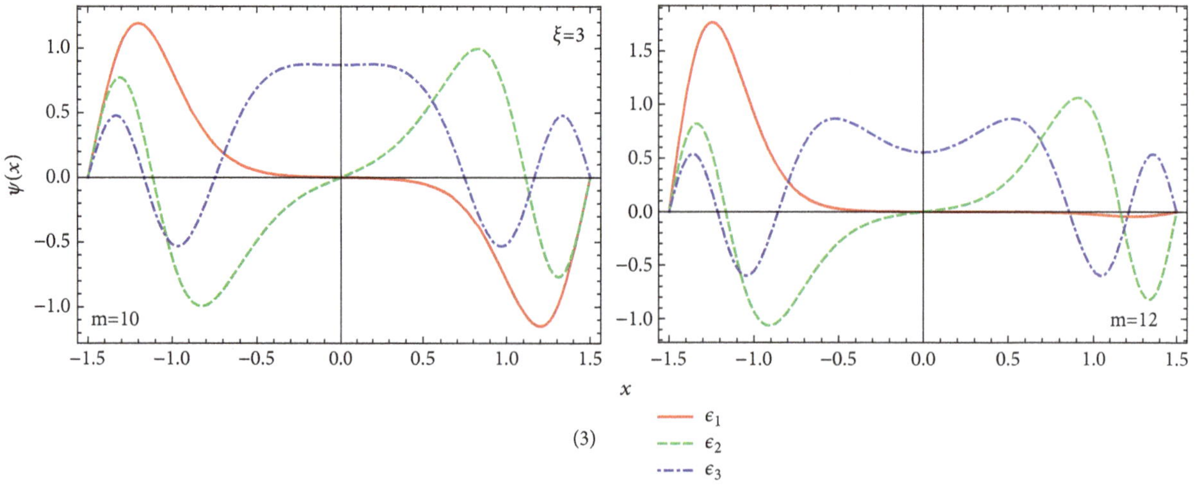

FIGURE 4: (Color online) The same as the above case but $m = 10, 12$.

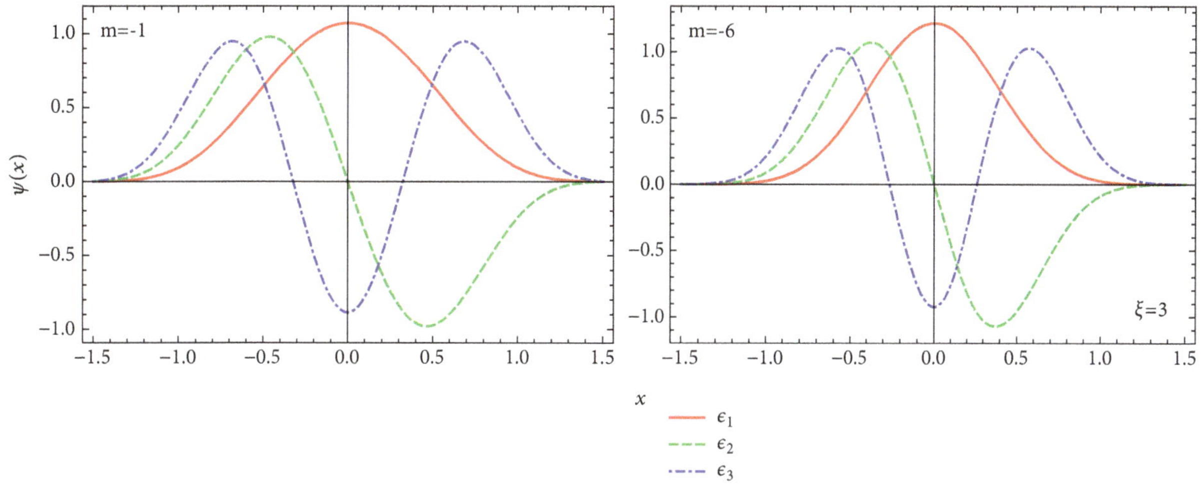

FIGURE 5: (Color online) The characteristics of the potential $V(z)$ as a function of the position z. We take $m = -1, -6$ and $\xi = 3$.

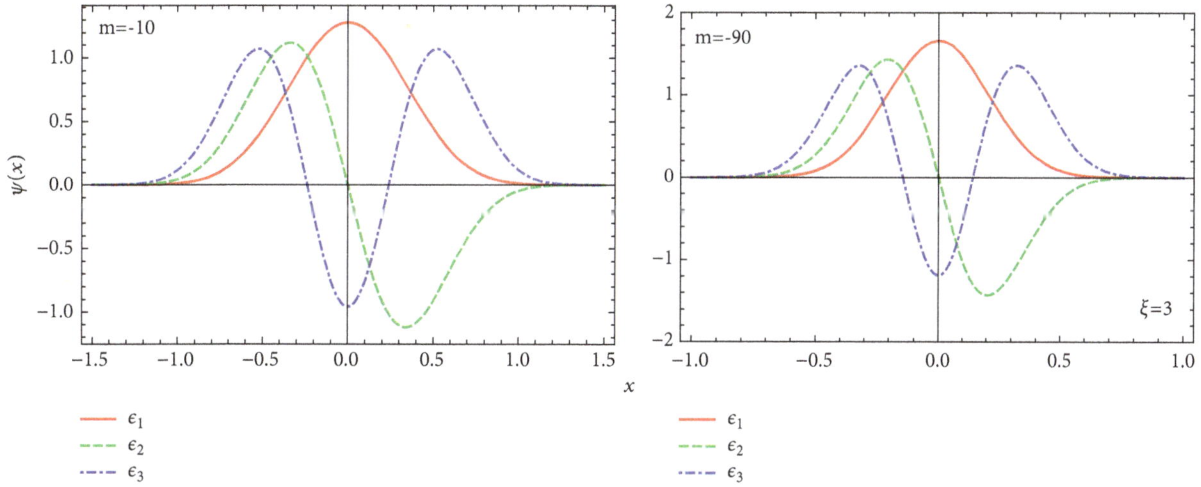

FIGURE 6: (Color online) The same as the above case but $m = -10, -90$.

cannot be given explicitly we have to solve the second-order differential equation (4) numerically. We denote the energy levels as ϵ_i $(i \in [1,6])$ in Table 1. We find that the energy levels ϵ_i decrease with the increasing m. Originally, we wanted to calculate the energy levels numerically by using powerful MAPLE, which includes some special functions such as the confluent Heun function that cannot be found in MATHEMATICA. As we know, the wave function is given by $\psi(z) = \exp(z\xi/2)H_c(\alpha, \beta, \gamma, \delta, \eta, z)$. Generally speaking, the wave function requires $\psi(z) \longrightarrow 0$ when $z \longrightarrow \infty$; i.e., $x \longrightarrow \infty$. Unfortunately, the present study is unlike our previous study [20, 21], in which $z \longrightarrow 1$ when x goes to infinity. The energy spectra can be calculated by series expansions through taking $z \longrightarrow 1$. On the other hand, the wave functions have a definite parity; e.g., for $m = 0$ some wave functions are symmetric. It is found that such properties are violated when the potential parameter m becomes larger as shown in Figure 4. That is, the wave functions for $m = 12$ are nonsymmetric. In addition, on the contrary to the case discussed by Razavy [18], in which he supposed the m is taken as positive integers, we are going to show what happens to the negative m case. We display the graphics in Figures 5 and 6 for this case. We find that the wave functions are shrunk towards the origin. This makes the amplitude of the wave function increase.

4. Conclusions

In this work we have studied the quantum system with the Razavy potential, which is symmetric with respect to the variable x and showed how its exact solutions are found by transforming the original differential equation into a confluent type Heun differential equation. It is found that the solutions can be expressed by the confluent Heun functions $Hc(\alpha, \beta, \gamma, \delta, \eta)$, in which the energy levels are involved inside the parameter η. This makes us calculate the eigenvalues numerically. The properties of the wave functions depending on m are illustrated graphically for a given potential parameter ξ. We have found that the even and odd wave functions with definite parity are changed to odd and even

TABLE 1: Energy levels of the Schrödinger equation with potential (1).

ν	ϵ_1	ϵ_2	ϵ_3	ϵ_4	ϵ_5	ϵ_6
$m = -6$	21.6608	35.7557	51.3448	68.3341	86.6500	106.233
$m = -5$	18.1891	31.3844	46.1503	62.3746	79.9715	98.8740
$m = -4$	14.6806	26.9167	40.8214	56.2549	73.1150	91.3249
$m = -3$	11.1259	22.3314	35.3346	49.9525	66.0599	83.5680
$m = -2$	7.51110	17.5996	29.6610	43.4412	58.7838	75.5860
$m = -1$	3.81463	12.6800	23.7644	36.6914	51.2639	67.3635
$m = 0$	0.00007	7.51170	17.6027	29.6729	43.4799	58.8919
$m = 1$	-3.99968	2.00200	11.1343	22.3606	35.4208	50.1750
$m = 2$	-8.32288	-3.99300	4.34771	14.7494	27.0959	41.2385
$m = 3$	-13.2815	-10.6927	-2.64788	6.87526	18.5501	32.1389
$m = 4$	-19.5196	-9.46859	-1.17161	9.87916	22.9677	38.0537
$m = 5$	-27.7547	-15.7094	-9.29612	1.24110	13.8439	28.5940
$m = 6$	-38.0314	-21.6913	-17.5131	-7.12621	4.89289	19.3065
$m = 7$	-49.9928	-28.2027	-25.9897	-14.8827	-3.78434	10.2625
$m = 8$	-63.3335	-35.8866	-21.7455	-12.1464	1.51447	17.5661
$m = 9$	-77.8339	-44.5255	-27.8571	-20.2355	-6.89162	8.76577
$m = 10$	-93.3024	-54.9017	-33.6970	-28.1690	-14.8944	0.229704
$m = 11$	-109.592	-65.743	-39.7373	-36.1005	-22.4007	-8.04337
$m = 12$	-126.580	-77.2416	-46.3335	-29.3139	-16.0647	1.06475

wave functions when the potential parameter m increases. This arises from the fact that the parity, which is a defined symmetry for very small m, is completely violated for large m. We have also noticed that the energy levels ϵ_i decrease with the increasing potential parameter m.

Conflicts of Interest

The authors declare that there are no conflicts of interest regarding the publication of this paper.

Acknowledgments

This work is supported by Project 20180677-SIP-IPN, COFAA-IPN, Mexico, and partially by the CONACYT project under Grant no. 288856-CB-2016 and partially by NSFC with Grant no. 11465006.

References

[1] L. I. Schiff, *Quantum Mechanics*, McGraw-Hill Book Co, New York, NY, USA, 3rd edition, 1955.

[2] L. D. Landau and E. M. Lifshitz, *Quantum Mechanics (Non-Relativistic Theory)*, Pergamon, New York, NY, USA, 3rd edition, 1977.

[3] D. ter Haar, *Problems in Quantum Mechanics*, Pion Ltd, Londonm, UK, 3rd edition, 1975.

[4] F. Cooper, A. Khare, and U. Sukhatme, "Supersymmetry and quantum mechanics," *Physics Reports*, vol. 251, no. 5-6, pp. 267–385, 1995.

[5] S. H. Dong, *Factorization Method in Quantum Mechanics*, Springer, Kluwer Academic Publisher, 2007.

[6] Z. Q. Ma and B. W. Xu, "Quantum correction in exact quantization rules," *Europhys. Lett*, vol. 69, no. 5, 685 pages, 2005.

[7] W. C. Qiang and S. H. Dong, "Proper quantization rule," *EPL*, vol. 89, no. 1, Article ID 10003, 2010.

[8] N. Rosen and P. M. Morse, "On the vibrations of polyatomic molecules," *Physical Review A: Atomic, Molecular and Optical Physics*, vol. 42, no. 2, pp. 210–217, 1932.

[9] M. F. Manning, "Energy levels of a symmetrical double minima problem with applications to the NH_3 and ND_3 molecules," *The Journal of Chemical Physics*, vol. 3, no. 3, pp. 136–138, 1935.

[10] M. Baradaran and H. Panahi, "Lie Symmetry and the Bethe Ansatz Solution of a New Quasi-Exactly Solvable Double-Well Potential," *Advances in High Energy Physics*, vol. 2017, Article ID 2181532, 8 pages, 2017.

[11] H. Konwent, "One-dimensional Schrödinger equation with a new type double-well potential," *Physics Letters A*, vol. 118, no. 9, pp. 467–470, 1986.

[12] H. Konwent, P. Machnikowski, and A. Radosz, "A certain double-well potential related to SU(2) symmetry," *Journal of Physics A: Mathematical and General*, vol. 28, no. 13, pp. 3757–3762, 1995.

[13] Q.-T. Xie, "New quasi-exactly solvable double-well potentials," *Journal of Physics A: Mathematical and General*, vol. 45, no. 17, Article ID 175302, 2012.

[14] B. Chen, Y. Wu, and Q. Xie, "Heun functions and quasi-exactly solvable double-well potentials," *Journal of Physics A: Mathematical and Theoretical*, vol. 46, no. 3, Article ID 035301, 2013.

[15] R. R. Hartmann, "Bound states in a hyperbolic asymmetric double-well," *Journal of Mathematical Physics*, vol. 55, no. 1, 012105, 6 pages, 2014.

[16] A. E. Sitniksky, "Exactly solvable Schrödinger equation with double-well potential for hydrogen bond," *Chemical Physics Letters*, vol. 676, pp. 169–173, 2017.

[17] D.-N. Le, N. D. Hoang, and V.-H. Le, "Exact analytical solutions of the Schrödinger equation for a two dimensional purely sextic double-well potential," *Journal of Mathematical Physics*, vol. 59, no. 3, 032101, 15 pages, 2018.

[18] M. Razavy and Am. J. Phys, "An exactly soluble Schrödinger equation with a bistable potential," *American Journal of Physics*, vol. 48, no. 4, 285 pages, 1980.

[19] S.-H. Dong, *Wave Equation in Higher Dimensions*, Springer, Berlin, Germany, 2011.

[20] S. Dong, Q. Fang, B. J. Falaye, G. Sun, C. Yáñez-Márquez, and S. Dong, "Exact solutions to solitonic profile mass Schrödinger problem with a modified Pöschl–Teller potential," *Modern Physics Letters A*, vol. 31, no. 04, p. 1650017, 2016.

[21] S. Dong, G. H. Sun, B. J. Falaye, and S. H. Dong, "Semi-exact solutions to position-dependent mass Schrödinger problem with a class of hyperbolic potential V_0tanh(ax)," *The European Physical Journal Plus*, vol. 131, no. 176, 2016.

[22] G. H. Sun, S. H. Dong, K. D. Launey et al., "Shannon information entropy for a hyperbolic double-well potential," *International Journal of Quantum Chemistry*, vol. 115, no. 14, pp. 891–899, 2015.

[23] N. Ainsworth-Vaughn, *Heun's Differential Equations*, A. Ronveaux, Ed., Oxford University Press, Oxford, UK, 1995.

[24] C. A. Downing, "On a solution of the Schrödinger equation with a hyperbolic double-well potential," *Journal of Mathematical Physics*, vol. 54, no. 7, 072101, 8 pages, 2013.

[25] P. P. Fiziev, "Novel relations and new properties of confluent Heun's functions and their derivatives of arbitrary order," *Journal of Physics A: Mathematical and General*, vol. 43, no. 3, article 035203, 2010.

[26] R. Hartmann and M. E. Portnoi, "Quasi-exact solution to the Dirac equation for the hyperbolic-secant potential," *Physical Review A*, vol. 89, Article ID 012101, 2014.

[27] D. Agboola, "On the solvability of the generalized hyperbolic double-well models," *Journal of Mathematical Physics*, vol. 55, no. 5, 052102, 8 pages, 2014.

[28] F.-K. Wen, Z.-Y. Yang, C. Liu, W.-L. Yang, and Y.-Z. Zhang, "Exact polynomial solutions of Schrödinger equation with various hyperbolic potentials," *Communications in Theoretical Physics*, vol. 61, no. 2, pp. 153–159, 2014.

Permissions

All chapters in this book were first published in AHEP, by Hindawi Publishing Corporation; hereby published with permission under the Creative Commons Attribution License or equivalent. Every chapter published in this book has been scrutinized by our experts. Their significance has been extensively debated. The topics covered herein carry significant findings which will fuel the growth of the discipline. They may even be implemented as practical applications or may be referred to as a beginning point for another development.

The contributors of this book come from diverse backgrounds, making this book a truly international effort. This book will bring forth new frontiers with its revolutionizing research information and detailed analysis of the nascent developments around the world.

We would like to thank all the contributing authors for lending their expertise to make the book truly unique. They have played a crucial role in the development of this book. Without their invaluable contributions this book wouldn't have been possible. They have made vital efforts to compile up to date information on the varied aspects of this subject to make this book a valuable addition to the collection of many professionals and students.

This book was conceptualized with the vision of imparting up-to-date information and advanced data in this field. To ensure the same, a matchless editorial board was set up. Every individual on the board went through rigorous rounds of assessment to prove their worth. After which they invested a large part of their time researching and compiling the most relevant data for our readers.

The editorial board has been involved in producing this book since its inception. They have spent rigorous hours researching and exploring the diverse topics which have resulted in the successful publishing of this book. They have passed on their knowledge of decades through this book. To expedite this challenging task, the publisher supported the team at every step. A small team of assistant editors was also appointed to further simplify the editing procedure and attain best results for the readers.

Apart from the editorial board, the designing team has also invested a significant amount of their time in understanding the subject and creating the most relevant covers. They scrutinized every image to scout for the most suitable representation of the subject and create an appropriate cover for the book.

The publishing team has been an ardent support to the editorial, designing and production team. Their endless efforts to recruit the best for this project, has resulted in the accomplishment of this book. They are a veteran in the field of academics and their pool of knowledge is as vast as their experience in printing. Their expertise and guidance has proved useful at every step. Their uncompromising quality standards have made this book an exceptional effort. Their encouragement from time to time has been an inspiration for everyone.

The publisher and the editorial board hope that this book will prove to be a valuable piece of knowledge for researchers, students, practitioners and scholars across the globe.

List of Contributors

Jun-Zhen Wang and Bao-Chun Li
Department of Physics, Shanxi University, Taiyuan, Shanxi 030006, China

Qin Chang and Na Wang
Institute of Particle and Nuclear Physics, Henan Normal University, Henan 453007, China

Ru-Min Wang
Institute of Theoretical Physics, Xinyang Normal University, Henan 464000, China

Jie Zhu
Institute of Particle and Nuclear Physics, Henan Normal University, Henan 453007, China
Institute of Theoretical Physics, Xinyang Normal University, Henan 464000, China

Majid Hashemi and Mahbobeh Jafarpour
Physics Department, College of Sciences, Shiraz University, Shiraz 71946-84795, Iran

O. J. Oluwadare
Department of Physics, Federal University Oye-Ekiti, Ekiti State, Nigeria

K. J. Oyewumi
Theoretical Physics Section, Department of Physics, University of Ilorin, Ilorin, Nigeria

M. Janbazi
Young Researchers and Elites Club, Shiraz Branch, Islamic Azad University, Shiraz, Iran

R. Khosravi
Department of Physics, Isfahan University of Technology, Isfahan 84156-83111, Iran

E. Noori
Young Researchers and Elite Club, Karaj Branch, Islamic Azad University, Karaj, Iran

Akshay Chatla and Bindu A. Bambah
School of Physics, University of Hyderabad, Hyderabad 500046, India

Sahithi Rudrabhatla
Department of Physics, University of Illinois at Chicago, Chicago, IL 60607, USA

Hesham Mansour and Ahmed Gamal
Physics Department, Faculty of Science, Cairo University, Giza, Egypt

Halil Mutuk
Physics Department, Faculty of Arts and Sciences, Ondokuz Mayis University, 55139 Samsun, Turkey

Ardian Nata Atmaja
Research Center for Physics, Indonesian Institute of Sciences (LIPI), Kompleks Puspiptek Serpong, Tangerang 15310, Indonesia

Ilham Prasetyo
Research Center for Physics, Indonesian Institute of Sciences (LIPI), Kompleks Puspiptek Serpong, Tangerang 15310, Indonesia
Departemen Fisika, FMIPA, Universitas Indonesia, Depok 16424, Indonesia

G. Aydın
Department of Physics, Mustafa Kemal University, 31034 Hatay, Turkey

José T. Lunardi
Department of Mathematics & Statistics, State University of Ponta Grossa, Avenida Carlos Cavalcanti 4748, 84030-900 Ponta Grossa, PR, Brazil

Luiz A. Manzoni
Department of Physics, Concordia College, 901 8th St. S., Moorhead, MN 56562, USA

Hang Zhou and Zhen-Hua Zhang
School of Nuclear Science and Technology, University of South China, Hengyang, Hunan 421001, China

Bo Zheng
School of Nuclear Science and Technology, University of South China, Hengyang, Hunan 421001, China
Helmholtz Institute Mainz, Johann-Joachim-Becher-Weg 45, D-55099Mainz, Germany

Ridhi Chawla and M. Kaur
Physics Department, Panjab University, Chandigarh, 160014, India

Lin-Fang Deng, Zheng-Wen Long and Ting Xu
Department of Physics, Guizhou University, Guiyang 550025, China

Chao-Yun Long
Department of Physics, Guizhou University, Guiyang 550025, China
Laboratory for Photoelectric Technology and Application, Guizhou University, Guiyang 550025, China

Ya-Qin Gao
Department of Physics, Taiyuan University of Science and Technology, Taiyuan, Shanxi 030024, China

Hai-Ling Lao and Fu-Hu Liu
Institute of Theoretical Physics & State Key Laboratory of Quantum Optics and Quantum Optics Devices, Shanxi University, Taiyuan, Shanxi 030006, China

T. A. Ishkhanyan
Russian-Armenian University, Yerevan 0051, Armenia
Institute for Physical Research, Ashtarak 0203, Armenia
Moscow Institute of Physics and Technology, Dolgoprudny 141700, Russia

A. M. Ishkhanyan
Russian-Armenian University, Yerevan 0051, Armenia
Institute for Physical Research, Ashtarak 0203, Armenia

Halil Mutuk
Physics Department, Ondokuz Mayis University, 55139 Samsun, Turkey

Hui-Ling Li and Wei Li
College of Physics Science and Technology, Shenyang Normal University, Shenyang 110034, China

Yi-Wen Han
School of Computer Science and Information Engineering, Chongqing Technology and Business University, Chongqing 400070, China

Huriye Gürsel and Ezzet Sakallı
Physics Department, Faculty of Arts and Sciences, Eastern Mediterranean University, Famagusta, North Cyprus, Mersin 10, Turkey

Vitaly Beylin and Vladimir Kuksa
Research Institute of Physics, Southern Federal University, 344090 Rostov-on-Don, Pr. Stachky 194, Russia

P. Pirinen and J. Suhonen
University of Jyvaskyla, Department of Physics, 40014, Finland

E. Ydrefors
Instituto Tecnológico de Aeronáutica, DCTA, 12228-900 São José dos Campos, Brazil

A. Gajos, E. Czerwi Nski, K. Dulski, N. Gupta-Sharma, K. Kacprzak, D. Kisielewska, G. Korcyl, T. Kozik, E. Kubicz, Sz. Niedfwiecki, M. PaBka, M. Pawlik-Niedfwiecka, J. Raj, Z. Rudy, S. Sharma, Shivani, M. Silarski, M. Skurzok, M. ZieliNski and P. Moskal
Faculty of Physics, Astronomy and Applied Computer Science, Jagiellonian University, 30-348 Cracow, Poland

C. Curceanu
INFN, Laboratori Nazionali di Frascati, 00044 Frascati, Italy

M. Gorgol, B. Jasi Nska and B. Zgardzi Nska
Department of NuclearMethods, Institute of Physics, Maria Curie-Skłodowska University, 20-031 Lublin, Poland

B. C. Hiesmayr
Faculty of Physics, University of Vienna, 1090 Vienna, Austria

A. Kap Bon
Faculty of Physics, Astronomy and Applied Computer Science, Jagiellonian University, 30-348 Cracow, Poland
Institute of Metallurgy and Materials Science of Polish Academy of Sciences, Cracow, Poland

P. Kowalski, L. Raczy Nski, R. Shopa and W. Wi Vlicki
Świerk Department of Complex Systems, National Centre for Nuclear Research, 05-400 Otwock Świerk, Poland

W. Krzemień
High Energy Department, National Centre for Nuclear Research, 05-400 Otwock Świerk, Poland

M. Mohammed
Faculty of Physics, Astronomy and Applied Computer Science, Jagiellonian University, 30-348 Cracow, Poland
Department of Physics, College of Education for Pure Sciences, University of Mosul, Mosul, Iraq

D. Bazeia, L. Losano and M. A. Marques
Departamento de Física, Universidade Federal da Paraíba, 58051-970 João Pessoa, PB, Brazil

R. Menezes
Departamento de Física, Universidade Federal da Paraíba, 58051-970 João Pessoa, PB, Brazil
Departamento de Ciências Exatas, Universidade Federal da Paraíba, 58297-000 Rio Tinto, PB, Brazil

Qian Dong and Shi-Hai Dong
Laboratorio de Información Cuántica, CIDETEC, Instituto Politécnico Nacional, UPALM, CDMX 07700, Mexico

F. A. Serrano
Escuela Superior de Ingeniera Mecánica y Eléctrica UPC, Instituto Politécnico Nacional, Av. Santa Ana 1000, México, D. F. 04430, Mexico

Guo-Hua Sun
Catedrática CONACYT, CIC, Instituto Politécnico Nacional, CDMX 07738, Mexico

Jian Jing
Department of Physics and Electronic, School of Science, Beijing University of Chemical Technology, Beijing 100029, China

Index

www.ingramcontent.com/pod-product-compliance
Lightning Source LLC
Chambersburg PA
CBHW080250230326
41458CB00097B/4229